Prestressed Concrete

FIFTH EDITION

Warner Foster Kilpatrick Gravina

Project Management Team Leader: Jill Gillies

Senior Production Manager: Lisa D'Cruz

Courseware Associate: Jessica Darnell

ISBN 978 0 6557 0639 7

Printed and bound in Australia by The SOS Print + Media Group

Pearson Australia
707 Collins Street
MELBOURNE VIC 3008
www.pearson.com.au

Table of Contents

Preface

Preface to the Fifth Edition

The prime purpose of this new edition of Prestressed Concrete is to take account of important changes that have been made in the new edition of the Australian Standard for Concrete Structures, which appeared in 2018 and Amendment 2 in 2021. Changes include modifications to the rectangular stress-block parameters for flexural strength calculations, modifications to the design clauses for shear and torsion, and changes in the safety coefficients for ultimate strength design which result in a slight reduction in the overall safety margins for design. We have also taken the opportunity to correct minor errors and to rearrange the material in Appendix B, which deals with the effects of creep and shrinkage in prestressed flexural members, to make it more directly applicable to the design process.

Robert Warner	Stephen Foster	Andrew Kilpatrick	Rebecca Gravina
Adelaide	Sydney	Bendigo	Melbourne

November, 2021

Preface to the Fourth Edition

Following the death of Ken Faulkes, who was an original co-author of previous editions of this book, Rebecca Gravina has joined the team of co-authors in the preparation of the fourth edition of *Prestressed Concrete.* In this new edition we have taken the opportunity to make corrections to the text and to extend and improve the treatment of creep and shrinkage effects in Appendix B.

Robert Warner	Stephen Foster	Rebecca Gravina
Adelaide	Sydney	Melbourne

December, 2016

Preface to the Third Edition

When the first edition of this book appeared, more than thirty years ago, prestressed concrete and reinforced concrete were considered to be separate and distinct materials of construction. In Australia, different design standards had to be used for the two materials. At that time, prestressed concrete was designed to remain uncracked under full working load, using elastic analysis and allowable stress concepts. The apparent gulf between reinforced concrete and prestressed concrete was exaggerated by severe restrictions that were placed on the use of partial prestressing by the then-current prestressed concrete standard, AS 1481. One of the main aims of the first edition of this book was to present a rational and unified approach to the analysis and design of prestressed concrete, and hence to encourage designers to choose freely from the full range of design options, including reinforced concrete and any appropriate level of prestress. A further aim of the book was to present rational ways of selecting prestress levels that would optimise service load behaviour and economy. This aim was reflected in its full title: *Prestressed Concrete: with emphasis on partial prestressing*[1].

Today, the Australian Concrete Structures Standard, AS 3600, adopts a unified, performance-based approach to design for all reinforced concrete structures and members, irrespective of whether or not prestress is used. This integrated approach has removed unnecessary restrictions on design and has provided significant advantages to innovative designers. This in turn has resulted in partial prestressing becoming the design norm, and the term prestressed concrete now means, simply, structural concrete with prestress.

The world-wide developments that occurred in the field of concrete structures also made possible a more elegant, unified treatment of the underlying theory. Structural concrete members may contain any mix of reinforcing and prestressing steel, with reinforced concrete members being at one end of this continuum. Such a unified approach to analysis and design was adopted in a previous text book, titled *Concrete Structures*[2].

Nevertheless, when we consider the processes of teaching and learning today, we find that many students and lecturers prefer to deal first with the basics of reinforced concrete before going on to consider the concepts of prestressed concrete. Furthermore, it is now usual for many undergraduate course programs in structural engineering to cover reinforced concrete but not prestressed concrete. Although used widely in practice, prestressed concrete has become an option for undergraduate students, and hence a subject to be studied by postgraduates and practising engineers. For these reasons, we have

1. Warner, R. F. and Faulkes, K. A., 1979, Prestressed Concrete: with emphasis on partial prestressing, Pitman Publishing, Melbourne.
2. Warner, R. F., Rangan, B. V., Hall, A. S. and Faulkes, K. A., 1998, Concrete Structures, Longman-Cheshire, Melbourne.

chosen to prepare a new edition of *Prestressed Concrete*, rather than *Concrete Structures*, in order to provide an up-to-date treatment of prestressed concrete. We are in fact reverting to the approach we originally adopted when the first editions of *Reinforced Concrete*[3] and *Prestressed Concrete*[1] appeared in 1977 and 1979.

The present book is thus intended as a companion to *Reinforced Concrete Basics*[4], which appeared recently, in 2007, and provides a thorough treatment of the analysis and design of reinforced concrete structures and members. This third edition of *Prestressed Concrete* maintains a basic and rational approach to the analysis and design of prestressed concrete. It builds on, but integrates with, the ideas of reinforced concrete. For detailed design, both books refer to the requirements of the new edition of the Australian Concrete Standard, AS 3600-2009.

The sequencing of material in this new edition follows that of the second edition. Chapter 1 presents a simple, introductory, non-mathematical overview of the field of prestressed concrete, including both analysis and design, and introduces the important design concepts of equivalent loads and load balancing. Chapters 2 and 3 provide additional preliminary information on material properties and codified design procedures. Chapters 4 to 6 deal with the behaviour and design of flexural members, while Chapters 7 to 9 fill in important information on non-flexural behaviour, namely shear, torsion, anchorage and losses. The final Chapters, 10 to 12, concentrate on design aspects for determinate and indeterminate beams and floor slabs.

Chapters 1, 4 to 6 and 10 together provide the basis for an introductory course on the analysis and design of prestressed concrete members. Some additional material from Chapters 7 to 9 is however needed to round out the treatment to introduce the topics of shear, losses and anchorage.

The contents of the book have been extensively revised, updated and rewritten to take account of the many developments in theory and practice that have occurred in the intervening years since the second edition. The treatment of creep and shrinkage effects in prestressed concrete flexural members, in Chapters 4 and 5 and Appendix B, has been simplified by using an approach

3. Warner, R. F., Rangan, B. V. and Hall, A. S., 1977, Reinforced Concrete, Pitman Publishing, Melbourne.
4. Foster, S. J., Kilpatrick, A. E. and Warner, R. F., 2021, Reinforced Concrete Basics, 3rd Ed., Pearson Education Australia, Melbourne.

that is based on the fundamental structural concepts of equilibrium, compatibility and elastic behaviour. The use of more complex visco-elastic analyses has thus been avoided. A greater emphasis has been placed on strut-and-tie modelling in Chapters 7 and 8 to deal with shear, torsion, and anchorage. This reflects recent developments in our knowledge and understanding of this important design tool.

The authors are deeply indebted to their friend and colleague Andrew Kilpatrick. It is only through his unstinting help that we have been able to complete the text.

It is with great sadness that we acknowledge the death of our long-time friend and colleague Professor Ken Faulkes. Ken was working with us on the final chapters of the book at the time of his death. His contributions have been invaluable. His vast knowledge and his experience, based on more than 50 years of research and practice in the field, are reflected throughout this text.

Robert Warner   Stephen Foster
1 June 2013

Notation

A_c = area of concrete
A_{eq} = combined equivalent area of steel and tendon
A_g = gross area of a non-transformed cross-section
A_p = area of prestressing steel
A_{pt} = area of prestressing steel on the tension side of the neutral axis at M_u
A_s = area of reinforcement
A_{sc} = area of longitudinal compressive reinforcement
A_{st} = area of longitudinal tensile reinforcement
A_{sv} = area of all legs of vertical shear reinforcement that cross a shear crack
$A_{sv.min}$ = minimum required area of shear reinforcement in a beam
a = a coefficient
= shear span
= clear distance between reinforcing bars
= distance between points of contraflexure
a_{sup} = length of a support for a flexural member
a_v = distance from the section at which shear is being considered to the face of the nearest support
b = width of a rectangular cross-section
b_{ef} = effective width of the flange of a T- or L-section
b_v = effective width of the web of a beam for shear
b_w = web width of a T- or L-section
C = compressive force
C_c = compressive force in the concrete of a cross-section
C_f = compressive force in the flange outstands of a T- or L-section
C_{sc} = compressive force in the steel reinforcement of a cross-section
C_w = compressive force in the web of a T- or L-section
C^* = compressive force at the strength limit state
c = cover of concrete to a reinforcing bar
D = overall depth of a member
D_s = overall depth of a slab

d	=	effective depth, from the extreme fibre in compression to the resultant of the steel forces (T_s and T_p), on the tension side of a cross-section.
d_c	=	depth from the extreme fibre in compression to the concrete compressive stress resultant
d_{eq}	=	depth in section of equivalent area A_{eq}
d_n	=	depth to the neutral axis in a cross-section in bending
	=	depth to the neutral axis of strain, if inelastic strains are present
d_{ne}	=	depth to the neutral axis of stress if inelastic strains are present
d_o	=	depth from the extreme fibre in compression to the centroid of the outer-most layer of longitudinal tensile reinforcement, but not less than $0.8D$
	=	diameter of a reinforcing bar
d_{om}	=	mean value of d_o around the punching shear perimeter u_e
d_p	=	depth from the extreme compressive fibre to the prestressing steel
d_s	=	depth from the extreme compressive fibre to the reinforcing steel
e	=	eccentricity of prestress force P
e_{eq}	=	eccentricity of equivalent area A_{eq} in section
E_c	=	mean secant modulus of elasticity of in situ concrete measured at an axial stress of 45 per cent of the peak stress
E_{co}	=	modulus of elasticity of concrete at time t_o
E_c^*	=	modulus of elasticity of concrete at time t^*
E_d	=	design action effect
E_{eq}	=	modulus of elasticity of equivalent area steel A_{eq}
E_p	=	mean modulus of elasticity of prestressing steel
E_R	=	reduced elastic modulus for concrete, in creep calculations
$E_{R\chi}$	=	age-adjusted effective modulus for concrete, used in creep calculations
E_s	=	mean modulus of elasticity of the steel reinforcement
e	=	eccentricity of a prestressing force from the section centroid
F	=	force
f_{cm}	=	mean compressive strength of a concrete cylinder
f_{cmi}	=	mean compressive strength of in situ concrete
f_p	=	sustained stress in prestressing tendon
f_{cmi}	=	mean compressive strength of in situ concrete
f_{ct}	=	tensile strength of concrete
$f_{ct.f}$	=	mean concrete tensile strength obtained from a flexure (modulus of rupture) test
f_{cv}	=	concrete punching shear stress capacity

f_{pb} = characteristic minimum breaking stress of prestressing steel
f_s = stress limit in reinforcing steel to control serviceability cracking
f_{su} = tensile strength of steel reinforcement (ultimate)
f_{sy} = characteristic yield strength of steel reinforcement
$f_{sy.f}$ = characteristic yield strength of steel reinforcement used as fitments (stirrups)
f_c' = characteristic compressive strength of a concrete cylinder at age 28 days
f_{cp}' = characteristic compressive strength of concrete at transfer
f_{ct}' = characteristic uniaxial tensile strength of concrete
$f_{ct.f}'$ = characteristic value of $f_{ct.f}$
G = service permanent action (dead load)
h = midpoint sag of a parabola
h_d = width of a bearing plate
h_s = depth of a bearing plate
h_x = midpoint sag of a parabolic cable in a slab in the x-direction
h_y = midpoint sag of a parabolic cable in a slab in the y-direction
I_{cr} = second moment of area of a cracked, transformed cross-section
I_{cs} = average second moment of area of the column strip in a slab
I_{wb} = average second moment of area of the wide-beam portion of a slab
I_{ef} = effective second moment of area
$I_{ef.av}$ = weighted average effective second moment of area for a flexural member
I_g = second moment of area of the gross concrete cross-section
k = a factor
k_{cs} = long-term deflection factor
k_u = neutral axis depth parameter for a cross-section, d_n/d at M_u
k_{uo} = neutral axis depth parameter in a cross-section, d_n/d_o, when M_{uo} is acting
$k_1, k_2, k_3, k_4, k_5, k_6$
= deflection coefficients or modifying factors
$E^*_{R\chi}$ = age-adjusted effective modulus at time t^*
j = time in days after prestressing
L = centre-to-centre distance of the supports for a flexural member
L_n = clear span, face-to-face of supports

L_{pa} = length of beam from the jack to the point at which friction loss is being calculated

L_x, L_y = shorter and longer spans, respectively, of a slab supported on all four sides

M = bending moment

M_{cr} = bending moment that causes a cross-section to begin cracking (including shrinkage)

M_{cro} = bending moment that causes a cross-section to begin cracking (excluding shrinkage)

M_{dec} = decompression moment (zero bottom fibre stress)

M_G = bending moment due to self-weight

M_{GQ} = bending moment due to self-weight plus live load

M_L, M_M, M_R

= bending moments at the left, middle and right ends of a beam or slab

M_o = bending moment corresponding to zero curvature

= decompression moment

M_p = section moment due to prestress (= $P.e$ in a determinate member)

= $M_1 + M_2$; total moment in a section in an indeterminate member due to prestress

M_G = bending moment due to dead load

M_Q = bending moment due to live load

M_u = ultimate bending capacity (strength) of a cross-section

M_{uo} = value of M_u in the absence of an axial force

M_y = bending moment at which the tensile reinforcement yields

M^* = bending moment at a cross-section due to the design load for the strength limit state

M^*_{max} = maximum value of M^* in a flexural member

M^*_v = bending moment to be transmitted from a slab to a column at the strength limit state

M_1 = primary moment due to prestress in a section of an indeterminate beam

M_2 = secondary (hyperstatic) moment due to prestress in a section of an indeterminate beam

N = axial force

n = modular ratio

P = prestressing force

= force acting normal to a cross-section

P_e = effective prestressing force

P_i	=	initial prestressing force, prior to time-dependent losses
P_T	=	total prestressing force in a slab per panel in one direction
P_u	=	breaking load of prestressing tendon
	=	capacity of prestressed section against failure at transfer
P_v	=	vertical component of prestressing force in a tendon
P_x	=	prestressing force per unit width of slab in *x*-direction
P_y	=	prestressing force per unit width of slab in *y*-direction
P^*	=	concentrated load for the strength limit state
p_{cw}	=	ratio of the area of compressive reinforcement to the area of the web, $A_{sc}/(b_w d)$
p_w	=	ratio of the area of the tensile reinforcement to the area of the web, $(A_{st}+A_{pt})/b_w d$
Q	=	service imposed action (live load)
R	=	resultant force
	=	design relaxation of prestressing tendon
R_b	=	basic relaxation of prestressing tendon
R_d	=	design capacity
R_u	=	nominal ultimate capacity
$R_{u.sys}$	=	mean value of the calculated capacity of a structural system
S	=	vertical component of prestressing force at an anchorage
s	=	spacing of fitments (stirrups) along a beam
T	=	tensile force
	=	average annual temperature
T_p	=	force in a prestressing tendon
T_s	=	force in the longitudinal tensile reinforcement
t	=	thickness (depth) of a slab
	=	thickness of the compression flange of a section
	=	time
t_h	=	hypothetical thickness of a member
t^*	=	time infinity
u	=	perimeter around which punching shear occurs
u_e	=	that part of the perimeter of a member that is exposed to a drying environment, plus half the perimeter of any internal voids
V	=	shear force
V_1	=	shear force carried by the concrete at web-shear cracking
V_{dec}	=	shear force in section at decompression moment

V_p	=	vertical component of prestressing force
V_s	=	tensile force in a vertical stirrup
V_u	=	flexural shear capacity of a beam containing shear reinforcement
V_{uc}	=	ultimate flexural shear capacity of a beam without shear reinforcement
	=	inclined shear cracking load
$V_{u.max}$	=	maximum flexural shear capacity of a beam
$V_{u.min}$	=	flexural shear capacity of a beam containing minimum shear reinforcement
V_{uo}	=	punching shear capacity of a slab without moment transfer
V_{us}	=	contribution by the shear reinforcement to the flexural shear capacity of a beam
V^*	=	flexural shear force at a cross-section due to the design load for the strength limit state
V^*_{max}	=	maximum value of V^* in a flexural member
W	=	concentrated load
W_p	=	equivalent concentrated load exerted by a tendon at a kink
W^*	=	concentrated load for the strength limit state
w	=	uniformly distributed load kN/m
w_b	=	uniformly distributed load to be balanced
w_G	=	uniformly distributed dead load
w_l	=	uniformly distributed long-term serviceability load, $w_G + \psi_l w_Q$
w_p	=	distributed equivalent load from a curved prestressing tendon
w_{px}	=	distributed equivalent load from a curved x-direction prestressing tendon
w_{py}	=	distributed equivalent load from a curved y-direction prestressing tendon
w_Q	=	uniformly distributed service live load
w_s	=	uniformly distributed short-term serviceability load $w_G + \psi_s w_Q$
w^*	=	uniformly distributed design load for the strength limit state
y_b	=	distance from the neutral axis to the bottom fibre of a cross-section
y	=	distance from the bottom fibre in an uncracked (gross) cross-section to its centroidal axis
Z	=	elastic section modulus of a cross-section
z	=	lever arm (distance) between the forces C and T of a cross-section in bending
α	=	an angle
	=	reduction factor for steel strain increment in unbonded tendons
α_1, α_2	=	compressive stress-block factors

	=	non-dimensional parameters used to determine creep and shrinkage curvatures
α_{tot}	=	total angular deviation of prestressing cable over length L_{pa}
β	=	beam section shear capacity parameter, $\beta_1\beta_2\beta_3$
	=	friction coefficient
β_x, β_y	=	coefficients for bending moments in slabs supported on four sides
β_1, β_2, β_3	=	parameters for the shear capacity of a beam cross-section
γ	=	compressive stress block factor
	=	unit weight of a material
	=	a coefficient
γ_c, γ_{rc}	=	unit weight of reinforced concrete
Δ	=	deflection of a member in bending
Δ_G	=	deflection due to self-weight
Δ_P	=	deflection due to prestress
Δ_{PG}	=	deflection due to self-weight plus prestress
Δ_s	=	short-term deflection due to the short-term serviceability load, $G + \psi_s Q$
$\Delta_{s.inc}$	=	difference between the short-term deflections due to the loads $G + \psi_s Q$ and G (incremental deflection)
$\Delta_{s.sus}$	=	short-term deflection due to the sustained serviceability load, $G + \psi_l Q$
Δ_{tot}	=	total long-term deflection
$\Delta_{tot.inc}$	=	total long-term (incremental) deflection after the attachment of deflection-sensitive partitions or finishes
ΔX	=	an increment in force
ΔX_c	=	increment in force in the concrete
$\Delta X_{c.c}$	=	increment in force in the concrete due to creep
$\Delta X^i_{c.c}$	=	increment in force in the concrete, due to creep, applied at time t_i
$\Delta X^*_{c.c}$	=	increment in force in the concrete, due to creep, applied at time t^*
$\Delta X^*_{c.sh}$	=	increment in force in the concrete, due to shrinkage, at time t^*
ΔX_p	=	increment in force in the tendon
$\Delta\varepsilon$	=	increment in strain
$\Delta\varepsilon_{cc}$	=	increment in creep strain in the concrete
$\Delta\varepsilon_{cc.a}$	=	increment in creep strain in the concrete at fibre a

$\Delta\varepsilon^{i}_{cc.a}$ = increment in creep strain in the concrete at fibre a, applied at time t_i

$\Delta\varepsilon^{*}_{cc.a}$ = increment in creep strain in the concrete at fibre a, applied at time t^*

$\Delta\varepsilon_{ce}$ = increment in elastic strain in the concrete

$\Delta\varepsilon_{p}$ = increment in strain in the tendon (all tendon strains are elastic)

$\Delta\varepsilon_{cc.p}$ = increment in creep strain in the concrete at the level of the tendon

$\Delta\varepsilon_{p.sh}$ = increment in strain in the tendon due to concrete shrinkage

$\Delta\kappa^{*}_{c}$ = creep curvature correction at t^* to allow for steel and tendon in section

ε = strain

$\varepsilon_{a.csh}$ = concrete strain due to creep and shrinkage in the top fibre of a section

ε_{c} = strain in concrete

ε_{cc} = creep strain in concrete

ε^{*}_{cc} = long-term value of creep strain ε_{cc}

$\varepsilon^{*}_{cc.eq}$ = long-term free creep strain in section at depth d_{eq}

ε_{ce} = elastic strain in concrete

ε_{cea} = elastic strain in the concrete, top fibre

ε_{ceb} = elastic strain in the concrete, bottom fibre

ε_{csh} = shrinkage strain in concrete

$\varepsilon_{c.csh}$ = concrete strain due to creep and shrinkage

ε_{ce} = initial elastic strain in concrete due to an applied stress

= elastic strain due to effective prestress in the concrete at tendon level

ε_{ci} = elastic strain due to initial prestress in the concrete at tendon level

ε_{cp} = concrete strain at tendon level

ε_{cs} = AS 3600 terminology for design shrinkage strain in concrete

ε^{*}_{cs} = long-term shrinkage strain in concrete

ε_{csd} = drying shrinkage strain

ε_{cse} = autogenous shrinkage strain

ε_{cu} = maximum compressive strain in the concrete of a cross-section in flexure

ε_{o} = strain at the peak stress of concrete in compression

ε_{o}, ε_{c} = compressive strain in concrete

ε_{p} = strain in the prestressing tendon

ε_{pe} = strain in prestressing tendon due to effective prestress

ε_{pi} = initial strain in prestressing tendon immediately after transfer

ε_{sc} = strain in the compressive steel reinforcement

ε_{sh} = total shrinkage strain for calculating deflection in a slab

ε_{st} = strain in the tensile steel reinforcement

ε_{sy} = yield strain of steel reinforcement

ε_{u} = concrete extreme fibre compressive strain at M_u

ε^*_{cc} = final creep strain in the concrete

ε^*_{cs} = final shrinkage strain in the concrete

ε_{cs} = design shrinkage strain according to AS 3600

η = friction loss factor

θ = an angle

= slope of prestressing cable

κ = curvature

κ_L, κ_M, κ_R

= curvatures at the left, middle and right ends of the beam

κ_o = initial elastic curvature at time t_o

$\kappa_c(t)$ = creep curvature at time t

κ^*_c = total long-term creep curvature at time t^*

κ^*_{co} = long-term free creep curvature at time t^*

κ^{**}_c = $\kappa^*_{co} + \Delta\kappa^*_c$ improved estimate of long-term creep curvature

κ_{sh} = shrinkage curvature

κ^*_{sh} = long-term shrinkage curvature

μ = coefficient of friction

ρ = the density of concrete

Σ = algebraic summation

σ = stress normal to a section

σ_1 = principal tensile stress

σ_{ca} = concrete top fibre stress

σ_{cb} = concrete bottom fibre stress

σ_{cbp} = concrete bottom fibre stress due to prestress

σ_{cc} = constant sustained concrete compressive stress

σ_{ce} = elastic stress due to effective prestress in the concrete at tendon level

σ_{ci}	=	elastic stress due to initial prestress in the concrete at tendon level
σ_{cs}	=	concrete (tensile) stress in a cross-section caused by shrinkage
σ_s, σ_{st}	=	tensile stress in steel reinforcement
σ_{sc}	=	compressive stress in steel reinforcement
σ_{scr}	=	stress in the tensile reinforcement obtained from a cracked section analysis
σ_{pa}	=	concrete top fibre stress due to prestress
	=	stress in prestressing cable at distance L_{pa} from the jack
σ_{pb}	=	concrete bottom fibre stress due to prestress
σ_{pj}	=	stress in prestressing cable at the jack
σ_{pu}	=	stress in prestressing steel at M_u
τ	=	shear stress
ϕ	=	capacity reduction factor
ϕ_{sys}	=	capacity reduction factor for a structural system
φ_{cc}	=	creep coefficient as used in AS 3600
$\varphi_{cc.b}$	=	basic value of φ_{cc}, for $t_o = 28$ days, as used in AS 3600
$\varphi(t, t_o)$	=	creep function; value at time t due to sustained stress applied at t_o
φ^*_{cc}	=	long-term value of $\varphi(t, t_o)$ at $t = t^*$, for stress applied at t_o
χ	=	aging coefficient for use with E_R
ψ_c	=	combination factor for imposed actions (live loads)
ψ_s	=	short-term live load factor used to determine the serviceability load
ψ_l	=	long-term live load factor used to determine the serviceability load

CHAPTER 1

Introduction

The basic ideas of prestressed concrete are introduced in this chapter. We explain what prestressing is and the advantages and disadvantages of prestressing concrete members. Methods of post-tensioning and pretensioning are explained. The chapter includes a short historical note on the development of prestressed concrete, from its beginnings at the end of the 19th Century.

1.1 Prestressed concrete

1.1.1 Plain concrete and reinforced concrete

When external load is applied to structural members such as beams and slabs, large regions in the member are subjected to tensile stress. Tensile stresses may also be induced in structural members by load-independent effects such as temperature gradients and imposed deformations due to foundation movement. Because of its very low tensile strength, plain concrete cannot be used to construct such members where significant tension is present.

The compressive strength of concrete is reasonably good and if small amounts of reinforcing steel are placed in strategic locations in the concrete to carry the internal tensile forces that develop, an effective load-carrying mechanism is created. The resulting composite material is **reinforced concrete**. The great advantage of concrete as a building material is that it is very cheap, and even when reinforcing steel is added in small quantities, the cost advantage is still sufficient to ensure that reinforced concrete is presently the most widely used structural material, both in Australia and world-wide.

The flexural behaviour of reinforced concrete is illustrated in Figure 1.1, where the beam is supported on a simple span of length L and has two equal point loads W acting at the third points of the span which are located at distance $a = L/3$ from the supports.

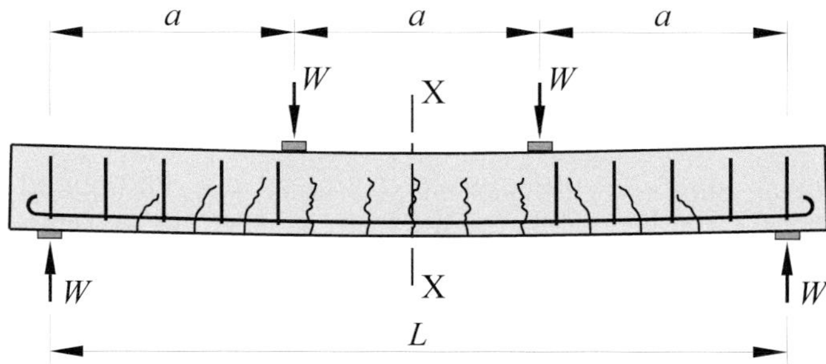

(a) Reinforced concrete beam under service load

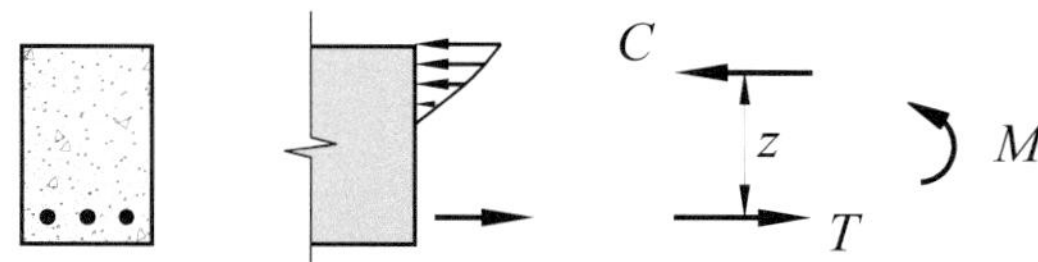

(b) Stresses and internal forces at section X-X

Figure 1.1 Reinforced concrete beam with external loads

Tensile stresses develop in the lower fibres of the beam as the first increments of load are applied. With increasing load, cracks soon appear throughout the mid-span region, where the moment is largest. At any cracked cross-section X-X between the load points (Figure 1.1(b)), the internal moment M is resisted by a tensile force T in the steel, located in the lower cracked region of the concrete, and an equal compressive force C in the intact compressive concrete above the crack. The steel is effective in carrying the tensile force in a cracked section, but it does not prevent or delay cracking of the tensile concrete. That is not its function. With increasing load the cracked region extends outwards towards each support, and the beam deflection increases. Under full service load, a well-developed pattern of fine cracks is present in the lower fibres of the beam.

With overload, the existing cracks widen and the cracked region extends even further outwards. At high overload, the steel reinforcement in the mid-span region yields. The cracks then widen even more and the deflection increases rapidly with only a very slight further increase in load. Eventually the ultimate moment M_u of the sections is reached in the mid-span region and the beam fails in flexure at its load capacity W_{max}. In the design of reinforced concrete flexural members, the aim is to achieve good service load behaviour, in particular with narrow crack widths and small deflections, and adequate strength to prevent premature failure.

1.1.2 Prestressed concrete

Prestressing is another way of circumventing the poor tensile strength of plain concrete: a system of permanent compressive stresses is introduced into the regions of a concrete member where tensile stresses will subsequently develop when the external service loads act. This pre-compression delays tensile cracking, and may even prevent it altogether at service loads. The downward deflection due to external load is also reduced. Prestressing is, thus, an effective way of improving the service-load behaviour of a reinforced concrete member. Compressive prestress in the concrete cross-sections is usually achieved by the use of highly stressed, high-strength tensile steel or fibre reinforced plastic (FRP) tendons that run through the length of the member. The tendons are permanently anchored to the concrete at each beam end. At each internal cross-section, the tensile force in the tendon produces equilibrating compressive stresses in the concrete.

To illustrate the use of prestressing, we return to the reinforced concrete beam in Figure 1.1 and we consider the effect of stressing two draped external steel prestressing tendons, placed on the side faces as shown in Figure 1.2(a). At the ends, the tendons are anchored to the concrete at the section mid-depth. In-span they are draped around cast-in-place pins located in the lower fibres of the concrete, directly under the load points. In the mid-span region the eccentricity of the tendons, relative to the section centroid, is e and the total tensile force in the two tendons is P. The prestressed tendons apply forces to the concrete at the pins and at the end anchors. This is shown in Figure 1.2(a).

The upward force at each pin, W_p, is slightly inclined from the vertical, and the force P at each end is slightly inclined from the horizontal. These forces are self-equilibrating. The overall effect of the prestress is to create an upwards camber in the beam, as in Figure 1.2(b). When the external loads W are applied there is a downwards deflection (Figure 1.2(c)), but it is reduced by the prior prestress.

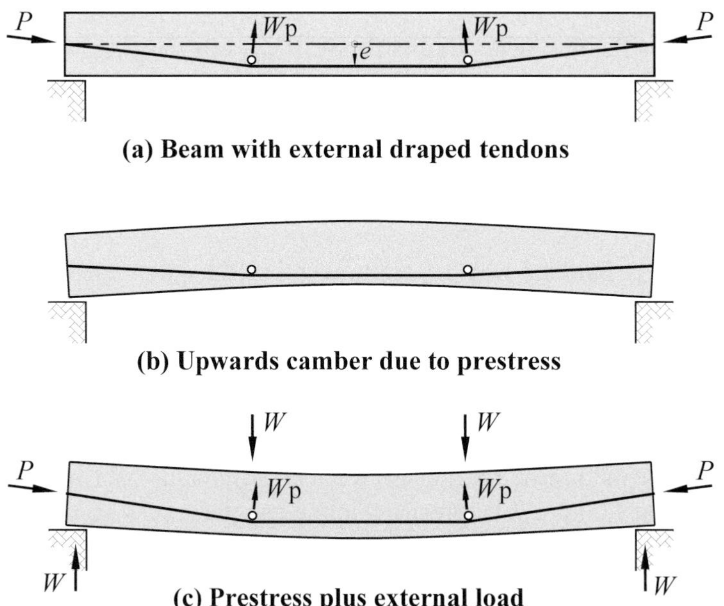

Figure 1.2 Prestressed concrete beam, draped tendons

To investigate the stresses in the concrete due to the prestress, we consider the free body to the left of section X-X in the central region, as shown in Figure 1.3. At X-X the tensile prestressing force P in the tendons is horizontal, and at eccentricity e. This induces compressive stresses in the concrete, which have a resultant force $C = P$ that also must act at eccentricity e. The eccentric force C is statically equivalent to a compressive force C acting at the centroid of the section, plus a negative moment $M_p = Ce$, as in Figure 1.4(a). The concrete stresses are thus the sum of a uniformly distributed compressive stress C/A_c and the bending stresses due to M_p. At the bottom and top fibres, the stresses are:

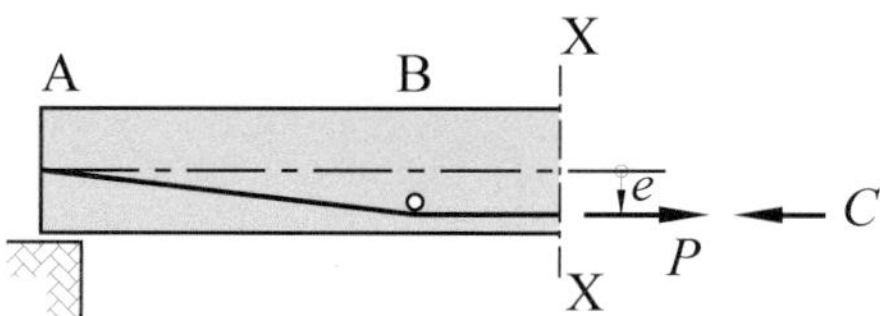

Figure 1.3 Equilibrating forces at section X-X

$$\sigma_{pb} = P\left[\frac{1}{A_c} + \frac{e}{Z}\right] \tag{1.1}$$

$$\sigma_{pa} = P\left[\frac{1}{A_c} - \frac{e}{Z}\right] \tag{1.2}$$

Here A_c is the area of the concrete cross-section and Z is its section modulus, which for a rectangular section is $bD^2/6$, where D is the section depth and b is the section width. The stress distribution in the section is shown in Figure 1.4(b). While the upper fibre stress σ_{pa} is shown as compressive, it will be tensile if the eccentricity e is sufficiently large, whereas the bottom fibre stress σ_{pb} is always compressive. The effect of any reinforcing steel in an uncracked section is very small and is ignored in Equations 1.1 and 1.2.

The external loads W induce a positive moment $M = Wa$ in the central region, with compressive stress in the upper fibres and tensile stress in the lower fibres. However, compressive stress is already present in the lower fibres of the section due to prestress. The resultant stresses at section X-X due to prestress plus external load are as shown in Figure 1.4(c).

Whether or not the resultant stresses remain compressive in the bottom fibres, as shown in Figure 1.4(c), depends on the magnitudes of the prestressing force P, the eccentricity e and the load-induced moment, M. In any case, cracking will be delayed, or possibly even prevented, by the prestress. Also, the initial upwards deflection due to the prestress (Figure 1.2(b)) reduces, and may eliminate completely, the downwards deflection due to the external load W.

X

X

e

C

=

X

X

C

$M_p = C.e = P.e$

(a) Stress resultants at X-X due to prestress

σ_{pa}

+

=

σ_{pb}

due to C + due to M_p = concrete stresses

(b) Stresses due to prestress

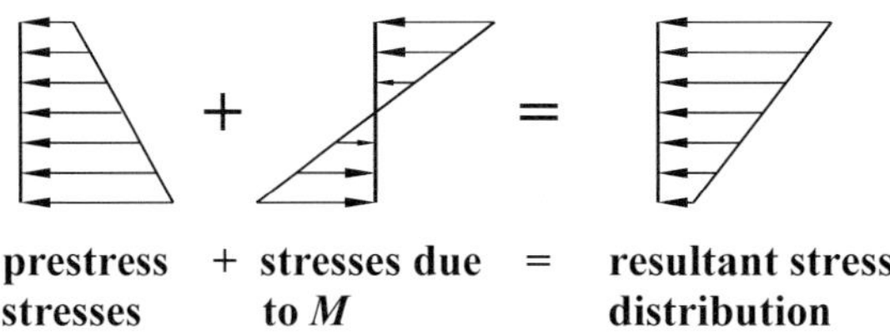

(c) Stresses due to prestress and external load

Figure 1.4 Concrete stresses at section X-X due to prestress and external load

In this example the prestressing tendons have been "draped". The downward eccentricity in the middle region produces a negative moment $M_p = Pe$, which opposes the moment M due to the external loads. In the outer regions the eccentricity e reduces progressively to zero. The bottom fibre compressive stress due to prestress, σ_{pb}, reduces to the value P/A_c at the end of the beam, while the tensile stress due to external loading reduces to zero.

Various other tendon arrangements, different to the one shown in Figure 1.2, can be used to reduce cracking and deflection. For example, Figure 1.5(a) shows a curved tendon located in a parabolically shaped duct that has been cast in the concrete. The tendon is tensioned against the ends of the hardened concrete and then anchored permanently. As we shall see shortly, this form of construction is known as **post-tensioning**. The force in the tendon is P and the maximum eccentricity at mid-span, as before, is e. The upwards deflection due to the prestress is shown in the Figure.

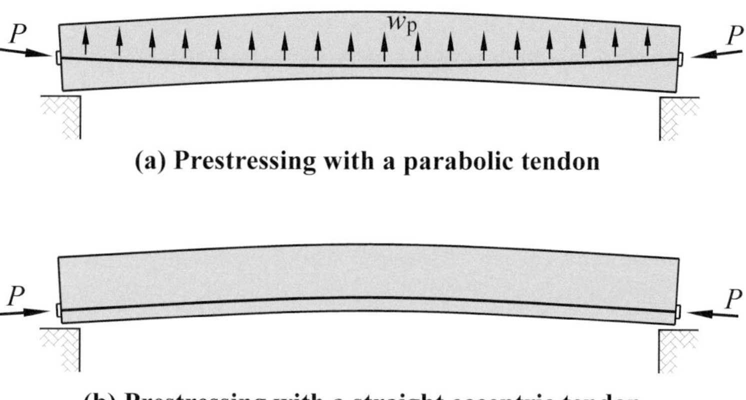

Figure 1.5 Parabolic and straight tendons

The self-equilibrating force system exerted on the concrete by the curved prestressing tendon is shown in Figure 1.5(a). It consists of slightly inclined end forces P at the anchorage points, and an upward distributed force w_p acting along the beam in a direction perpendicular to the curved cable. If the cable curvature is constant (which is the case if the shape is parabolic) then it can be shown that w_p is a uniformly distributed force.

At mid-span the tendon force P is horizontal. It is equilibrated by a horizontal compressive force C in the concrete. The distribution of compressive concrete stresses here is therefore the same as that shown already in Figure 1.4(b). Although the eccentricity of the tendon decreases towards the supports, the prestressed tendon induces compressive stresses in the lower fibres at each section so that cracking is delayed or prevented when the load is applied. The total downwards deflection is also reduced.

In Figure 1.5(b) yet another tendon shape is shown. This time the tendon is straight, with the eccentricity e constant along the full length of the beam.

This profile is typical when **pre-tensioning** is employed. The straight tendon induces the same stress distribution in every section along the beam, and this is as shown in Figure 1.4(b). There is a uniform compression of P/A_c and bending stresses due to the moment $M_p = P\,e$. The overall effect is again to induce compressive stresses in the lower fibres and hence to delay or prevent cracking. The initial upwards camber acts to reduce or prevent deflection under external service load. However, the negative prestressing moment is now constant along the beam, and so becomes greater than the positive applied moment in the outer regions near the beam ends. In the design of such beams, care must be taken to ensure that the negative moment due to prestress is not excessive in the end regions.

For the beam in Figure 1.5(b), the self-equilibrating forces due to the straight tendon consists simply of equal and opposite horizontal forces P applied with eccentricity e at each end of the beam. This is statically equivalent to axial end forces P, plus a pair of negative end moments $M_p = -Pe$ at each end.

Figures 1.2 to 1.5 show how various tendon shapes are used to induce an equilibrating system of forces that act on the concrete. These forces act at the ends of the tendon where it is anchored to the concrete, and at any point along the span where the tensioned tendon changes direction. In particular, a concentrated force W_p acts at a kink in the tendon (as in Figure 1.2) and a distributed transverse force w_p acts over any length of member in which the tendon is curved (as in Figure 1.5(a)).

The forces exerted by the prestressed tendon on the beam can be thought of as **equivalent loads**. For example, in Figure 1.2 the equivalent loads are the upwards acting point loads W_p and the inclined end forces P, while in Figure 1.5(a) the equivalent loads consist of a uniformly distributed equivalent load w_p and the inclined end forces P.

The concept of equivalent loads gives us a simple but very useful view of the effect of prestress on the behaviour of members, both statically determinate and statically indeterminate. It also provides a convenient method for evaluating the stresses that are produced in a beam by the prestress. Furthermore, a simple but extremely useful design technique, called **load balancing**, can be developed from the equivalent load concept. Ideas of equivalent loads and load balancing are discussed in detail in Chapter 4.

THE IDEA OF PRESTRESSING

The idea of prestressing has wider applicability than in the field of prestressed concrete. A simple form of prestressing has been used by coopers for centuries to construct wine barrels by forcing heated metal tension bands over wooden staves. The precompression induced when the bands cool prestresses the staves together and prevents leaking. A variation of this technique is used today in the construction of large circular prestressed concrete liquid-retaining tanks. The procedure involves winding prestressing tendons around precast vertical concrete 'staves'.

One of the first suggestions to introduce prestress into structural concrete was made by P H Jackson in 1886 in San Francisco. A patent taken out in Berlin in 1888 by Doehring anticipated the idea of the production method which uses a pretensioning bed. Various proposals and tests followed, but this early developmental work was unsuccessful because mild steel reinforcing bars were used as the prestressing medium.

In the 1920s, R H Dill in the USA recognised that high strength wire could be used to produce a satisfactory prestressed member. However, the first successful practical designs in prestressed concrete were carried out in Europe by Eugene Freyssinet in the 1930s, when the time-dependent creep and shrinkage behaviour of concrete came to be better understood. In the United States, prestressed concrete was first used in the construction of circular water tanks. In the late 1930s the Preload Corporation developed the technique of winding wires around cast-in-place circular concrete walls.

Shortly after World War 2, Freyssinet designed a number of successful and highly acclaimed bridges in France, which led to wide acceptance of prestressed concrete. An upsurge in interest in prestressed concrete at that time can be attributed in part to the scarcity and high cost of steel and other structural materials in those post-war years. In the United States, prestressed concrete was first used in bridge construction in the late 1940s. Interesting historical information on the development of prestressed concrete, on personalities involved in its early development, and on the range of structures previously constructed in prestressed concrete, is to be found in the T Y Lin Symposium on Prestressed Concrete, reported in the Prestressed Concrete Journal (Lin, 1976).

1.2 Prestressing as a design option

Prestressed concrete, like reinforced concrete, takes advantage of the compressive strength of concrete, while circumventing its weakness in tension. Prestressed concrete is usually made from concrete of medium to high strength, with a small quantity of very high strength prestressing steel tendon. Ordinary non-prestressed reinforcing steel is also included in the member, both as subsidiary longitudinal reinforcement, that becomes effective after cracking, and as transverse stirrups to improve shear strength.

As a design option, prestressing has associated with it a number of advantages, and some disadvantages, that the designer always needs to keep in mind.

1.2.1 Advantages of prestressing

Improved service load behaviour

We have seen how prestressing can improve the service load behaviour of a concrete member. Even quite moderate levels of prestressing can reduce crack widths to very small values under working load. Deflections can be controlled to remain within a desired range or entirely eliminated, for a chosen service load, by introducing the appropriate level of prestress. It should be noted that prestress does not in itself have any significant effect on flexural strength although the presence of prestressing steel in the tensile region of an under-reinforced beam contributes to its moment capacity.

Efficient use of high strength steel and concrete.

Prestressed concrete makes possible the efficient and economic use of both very high strength steel and high strength concrete in an efficient structural medium. High strength steel is not efficient if used as ordinary reinforcement. This is because unacceptably large deflections and excessively wide cracks may occur in the member under service load, long before the steel is stressed even into its working load range.

More slender concrete structures with larger spans.

With the effective use of high strength materials together with the control of deflections and crack widths by prestressing, it becomes possible on the one hand to produce concrete structures that are more slender than reinforced concrete structures, or, on the other hand, to use significantly larger spans than would otherwise be possible. Prestressing is a particularly useful design option for large-span structures where self-weight is a high proportion of the total load and also in structures subjected to large imposed dead load. It is particularly advantageous when used in medium-span and long-span bridges.

Improved recovery after overload

Prestress ensures that there is crack closure and good deflection recovery following an unexpected short-term overload, even when the overload is quite severe. This is not the case with reinforced concrete construction.

Improved strength in shear and torsion

The presence of compressive prestress in the concrete delays the formation of inclined cracks as well as vertical cracks. Prestress can therefore be used to reduce or eliminate shear and torsion problems, for example by increasing the inclined cracking load to a level that exceeds the flexural strength of the member. Even if inclined cracking does occur, the shear and torsional capacity of the member is improved by the presence of prestress. The capacity of a member in shear and torsion can be increased substantially by introducing special vertical prestress.

Improved fatigue resistance

Fatigue failure in a concrete flexural member is far more likely to occur in the tensile steel than in the compressive concrete. The fatigue resistance therefore depends largely on the fatigue properties of the tensile steel, and on the amplitude of the stress cycles that occur in the tensile steel due to the load cycles. Prestress increases greatly the minimum stress level and so reduces the amplitude of the tensile stress cycle. The fatigue resistance of a member can be greatly improved by introducing (or increasing the level of) longitudinal prestress.

Other uses

Prestressing has a variety of useful applications apart from the construction of slender concrete structures and members. For example, it is used in the repair and refurbishment of existing structures of concrete, steel and even timber.

1.2.2 Some cautionary aspects of prestressing

Prestress improves the structural performance of a concrete member in various ways, but if it is used inappropriately, or without proper understanding, serious problems can occur. If, for example, high prestress is applied at a large eccentricity to a flexural member, the camber (upward deflection) may become unacceptably large and it will increase with time because of concrete creep. Excessive camber can be a form of unserviceability. Cracking may also occur in the upper surface of the member during prestressing, before the full service load acts. In extreme situations, if an excessively high prestress is used, catastrophic failure can occur by crushing of the concrete during the prestressing operation.

Crushing of the concrete in the anchorage regions in post-tensioned members is a potential problem that can arise during construction. Very high, localised forces are induced in the concrete directly behind the mechanical anchorages. Alternatively, longitudinal cracks can occur in the anchorage region, in both post-tensioned and pre-tensioned members, and extend into the span and effectively destroy the member.

The production of prestressed concrete requires special equipment and refined construction operations. The services of a skilled labour force are also required. High-quality materials need to be used, with careful attention paid to quality assurance. Thus there are additional costs involved when prestressing is chosen as a design option. The disadvantages must be considered carefully and weighed against the structural advantages and cost savings. The decision to use prestressing needs to be based on a proper analysis of overall cost effectiveness and structural efficiency.

1.3 Use of high-strength tendons and cables

The advantages of prestressing were well appreciated by the early pioneers of reinforced concrete construction. Ingenious attempts to introduce prestressing into concrete were made in the late 19th Century, when reinforced concrete was still being developed as a building material. In early experiments, prestress was applied to mild steel reinforcement but this was not successful because the prestressing force was completely lost due to the large inelastic deformations that occur in concrete over time as a result of creep and shrinkage. It was not appreciated then that concrete undergoes such large, long-term inelastic strains.

Progressive loss of prestress due to creep and shrinkage

To illustrate the effect of creep and shrinkage we consider the concrete member shown in Figure 1.6, of length L and axially prestressed by means of a post-tensioned tendon in a duct at the centroid of the section. The initial condition of zero stress prior to prestressing is represented in Figure 1.6(a). The tendon is tensioned by jacking against the right end of the concrete to produce a tensile force P in the tendon and an equal and opposite compressive axial force C in the concrete. The tendon ends are then anchored to the concrete. Just after the prestressing operation the deformations consist of (a) an initial tensile tendon strain of $\varepsilon_{po} = P/A_pE_p$, where A_p and E_p are the area of the tendon and its elastic modulus, respectively, and (b) an initial compressive elastic strain in the concrete of $\varepsilon_{co} = P/A_cE_c$, where A_c and E_c are the area and elastic modulus of the concrete. The corresponding deformations are shown in Figure 1.6(b).

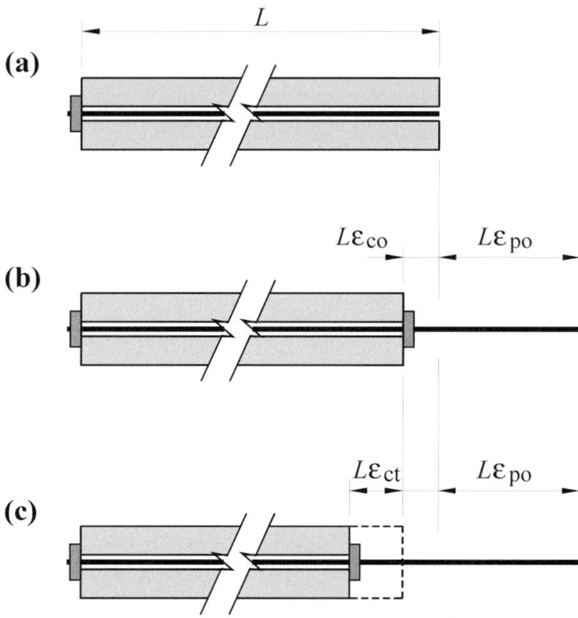

Figure 1.6 Strains during and after post-tensioning, axially placed tendon

Following stressing, the concrete continues to shorten due to creep and shrinkage. The tensioned tendon shortens by an equal amount, so that the prestressing force decreases with time. The situation after a long period of time is shown in Figure 1.6(c), where the final inelastic concrete strain is ε_{ct}. The decrease in tensile strain in the tendon is equal to ε_{ct} and the loss of prestress is $\Delta P = A_p E_p(\varepsilon_{ct})$ in both the tendon and the concrete.

It is instructive to consider typical numerical values. If mild steel bars are used, with a yield stress of 250 MPa and an initial level of prestress of say 200 MPa, the initial elastic steel strain is about 1000 microstrain, i.e. 0.001. The long-term shrinkage strain of concrete in Australia is in the order 0.0006 to 0.0010. Assuming a shrinkage strain of, say, 0.0008, we see that the prestressing force drops sharply over time to just one-fifth of its original value. When we add long-term creep, the result is an almost total loss of prestress. Prestressing losses are always significant, but with mild steel as the prestressing medium, the prestress can be completely lost within weeks.

Use of very high strength prestressing steel to manage losses

It was not until very high strength prestressing steels were introduced in the 1930s that loss of prestress due to creep and shrinkage could be limited, and practical success was achieved in the manufacture of prestressed concrete. At present, prestressing tendons with an ultimate strength of about 1750 MPa are in common use. An initial prestress to 75 per cent of the ultimate strength, say to 1300 MPa, gives an initial strain of about 0.007. With creep and shrinkage losses of, say, 0.001, the loss of prestress is still significant, in the order of 15 per cent, but the residual tension in the steel is ample to maintain good permanent prestress in the concrete.

Leonhardt (1980) used the word 'springiness' to describe the large extension needed in the tendon during the prestressing operation, relative to the small contraction of the concrete. The difference in strains needed in the two materials is almost an order of magnitude. It is important to have this springiness present if the prestressing process is to be successful.

Types of high-strength prestressing steel

Prestressing steel in use today comes in the form of small diameter, **high strength steel wire** and large diameter **high strength alloy bar**. To increase the capacity of the wire product, individual wires are often fabricated into **multi-wire strand**, typically with seven wires per strand. To further increase the tensile capacity of the prestressing unit, many strands can be used together to form a single post-tensioning **cable**. Details of the properties of wires, strands and cables, as well as information on creep and shrinkage of concrete, will be presented in Chapter 2. The terms **tendon** and **cable** are both used for the prestressing steel in a prestressed member. Depending on context, these terms may indicate either an individual component, for example a cable in a single duct, or the entire assemblage.

Development work has been going on for some years to produce prestressing tendons from materials other than steel, and we can expect to see their increased use in the near future. In this text, however, attention is focused on the use of steel tendons.

1.4 Methods of prestressing

In Figures 1.2 to 1.5 we have seen how a range of different tendon shapes can be used to prestress a concrete beam and so improve its working load behaviour. We now look at some of the methods that are used to introduce prestress into concrete members and structures:

Post-tensioning

In general, the term **post-tensioning** refers to construction processes in which the cable is tensioned *after* the concrete has been placed and hardened. The cables are typically contained in ducts that are cast in the concrete member. They are prestressed by jacking their ends against the concrete. A simple example of a post-tensioned beam with a single curved cable was shown in Figure 1.5(a).

In large members, such as bridge girders, it may be necessary to use many individual cables to achieve a sufficiently high prestressing force. Each cable may follow a separate path through the member, in its own duct. The individual paths are chosen so as to avoid congestion, and may even follow a three dimensional path, curved both in plan and elevation. The paths of the individual cables are chosen so that, when acting together, the resultant prestressing force has the profile required to counter the tensile stresses at various cross-sections along the member that will develop when external load is applied.

Various losses occur in the prestressing force in post-tensioned members, both during the prestressing operation and subsequently, and these always need to be evaluated. Progressive losses over time occur due to creep and shrinkage of the concrete as already explained in relation to Figure 1.6. Time dependent loss of prestress also occurs due to the process of **relaxation** in the tensioned prestressing tendon. Loss of prestress occurs along the tendon during prestressing because of **friction** between the tendon and the duct. The maximum force exists in the tendon at the end (or ends) where jacking takes place, but reduces progressively along the tendon due to friction between the tendon and the duct. The greater the curvature in the duct, the greater is the friction loss. Friction loss is reduced by jacking at both ends of the tendon, either sequentially or simultaneously. An immediate loss of prestress also occurs near the end of the jacking operation due to **slip**, when the force in the jack is transferred to the anchorage grips.

After the post-tensioning operation is complete, the stressed cable sits in its duct, unbonded to the surrounding concrete. This would adversely affect service load behaviour. However, bond between the cable and the inner duct surface, and hence the surrounding concrete, can be achieved by subsequent pressure grouting of the duct. Grouting also helps protect the cables against corrosion. If the cables are left unbonded, additional reinforcement may be required in the member to control cracking and also to provide adequate flexural strength.

Post-tensioning is particularly useful for the in situ construction of large flexural components such as long-span floor slabs and short transfer beams. It is used frequently in large bridge girders. Post-tensioning also plays an important role in the development of innovative methods of bridge construction, such as the segmental and launching methods which eliminate the need for temporary shoring.

Pretensioning

The term **pretensioning** refers to a process in which the prestressing steel is tensioned *before* the concrete is cast. Pretensioned members tend to be smaller than post-tensioned members and are usually constructed under factory conditions in a prestressing "bed". Individual wires or strands are used to provide the necessary prestressing force. These are tensioned and anchored temporarily at the ends of the bed, as indicated in Figure 1.7(a). The formwork is then set up around the tendons, the concrete is cast, as in Figure 1.7(b), and when it has attained sufficient strength, the prestressing force is released from the abutments, Figure 1.7(c). Within the member, the tension in the tendon is transferred to the surrounding concrete, provided of course that the bond is adequate.

In the pretensioning process an initial **elastic loss** of prestress occurs because of the initial elastic compressive strain in the concrete. This can be appreciated from Figure 1.6, where, in the case of pretensioning, the total tendon strain ε_{po} exists after pre-tensioning but prior to transfer. Just after transfer the strain in the tendon is $(\varepsilon_{po} - \varepsilon_{co})$ and so the initial elastic loss is $A_p E_p(\varepsilon_{po} - \varepsilon_{co})$. The further loss in prestress over time is $A_p E_p \varepsilon_{ct}$. The total long-term strain in the tendon reduces to $(\varepsilon_{po} - \varepsilon_{co} - \varepsilon_{ct})$.

In factory pretensioning the abutments are usually far enough apart to allow the simultaneous manufacture of large numbers of identical units of small to medium size, such as railway sleepers or small-span bridge planks. Alternatively, if a single larger member is to be pretensioned, the tendons may be stressed directly against specially strengthened formwork.

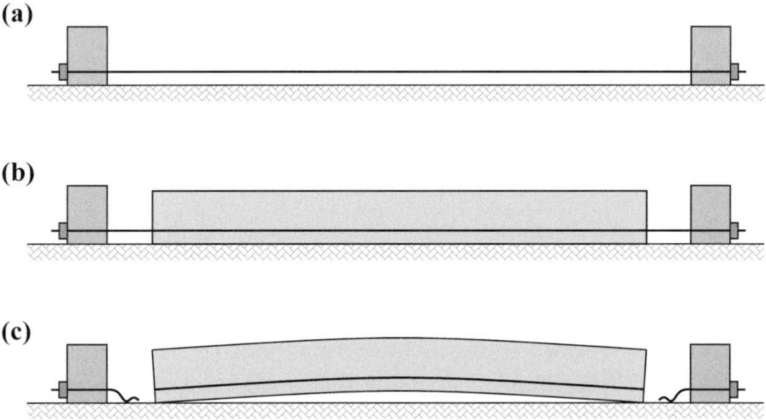

Figure 1.7 Pretensioning process

Although the tendons in a pretensioned member must be piece-wise straight, kinks and draped shapes can be used to vary the eccentricity along the beam. For example, the tendon shape shown in Figure 1.2 can be achieved in a pretensioning bed. The crossbars or pins are held temporarily and rigidly in place by attaching them to specially stiffened formwork or to the floor of the bed.

The transfer of prestress to the concrete is carried out as soon as possible to speed up the pretensioning process. Steam curing is employed to allow the manufacturing process to take place within a 24 hour cycle. The concrete is not then fully mature at transfer, and losses due to the immediate elastic compression of the concrete and to subsequent creep and shrinkage are somewhat higher than in the case of post-tensioned construction.

In pretensioned members, the tensioned prestressing steel is anchored to the concrete in the end regions through bond. The build up in prestressing force from zero at the end face to the full value P occurs over a short distance, called the **anchorage length**. Pretensioning wires, which may be as small as 5mm in diameter, are crimped or indented in order to improve the bonding characteristics. Alternatively the prestressing steel may be in the form of strands with good bond characteristics.

Prestressing with external tendons and cables

In Figure 1.2 prestress is applied to the beam by means of external tendons. External unducted post-tensioning cables are often used for the construction of large concrete hollow-box bridge girders. The cables, being external, are not protected from corrosion by the concrete, but advantages of this form of construction include ease of inspection and ease of replacement of damaged or corroded cables. External post-tensioned tendons are used to repair damaged or defective structural members of steel, concrete and timber.

External prestressing without high tensile tendons

It is of course possible to introduce a prestressing force into a concrete member without using highly stressed tendons. In Figure 1.8 an eccentric compressive force is induced in the concrete member by jacking against rigid external supports. External prestressing has a number of serious disadvantages. It is potentially dangerous and must be used with great care, but preferably not at all.

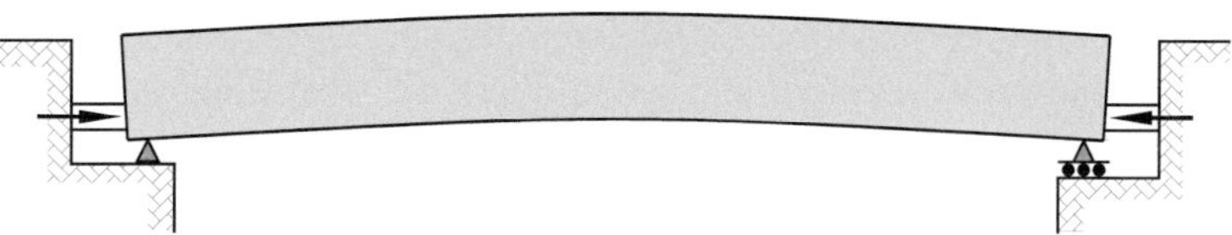

Figure 1.8 Prestressing by external forces

This system of prestressing has virtually no in-built 'springiness' in the sense discussed above in Section 1.3. Small creep and shrinkage in the concrete member can result in a rapid and complete loss of prestress. Provision therefore must be made for almost continuous re-jacking throughout the life of the structure. Even so, any slight foundation movement could lead to a complete loss of prestress.

If the beam in Figure 1.8 were of plain concrete, it would have no reserve of strength. Any unexpected overload could result in sudden catastrophic collapse. Loss of prestress is almost unavoidable, and would also result in catastrophic collapse. Reinforcing steel would be needed to ensure robustness, satisfactory overload behaviour and adequate ultimate strength. In contrast to the situation in Figure 1.8, the presence of a prestressed tendon (even unbonded) in a member contributes to ductile overload behaviour and to improving the load carrying capacity of the member.

1.5 Anchorage and bond of tendons

Anchorage

For prestressing to be effective, the tensile force in the prestressing steel has to be anchored effectively and permanently to the concrete. Special physical anchors are employed in post-tensioned construction to transfer the prestressing force from the ends of the cable into the adjacent concrete. Not only must slip of the cable be minimised, but the very large concentrated forces have to be distributed through the end region of the member in a manner that does not lead to splitting or cracking of the concrete. Special transverse reinforcement is provided in the concrete behind the physical anchor to contain and prevent the spread of cracks. Various patented anchorage devices are offered by specialist prestressing companies as a package, together with jacking equipment, anchorage zone reinforcement and cable ducts. In Australia, such firms may also undertake the detailed design of the cables and anchorages after the prestressing force and cable profile have been determined by the structural designer. Besides supplying the appropriate hardware (anchorages, jacks etc.) these specialist firms also carry out the installation and final prestressing, anchorage and grouting operations on site. Details of the anchorages available can be obtained from the supply companies.

Bond

In post-tensioned construction effective bond may or may not exist between the prestressing steel and the adjacent concrete. The structural behaviour of the member is affected by the bond condition. Good bond leads to improved behaviour in the post-cracking range, with cracks remaining smaller and more evenly distributed. The ultimate strength of the member is also improved by good bond.

In Australia, post-tensioning ducts are usually grouted in order to provide bond between the tendon and the duct. A rippled outer surface of the duct improves bond with the surrounding concrete. In the United States, the additional cost and inconvenience of the grouting process has led to the development of an unbonded, post-tensioned method of prestressing large-span floor slabs. The method is also popular in Europe. In any unbonded construction it is important that bonded reinforcing steel be included in the concrete to control cracking and ensure good overload behaviour.

In pretensioned construction, anchorage is achieved through direct bond between the previously tensioned tendon and the subsequently placed surrounding concrete. The adequacy of the bond in the end regions must be checked as part of the design calculations. In pretensioned construction the tendons are normally bonded along the full length of the member.

Debonding

When straight tendons are employed at a constant eccentricity, local overstressing in compression can occur in the upper concrete fibres near the ends of the member. To avoid this, some individual wires or strands can be debonded near the ends of the beam. **Debonding** is achieved by coating the strand with grease, or with a cover sleeve that prevents contact between the stressed strand and the concrete.

1.6 Cable profile and level of prestress

The magnitude of the prestressing force P and its profile, that is its varying eccentricity along the member $e(x)$, are the prime design parameters that affect behaviour at service load and overload, and are dealt with in later chapters. The losses in P are important and are dealt with in Chapter 9. The initial and final values of the prestressing force both have to be taken into account in the design calculations.

1.6.1 Cable profile

In this text we will use the term **cable profile** to mean the line of action of the resultant prestressing force as it passes along the length of the member. In the case of a small pretensioned member this may be simply a straight line at constant e. In large post-tensioned members e can vary from section to section, as already suggested in Figures 1.2 to 1.5.

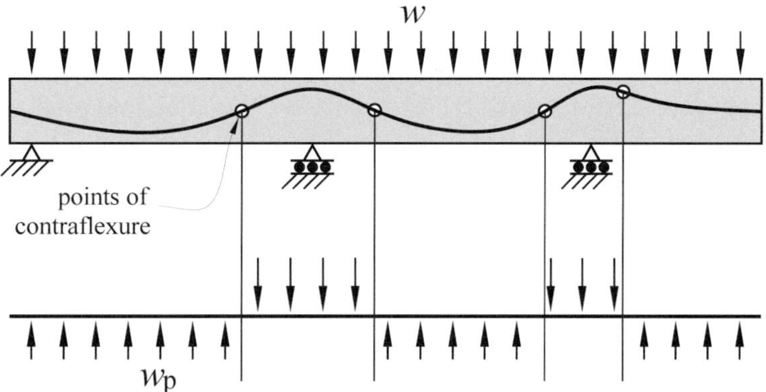

Figure 1.9 Cable profile, continuous girder

In continuous beams, the eccentricity of the cable profile rises to a maximum upward (negative) value in the regions of peak negative moment around the intermediate supports. A vertical kink at an interior support would be theoretically ideal but is not practically achievable in post-tensioned construction. Instead, a sharp negative curvature is provided over the small length of beam around the intermediate supports, as shown in Figure 1.9.

1.6.2 Level of prestress

In early design applications in the 1930s and 1940s in Europe, designers used a prestressing force P that was large enough to eliminate all cracking under service load conditions. Very little supplementary ordinary reinforcing steel was then needed to satisfy the strength and the serviceability requirements, and very little was in fact used. This construction is said to be **fully prestressed**. The early successful designs in prestressed concrete, notably those of Eugene Freyssinet, set this pattern for some decades after the Second World War. Fully prestressed concrete fitted well into the permissible stress design philosophy of the time, and was the normal approach to the design of prestressed concrete structures until well into the latter half of the 20th Century. Reinforced concrete and prestressed concrete at that time were treated as different forms of construction. In Australia, there were separate design standards for reinforced concrete and prestressed concrete.

Nevertheless, the possibility of a mixed, intermediate form of construction had long been attractive to some engineers. As early as 1939, Emperger suggested the use of a small number of prestressed tendons in reinforced concrete members to improve service load behaviour (Emperger, 1941) Published test results showed the validity of this design concept. The term **partial prestressing** was used by Abeles (1967) for such intermediate forms of construction. A member was said to be partially prestressed if the precompression was not sufficient to prevent cracks from forming under the design service load. In order to ensure adequate strength, such members required some ordinary reinforcement in the tensile regions. Transverse shear reinforcement was also needed, similar to that in a reinforced concrete member. Such construction could be described as prestressed reinforced concrete. The financial and structural advantages of using lower levels of prestress, together with some reinforcing steel, gradually came to be appreciated by designers and this approach was adopted in Europe in the 1970s, particularly in Switzerland, and since then has become widespread. The structural advantages include good ductility and robustness. Fully prestressed construction also lost favour because of brittle overload behaviour due to insufficient reinforcing steel.

The introduction of limit states concepts and performance-based design criteria for the codified design of concrete structures has also led to a better understanding of the advantages of using a combination of reinforcing steel and prestressing tendons in a member. Today, the level of prestress is treated as a design variable, to be chosen to suit the specific situation, and this approach is used in Chapters 11 and 12 in procedures for the design of beams and slabs.

This unified approach is reflected in AS 3600, which has been a common code for prestressed concrete and reinforced concrete for several generations. AS 3600 refers simply to reinforced concrete structures on the one hand (i.e. structures without prestress) and to concrete structures with prestress on the other hand. In the latter case, full prestressing is still an option, but at one end of the spectrum of possibilities. It is appropriate in special circumstances, for example if all cracking must be eliminated. This requirement may apply to radiation shielding structures and to leak-proof liquid containers that are subjected to severe temperature fluctuations. Even in such special structures, the inclusion of some ordinary reinforcement is highly desirable to provide ductility and robustness.

1.7 References

Abeles, P. W., 1967, Design of Partially Prestressed Concrete Beams, *ACI Journal*, Proc. Vol. 64, No 10, (Title 64-58), Oct. pp. 669-677.

Anon, 1977, The Mother of Invention, *50th Anniversary Edition of the Constructional Review*, Vol 50, No 4.

AS 3600:2018, *Concrete Structures*, Standards Australia, Sydney, Australia.

AS/NZS 1170.0:2002, *Structural Design Actions, Part 0: General principle*s, Standards Australia, Sydney, Australia.

AS/NZS 1170.1:2002, *Structural Design Actions – Part 1: Permanent, imposed and other actions*, Standards Australia, Sydney, Australia.

AS/NZS 1170.2:2011, *Structural Design Actions – Part 2: Wind actions*, Standards Australia, Sydney, Australia.

AS/NZS 1170.4:2002, *Structural Design Actions – Part 4: Earthquake actions in Australi*a, Standards Australia, Sydney, Australia.

Blockley D., 1980, *The Nature of Structural Design and Safet*y, Ellis Horwood, Chichester.

Emperger, Friedrich I. E., von, 1941, *Stahlbeton mit vorgespannten Zulagen aus hochwertigem Stahl* (Reinforced concrete with prestressed reinforcement made of high quality steel), Beton und Eisen.

Leonhardt F., 1980, *Vorlesungen über Massivbau, Teil 5, Spannbeton*, Springer-Verlag, Berlin (*Lectures on Structural Concrete, Part 5: Prestressed Concrete*).

Leonhardt F., 1964, *Prestressed Concrete Design and Construction*, English Edition, Wilhelm Ernst, Berlin.

Libby J. R., 1971, *Modern Prestressed Concrete*, Van Nostrand Reinhold, New York.

Lin, T. Y., 1976, Symposium on prestressed concrete, Past, Present and Future, University of California, 5th June 1976, *Journal of Prestressed Concrete Institute, Special Commerorative Issue*, Sept.

Thürlimann B, 1971, A Case for Partial Prestressing, *Proceedings Structural Concrete Symposium*, University of Toronto, Canada, May.

CHAPTER 2

Properties of materials

The behaviour of a prestressed concrete member largely reflects the behaviour and properties of the component materials: concrete, reinforcing steel, and prestressing tendon. In this chapter we look at the behaviour of these materials and their properties, insofar as they affect structural behaviour. Reinforcing steel and prestressing tendons behave in a relatively simple manner structurally, but concrete is a more complex material, partly because of the effects of shrinkage and creep. Design information from AS 3600 is summarised here.

2.1 Introduction

The structural behaviour of reinforcing steel and prestressing steel is relatively straightforward, as both materials act in an approximately elastic-plastic manner. Concrete, however, is a more complex material to deal with. It has relatively good compressive strength, but its tensile strength is very low.

Under small, short-term, compressive stresses, the response of concrete is close to being linear and elastic. Its elastic modulus increases as it ages, so that it is necessary to distinguish between the stiffness of the young concrete when prestress is applied, and later when increments of external loads act. Its behaviour becomes increasingly non-linear in the overload range. Both the stiffness and the strength properties of concrete depend on a wide range of factors, including mix details, curing conditions, age, previous history of ambient temperature and humidity, and prior stress history.

Even if concrete remains unstressed, large compressive strains develop gradually over time. This phenomenon is known as **shrinkage**. It is related to moisture content and moisture movement in the concrete, and loss of moisture to the atmosphere. When concrete is subjected to sustained stress, an additional strain develops over time, in addition to the elastic and shrinkage strains, and this is called **creep**. The long-term creep strain can be around three times larger than the initial elastic strain. Although creep and shrinkage are physically inter-related, in routine design calculations they are treated as being independent: creep is the *time-dependent* and *stress-dependent* component of strain, and shrinkage is the *time-dependent* but *stress-independent* component. Shrinkage and creep play an important role in prestressed concrete design, firstly because large additional deformations and deflections develop in a member over time, and, secondly, because they bring about a significant long-term loss of prestress. Loss of prestress due to creep and shrinkage was discussed briefly in Chapter 1.

While variability is inherent in all concrete properties, it is especially large for creep and shrinkage of concrete. Generalised creep and shrinkage data should be used with care in specific design situations, irrespective of whether the data come from textbooks, design handbooks or codes of practice. Testing to determine reliable design properties of the concrete is always a preferred option to be considered, especially for larger construction projects.

2.2 Properties of prestressing steel

Three main forms of high-strength steel are used as prestressing tendon:

- wires,
- multi-wire strands, and
- high-strength alloy bars.

The Australian and New Zealand Standard AS/NZS 4672 deals with steel prestressing materials. Important additional design information is provided in Table 3.3.1 of AS 3600.

A typical stress-strain curve for prestressing steel is shown as curve (a) in Figure 2.1. Although there is no distinct yield point, AS 3600 allows an assumed yield stress to be taken as the 0.1 per cent proof stress. This is shown as (b) in Figure 2.1. In the past it has been common practice to take the 0.2 per cent proof stress as the yield stress, which is somewhat higher. If demonstrating adequate strength becomes a critical issue in design calculations, an alternative to taking the 0.1 per cent proof stress as the yield stress is to base the strength calculations on a bi-linear approximation to the stress-strain curve, as indicated by curve (c) in Figure 2.1.

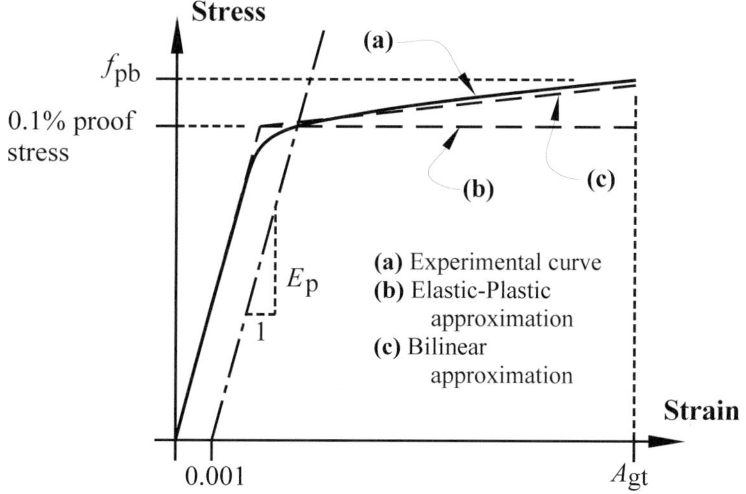

Figure 2.1 Stress-strain relationship for prestressing steels

Some important characteristic properties of prestressing steel are summarised below in Table 2.1. The characteristic minimum **breaking stress**, denoted by f_{pb}, is a practical lower bound to the tensile strength of any prestressing steel and is used in strength design calculations. For prestressing materials, the characteristic strength is defined as the 5 per cent fractile. This means that if strength tests were carried out on many specimens, only five per cent of the results would *not* reach the value f_{pb}. In Table 2.1, the limit of **uniform elongation**, A_{gt}, is the strain at the breaking stress, f_{pb}, or the characteristic minimum breaking load, P_u.

TABLE 2.1 Typical material data for prestressing steels (AS/NZS 4672.1)

Type of tendon	Dia. (mm)	Area (mm^2)	E_p (GPa)	f_{pb} (MPa)	P_u (kN)	A_{gt}
Stress relieved wire	5	19.9	205	1 700	33.8	0.035
	7	38.5	205	1 670	64.3	0.035
7-wire strand						
ordinary	12.7	98.6	195	1 870	184	0.035
	15.2	143	195	1 790	250	0.035
compact	15.2	165	195	1 820	300	0.035
	18.0	223	195	1 700	380	0.035
High strength alloy bar	26	562	200	1 030	579	0.060
	32	840	200	1 030	865	0.060
	40	1 232	200	1 030	1 269	0.060
	56	2 428	200	1 030	2 501	0.060

2.2.1 Prestressing wires

Wires are manufactured from medium- to high-carbon steel by cold drawing. Hard-drawn wire is usually stress relieved by a process of straightening and low temperature heat treatment, which improves mechanical properties such as ductility. A further process of stabilisation, by simultaneous stretching and heat treating, is sometimes used to improve the stress-relaxation properties of the steel. Steel prestressing wires may be indented or crimped to improve the bond and anchorage properties. Sizes vary from 2 mm up to 8 mm diameter, but 5 mm and 7 mm diameter wires are most frequently used in Australia. The stress-strain curve in Figure 2.1(a) is typical of prestressing wire, with the material mechanical properties as given in Table 2.1. If the elastic-plastic stress-strain curve (b) is to be used, the yield stress is taken as $0.80f_{pb}$ and $0.83f_{pb}$ for as-drawn and stress-relieved wire, respectively. The modulus of elasticity of wire, both as-drawn and stress-relieved, is taken to be 205 GPa, with a variability of ±10 GPa, unless determined otherwise by test.

2.2.2 Prestressing strand

In order to cater for the large prestressing forces needed in much structural work, prestressing wires are fabricated into multi-wire strands. In Australia both 7-wire and 19-wire strands are produced. In the commonly used 7-wire strand, six outer wires are twisted tightly into a helix around a single core wire, with a pitch of about 12 to 16 times the nominal strand diameter. The strand is usually stress-relieved to improve the mechanical properties. Stress-relieved 7-wire strands come in various sizes, ranging from 9.5 mm to 18 mm diameter, and in **ordinary** or **compact** form. Compact strands are run through a die after twisting to make the cross-sectional shape smaller and, hence, to reduce the nominal strand diameter for a given cross-sectional area.

Characteristic minimum breaking strengths of strands are given in Table 3.3.1 of AS 3600, and some typical values are included here in Table 2.1. Values for the cross-sectional areas of strands should be regarded as nominal, as areas vary slightly from time to time, depending on the amount of wear that has taken place in the manufacturing dies.

The apparent elastic modulus of prestressing strand is slightly lower than for a single wire. This is because of its spiral configuration and the tendency for the outer wires to move relative to each other and to straighten very slightly when tensioned. In AS 3600 the modulus of elasticity of stress-relieved strand is given as 200 ±5 GPa. Design data provided by prestressing companies quote the elastic modulus as varying between 180 and 205 MPa. Stress-strain curve (a) in Figure 2.1 is representative of prestressing strand, with the material mechanical properties as provided in Table 2.1. If the elastic-plastic stress-strain model is used, the yield stress is taken as $0.82f_{pb}$.

If 19-wire strands (or indeed any other types of tendon) are to be used, it is highly advisable to determine the design properties by test, or from the manufacturer. Large-format tendon, such as 19-wire strands, should not be used in pretensioning work.

2.2.3 High-strength alloy-steel bars

High-strength alloy-steel bars are usually hot-rolled, sometimes with ribs introduced during the rolling process to improve bond. The diameter varies between 26 mm and 75 mm. The characteristic minimum breaking strength of bars of all diameters is stipulated by AS 3600 as 1030 MPa. The yield stress is taken as the 0.1 per cent proof stress, or as $0.81f_{pb}$ for super-grade bars and $0.89f_{pb}$ for ribbed bars. The bilinear approximation, shown as curve (c) in Figure 2.1, can also be used for alloy bar if strength calculations become an important issue. The elastic modulus E_p for alloy bar is set at 205 ±10 GPa in AS 3600, while it is 200 ±10 GPa in AS/NZS 4672.1–2007.

2.2.4 Stress relaxation in prestressing tendons

Prestressing tendons suffer some loss of stress due to **relaxation** when held at a sustained strain over an extended period of time. A small amount of relaxation occurs in the tensioned tendons in any prestressed construction and this may need to be allowed for in the design calculations. Values commonly quoted for relaxation loss are between 2 and 4 per cent. AS 3600 and AS/NZS 4672.1 distinguish between two grades of materials: Relax 1 (formerly referred as **normal-relaxation**), and Relax 2 (formerly referred as **low-relaxation**).

TABLE 2.2 Relaxation data for prestressing steels (AS/NZS 4672.1)

Type of tendon	Initial force as percentage of f_{pb}	# Basic Relaxation R_b (max %)	
		Relax 1 (normal-relaxation)	Relax 2 (low-relaxation)
Stress relieved wire	80	10.0	3.0 #
	70	6.5	2.0
	60	3.5	1.0
7-wire strand	80	12.0	3.5 #
	70	8.0	2.5
High strength alloy bar	70	–	4.0 #
	60	–	1.5

Note: # Values of the Basic Relaxation, R_b, are used in Equation 2.1 below.

A calculation procedure for relaxation is given in AS 3600 that applies to all types of tendon. The **basic relaxation** of a tendon, R_b, is defined by AS 3600 as that measured at a temperature of 20 degrees Celsius after 1000 hours of a stress sustained at 0.8 of the characteristic minimum breaking stress, f_{pb}, for strand and $0.7f_{pb}$ for stress-relieved wire and steel-alloy bars. A standard test procedure is prescribed in AS/NZS 4672.1. In lieu of tests, values of the basic relaxation for prestressing steels are included in Table 2.2. The **design relaxation** for a tendon, R, is calculated from the basic relaxation as:

$$R = k_7 k_8 k_9 R_b \tag{2.1}$$

The factor k_7 takes account of the time over which relaxation occurs. It is calculated as:

$$k_7 = \log(5.4\, j^{1/6}) \tag{2.2}$$

where j is the time in days after prestressing (i.e. the number of days over which relaxation occurs).

TABLE 2.3 Values for relaxation coefficient k_8

σ_p / f_{pb}	k_8	
	Wire and strand	**Alloy bar**
0.40	0.0	0.0
0.70	0.75	1.0
0.80	1.0	1.83
0.85	1.75	2.25

The term k_8 takes account of the stress σ_p that is sustained in the tendon. Values are given in Table 2.3. The factor k_9 takes account of the average annual temperature ($T\,^{\circ}$C) at the location of the structure and is given by:

$$k_9 = T/20 \geq 1.0 \tag{2.3}$$

Any high temperatures that may occur during the life of the member and during construction (such as during steam curing) need to be allowed for. Losses due to steel relaxation are discussed further in Chapter 9 of this book.

2.3 Properties of reinforcing steel

In Australia, reinforcing steel is classified using two **ductility classes**: **normal** (N) and **low** (L). This classification depends on the **characteristic uniform strain at fracture**, ε_{su}. For Class N, ε_{su} is at least 0.05, while for Class L the value must be at least 0.015. In addition, a high ductility class (E), is available for earthquake resistant design. It is commonly used in New Zealand. Stainless steel bars are also available, either plain or ribbed, for which BS 6744 applies.

Standard AS/NZS 4671 applies to reinforcing steel. AS 3600 also provides specific design information, such as the **characteristic yield strength** f_{sy} and the **elastic modulus** E_s. Additional design information is available in publications provided by the manufacturers.

The types of reinforcing steel commercially available in Australia at the present time include:

- plain (undeformed) mild steel bar;
- deformed steel bar (Class N);
- hard-drawn wire (Class L); and
- reinforcing steel: plain, deformed or indented (Classes N and L).

Designated grades are used in AS 3600 to indicate characteristic yield strength, f_{sy}, and ductility. Plain round bar, with a value for f_{sy} of 250 MPa and with Class N ductility, is designated as R250N. There are two classes for deformed bar with f_{sy} = 500 MPa: D500L and D500N. It should be noted that D500L bars are only used for fitments. Welded wire mesh also has two grades, D500L and D500N. Detailed specifications for the structural properties of reinforcing steel are included in AS/NZS 4671, Steel Reinforcing Materials. According to AS 3600, the modulus of elasticity and **coefficient of thermal expansion** for all reinforcing steels may be taken 200×10^{3} MPa and 12×10^{-6} per degree Celsius, respectively, or that determined by test.

The areas of Class N reinforcing bars are given in Table 2.4. Useful information concerning the properties of reinforcement is provided in the Reinforcement Detailing Handbook (CIA, 2010).

TABLE 2.4 Details of normal ductility reinforcement (f_{sy} = 500 MPa)

Grade and diameter (mm)	N10	N12	N16	N20	N24	N28	N32	N36
Nominal area (mm^2)	80	110	200	310	450	620	800	1020

2.4 Strength properties of concrete

The concrete strengths used in prestressed concrete construction are usually somewhat higher than for reinforced concrete, with cylinder strengths ranging typically upwards from 40 MPa. Additives and low-pressure steam curing allow high concrete strengths to be obtained at early ages in order to achieve early transfer of prestress to the concrete in pretensioning construction.

2.4.1 Characteristic compressive concrete strength, f_c'

The **characteristic compressive strength**, f_c', is used in structural design as a measure of the strength of concrete in uniaxial compression at 28 days. It is defined in statistical terms as the strength at 28 days, assessed by standard cylinder tests, that is exceeded by 95 per cent of the concrete tests. The tests are carried out over a short period of time in accordance with the procedure specified in Australian Standard AS 1012.9. In the statistical evaluation of concrete strength, a normally distributed population is assumed. The difference between f_c' and the **mean strength,** f_{cm}, depends on the standard deviation s:

$$f_c' = f_{cm} - 1.64s \tag{2.4}$$

The standard deviation s varies with the degree of quality control and also with the grade of concrete. Typical values of s are in the range of 4 to 5 MPa, but reduce to 3 MPa for very good quality control. Conservative values for s vary from 5 to 6 MPa. The margin between f_c' and f_{cm} ranges from about 5 MPa for $s = 3$ MPa up to about 10 MPa for $s = 6$ MPa. In AS 3600, the

mean strength of in situ concrete, f_{cmi}, is taken to be 90 per cent of the mean cylinder strength, f_{cm}.

In prestressed concrete design, both the 28-day characteristic strength, f_c', and the strength of concrete at transfer, f_{cp}, need to be specified. AS 3600 does not impose any limits for minimum concrete strength at transfer but it does require that the strength of the entire member at transfer be checked against failure due to concrete crushing.

2.4.2 Gain in compressive strength with time

A rapid early gain in concrete strength with time is of advantage in pretensioned concrete work, and Type HE, high early strength cement, is used for this form of construction. Low-pressure steam curing is also employed in order to gain sufficient strength for transfer at an age of 24 hours.

The ratio of the standard 28 day strength to the 7 day strength depends on the chemical composition and fineness of the cement, and on ambient temperature and curing details. Figure 2.2 gives a broad indication of strength gain with time under average conditions, where $f_c(t)$ is the mean strength of concrete at age t after casting, and $f_c(28)$ is the 28 day mean strength. The *fib* Model Code (2013) gives the strength gain of concrete with time as:

$$f_c(t) = \exp\left(s \cdot \left(1 - \sqrt{\frac{28}{t}}\right)\right) f_c(28) \tag{2.5}$$

where $s = 0.38$ for general purpose (Type GP) cement and $s = 0.25$ for high early strength (Type HE) cement. Methods for increasing the rate of strength gain to speed up factory turn-around or field construction include adding accelerators to the concrete mix as well as the use of steam curing.

Care is needed in the use of accelerators, because other material properties may be adversely affected. Calcium chloride can have a severely adverse effect on the shrinkage characteristics of the concrete as well as on the rate of corrosion of the steel. When low-pressure steam curing of concrete is to be undertaken for pretensioning, preliminary testing of the concrete is highly advisable.

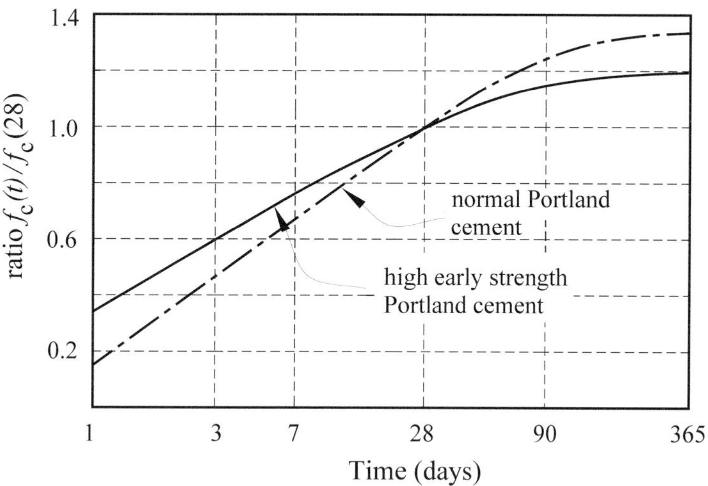

Figure 2.2 Strength gain with time, concrete

2.4.3 Tensile strength

Several different methods are used to determine the tensile strength of concrete, including a direct tension test, in which the specimen is subjected to pure tension, and indirect methods such as the Brazil (splitting) test and the modulus of rupture test. In the indirect methods, the stress varies over the cross-section of the test specimen and differing tensile strengths are obtained. AS 3600 differentiates between the **mean strength in uniaxial tension**, f_{ct}, the mean measured **flexural tensile strength**, $f_{ct.f}$, obtained from a modulus of rupture test, and the mean measured **splitting strength**, $f_{ct.sp}$.

For design calculations, AS 3600 allows the characteristic uniaxial tensile strength, f'_{ct}, and the characteristic flexural tensile strength, $f'_{ct.f}$, to be evaluated approximately from the characteristic compressive strength f'_c as follows:

$$f'_{ct} = 0.36\sqrt{f'_c} \tag{2.6}$$

$$f'_{ct.f} = 0.60\sqrt{f'_c} \tag{2.7}$$

The mean and upper characteristic values are obtained by multiplying these values by 1.4 and 1.8, respectively.

2.5 Short-term deformation of concrete

2.5.1 Elastic modulus

When subjected to short-term compressive stresses in the working load range, concrete's response is almost linear-elastic. According to AS 3600, the mean elastic modulus at a specific age, E_c, can be calculated from the mean compressive strength of the in situ concrete, f_{cmi}, at the same age. There are two expressions for the elastic modulus:

$$\text{for } f_{cmi} \leq 40 \text{ MPa}: \quad E_c = \rho^{1.5} \times [0.043\sqrt{f_{cmi}}] \text{ MPa} \tag{2.8}$$

$$\text{for } f_{cmi} > 40 \text{ MPa}: \quad E_c = \rho^{1.5} \times [0.024\sqrt{f_{cmi}} + 0.12] \text{ MPa} \tag{2.9}$$

where f_{cmi} is in MPa and the density of the concrete, ρ, is in kg/m^3.

For normal weight concretes, the density can be taken to be 2400 kg/m^3 and for f_{cmi} not greater than 40 MPa, Equation 2.8 reduces to:

$$E_c = 5060\sqrt{f_{cmi}} \quad \ldots\ldots \quad f_{cmi} \leq 40 \text{ MPa} \tag{2.10}$$

In the equations above, E_c can be ± 20 per cent from the calculated value.

2.5.2 Stress-strain relation

A fundamental material property is the relation between strain and stress. For concrete, expressions for the stress-strain relation should allow for variations in the initial slope (elastic modulus) of the curve, because it is affected by age and compressive strength. Various stress-strain curves have been proposed for concrete in compression. One such equation is a modified form of the Popovics (1970) model, subsequently developed by Thorenfeldt et al. (1987) and Collins and Porasz (1989) and expressed as:

$$f_c = f_{co}\frac{n\eta}{n - 1 + \eta^{nk}} \tag{2.11}$$

where f_{co} is the strength of the in situ concrete. AS 3600 adopts the value $f_{co} = 0.9f_c'$ for strength design, and for serviceability design $f_{co} = f_{cmi}$. In Equation 2.11, $\eta = \varepsilon_c/\varepsilon_o$ where ε_c is the concrete strain and ε_o is the strain corresponding to the peak in situ strength, f_{co}. Also, $n = E_c/(E_c - E_{cp})$ and

E_{cp} is the secant modulus corresponding to the peak in situ strength ($E_{cp} = f_{cmi}/\varepsilon_o$).

The term k in Equation 2.11 is a decay factor that defines the post peak response. It increases with concrete strength. A suitable expression for k is given by Collins and Porasz (1989) in the form of the step function:

$$\text{for } \varepsilon \le \varepsilon_o \qquad k = 1.0 \tag{2.12}$$

$$\text{for } \varepsilon > \varepsilon_o \qquad k = 0.67 + f_{co}/62 \le 1.0 \tag{2.13}$$

where f_{co} is in MPa.

The strain ε_o, corresponding to the peak stress, depends on the aggregate and mix properties. For Australian concretes it is approximately (Attard and Stewart, 1998):

$$\varepsilon_o = 4.11 f_{cmi}^{3/4}/E_{co} \tag{2.14}$$

where both f_{cmi} and the initial tangent modulus, $E_{co} \approx 1.1E_c$, are in MPa. For normal strength concretes (25 to 50 MPa), ε_o is of the order of 0.002, but it increases to about 0.0028 for concrete strengths of 100 MPa.

The shape of the stress-strain relationship for concrete in uniaxial compression depends on the mean in situ concrete strength f_{cmi}. Figure 2.3 shows curves derived from Equations 2.11 to 2.14, with strengths ranging from 20 MPa up to 100 MPa.

There is a slow, less-than-proportional increase in initial elastic modulus with increasing stress. However, the shape of the stress-strain relation also depends on the type of aggregate and on the rate of application of load. For this reason, it is desirable to use experimental data for accurate calculations of structural deformations.

In AS 3600 various **standard strength grades** are specified. These are preferred values of the characteristic compressive strength for use in design. The grades are: 20, 25, 32, 40, 50, 65, 80 and 100. These figures are the values of f_c' in MPa. Table 2.5 gives values of some properties for the standard grades.

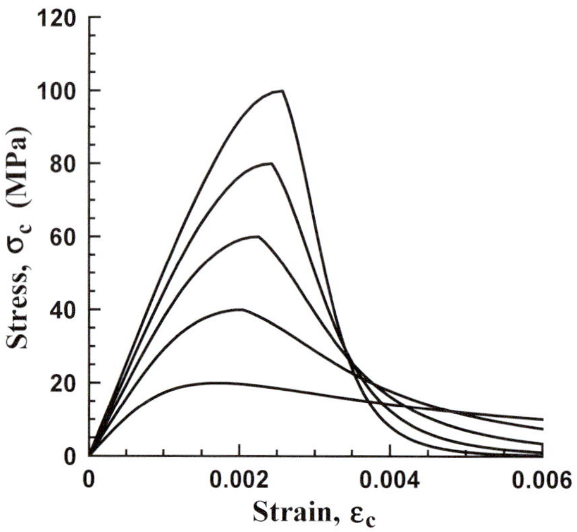

Figure 2.3 Stress-strain curves, concrete in compression

TABLE 2.5 Properties of standard grade concretes

Standard strength grade f'_c (MPa)	Mean in situ comp. strength f_{cmi} (MPa)	Mean in situ elastic modulus E_c (MPa)	Flexural tensile strength $f'_{ct.f}$ (MPa)	Uniaxial tensile strength f'_{ct} (MPa)	Modular ratio n
20	22	24 000	2.7	1.6	8.3
25	28	26 700	3.0	1.8	7.5
32	35	30 100	3.4	2.0	6.6
40	43	32 800	3.8	2.3	6.1
50	53	34 800	4.2	2.5	5.7
65	68	37 400	4.8	2.9	5.3
80	82	39 600	5.3	3.2	5.0
100	99	42 200	6.0	3.6	4.7

2.6 Shrinkage of concrete

The magnitude of the shrinkage strain in concrete depends on the age *t,* on the initial amount of moisture contained in the concrete and on the moisture present in the surrounding atmosphere. Some shrinkage occurs even when the atmosphere is at 100 per cent humidity. This is called **endogenous shrinkage** (or autogenous or chemical shrinkage). There is additional shrinkage if the humidity in the atmosphere is less than 100 per cent, and this is called **drying shrinkage**. At time *t* after setting, the total shrinkage is:

$$\varepsilon_{cs}(t) = \varepsilon_{csd}(t) + \varepsilon_{cse}(t) \tag{2.15}$$

The final subscripts **d** and **e** refer here to the terms drying shrinkage and endogenous shrinkage, respectively. In normal strength concretes, drying shrinkage forms the major component of the shrinkage, but endogenous shrinkage becomes more important in high strength concretes.

It is important to observe how time is measured when dealing with creep and shrinkage. In equations for endogenous shrinkage, *t* is the time from the first set of the concrete. For drying shrinkage calculations, *t* is measured from the end of curing. For creep due to a stress increment, time is measured from the time when the stress increment is applied.

Under reasonably constant ambient conditions, shrinkage strain increases with time to approach a long-term limiting value of ε^*_{cs} (also written as ε^*_{sh}) as in Figure 2.4(a). This can be represented by a time function $f(t)$, which is zero at time zero (at the setting of the concrete, or at the end of curing) and approaches unity as *t* approaches infinity:

$$\varepsilon_{cs}(t) = \varepsilon^*_{cs} \cdot f(t) \tag{2.16}$$

In practice, changes in ambient conditions can occur over relatively short time periods, so that the rate of increase of shrinkage strain is irregular. Figure 2.4(a) is thus a highly idealised, 'average' curve, which nevertheless is used to evaluate the overall structural effects of shrinkage.

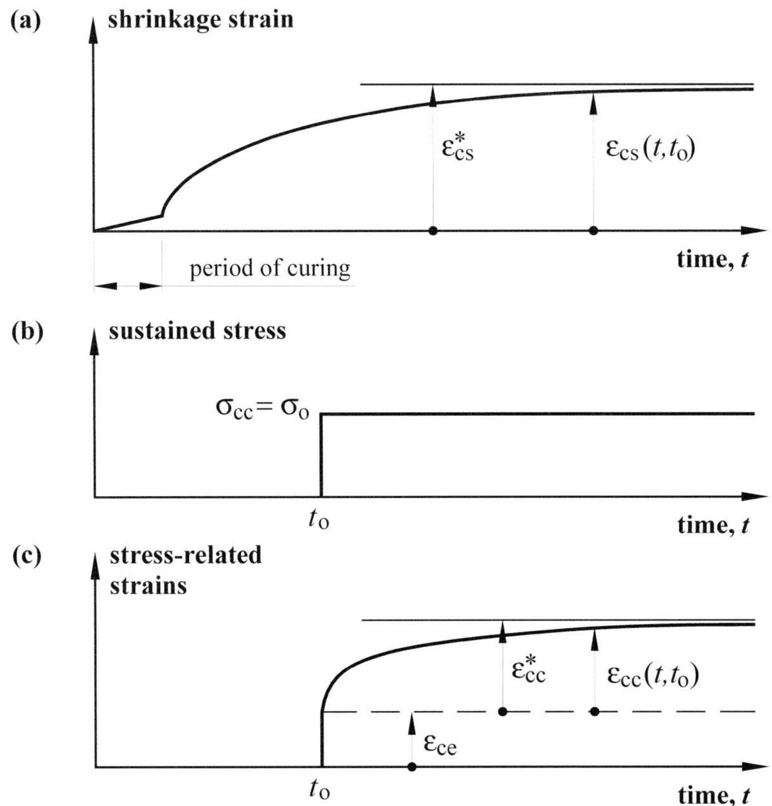

Figure 2.4 Development of creep and shrinkage strain with time

In Figure 2.4(a) the total shrinkage strain is shown on a time axis that commences after casting at first set of the concrete. In Figure 2.4(b) the sustained stress σ_{cc} is applied when the concrete is at age t_o on the time axis. In Figure 2.4(c), the initial elastic strain, ε_{ce}, occurs at t_o, when σ_o is applied. The creep strain $\varepsilon_{cc}(t, t_o)$ develops progressively from time t_o. To avoid confusion, we will use italics below to make clear which time scale that is being referred to.

AS 3600 data for concrete shrinkage

In lieu of special shrinkage tests, AS 3600 provides data to evaluate a **design shrinkage strain,** ε_{cs}. This is the sum of the endogenous and drying components, ε_{cse} and e_{csd}, as in Equation 2.14.

The endogenous component (which occurs equally in fully moist or dry environments) is calculated by AS 3600 as:

$$\varepsilon_{cse} = \varepsilon^*_{cse}(1.0 - e^{-0.07t}) \tag{2.17}$$

Here, t is the time in days *after the concrete has set*. The term ε^*_{cse} is the final value of the endogenous strain, and is given by AS 3600 as:

$$\text{for } f_c' \leq 50\text{ MPa} : \varepsilon^*_{cse} = (0.07f_c' - 0.5) \times 50{\times}10^{-6} \tag{2.18}$$

$$\text{for } f_c' > 50\text{ MPa} : \varepsilon^*_{cse} = (0.08f_c' - 1.0) \times 50{\times}10^{-6} \tag{2.19}$$

The drying shrinkage at an intermediate time t *after the commencement of drying* is expressed as:

$$\varepsilon_{csd} = k_1 k_4\, \varepsilon_{csd.b} \tag{2.20}$$

where $\varepsilon_{csd.b}$ is a **basic drying shrinkage** strain, and is calculated as:

$$\varepsilon_{csd.b} = (0.9 - 0.005f_c') \times \varepsilon^*_{csd.b} \tag{2.21}$$

where $\varepsilon^*_{csd.b} = 800\text{x}10^{-6}$, or determined by testing. The term k_4 takes account of the environmental conditions, with values given in Table 2.6.

TABLE 2.6 Values for environment coefficient k_4, used to calculate both shrinkage and creep strains

Environment	k_4
Arid environment	0.70
Interior environment within a building structure	0.65
Temperate inland environments	0.60
Tropical environments and environments within 50 km of the coast	0.50

The term k_1 in Equation 2.20 allows for the time over which drying occurs, and the size and shape of the cross-section. The values of k_1 depend on the time *t since commencement of drying* and on the **hypothetical thickness** of the member, t_h, which is calculated as:

$$t_h = 2A_g/u_e \tag{2.22}$$

The term A_g is the gross cross-sectional area of the member and u_e is the **exposed perimeter**, taken as the portion of the perimeter of the section that is exposed to the atmosphere, plus the perimeter of any closed voids contained in the cross-section. For a rectangular cross-section of overall dimensions b and D, with all sides exposed, the hypothetical thickness is:

$$t_h = bD/(b+D) \tag{2.23}$$

If the member is a thin wall or slab, with $b >> D$, t_h becomes equal to the thickness D.

Frequently in design practice only the long-term shrinkage, ε^*_{cs}, is needed. This is the final value at the end of the shrinkage process (usually taken to be at about 30 years). Final, long-term values of k_1 are expressed in this text as k_1^* and are given in Table 2.7 for t_h values between 50 and 400 mm.

TABLE 2.7 Values of k_1^* for evaluating long-term shrinkage strains at time t^* = 30 years

t_h (mm)	50	100	200	400
k_1^*	1.73	1.51	1.22	0.93

If intermediate values of k_1 are needed, they can be obtained from Figure 2.5 below, in which the ratio k_1/k_1^* is plotted against time since the commencement of drying, t, for various values of the hypothetical thickness t_h. Values for k_1 are determined from:

$$k_1 = \frac{\alpha_1 t^{0.8}}{t^{0.8} + 0.15 t_h} \tag{2.24}$$

where:

$$\alpha_1 = 0.8 + 1.2\,e^{-0.005 t_h} \tag{2.25}$$

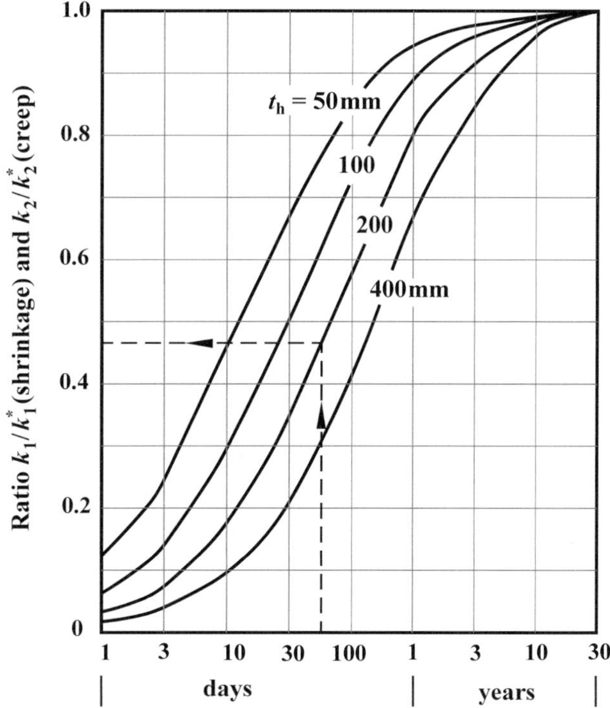

Figure 2.5 Ratios of k_1 and k_2 at time t relative to their final value at time equal to 30 years.

It is not immediately obvious from AS 3600 that the relationship between shrinkage at time t after the onset of drying and creep at time t after loading (discussed in the next Section), relative to their final values, are treated identically. This is evident in Figure 2.5, which, in the authors' opinion, is an improved representation of the influence of time on shrinkage and creep as presented in AS 3600. The plotted ratios are given by:

$$\frac{k_1}{k_1^*} = \frac{k_2}{k_2^*} = \frac{\alpha t^{0.8}}{t^{0.8} + 0.15 t_h} \tag{2.26}$$

where $\alpha = 1 + 8.8 \times 10^{-5} t_h \approx 1.0$. Taking α as unity gives a less than four per cent error at $t_h = 400$ mm, and reduces for lower t_h values, and well within the variability of the outcomes, which is of the order of ± 30 per cent.

For shrinkage, the influence of thickness on the transport of moisture through diffusion, leading to moisture loss from the surfaces, is readily observed from Figure 2.5. For example, an element with $t_h = 50$ mm will reach 90 per cent of its final shrinkage after approximately six months, whereas for a member with $t_h = 400$ mm, this takes about 5 years. For a typical example of a 200 mm thick suspended slab, 50 per cent of the final shrinkage occurs after about two months.

To evaluate the long-term value of the design shrinkage strain, ε_{cs}^*, the component ε_{cse}^* is determined from Equations 2.18 and 2.19 and the component ε_{csd}^* is evaluated using Equation 2.20 with the long-term value of k_1^* obtained from Table 2.7. The components are then added as per Equation 2.15. Typical final design shrinkage strains (at 30 years) are listed in Table 3.1.7.2 in AS 3600 for various concrete strengths, environmental conditions, and hypothetical thicknesses. Values range from around 800 to 350 microstrain.

To evaluate the design shrinkage strain ε_{cs} at intermediate times, the component ε_{csd} is determined from Equation 2.20 using the appropriate value of k_1 obtained from Equation 2.24 or Figure 2.5. The component ε_{cse} is obtained from Equation 2.17. These components are added together as in Equation 2.15.

Non-uniform shrinkage

The AS 3600 data for evaluating shrinkage are based on the assumption that shrinkage is uniform throughout the member, with equal strain in all fibres. In real situations, the moisture conditions within the concrete vary, with less moisture at the surfaces (where drying occurs), and a reserve of moisture in the interior. Shrinkage in a section is therefore non-uniform: lower in the interior regions, but much higher at and near the surfaces. This results in tensile stresses in the surface regions of the concrete, and equilibrating compressive stresses in the interior. Non-uniform shrinkage is inevitable in practice. It often leads to load-independent cracking on the surfaces of concrete members.

2.7 Creep of concrete under constant stress

In Figure 2.4(b) a constant sustained stress σ_o is applied to the concrete at age t_o after casting. Creep strain, like shrinkage strain, is taken to be positive if compressive. The creep strain depends on the stress level, the time of loading, $(t - t_o)$ and the age of the concrete at first loading, t_o, as well as on the ambient conditions and the details of the concrete mix. It is written as $\varepsilon_{cc}(t, t_o)$, and the final, maximum, long-term creep strain as $\varepsilon^*_{cc}(t_o) = \varepsilon_{cc}(t^*, t_o)$.

creep function

In the working load range, creep strain is taken to be proportional to the sustained stress σ_o. The elastic strain, ε_{ce}, at time t_o is also proportional to stress:

$$\varepsilon_{ce}(t_o) = \sigma_o / E_c(t_o) \tag{2.27}$$

The ratio of creep strain to elastic strain is therefore *independent* of the sustained stress σ_o, and the **creep function** $\varphi(t, t_o)$ is:

$$\varphi(t,t_o) = \varepsilon_{cc}(t,t_o)/\varepsilon_{ce}(t_o) \tag{2.28}$$

The term t_o is a reminder that the function depends on the age of the concrete when the sustained stress is first applied. The final value of the function at time infinity, t^*, is written as:

$$\varphi^*_o = \varphi(t^*,t_o) = \frac{\varepsilon_{cc}(t^*, t_o)}{\varepsilon_{ce}(t_o)} \tag{2.29}$$

Being independent of stress, φ^*_o is a useful measure of the potential creep of the concrete when loaded at age t_o. Physically, φ^*_o can be thought of as the long-term creep strain observed in a constant-stress creep test, divided by the initial elastic strain observed in the specimen when it is first loaded at age t_o.

The creep strain is evaluated from the creep function as:

$$\varepsilon_{cc}(t, t_o) = \varepsilon_{ce}(t_o)\,\varphi(t, t_o) \tag{2.30}$$

and the long-term creep strain is:

$$\varepsilon^*_{cc}(t_o) = \varepsilon_{ce}(t_o)\,\varphi^*_o \tag{2.31}$$

Specific creep

An alternative method of treating creep strain is to use the specific creep, $C(t, t_o)$, which is defined as the creep strain resulting from a unit sustained stress, first applied at age t_o. The creep strain due to a constant sustained stress σ_o is:

$$\varepsilon_{cc}(t, t_o) = \sigma_o C(t, t_o) \tag{2.32}$$

The specific creep and the creep function are related through the elastic modulus by:

$$\varphi(t, t_o) = E_c(t_o)\,C(t, t_o) \tag{2.33}$$

Effective concrete modulus $E_R(t, t_o)$

If the stress remains constant (or nearly constant) over time, the creep strain and the elastic strain can be lumped together and the total stress-dependent strain can be calculated using an 'effective' modulus, which gradually decreases with time. The reduced modulus is:

$$E_R(t,t_o) = \frac{E_c(t_o)}{1 + \varphi(t,t_o)} \tag{2.34}$$

The linear (but time-dependent) relation between total strain and stress is expressed as:

$$\varepsilon_c(t,t_o) = \frac{\sigma_o}{E_R(t,t_o)} \tag{2.35}$$

Although there is little advantage in using Equations 2.34 and 2.35 in constant stress creep calculations, a modification of the effective modulus, called **the age-adjusted effective modulus**, is used to deal with creep under varying stress, as we shall see shortly in Section 2.8.2.

AS 3600 creep data

Data are provided in AS 3600 for evaluating creep strains under conditions of constant sustained stress and constant ambient conditions. The data have been derived from Australian tests. However, field values can differ by 30 per cent, or more, from the AS 3600 values. When creep and shrinkage are significant design considerations, data from tests should be used.

In AS 3600 terminology, the quantity $\varphi(t,t_o)$ is referred to as the **creep coefficient**. A **design creep coefficient**, $\varphi_{cc,}$ is evaluated from a **basic creep coefficient**, $\varphi_{cc.b}$, and various modifying factors:

$$\varphi_{cc} = k_2 k_3 k_4 k_5 k_6 \varphi_{cc.b} \tag{2.36}$$

The basic creep coefficient, $\varphi_{cc.b}$, is the final value of the creep coefficient, φ_o^*, when a sustained stress of $0.4f_c'$ is applied at 28 days, in ambient conditions like those that would exist in an enclosed building. The preferred method for evaluating $\varphi_{cc.b}$ is by test, in accordance with AS 1012.16–1996. In the absence of test data, AS 3600 provides numerical values, which are reproduced here in Table 2.8.

TABLE 2.8 Values of $\varphi_{cc.b}$

Concrete strength (MPa)	25	32	40	50	65	80	100
$\varphi_{cc.b}$	4.2	3.4	2.8	2.4	2.0	1.7	1.5

In Equation 2.36, the factor k_2 accounts for the increase in creep strain with duration of loading. It also takes account of the cross-sectional shape (as represented by the hypothetical thickness t_h) and is comparable with the term k_1 previously used to evaluate shrinkage strains. Numerical values of k_2 are presented in graphical format in Figure 3.1.8.3 of AS 3600.

Often the designer will only need to consider final long-term creep values. The maximum value of k_2 at t^* is denoted here as k_2^*, with values provided in Table 2.9 for a range of hypothetical thickness (t_h) values.

TABLE 2.9 Values of k_2^* for evaluating long-term creep

t_h (mm)	50	100	200	400
k_2^*	1.74	1.49	1.20	1.01

Intermediate values of k_2 can be obtained from the expression:

$$k_2 = \frac{\alpha_2 t^{0.8}}{t^{0.8} + 0.15 t_h} \tag{2.37}$$

where t is the time in days since loading, and:

$$\alpha_2 = 1.0 + 1.12 e^{-0.008 t_h} \tag{2.38}$$

As described above, Figure 2.5 plots values for the ratio k_2/k_2^* against time t for a range of t_h values, where the influence of thickness of the section can be readily observed. For example, at time equal to 100 days, a section of t_h = 400 mm will have achieved approximately 40 per cent of its final creep strain, whereas for t_h = 50 mm, this value nears 85 per cent.

In AS 3600 the expression:

$$k_3 = 2.7/[1 + \log(t_o)] \tag{2.39}$$

is referred to as the maturity factor, and is used to evaluate the maturity of the concrete at the age when the sustained stress is first applied, t_o. Values may be obtained directly from Table 2.10. It should be noted that in AS 3600 the term τ is used instead of t_o for the age at first loading.

TABLE 2.10 Values of creep maturity coefficient, k_3

t_o (days)	3	7	28	365	10,000
k_3	1.83	1.46	1.10	0.76	0.54

Note: In interpolating intermediate values, the non-linearity of the time scale should be allowed for.

Values for k_4 are the same as for shrinkage, and are given in Table 2.6.

The term k_5 is a modifier for high strength concrete and is unity for concrete strengths less than or equal to 50 MPa. For strengths greater than 50 MPa but not exceeding 100 MPa:

$$k_5 = 2.0 - \alpha_3 - 0.02(1.0 - \alpha_3) f'_c \tag{2.40}$$

$$\alpha_3 = 0.7/(k_4 \alpha_2) \tag{2.41}$$

where k_4 is given in Table 2.6 and α_2 by Equation 2.38.

The term k_6 takes into account non-linear creep that occurs in the case of high applied sustained stress. It is taken as unity unless the sustained stress is greater than $0.45 f_{cmi}$, in which case it is evaluated from:

$$k_6 = e^{1.5(\sigma_o / f_{cmi} - 0.45)} \tag{2.42}$$

EXAMPLE 2.1

Calculation of Creep and Shrinkage Strains

Estimate the total long-term compressive strain due to a sustained stress of 10 MPa in an axially loaded, plain concrete compression member with cross-section 400 mm square. The environment is temperate inland. For the concrete, f'_c = 32 MPa at 28 days. Two load cases are to be considered: (a) the load is applied at age 14 days; (b) the load is applied at age 180 days.

Other data: Type GP cement; 7-day curing period; all sides are exposed.

SOLUTION

(a) Loading at 14 days

Elastic strain:

For $f'_c = 32$ MPa, Table 2.5 gives the mean in situ strength of the concrete at $t = 28$ days as $f_{cmi} = 35$ MPa. From Equation 2.5, for age $t = 14$ days:

$$f_{c.14} = \exp\left(0.38 \cdot \left(1 - \sqrt{\frac{28}{14}}\right)\right) \times 35 = 30 \text{ MPa}$$

and from Equation 2.8:

$$E_{c.14} = 5060 \times \sqrt{30} = 27.7 \times 10^3 \text{ MPa}$$

The elastic increment of strain at 14 days is:

$$\varepsilon_{ce.14} = 10/27.7 \times 10^3 = 360 \times 10^{-6}$$

Shrinkage strain:

The endogenous shrinkage is determined from Equation 2.18, which gives $\varepsilon^*_{cse} = 87 \times 10^{-6}$.

To evaluate the drying shrinkage, we calculate the hypothetical thickness (Equation 2.23) as:

$$t_h = \frac{2 \times 400 \times 400}{4 \times 400} = 200 \text{ mm}$$

From Equation 2.21: $\varepsilon_{csd.b} = 592 \times 10^{-6}$

From Table 2.6, $k_4 = 0.6$, and from Table 2.7, $k^*_1 = 1.22$

From Equation 2.21: $\varepsilon_{csd} = 1.22 \times 0.6 \times 592{\times}10^{-6} = 433{\times}10^{-6}$

The final design shrinkage strain is:

$$\varepsilon^*_{cs} = (87 + 433) \times 10^{-6} = 520{\times}10^{-6}$$

Creep strain:

For $f'_c = 32$ MPa, Table 2.8 gives the basic value $\varphi_{cc.b} = 3.4$.

From the shrinkage calculation we have $t_h = 200$ mm and $k_4 = 0.6$.

From Table 2.10, for $t_o = 14$ days and using interpolation, we obtain $k_3 = 1.26$.

From Table 2.9, $k^*_2 = 1.20$. In this example, $k_5 = k_6 = 1.0$. Hence:

$$\varphi^*_{cc} = 1.20 \times 1.26 \times 0.6 \times 1.0 \times 1.0 \times 3.4 = 3.08$$

The final creep strain is:

$$\varepsilon^*_{cc} = \varphi^*_{cc} \times \varepsilon_{ce.14} = 3.08 \times 360{\times}10^{-6} = 1110{\times}10^{-6}$$

Total strain:

At time t^*, with first loading at $t_o = 14$ days, the total strain is:

$$\varepsilon^*_c = \varepsilon_{ce.14} + \varepsilon^*_{cs} + \varepsilon^*_{cc}$$
$$= (360 + 520 + 1110) \times 10^{-6} = 1990{\times}10^{-6}$$

(b) Loading at 180 days

Elastic strain:

As before, at age 28 days $f_{cmi} = 35$ MPa and from Equation 2.5 the 180 day strength is calculated as:

$$f_{c.180} = \exp\left(0.38 \cdot \left(1 - \sqrt{\frac{28}{180}}\right)\right) \times 35 = 44 \text{ MPa}$$

From Equation 2.9 for $\rho = 2400$ kg/m^3:

$$E_{c.180} = 2400^{1.5}[0.024\sqrt{44} + 0.12] = 32.8\times10^{3} \text{ MPa}$$

The elastic component of the strain at 180 days is:

$$\varepsilon_{ce.180} = 10/32.8\times10^{3} = 305\times10^{-6}$$

Shrinkage strain:

The final shrinkage strain is independent of load and time of loading, and from (a) above: $\varepsilon^*_{cs} = 520\times10^{-6}$.

Creep strain:

From Table 2.10 and with interpolation we obtain $k_3 = 0.83$ for $t_o = 180$ days. Other k values are unchanged from the previous calculation for 14 days. The final creep coefficient is:

$$\varphi^*_{cc} = 1.20 \times 0.83 \times 0.6 \times 1.0 \times 1.0 \times 3.4 = 2.03$$

and the final creep strain is:

$$\varepsilon^*_{cc} = \varphi^*_{cc} \times \varepsilon_{ce.180} = 2.03 \times 305\times10^{-6} = 619\times10^{-6}$$

It is observed that this is considerably lower than for the example of loading at 14 days, where the creep strain was 1110 microstrain.

Total strain:

With first loading at $t = 180$ days, the total strain is:

$$\varepsilon^*_{c} = \varepsilon_{ce.180} + \varepsilon^*_{cs} + \varepsilon^*_{cc}$$

$$= (305 + 520 + 619)\times10^{-6} = 1444\times10^{-6}$$

2.8 Concrete creep under varying stress

In practice, the sustained stress that causes creep is rarely constant. Variations in stress over time occur because of changes in the external loads. A gradual decrease in stress also occurs in the concrete in a prestressed cross-section. This is because of the gradual redistribution of internal stresses, which is due to the build-up of compressive creep and shrinkage strains in the concrete, and transfer of compressive prestress from the concrete to the adjacent compressed steel and tendon. This is effectively a feedback effect, in which the decrease in concrete stress is itself initiated by the creep that originates from the stress.

2.8.1 Superposition approach for discrete changes in stress

If the changes in stress occur in discrete increments at known times, then the creep strain can be evaluated independently for each stress increment and the results can be superposed. The stress history in Figure 2.6 consists of stress σ_o applied at t_o, and an increment $\Delta\sigma_1$ at t_1, where $\Delta\sigma_1 = \sigma_1 - \sigma_o$. For $t > t_1$, the elastic and creep strains are, respectively:

$$\varepsilon_{ce}(t) = \varepsilon_{ceo} + \varepsilon_{ce1} = \frac{\sigma_o}{E_c(t_o)} + \frac{\Delta\sigma_1}{E_c(t_1)} \tag{2.43}$$

$$\varepsilon_{cc}(t) = \varepsilon_{cc}(t,t_o) + \varepsilon_{cc}(t,t_1) = \frac{\varphi(t,t_o)\sigma_o}{E_c(t_o)} + \frac{\varphi(t,t_1)\Delta\sigma_1}{E_c(t_1)} \tag{2.44}$$

If the stress history consists of many increments, $\Delta\sigma_1$, $\Delta\sigma_2$, ... $\Delta\sigma_i$, ... $\Delta\sigma_n$, applied at times t_1, t_2, ... t_i, ..., t_n, then:

$$\varepsilon_{ce}(t) = \varepsilon_{ceo} + \sum_{i=1}^{n} \varepsilon_{ce}(t_i) = \frac{\sigma_o}{E_c(t_o)} + \sum_{i=1}^{n} \frac{\sigma_i}{E_c(t_i)} \tag{2.45}$$

$$\varepsilon_{cc}(t) = \varepsilon_{cc}(t,t_o) + \sum_{i=1}^{n} \varepsilon_{cc}(t,t_i) = \frac{\varphi(t,t_o)\sigma_o}{E_c(t_o)} + \sum_{i=1}^{n} \frac{\varphi(t,t_i)\Delta\sigma_i}{E_c(t_i)} \tag{2.46}$$

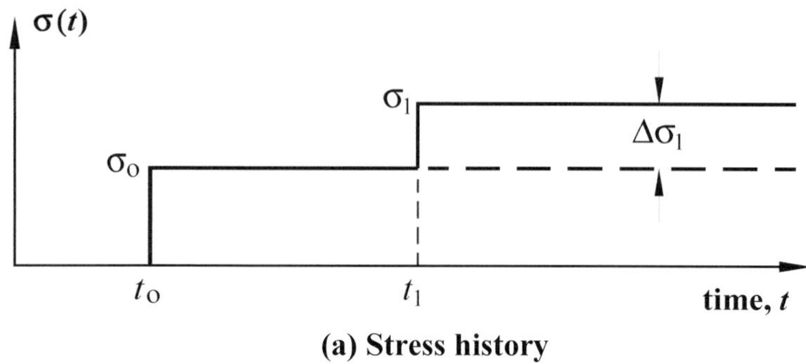

(a) Stress history

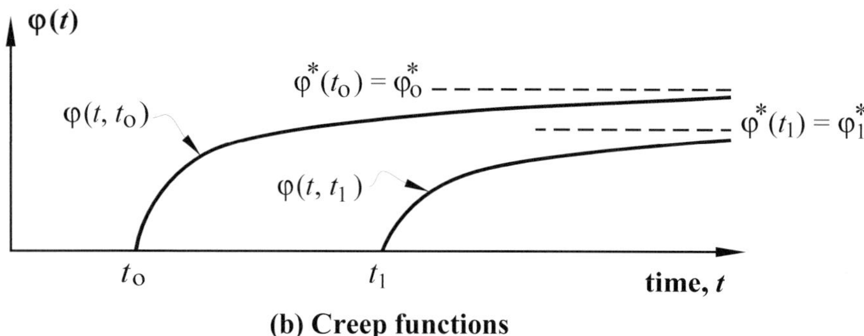

(b) Creep functions

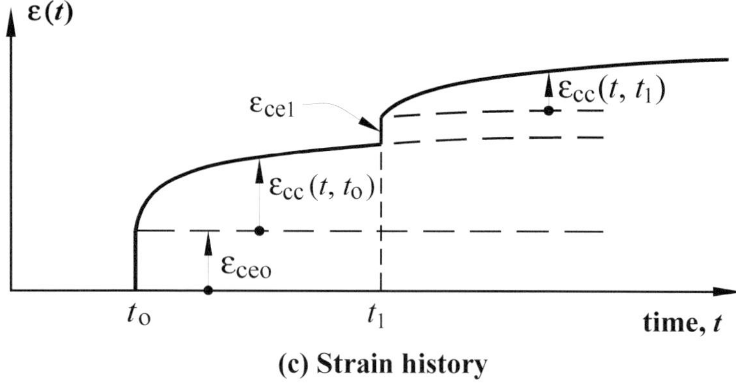

(c) Strain history

Figure 2.6 Creep under a two-step load history

Creep recovery

If the stress history includes large stress decrements as well as increments, the superposition principle cannot be relied on to give accurate results, because the creep recovery that follows unloading after a period of sustained loading is greater than that predicted by superposition. This is illustrated in Figure 2.7, where σ_o is applied at time t_o and removed at t_1. Note that in Figure 2.7(b) full elastic recovery at t_1 does not occur, as the elastic modulus has increased over the time interval.

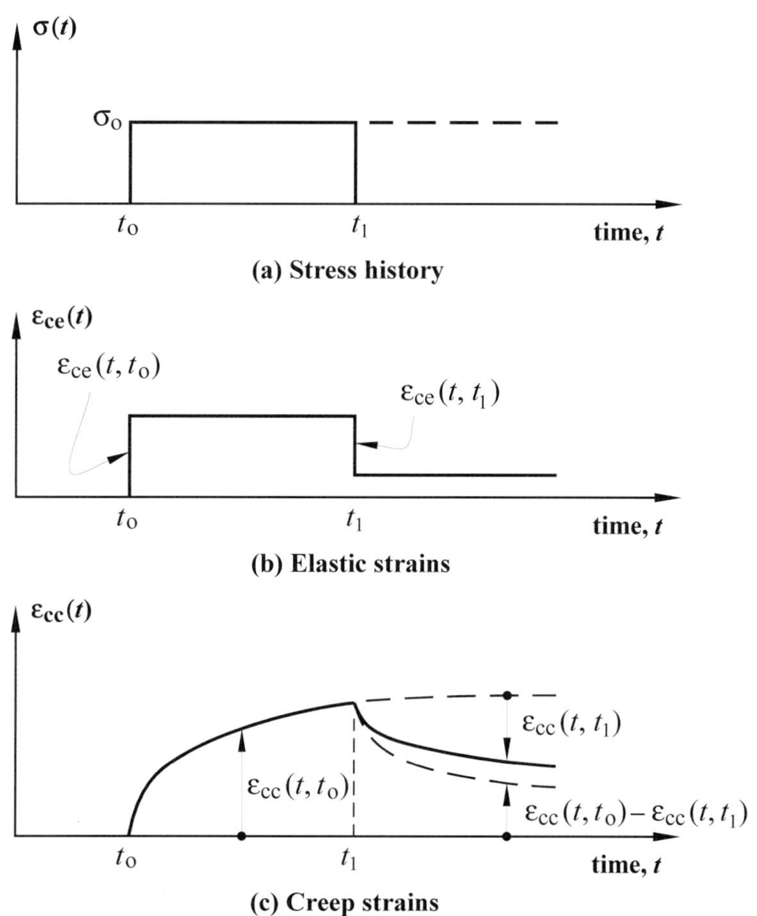

Figure 2.7 Creep recovery

In Figure 2.7(c), the full line represents behaviour as predicted by superposition. The dashed lower line that represents observed behaviour is considerably lower.

Nevertheless, the superposition approach is used in practice, even when there is creep recovery, in order to achieve a relatively simple analysis with admittedly approximate results. Further information on creep recovery is available in the literature, for example from Yue and Taerwe (1992).

2.8.2 Creep under gradually varying stress

As previously noted, concrete creep usually occurs in prestressed members under a gradually decreasing stress. For an accurate study of such effects, the entire previous history of stress and strain has to be considered, with the use of complex memory models. This is not a practical proposition for routine design. Approximate methods of analysis are therefore used that are based on the superposition assumption and the creep properties of concrete under constant stress. In Appendix B, details are given of several analytic methods for evaluating loss of prestress and long-term deformations and deflections. They are also discussed briefly in Chapter 4.

The age-adjusted effective modulus

One frequently used treatment of creep under varying stress adopts a modified form of the effective modulus idea, mentioned previously in Section 2.7.

If the sustained stress is constant, the effective modulus, $E_R(t, t_o)$, previously defined by Equation 2.34, provides a good estimate of the total stress-dependent strain. It is inaccurate if there is a significant stress variation over time, but improved accuracy can be achieved by modifying $E_R(t, t_o)$ to include an "*ageing coefficient*" $\chi(t, t_o)$ as follows:

$$E_{R\chi}(t,t_o) = \frac{E_c(t_o)}{1 + \chi(t, t_o)\varphi(t,t_o)} \tag{2.47}$$

This modified modulus is called the **age-adjusted effective modulus**. The ageing coefficient is evaluated so that it allows approximately for variations in stress. The concept of an age-adjusted effective modulus was introduced by Bazant in 1972, that was developed from the original idea of Trost (1967).

The ageing coefficient is unity if the stress is constant. The age-adjusted effective modulus is not a true material property because its value depends on the load history. Its use is explained in Appendix B, where values are given for the ageing coefficient and hence for the age-adjusted effective modulus for use in practical applications.

EXAMPLE 2.2

Calculation of Creep with Stress Increments

An axial compressive stress is applied to the $400\ \text{mm} \times 400\ \text{mm}$ square section member of Example 2.1 in four equal increments of 2.5 MPa at age 14, 72, 130 and 180 days after casting. After 180 days the stress of 10 MPa is maintained for the life of the member. Calculate the total long-term compressive strain using (a) an incremental approach and (b) an approximation.

Solution

(a) Incremental calculations

Values of k_3 for 14, 72, 130 and 180 days are obtained from Table 2.10, using interpolation, as: 1.26, 0.95, 0.87 and 0.83. The basic creep coefficient is $\varphi_{cc.b} = 3.4$. Calculations of elastic and creep strains are given in Table 2.11.

In Example 2.1, the shrinkage strain was calculated as $520{\times}10^{-6}$. The total long-term strain is:

$$\varepsilon_c^* = (321 + 520 + 776) \times 10^{-6} = 1617{\times}10^{-6}$$

(b) Approximate calculation

An approximate value can be obtained by averaging results for loads applied at 14 days and at 180 days. From Example 2.1 we estimate the total long-term strain as:

$$\varepsilon_c^* = (1990 + 1444) \times 10^{-6}/2 = 1717{\times}10^{-6}$$

This is only six per cent different to the result in (a) and is more than adequate, especially given uncertainties in the creep data.

TABLE 2.11 Strain calculations for Example 2.2

Age, t (days)	$f_{cmi}(t)$ (MPa)	$E_c(t)$ (MPa)	$\varepsilon_{ce}(t)$ (10^{-6})	φ^*_{cc}	ε_{cc} (10^{-6})
14	30	27 700	90	3.08	277
72	40	32 000	78	2.32	181
130	43	32 600	77	2.13	164
180	44	32 800	76	2.03	154
			Σ 331		**Σ 776**

2.9 References

AS/NZS 4671–2001, *Steel reinforcing materials*, Standards Australia, Sydney, Australia.

AS/NZS 4672.1–2007, *Steel prestressing materials - General Requirements*, Standards Australia, Sydney, Australia.

AS 1012.9–1999, *Methods of testing concrete - Determination of the compressive strength of concrete specimens*, Standards Australia, Sydney, Australia.

AS 1012.16–1996, *Methods of testing concrete - Determination of creep of concrete cylinders in compression*, Standards Australia, Sydney, Australia.

AS 3600–2018, Australian Concrete Structures Standard, Standards Australia, Sydney, Australia.

Attard, M.M. and Stewart, M.G. 1998, An improved stress block model for high strength concrete, *Research Report No. 154.10.1997*, Department of Civil, Surveying and Environmental Engineering, University of Newcastle, Newcastle, Australia.

Bazant, Z. P., 1972, Prediction of concrete creep effects using age-adjusted effective modulus method, *ACI Journal Proc.*, Vol. 69, No. 4 (April), pp 212-217.

BS 6744:2016, Stainless steel bars - reinforcement of concrete - requirements and test methods, British Standards Institution, London, United Kingdom.

CIA Z6, 2010, *Reinforcement Detailing Handbook*, Concrete Institute of Australia.

Collins, M.P., and Porasz, A., 1989, Shear design for high-strength concrete, Comite Euro-International du Beton, *Bulletin d'Information No. 193*, pp. 77-83.

fib Model Code for Concrete Structures 2010, 2013, *International Federation for Structural Concrete (fib)*, Lausanne, Switzerland, Wilhelm Ernst & Sohn, Berlin, Germany, 402 pp.

Thorenfeldt, E., Tomaszewicz, A., and Jensen, J.J., 1987, Mechanical properties of high strength concrete and application in design, *Proceedings of the Symposium on Utilization of High Strength Concrete*, Tapir, Trondheim, Norway, pp. 149-159.

Trost, H., 1967, Auswirkungen des Superpositionsprinzips auf Kriech- und Relaxationsprobleme (Application of the superposition principle to creep and shrinkage problems in concrete and prestressed concrete), *Beton und Spanbeton*, Vol 62, No 10, pp 230- 38, No 11, pp 261-69.

Yue, L.L. and Taerwe, L, 1992, Creep recovery of plain concrete and its mathematical modelling, *Magazine of Concrete Research*, 44(161):281-290.

CHAPTER 3

Methods of design and analysis

In this chapter we first look briefly at the structural design process and then specifically at the design objectives and design criteria that apply to prestressed concrete. The various design checks specified in AS 3600 to ensure the serviceability and strength of prestressed concrete are discussed, together with the methods of analysis needed to make these checks. The chapter ends with a short introduction to strut-and-tie modelling.

3.1 The structural design process

The overall aim in structural design is to choose the details of a proposed structural system that will allow it to be constructed, while ensuring that it will:

- perform satisfactorily under normal, day-to-day service conditions;
- have sufficient reserves of strength to resist overloads, accidental loads and any environmental extremes that may reasonably be expected to occur during its service life;
- be easy to construct;
- be durable and robust; and
- be economically competitive.

These design objectives apply to all structures, irrespective of the materials and methods of construction, and can be summarised using the terms: serviceability, adequate strength, ease of construction, durability, robustness, and economy.

Structural design is a complex process which is of necessity iterative. It begins with the choice, on a trial basis, of the structural form and geometry of the system, and the structural details of individual members. Checks are then made to see whether the design objectives are met, and adjustments are made progressively to the design until all the objectives are met.

The design of a complex structural system is carried out in several stages, which are often referred to as **concept design**, **preliminary design** and, finally, **detailed design**. The design process has been explained in companion texts (Warner et al., 1988; Foster et al., 2021; and Dandy et al, 2018). In this book, we focus attention on the detailed stage of design for prestressed concrete members and structures.

In detailed design, the structural objectives are quantified by means of **design criteria**, which ensure that acceptable structural behaviour is achieved under a range of specified design loads. In Australia, the design criteria are specified in the Australian Standard for Concrete Structures, AS 3600. The design loads, and their combinations, for serviceability and strength design are specified in Standards AS/NZS 1170, Parts 0 to 3, and AS 1170 Part 4.

3.2 Design objectives and design criteria for prestressed concrete

We now consider the structural design objectives that apply specifically to prestressed concrete, together with the relevant design criteria that are specified in AS 3600.

3.2.1 Serviceability

For a structure to be serviceable, it must perform adequately under normal working conditions throughout its service life. The design criteria for serviceability will depend on the intended use of the structure. For prestressed concrete beams and flat slab floors, the prime requirements are that cracking and deflections do not become excessive. Similar serviceability requirements apply to other prestressed flexural members. For some special structures there will be additional requirements, such as non-leakage from containment vessels. In bridges, vibration control is an important serviceability requirement.

Serviceability problems arise in structures not only as the result of external loads, but also because of load-independent effects such as temperature fluctuations, temperature gradients, foundation movements and other imposed deformations. Such effects can induce cracking and deflections that are more severe than those due to the applied loads. Other relevant serviceability problems in prestressed concrete members relate to the prestress itself. For example, unacceptably large upward deflections can result from the use of an excessively high level of eccentric prestress.

Both **direct and indirect design criteria** can be used to specify minimum serviceability requirements. In the deflection control of beams, for example, a direct criterion specifies an upper allowable limit on deflection, whereas an indirect criterion might specify limits on the span-depth ratios, these limiting ratios being chosen so that the deflections will not become excessive. Indirect criteria tend to be used when direct criteria are difficult to formulate or apply, and sometimes to simplify the design calculations.

Serviceability criteria for prestressed concrete tend to be direct, with upper limits placed on deflections and crack widths. To check that such criteria are satisfied, reliable analytic methods are needed to predict structural behaviour. In Chapters 4 and 5, methods are presented for analysing the behaviour of uncracked and cracked prestressed concrete members under service load conditions. The specific design criteria for serviceability are discussed in detail in Chapters 10, 11 and 12 that focus on design procedures.

3.2.2 Strength

Various **strength design criteria** are prescribed in AS 3600 to ensure that the structure can safely resist a range of overload conditions. In the strength design of prestressed concrete beams, girders and slabs, the prime considerations are flexural failure and shear and torsion failure. Other important failure modes that need to be considered include crushing and splitting of the concrete in the end block regions behind the anchorages, and catastrophic failure during construction due to over-prestressing.

Methods of analysis for flexural strength are dealt with in Chapter 6 of this book. Shear and torsion are treated in Chapter 7, while anchorage and end block design are considered in Chapter 8.

3.2.3 Durability

Durability is the property that allows a structure to continue to perform in a satisfactory manner throughout its intended life, without requiring excessive maintenance or repair costs. Severe and rapid deterioration of some poorly constructed concrete structures in various parts of the world has recently emphasised the need to achieve good durability. Although our knowledge of the processes of deterioration of materials has improved as the result of recent research, reliable quantitative methods for evaluating durability have not yet been developed. Direct design criteria do not exist and an indirect approach has to be taken. At present, good durability is achieved indirectly, mainly by ensuring that the concrete is dense and of good quality, and that all steel is protected by an adequate cover of concrete. Minimum requirements for concrete quality and cover, contained in Section 4 of AS 3600, are the prime measures used to ensure durability, in lieu of direct design criteria.

3.2.4 Robustness

A structure is robust if it can withstand, within reason, unexpected adverse load and load-independent effects. For example, the possibilities of blast and collision require robustness in a structure. Methods of analysis for robustness have not yet been adequately developed. Indeed, a satisfactory, generally applicable definition of robustness still has to be formulated. AS 3600 deals qualitatively with robustness, in Clause 2.1.3. In AS 1170.0 a robustness test for multi-storey structural systems is outlined, in which limits are placed on the lateral deflections induced by specified horizontal loads. Unfortunately, the situation is more complex than this. For example, upward loads often occur in unusual situations involving blast. Robustness is not treated quantitatively in this text.

3.2.5 Ease of construction and economy

Ease of construction and economy are very important design objectives, but differ in nature to those previously discussed, and do not lend themselves to treatment with quantitative design criteria. Economy is a convenient comparative measure for evaluating and ranking alternative trial designs. In optimal design, cost is often chosen as the objective function to be minimised. Ease of construction can be treated indirectly as a comparative cost, which is included in the total cost. While comments will be made on constructability and economy in the following chapters, they are not treated as the other design objectives.

3.3 Design criteria and structural reliability

The choice of design criteria for inclusion in national design standards is made by code-writing committees. The formulation of appropriate design criteria requires good judgment and professional experience. An element of subjectivity can also be involved. Extensive, properly documented past experience, from both successful and unsuccessful designs, provides a good practical basis for quantifying design criteria.

In formulating design criteria for incorporation into codes of practice, code-writing committees recognise that absolute safety against failure or unserviceability is neither attainable nor even desirable. No matter how carefully and conservatively the design and construction of a building structure are carried out, the possibility of failure is unavoidable, whether due to gross human error or natural catastrophe. The aim of structural design is *not* to achieve absolute safety at any cost, but rather to ensure a very high level of reliability, at an acceptably low risk of failure or unserviceability, and at an acceptable cost to the community. Questions concerning safety and risk in structural engineering are not discussed in detail here. A thorough treatment of these important and fundamental matters is provided by Blockley (1980). Further information is available on reliability, as treated by codes and standards in Australia (Warner et al., 1988; Foster et al., 2015, Foster et al., 2016 and Stewart et al., 2016).

It is important for designers to understand that design criteria specify *minimum* requirements for structural behaviour, and that these criteria should be evaluated critically in regard to each specific structure being designed. They can and should be modified and made more stringent when this is appropriate.

EARLY PRESTRESSING IN AUSTRALIA

One of the first structural applications of prestressed concrete in Australia was in the construction of a precast multi-storey framed building to serve as an ice tower at the site of Warragamba Dam, near Sydney. This experimental building was constructed by the Sydney Water Board in the early 1950s. The precast beams and columns were initially post-tensioned and subsequently post-tensioned together after erection. One of the reasons given for the use of prestressed concrete in this structure was the scarcity of structural steel at the time.

An application of prestressed concrete in Australia that predated the Warragamba ice tower occurred in the early 1940s, when several kilometres of prestressed reinforced concrete pipeline were laid for the Lithgow water supply scheme. Large-scale structural use of prestressed concrete in Australia began in the 1950s with pretensioned joists and, a little later, the first prestressed concrete highway bridges. Detailed information on the subsequent development and use of prestressed concrete in Australia in the construction of buildings and structures as diverse as the Rip Bridge, the Australia Square Tower and the Sydney Opera House are to be found in the 50th anniversary edition of the Constructional Review in 1977.

3.4 AS 3600 design checks for prestressed concrete

In AS 3600, a unified approach is taken to the design of reinforced concrete and prestressed concrete structures. These are not treated as separate forms of construction. Rather, a common design basis is provided, which focuses on performance requirements and the design criteria relating to strength and serviceability, irrespective of whether prestressing is used. The designer can thus seek the best solution for each specific design problem by treating prestressing as a design option.

Current and previous editions of AS 3600 reflect recent international developments in structural design. In particular, the move towards design for performance can be seen in AS 3600 as well as in other national and international codes, going back to the 1978 CEB-FIP model code (CEB-FIP, 1978). On the other hand, AS 3600 does not emphasise the concept of limit state, even though it uses important limit states concepts, including partial safety coefficients that have been derived from reliability studies. The terminology in AS 3600 tends to follow that of the ACI code, and AS 3600 could be described as a strength and serviceability code.

3.4.1 AS 3600 strength checks

In the current standard, AS 3600, the general strength design criterion is:

$$R_{\mathrm{d}} \geq E_{\mathrm{d}} \tag{3.1}$$

The term R_{d} on the left side is the **design strength**, and E_{d} on the right hand side is the **design action effect**.

This generalised formulation is used because five different, alternative strength check methods are allowed by AS 3600. Each method requires a different type of strength analysis. The methods of analysis are:

1. an elastic analysis of the structure under the collapse load, used with inelastic ultimate strength calculations for individual, critical local cross-sections;
2. a non-linear collapse load analysis of the entire structure;
3. a non-linear finite element analysis of the entire structure;
4. a linear finite element analysis of the entire structure;
5. a strut-and-tie analysis of failure conditions in local regions and members in the structure.

The first of these methods is most commonly used, and forms the analytic basis for **ultimate strength design**, which will be the prime focus of attention in this book. The last-mentioned method, **strut-and-tie design**, is also frequently used in the design of members for shear and torsion and for the design of anchorage regions. Strut-and-tie modelling is explained briefly in Section 3.7 of this chapter and in more detail in a companion text to this book (Foster, et al., 2021). The ultimate strength method is dealt with in Chapter 6, specifically in relation to flexural strength design, while the strut-and-tie method is used in Chapter 7 for shear and torsion design, and in Chapter 8 for anchorage design.

In ultimate strength design, the stress resultants (moments, shears, torsion and axial force) in local critical sections and regions are determined by an elastic analysis of the structure under design ultimate load conditions. These stress resultants are the design action effect E_d in Equation 3.1. It should be noted that AS 3600 allows some redistribution of the elastic moments to be made in evaluating moments and shears in indeterminate flexural members. This is explained in Chapter 11.

In ultimate strength design, the design strength R_d is obtained by applying a capacity reduction factor ϕ to the ultimate strength of the section, R_u:

$$R_d = \phi R_u \tag{3.2}$$

The factor ϕ is a safety coefficient that allows for adverse variations in strength. Values for ϕ are specified in Section 2 of AS 3600 for the different stress resultants. It should be noted that the ϕ values in AS 3600–2018 are slightly higher than in the previous edition, AS 3600–2009. This results in a small increase in efficiency (of about 6 to 8 per cent) when designing for strength for certain conditions. The reasons for this change are explained by Foster et al. (2015, 2016) and Stewart et al. (2016). Some typical values of ϕ for Class N reinforcement are given in Table 3.1.

The design loads and load combinations to be used in the strength checks are specified in the loading standard, AS/NZS 1170.0. Various possible combinations need to be considered in turn, in order to identify the most adverse condition. A distinction is made between situations when live load acts for a short time period and when it acts permanently. The main load combinations and load factors specified in AS/NZS 1170.0 are as follows:

TABLE 3.1 Values of ϕ for strength design with Class N reinforcement

Stress resultant	Value of ϕ
Pure bending (for ductile members)	0.85
Pure axial compression	0.65
Bending with axial compression	
(i) for $N_u \geq N_{ub}$	$0.65k_\phi$
(ii) for $N_u < N_{ub}$	$0.65k_\phi + (\phi - 0.65k_\phi)(1 - N_u / N_{ub})$
Shear and torsion:	
(i) where minimum Class N fitments are provided, other than shear limited by web crushing;	0.75
(ii) otherwise:	0.70
Strut and tie modelling:	
(i) concrete in compression;	0.65
(ii) steel in tension	0.85
Bearing	0.60

Note: For non-ductile sections in flexure, values used for ϕ are below 0.85; see AS 3600 for details. For short columns with $Q/G \geq 0.25$, $k_\phi = 1.0$, otherwise $k_\phi = 12/13$.

dead load acting alone:	$1.35G$	(3.3)
dead load + live load:	$1.2G + 1.5Q$	(3.4)
dead load + long-term live load:	$1.2G + 1.5\psi_1 Q$	(3.5)
dead load + live load + wind:	$1.2G + \psi_c Q + W_u$	(3.6)
dead load + wind action reversal:	$0.9G + W_u$	(3.7)
dead load + earthquake + live load:	$G + E_u + \psi_c Q$	(3.8)

In these expressions, G is the dead load (or permanent action), Q is the live load (or imposed action), W_u is the wind load for strength design, calculated in accordance with the Wind Loading Standard, AS/NZS 1170.2, and E_u is the earthquake load.

The load combination factors, ψ_l and ψ_c, are applied to the live load in Equations 3.5 and 3.6. The factor ψ_l determines the long-term sustained component of the live load for dead and live load combinations, whereas ψ_c is used for the combinations that include wind and earthquake. Values of ψ_l generally vary from 0.4 for office floors and parking areas up to 0.6 for storage areas, but a value of unity is used in extreme situations, for example when heavy machinery is permanently installed. Values for ψ_c normally vary between 0.4 and 0.6 but may reach 1.2 if there is heavy machinery installed. For other special situations not mentioned here, for example involving snow loads, earth pressure and water pressure, reference should be made to Section 4.2 of AS/NZS 1170.0.

The critical load combinations given in AS/NZS 1170.0 are intended for all materials of construction and do not take account of prestressing. However, Clause 2.4 in AS 3600 specifies that the prestressing effect, denoted by P, is to be included in all relevant load combinations, with a load factor of unity, for both serviceability design and strength design, except for the special condition which exists during construction when prestress is first transferred to the concrete. At transfer, the more severe of the following is to be used:

$$\text{dead load + prestressing effect:} \qquad 1.15G + 1.15P, \text{ or} \tag{3.9}$$

$$0.9G + 1.15P \tag{3.10}$$

Equation 3.10 applies if the dead load counteracts the adverse effects of prestressing. An example of its use would be to check against excessive tensile stresses in the upper fibres of a beam when only self weight is acting (prior to the application of subsequent loads), or for a check against excessive upwards deflection due to prestress.

Collapse load methods of analysis and design will not be of prime concern in this book; nevertheless, it is instructive to see how Equation 3.1 applies to these methods. Considering by way of example the second method listed above, we note that the non-linear collapse method of analysis provides an estimate of the load capacity of the entire structure, for a chosen load combination. This now becomes the term R_d in Equation 3.1. The design action effect, E_d, is the specific design load combination itself, which the structure must be able to carry. Equation 3.1 states that the load capacity of the structure, R_d, as determined by collapse load analysis, must be at least as large as the design load $E_{d.}$ This strength design requirement must be satisfied for each of the relevant load combinations in Equations 3.3 to 3.10.

3.4.2 AS 3600 serviceability checks

AS 3600 uses both direct and indirect design checks for the serviceability requirements. Direct checks for deflections and crack widths can be made by calculating deflections and crack widths using appropriate analytic methods, and comparing values with acceptable limiting values specified in AS 3600.

An alternative indirect procedure in AS 3600 for crack control involves checking the stresses in the reinforcing steel and the stress increments in the prestressing steel in regions close to the tensile face of the member, and comparing these with designated limiting values. Some additional detailing requirements for the reinforcement have to be met when this indirect method is used.

Indirect checks on the deflections of reinforced concrete flexural members are allowed by AS 3600. These are made by comparing the span-to-depth ratio of a proposed design against specified limits which, generally, have been found in the past to yield acceptable deflection control. Unfortunately, AS 3600 does not allow indirect deflection checks for prestressed concrete beams or slabs. It specifically requires deflection calculations to be made. This means that short-term and long-term deflections always have to be evaluated. Appropriate deflection calculation methods are discussed in Chapters 4 and 5.

The service load combinations to be considered for serviceability design are specified in AS/NZS 1170.0. They involve the following load terms and factors: G, $\psi_s Q$, $\psi_l\, Q$, W_s, and E_s. The terms W_s and E_s are the wind and earthquake actions, respectively. The factors ψ_s and ψ_l are used for short-term and long-term live-load applications, respectively. Values of ψ_s given in Table 4.1 of AS/NZS 1170.0 vary between 0.7 and 1.0, depending on the type of member, while those for ψ_l vary from 0.4 up to 1.0.

3.5 The critical stress method of design

In the early days of prestressed concrete, just after the Second World War and well before the development of limit states concepts, a method of design was employed which involved checking the stresses in the top and bottom fibres of critical cross-sections of flexural members under working loads to ensure that specified allowable limiting stresses were not exceeded. This allowable stress design procedure was used generally for structures in reinforced concrete and steel, as well as for prestressed concrete. At that time, prestressed members were highly prestressed. This limiting stress method of design is referred to here as the **critical stress method**. The use of the method was already in decline in the 1980s. A full explanation of the method was given in an appendix to the second edition of this book (Warner and Faulkes, 1988).

The critical stress method is considerably more complex than the current approach used in this book. It has the disadvantage that it does not deal directly with structural performance. Separate, additional checks must therefore be made, following the critical stress calculations, to ensure that the current code requirements of strength and serviceability are met. The check on critical extreme fibre stresses is in fact redundant.

In rare situations a design with high prestress may be required, for example to keep a member crack-free both at transfer and under full service live load. Even in such extreme cases, the design approach proposed in this book is applicable and appropriate. It is simpler to apply, and it takes direct account of the relevant performance requirements.

3.6 Methods of analysis

3.6.1 Analysis for overload and strength

As previously explained, AS 3600 allows for five alternative methods of analysis to be used to carry out strength checks. While there are design situations in which each of these methods will be appropriate, strength checks based on ultimate strength theory are used in routine design situations and are the most commonly used. A linear elastic analysis allows the stress resultants in critical sections at high overload to be evaluated.

The basic assumptions made in the ultimate strength design of continuous members, and indeterminate structures generally, are that (a) the overall system behaves elastically in the overload range up to the point of failure, and that (b) local behaviour at ultimate strength is non-linear and inelastic. Although these assumptions may seem to be self-contradictory, they lead to safe and conservative results provided that the structure has adequate ductility to deform under high overload and so allow the internal stress resultants to redistribute to the extent needed. This is the reason why some redistribution of moments in indeterminate structures, away from the calculated elastic distribution, is allowed when the ultimate strength method is employed. As we will see in Chapter 11, AS 3600 allows up to 30 per cent redistribution of the elastically calculated moments in strength design. The theory of plasticity provides theoretical justification of the method, provided the structure is ductile.

The strut-and-tie approach is used for design in Chapters 7 and 8 for dealing with shear and torsion and anchorage design. Strut-and-tie concepts are explained in detail in a companion text (Foster, et al., 2021), and more briefly here in Section 3.7 of this chapter.

3.6.2 Analysis for short-term service load behaviour

Short-term deflections and crack widths are evaluated using analyses of structural behaviour under short-term service loading. For uncracked members, the methods of analysis presented in Chapter 4 are straight-forward because they are based on the assumption of linear elastic behaviour. The analytic methods in Chapter 5 for post-cracking behaviour are necessarily more complicated than for uncracked members.

3.6.3 Analysis for long-term behaviour under service load

Methods for analysing the time-varying behaviour of prestressed concrete members are used to evaluate the long-term losses in prestress and the additional long-term deformations and deflections that result from creep and shrinkage of the concrete. The analysis of time-varying behaviour becomes complicated because gradual changes occur in the concrete stresses that initiate the creep. Strictly speaking, the rate of creep in a fibre at any time instant depends on the entire pre-history of stress and strain, and an accurate analysis would require the use of memory models for concrete. Such analyses are quite complex. In practice, simplified and approximate methods are used to evaluate time-varying structural behaviour. Three methods are described in this book: a step-by-step method; a single-step method, modified by using the age-adjusted effective elastic modulus; and approximate calculations using simplified closed-form equations. These methods are briefly introduced at the end of Chapter 4, but full details are provided in Appendix B, together with numerical examples of their use. They are used in design examples in Chapters 10 and 11.

3.7 Strut-and-tie modelling and stress-fields

3.7.1 Strut-and-tie modelling

In strut-and-tie modelling, the internal load paths in any region of a reinforced concrete or prestressed concrete member are assumed to consist of:

- concrete struts in compression,
- steel tension ties (of reinforcing steel or prestressing tendon), and
- nodes, or joints, where the ends of the ties and struts meet, and where external loads may act.

Strut-and-tie modelling is mainly used to study the overload behaviour and strength of such regions. It can also be used for the strength design of non-flexural regions in a member, or indeed of an entire member. AS 3600 allows the use of strut-and-tie modelling for the strength design of members in shear and torsion, and it requires that the method be used for the design of all non-flexural members.

The origins of the strut-and-tie method are to be found in the truss analogy, which was developed in the late 1800s for the design of beams in shear. In the 1920s the truss analogy was extended to deal with torsion in reinforced concrete members. The strut-and-tie concepts were used intuitively by designers for many years to deal with complex design problems in reinforced concrete.

However, a new and more fundamental approach was initiated by Thürlimann and his associates in the 1960s in Zürich, Switzerland, when they applied the concepts of plasticity theory to concrete structures and developed a sound theoretical basis for strut-and-tie modelling. In the 1980s the concept of a strut-and-tie approach for the detailing of non-flexural regions of members was promoted by Marti (1985a, 1985b) and Schlaich et al. (1987).

Since then, strut-and-tie modelling has gained increasing acceptance and now appears in most national codes and standards. Strut-and-tie modelling makes use of the lower bound theory of plasticity so that the designer can *choose* a suitable load path and then design and detail the structure such that this load path is sufficiently strong to carry the applied loads through to the supports. Since the method is based on lower bound plasticity, in theory it errs on the 'safe' side.

While there can be no doubt that the strut-and-tie method is a powerful design tool, some precautionary words are needed. In order to achieve the load path demanded by the designer it may be necessary for the structure to undergo significant redistribution of the internal forces, and large deformations, which it may not be capable of doing due to ductility limitations of the concrete and steel. The designer therefore needs to keep this in mind when choosing the load path. Care is also needed in detailing to ensure that:

(i) the member is capable of delivering the redistributions demanded of it without undue distress to itself or adjacent components, and

(ii) the region defined by the strut-and-tie model is detailed with consideration of the full stress-field that it represents. This is discussed further in the following section.

Although strut-and-tie modelling can be applied to all components of a structure, it is usually used in regions near statical or geometrical discontinuities, referred to as D- (Disturbed or Discontinuity) regions. An example of a D-region is the end zone of a prestressed concrete member.

In the beams in Figure 3.1, D-regions occur near the supports, while the B- (Bernoulli) regions occur near mid-span. The choice of the length of the D-regions is somewhat arbitrary. For any selection, equilibrium has to be satisfied in the design and, provided the ductility demands of the structure can be met, the designer has considerable freedom in the selection of the load path. As we will see in Chapter 8, Saint Venant was the first to show for an elastic system that the length over which a point disturbance remains significant is approximately the height (H) of the section at which it occurs. This is approximately true also for reinforced concrete, which is highly non-linear, although numerical models show the disturbance can be taken over a greater length, up to $1.5H$ (Foster and Rogowsky, 1997a, 1997b).

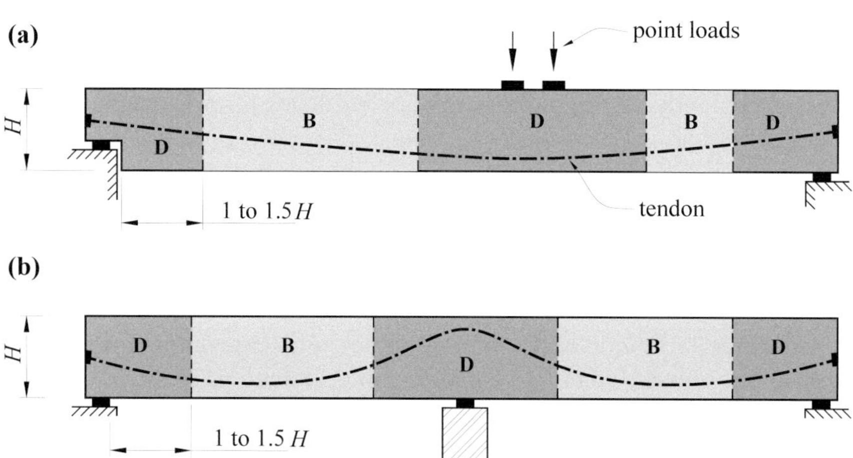

Figure 3.1 B- and D-regions in beams

3.7.2 Stress-fields

Both stress-fields and strut-and-tie models indicate how internal forces are transferred through a structural member to adjacent supporting members, and can also show how externally applied loads are transferred through an entire structural system into its supports.

Stress-fields show in more detail than a strut-and-tie model the flow of stresses through local regions of a member. There are various types of stress fields, including the following:

- compression band: a straight, parallel field of compressive stress of uniform intensity along its axis;

- tension band: a straight, parallel field of tensile stress of uniform intensity along its axis;
- compression fan: a diverging, fan-shaped field of compressive stress with uniform intensity on a transverse cross-section, but of varying intensity along the axis of the fan;
- compression chord: a straight, parallel field of compressive stress, of varying intensity along its axis;
- tension chord: a straight, parallel field of tensile stress, of varying intensity along its axis;
- node: a small region subjected to biaxial or triaxial stresses, and connecting the ends of several other fields.

Some typical stress-fields within a structural member are shown in Figure 3.2(a) and (c).

Compressive stress-fields can be replaced by equivalent straight struts, and tensile stress-fields by equivalent straight ties, each located at the centroids of the stress field that they are replacing. The struts and ties carrying the same forces as those represented by stress-field, when the stresses are integrated through their cross-section. Figure 3.2(b) shows the equivalent strut-and-tie model that represents the stress-fields in Figure 3.2(a). In Figure 3.2(a) the vertical (tensile) stress field is a representation of closely spaced stirrups, with the equivalent stress equal to the total force in the stirrups divided by the cross-sectional area over which the stirrups are spread.

As demonstrated, stress-fields and their equivalent strut-and-tie models are closely linked. While the strut-and-tie model is a useful tool for showing internal force paths, the stress-field provides an insight for detailing, especially the detailing of bursting reinforcement. For example, the strut-and-tie model of Figure 3.2(c) indicates that the bursting reinforcement may be concentrated around the tie and anchored past the nodes, but this is not the case. Reflecting the stress-field that our model replaces, it is seen that the resulting reinforcement should be distributed through the stress-field, centred about the tie, and must be extended through the full depth of the section and anchored beyond. Thus detailing of the element requires careful consideration of the full stress field.

,

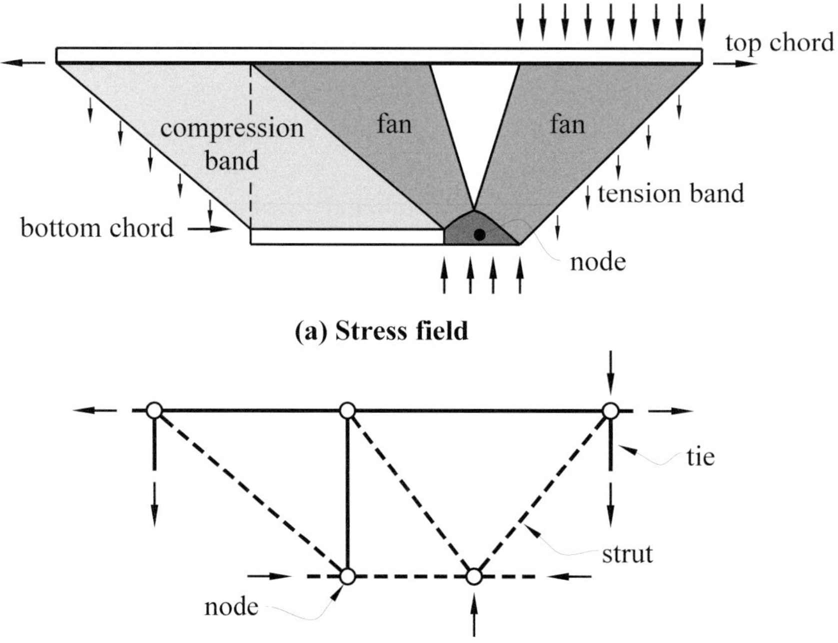

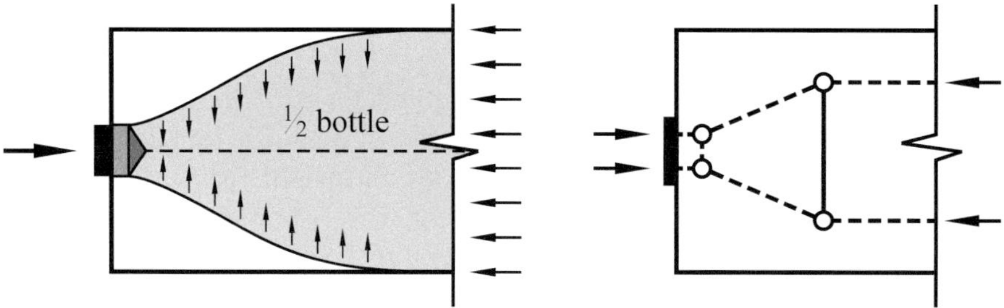

Figure 3.2 Basic elements for stress-field analysis

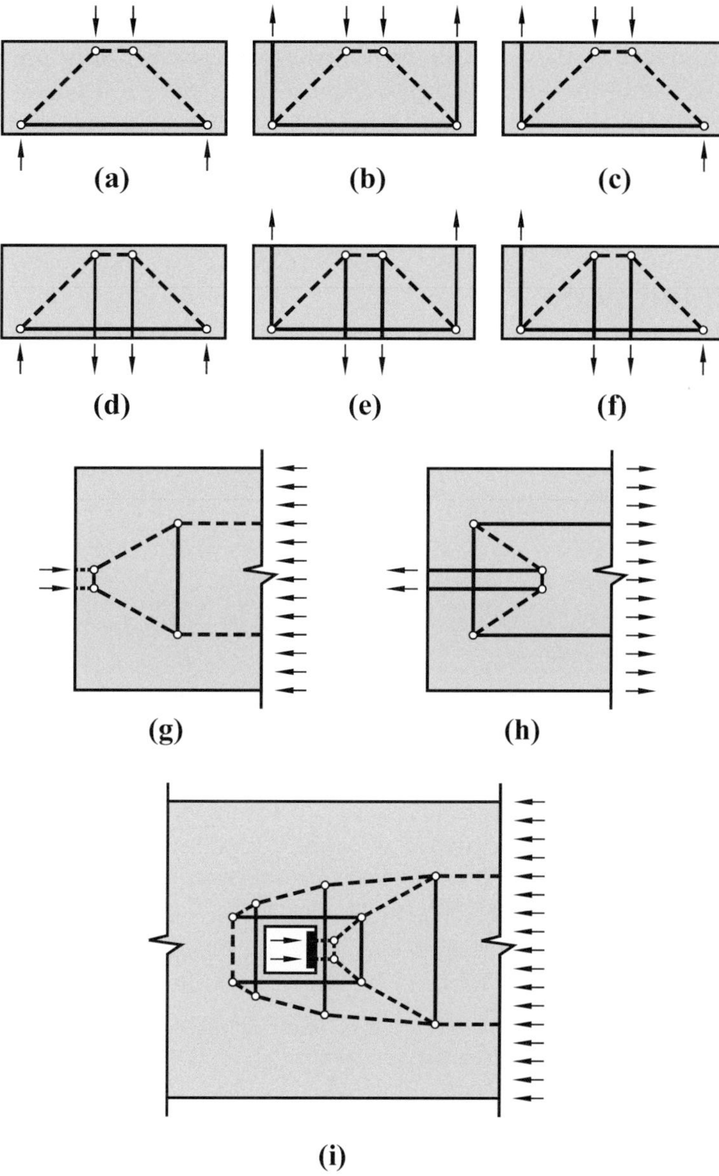

Figure 3.3 Basic strut-and-tie models

Some common strut-and-tie models that have been adopted for the detailing of steel reinforcement in member D-regions are shown in Figure 3.3, where Figure 3.3(a) to (h) were proposed by Marti (1991). A more complex, indeterminate model is shown in Figure 3.3(i) for the case where prestress is introduced within the span through a '*pocket*' or a '*blister*' (Rogowsky and Marti, 1991). In such a case, the disturbed region behind the anchor also requires consideration.

3.8 References

AS 3600–2018, *Concrete Structures*, Standards Australia, Sydney, Australia.

AS 1170.4, 2007, *Structural design actions - Earthquake actions in Australia*, Standards Australia, Sydney, Australia.

AS/NZS 1170.0, 2002, *Structural design actions, Part 0: General principles*, Standards Australia, Sydney, Australia.

AS/NZS 1170.1, 2002, *Structural design actions, Part 1: Permanent, imposed and other actions*, Standards Australia, Sydney, Australia.

AS/NZS 1170.2, 2011, *Structural design actions, Part 2: Wind actions*, Standards Australia, Sydney, Australia.

AS/NZS 1170.3, 2003, *Structural design actions - Snow and ice actions*, Standards Australia, Sydney, Australia.

Blockley, D.I., 1980, *The Nature of Structural Design and Safety*, Ellis Horwood, Chichester, UK.

CEB-FIP, 1978, *International System of Unified Standard Codes of Practice for Structures: Volume 2: CEB-FIP Model Code for Concrete Structures* (CEB Bulletin 125, International Federation for Structural Concrete, fib, Lausanne, Switzerland).

Dandy, G., Daniel, T., Foley, B. and Warner, R., 2018, *Planning and Design of Engineering Systems*, 3rd Ed, CRC Press, Taylor & Francis, Milton Park, UK, 2018.

Foster, S J, Kilpatrick, A E and Warner, R F, 2021, *Reinforced Concrete Basics*, 3E, (3rd Ed) Pearson, Melbourne, Australia., 589 pp.

Foster S.J., and Rogowsky, D.M. 1997a, Bursting Forces in Concrete Members resulting from In-plane Concentrated Loads, *Magazine for Concrete Research*, Vol. 49, No. 180, pp. 231-240.

Foster S.J., and Rogowsky, D.M. 1997b, Splitting of Concrete Panels under Concentrated Loads, *Structural Engineering and Mechanics*, Vol. 5, No. 6, pp 803-815.

Foster, S. Stewart, M., Sirivivatnanon, V., Loo, K.Y.M., Ahammed, M., Ng, T.S. and Valipour. H., A Re-Evaluation of the Safety and Reliability Indices for Reinforced Concrete Structures Designed to AS3600, UNICIV Report R-464, School of Civil and Environmental Engineering, UNSW Australia, November 2015, 201 pp.

Foster, S., Stewart, M., Loo, M., Ahammed, M. and Sirivivatnanon, V., Calibration of Australian Standard AS3600 Concrete Structures: Part I Statistical Analysis of Material Properties and Model Error, Australian Journal of Structural Engineering (AJSE), Vol., 17, Issue 4, 2016, pp.242-253.

Marti, P. 1985a, Basic Tools of Reinforced Concrete Beam Design, *ACI Journal Proceedings*, Vol. 7, No. 1, Jan.-Feb., pp. 46-56.

Marti, P. 1985b, Truss Models in Detailing, *Concrete International*, American Concrete Institute, Vol. 7, No. 12, Dec. pp. 66-73.

Marti, P. 1991, Dimensioning and Detailing, IABSE Colloquium 'Structural Concrete', Stuttgart, IABSE Vol. 62, pp. 411-443.

Rogowsky D.M., and Marti P. 1991, *Detailing for Post-Tensioning*, No. 3, VSL Report Series, VSL International.

Schlaich, J., Schäfer, K., and Jennewein, M. 1987, Toward a Consistent Design of Structural Concrete", Special Report, *PCI Journal*, Vol. 32, No. 3, May-June, pp. 74-150.

Stewart, M., Foster, S., Ahammed, M. and Sirivivatnanon, V., Calibration of Australian Standard AS3600 Concrete Structures: Part II Reliability Indices and Changes to Capacity Reduction Factors, Aust. J. of Struct. Eng. (AJSE), Vol., 17, Issue 4, 2016, pp. 254-266.

Warner, R.F., Rangan, B.V., Hall, A.S. and Faulkes, K.A., 1998, *Concrete Structures*, Longman Cheshire, Pearson Education, Melbourne.

Warner, R.F. and Faulkes, K.A., 1988, *Prestressed Concrete*, 2nd Ed., Longman Cheshire, Melbourne, Australia.

CHAPTER 4

Flexural behaviour of uncracked members

The behaviour of uncracked prestressed beams under short-term and sustained loading is described, and methods are then presented for calculating the stresses, strains and deflections in such members. The important design concepts of equivalent load and load balancing are also introduced in this chapter.

4.1 Introduction

The short-term response of an uncracked prestressed beam to external load is very close to being linear and elastic. Over time, however, creep and shrinkage induce large increases in the strains and deformations in the concrete, together with a redistribution of stresses in the concrete, steel and tendon, and a significant loss of prestress, as well as a large increase in deformations and deflections. The methods presented here for calculating stresses, strains, deformations and deflections under both short-term and sustained loadings are used in later chapters in the design of statically determinate and indeterminate beams and slabs.

4.2 Short-term behaviour of uncracked beams

In the cross-section of an uncracked prestressed concrete beam, the short-term stresses in the concrete are affected to only a minor extent by the presence of the reinforcing steel, the tendon and the duct. A simplified elastic analysis can therefore be based on the gross concrete area A_g and its second moment of area I_g. In some special situations a more accurate treatment may be required. If so, the refined analysis in Appendix A takes account of the presence of the duct, the tendon, and of tensile and compressive reinforcement.

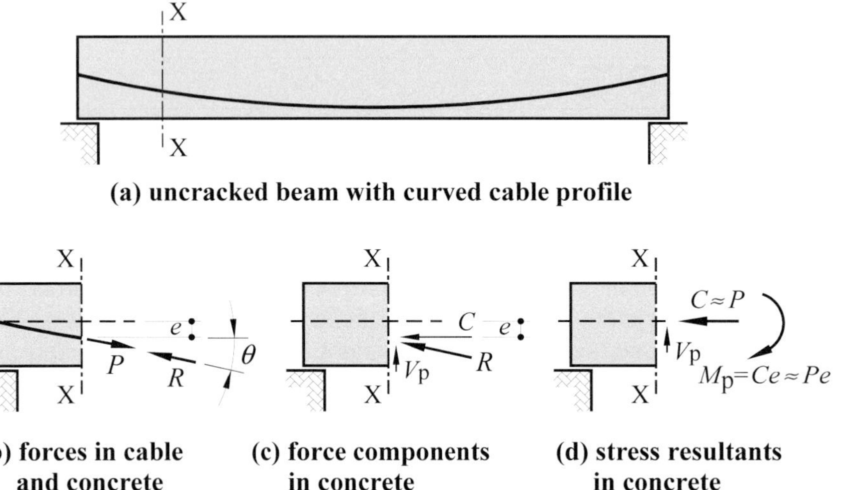

Figure 4.1 Stress resultants in concrete due to prestress

4.2.1 Stresses due to prestress acting alone

We first investigate the stresses due to prestress alone in the post-tensioned beam in Figure 4.1(a), which was previously discussed in Chapter 1 and was shown in Figure 1.3. We consider the portion of the beam to the left of section X-X. The forces on this free body all act at section X-X and they consist of the tensile prestressing force P in the cable, inclined at angle θ to the horizontal and located at eccentricity e below the centroidal axis, and an equal and opposite equilibrating compressive resultant force R in the concrete. The components of R are a horizontal force, $C = R\cos\theta$ and a vertical shear force $V_p = R\sin\theta$ (Figure 4.1(c)). The force C is the resultant of the longitudinal concrete stresses in X-X and V_p is the resultant of the transverse shear stresses. The angle θ is usually small and we use the simplifying approximations: $\sin\theta = \tan\theta = \theta$ and $\cos\theta - 1$. The vertical and horizontal components of R become $V_p = R\theta$, and $C = R = P$. We are only concerned here with the longitudinal stresses, but we will consider the shear stresses further in Chapter 7.

The compressive force C at eccentricity e is statically equivalent to an axial force of equal magnitude, acting at the centroid of the section, plus a moment $M_p = -Ce$, as shown in Figure 4.1(d). Provided the section X-X is not too close to the end of the beam, the longitudinal concrete stresses vary linearly from a maximum compressive stress σ_{cb} in the bottom fibre to a minimum in the top fibre, σ_{ca}, which may be either tensile or compressive, depending on the magnitude of the eccentricity e. These stresses are shown in Figure 4.2. With $P = C$, the concrete compressive stress at any depth y below the centroidal axis can be written as:

$$\sigma_{cy} = \frac{P}{A_g} + \frac{Pey}{I_g} \tag{4.1}$$

The eccentricity e of the prestress is positive when the tendon is located below the centroidal axis. As already noted, A_g and I_g are the area and the second moment of area of the gross concrete cross-section, respectively.

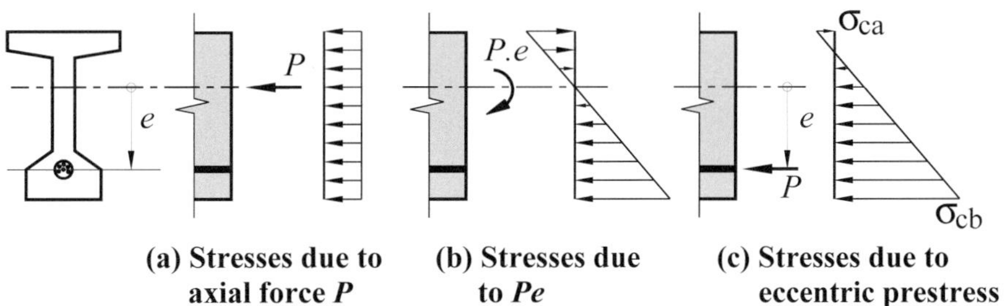

Figure 4.2 Stresses due to prestress only

In statically determinate members, the longitudinal stresses due to prestress at a cross-section are unaffected by the cable eccentricity at other sections along the beam, or by the cross-section details at other sections. This is not necessarily true for indeterminate members. We shall see in Chapter 11 how applying prestress to a continuous beam can induce deflections that are incompatible with the support conditions. This induces additional secondary reactions, known as hyperstatic reactions, and hence additional moments and additional stresses. Equation 4.1 is necessary but not sufficient to calculate the stresses due to prestress in statically indeterminate beams.

Although the equal forces C and P decrease over time due to shrinkage and creep losses, Equation 4.1 provides a good estimate of the concrete stresses at any time provided the appropriate (reduced) value of prestress is used. However, the strains in the section cannot be obtained from the stresses using the elastic modulus E_c. Such a calculation would ignore large inelastic strains and could lead to serious error.

4.2.2 Initial stresses due to prestress plus applied load

A positive moment M in the section due to external load produces tensile stress in the lower fibres of the concrete. With tension taken as negative, the elastic concrete stress at depth y below the centroidal axis is:

$$\sigma = -My/I_g \tag{4.2}$$

The stresses due to the prestress in Figure 4.3(a) and those due to the moment M in Figure 4.3(b) are added to obtain the resultant stresses in Figure 4.3(c). In the lower fibres, the compressive stress due to prestress is reduced when M acts, while in the upper fibre it is increased. The resultant compressive force C in the concrete thus acts at a distance z above the line of action of the tensile force P, as shown in Figure 4.3(d).

The horizontal forces in the section, P and C, are equal and comprise a couple which is equal to the applied moment M:

$$M = Cz = Pz \tag{4.3}$$

As M increases, P and C remain essentially constant in magnitude, but the line of action of C moves progressively upwards in the section, and the lever arm z increases proportionally as M increases:

$$z = M/P \tag{4.4}$$

We are ignoring the very small increment in the force P that in fact occurs because of the increase in tensile strain in the lower fibres of the beam as M increases, but this remains very small provided cracking does not occur.

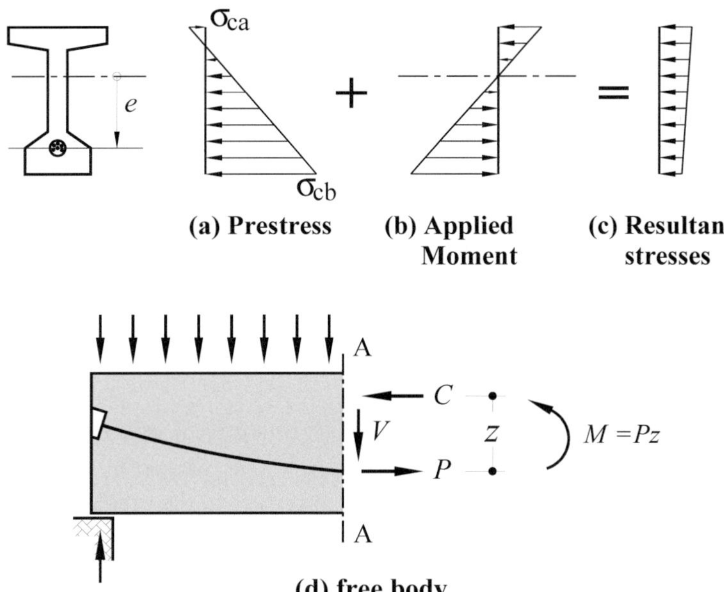

Figure 4.3 Stresses in the concrete due to prestress and applied moment

An important difference therefore exists between the working load behaviour of an uncracked prestressed concrete beam section and that of a cracked reinforced concrete beam section. In the cracked reinforced concrete section subjected to increasing moment, the neutral axis position remains constant at depth d_n below the top fibre (Foster et al., 2021). The value of d_n depends on the dimensions of the section and the amount of reinforcement but not on M. The internal forces C and T, and hence the steel and concrete stresses, increase proportionally as the moment M increases, while the lever arm z remains constant.

In contrast to the reinforced concrete section, in the prestressed concrete section, the internal forces C and P (Figure 4.3(d)) remain essentially constant, and an increase in M is accommodated by an increase in the lever arm. The compressive stress block in the concrete changes shape progressively as M increases, with the compressive stresses increasing in the upper fibres and decreasing in the lower fibres. The total compressive force C, however, and the tensile force P remain practically unchanged, even when large increases occur in the moment. This can be highly beneficial when fatigue is a design consideration.

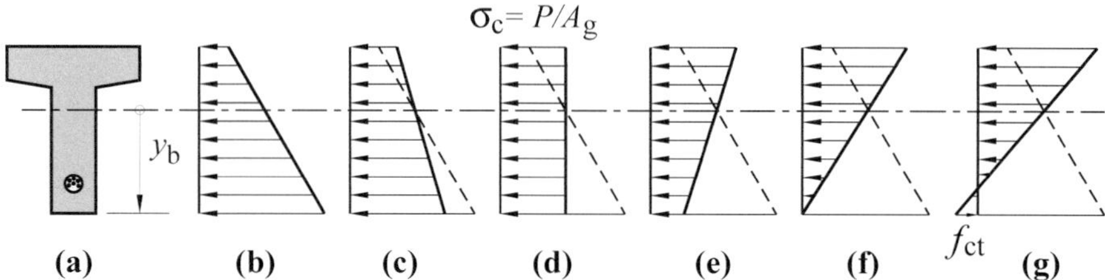

Figure 4.4 Stress distributions with increasing *M*

The changing shape of the concrete stress block with increasing M is shown in Figure 4.4. The initial prestress with $M = 0$ is shown in Figure 4.4(b). In the final stage (Figure 4.4(g)), where M approaches the cracking moment, the tensile stress in the bottom concrete fibre approaches the concrete's tensile strength, f_{ct}. Note that the concrete stress at the section centroid remains constant at P/A_g while M increases. The stress line rotates about this value at the centroidal axis, with the bottom fibre stress decreasing and then going into tension. Three key stages in Figure 4.4 are of particular interest:

Uniform compressive stress, $M = Pe$, Figure 4.4(d)

When the positive applied moment is equal in magnitude to the negative prestressing moment, $M = M_p = Pe$, the concrete stress is uniformly distributed over the section at a value of P/A_g. The elastic strains are also uniformly distributed and so there is zero elastic curvature in the section. The negative curvature due to prestress is equal and opposite to the positive curvature due to M, and the effects of prestress and load are in this sense **balanced**.

Decompression, Figure 4.4(f)

When the stress in the bottom fibre is equal to zero, the condition is called **decompression**. The decompression moment is calculated as follows:

$$M_{dec} = \frac{\sigma_{cbp} I_g}{y_b} \tag{4.5}$$

where σ_{cbp} is the initial compressive stress in the bottom fibre due to pre-stress. To evaluate σ_{cbp} we use Equation 4.1 with $y = y_b$:

$$\sigma_{cbp} = \frac{P}{A_g} + \frac{Pey_b}{I_g} \tag{4.6}$$

Cracking, Figure 4.4(g)

Cracking occurs when the bottom fibre stress, now tensile, equals the tensile strength of the concrete. However, the cracking moment is reduced because additional tensile stress is induced in the tensile fibres by prior shrinkage and creep. This is discussed in detail in Chapter 5.

EXAMPLE 4.1

CALCULATION OF STRESSES IN AN UNCRACKED SECTION

A prestressed concrete beam with a rectangular cross-section 800 mm deep and 400 mm wide has a span of 10 metres. It is post-tensioned by a cable that has a cross-sectional area of 1000 mm^2 and an eccentricity e that varies parabolically from zero at the ends to a maximum of 250 mm at mid-span. The prestressing force is 1200 kN. We wish to determine for the mid-span section:

- the stresses due to prestress acting alone;
- the decompression moment; and
- the stresses due to prestress, self-weight and a uniformly applied live load of 30 kN/m.

All loads act shortly after prestressing, so that the one value of E_c is used throughout these calculations.

Other Data

$$A_g = 320\times10^3 \text{ mm}^2;\ I_g = 17.07\times10^9 \text{ mm}^4;$$

$$Z_{bot} = Z_{top} = I_g/(D/2) = 17.07\times10^9/400 = 42.67\times10^6 \text{ mm}^3$$

The unit weight of reinforced concrete is taken as 25 kN/m^3, so that the self-weight is $w_G = 25 \times 0.32 = 8.0$ kN/m.

SOLUTION

Stresses due to prestress alone

At mid-span the prestress exerts a horizontal axial compression on the concrete of $P = 1200$ kN and a moment of:

$$Pe = -1200\times10^3 \times 250 = -300\times10^6 \text{ Nmm} = -300 \text{ kNm}$$

The axial force P produces a uniform compressive stress of:

$$\sigma_c = 1200\times10^3/320\times10^3 = 3.75 \text{ MPa}$$

The extreme fibre stresses due to the moment Pe are:

$$\sigma_c = \pm300\times10^6/42.67\times10^6 = \pm7.03 \text{ MPa}$$

The extreme fibre stresses due to prestress are thus:

$$\sigma_{ca} = 3.75 - 7.03 = -3.28 \text{ MPa} \quad \text{(tension)}$$

$$\sigma_{cb} = 3.75 + 7.03 = +10.78 \text{ MPa} \quad \text{(compression)}$$

Decompression moment

For zero bottom fibre stress, the moment from the externally applied loads must overcome the compressive stress of 10.78 MPa created at the bottom fibre by the prestress. Hence:

$$M_{dec} = \sigma_{cb} Z = 10.78 \times 42.67\times10^6 = 460\times10^6 \text{ Nmm} = 460 \text{ kNm}$$

Stresses due to prestress, self-weight and live load

Self-weight moment: $M_G = 8.0 \times 10^2/8 = 100$ kNm

Live load moment: $M_Q = 30.0 \times 10^2/8 = 375$ kNm

Total applied moment: $M = 100 + 375 = 475$ kNm

The extreme fibre stresses due to the applied moment are:

$$\sigma_{ca} = +475\times10^6/42.67\times10^6 = +11.13 \text{ MPa} \quad \text{(compression)}$$

$$\sigma_{cb} = -\sigma_{ca} = -11.13 \text{ MPa} \quad \text{(tension)}$$

The total stresses due to the applied moment plus the prestress are thus:

$$\sigma_{ca} = +11.13 - 3.28 = +7.85 \text{ MPa} \quad \text{(compression)}$$

$$\sigma_{cb} = -11.13 + 10.78 = -0.35 \text{ MPa} \quad \text{(tension)}$$

4.3 Equivalent load concept

In Chapter 1 we saw how a stressed tendon exerts an equilibrating system of forces on the prestressed beam. These forces act where the tendon is anchored to the concrete, and also wherever the tendon changes direction. If there is a sharp change in direction, i.e. a kink, the tendon exerts a concentrated transverse force on the concrete. Where the tendon is curved over a length of beam, it exerts a distributed transverse force on the concrete. These forces are illustrated in Figure 4.5. While the tendon exerts a local force on the concrete, the concrete in turn exerts an equal and opposite force on the tendon.

The forces exerted on the concrete by the tendon can be thought of as a self-equilibrating system of **equivalent loads**. It follows that the stresses in the concrete beam due to the prestress can be determined by considering the effect of these equivalent loads on the beam. This gives us an alternative method to the one explained above in Section 4.2 for determining the stresses due to prestress. The concept of equivalent loads also gives us a simple overview of the effects of prestress on both statically determinate and indeterminate beams. It provides the basis for a design concept called **load balancing**, which will be discussed shortly in Section 4.4 and will be used for design in Chapters 10, 11 and 12.

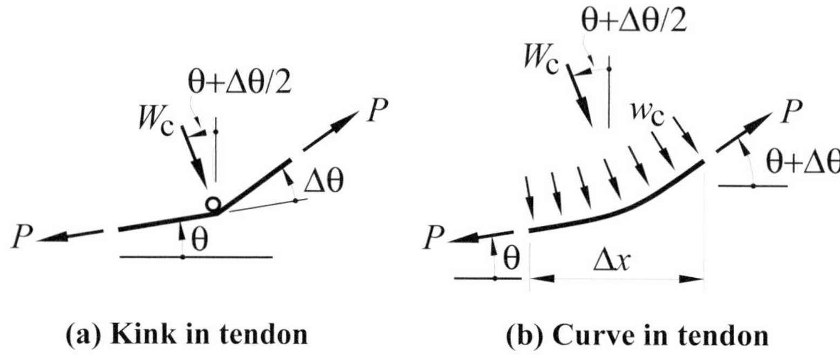

(a) Kink in tendon **(b) Curve in tendon**

Figure 4.5 Forces at a change in direction of the tendon

Equivalent load at a kink in the tendon

In Figure 4.5(a) the kink in the tendon has an angle of $\Delta\theta$. In reality, very high compressive stresses would be induced locally in the concrete at such a sharp kink in a stressed tendon. In practice a steel pin, or equivalent, is used to prevent local crushing of the concrete. The slope of the tendon is θ to the left of the kink and $\theta + \Delta\theta$ to the right. Ignoring any small loss of prestress due to friction, the tensile force in the tendon on each side of the kink is P and these are equilibrated by a transverse force W_c, which the concrete exerts through the pin onto the tendon. The tendon, in turn, exerts a force W_p through the pin onto the concrete. W_p is equal and opposite to W_c. For equilibrium in the transverse direction we have $W_c = 2P\sin(\Delta\theta/2)$. From Figure 4.5(a) we see that the line of action of W_c is at an angle of $(\theta + \Delta\theta/2)$ to the vertical. As we have seen, kinked tendons are used in pretensioned members.

In practical situations the kink angle $\Delta\theta$ is very small and the angle θ is also small. As a simplification we therefore set $\sin\theta = \tan\theta = \theta$ and write:

$$W_p = W_c = P\,\Delta\theta \qquad (4.7)$$

As an approximation, we also assume that the equivalent load W_p is vertical.

Equivalent load in a curved region of tendon

In Figure 4.5(b) a short length of tendon is curved, with a change in slope of $\Delta\theta$ over distance Δx. The situation is comparable to that in Figure 4.5(a), except that the transverse force acting on the concrete is now distributed and has the value w_p kN/m. The local concrete stresses are now smaller, and there is little danger of local crushing. The resultant force acting on the concrete over the length Δx is $W_p = w_p \Delta x$. The magnitude of w_p depends on the force P and the curvature in the cable: $w_p = P \times \kappa$. If the tendon has a parabolic shape then κ is constant along the beam and so w_p is uniformly distributed. We will consider parabolically shaped tendons in particular because these are frequently used in the design of post-tensioned members.

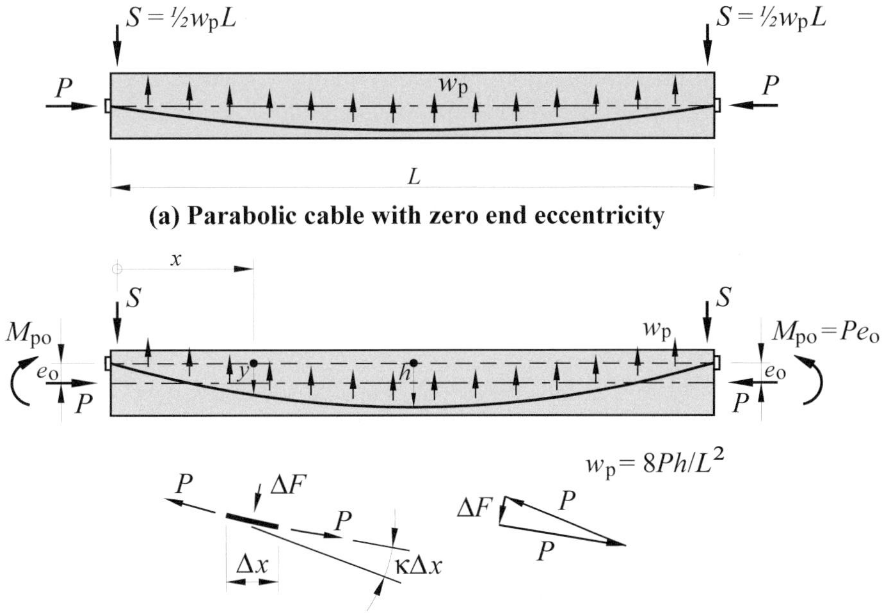

Figure 4.6 Equivalent loads for parabolic cables

Equivalent distributed load for a tendon of parabolic shape

Figure 4.6(a) shows a beam of span L with a parabolic cable.The cable eccentricity e is zero at each end of the beam, with a maximum value at mid-span. The difference between the mid-span eccentricity and the mean of the end eccentricities of a parabolic cable is called the **sag**. This is denoted as h. In Figure 4.6(a), h is also the maximum value of e. The eccentricity e at any cross-section is a function of the distance x along the beam:

$$e(x) = 4h\left[\frac{x}{L} - \left(\frac{x}{L}\right)^2\right] \tag{4.8}$$

The slope of the tendon is the first derivative of $e(x)$:

$$\theta(x) = \frac{de(x)}{dx} = \frac{4h}{L}\left(1 - \frac{2x}{L}\right) \tag{4.9}$$

The curvature in the tendon, κ, is the rate of change of slope, i.e. the second derivative of $e(x)$, which is $-8h/L^2$ and is constant along the span. A parabolic tendon thus induces a uniformly distributed load of value:

$$w_p = P\kappa = -8Ph/L^2 \tag{4.10}$$

As an approximation, we can also assume that this distributed force acts vertically.

The slope of the tendon at the end of the beam (at the anchorage) is $4h/L$, so that the transverse end force exerted on the concrete through the anchorage is $S = 4Ph/L$, while the horizontal force is taken as an approximation to be P.

In Figure 4.6(b) the tendon is also parabolic, with constant curvature along the span. The sag is again h and the equivalent load w_p is uniformly distributed (Equation 4.10). However, the end eccentricities are $-e_o$. The equivalent loads at the beam ends now consist of a vertical force S and a horizontal force P at eccentricity e. The horizontal eccentric end force is statically equivalent to an axial force P acting at the centroid, plus positive end moments $M_{po} = Pe_o$, as shown in Figure 4.6(b).

Equivalent loads in non-prismatic members

If the depth of a member varies, as for example at a haunch, care must be exercised in evaluating the stresses due to prestress. To illustrate the situation, the depth of the beam in Figure 4.7(a) changes with distance along the span.

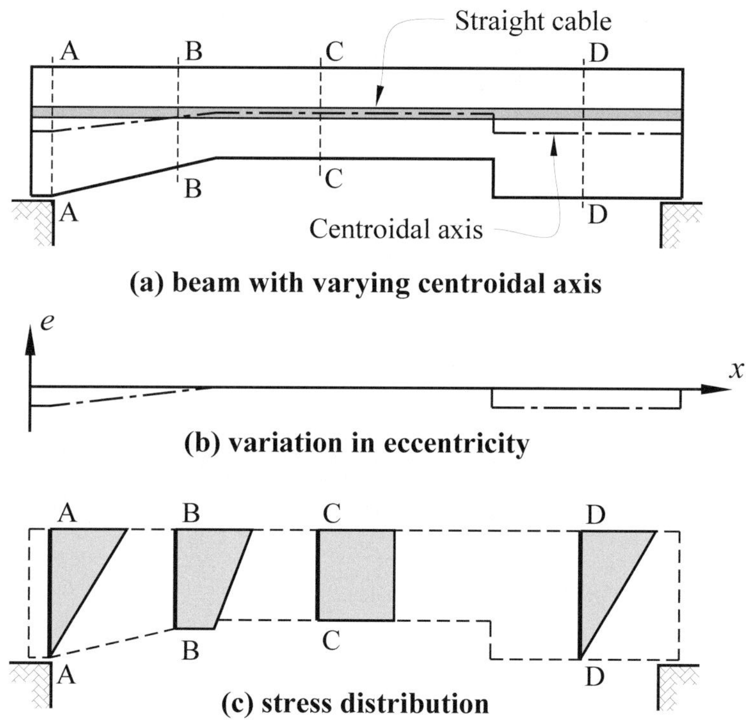

Figure 4.7 Non-Prismatic Members

In the haunch on the left side, the depth of the centroidal axis varies linearly. Over the middle region, the centroidal axis is at constant depth, but it increases abruptly where the overall depth suddenly increases. The tendon is itself straight, and the equivalent loads consist of the horizontal opposing end forces and the resultant moments due to their eccentricity. Nevertheless, the varying location of the centroidal axis, indicated in Figure 4.7(b), results in varying eccentricities of prestress and consequently varying stress distributions due to prestress. These are shown in Figure 4.7(c).

Stress calculations using the equivalent load method

In the equivalent load method, the forces exerted on the concrete by the tendon are treated as an external self-equilibrating load system. In the case of a post-tensioned member it is easy to picture how the tendon, located in a duct and physically separated from the concrete, exerts forces on the concrete. These forces, or equivalent loads, act on the concrete beam and any reinforcement contained in it but not including the tendon itself. This is shown in Figure 4.8(a). Just after prestressing, the stressed tendon is an external agent that exerts forces on the concrete beam.

Following the post-tensioning operation, the duct containing the tendon is often grouted. This introduces bond between the tendon and the adjacent concrete, and from this time on, the tendon will be an integral component of the beam. When additional load is applied, the tendon and the adjacent concrete now deform together, and the force in the tendon increases slightly from P to $(P + \Delta P)$. Even if the grouting is imperfect, there is still interaction between the tendon and the concrete and the tendon force changes as external load is applied to the member. The tendon is now an integral part of the member.

Nevertheless, simplifications and approximations are introduced into the design calculations which tend to blur the actual behaviour. As we have already seen, the calculations of concrete stresses in an uncracked section are based on the gross concrete section of a post-tensioned member (thus ignoring the duct, the tendon and any reinforcement).

The fundamentals are slightly different in the case of pretensioned members. In Figure 4.8(b), the strands or wires are placed in the prestressing bed and tensioned before the concrete is added. The stretched tendon is held in position by concentrated forces at A, B, C and D. The rigid prestressing bed is used to apply these forces to the tendon. After the concrete has been placed and has hardened, the tendon is fully bonded to the unstressed concrete and has become part of the beam. At transfer, the forces at A, B, C and D are 'released'. Statically, this is equivalent to applying equal and opposite forces to those in Figure 4.8(b) to the *entire member*, now including the tendon. Figure 4.8(c) shows the release forces, with $W_A = -R_A$, $W_B = -R_B$, $W_C = -R_C$ and $W_D = -R_D$. Nevertheless, in simplified calculations the stresses in the concrete due to prestress are calculated in the same way as for a post-tensioned member using the gross concrete section.

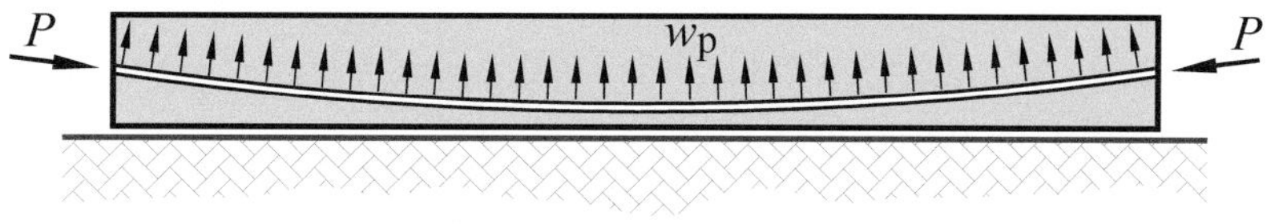

(a) Post-tensioned beam

(b) Pre-tensioned beam: forces acting on tendon before casting

(c) Pre-tensioned beam: at transfer

Figure 4.8 Equivalent loads for (a) post-tensioned beams and (b) and (c) pre-tensioned beams (hog not shown)

The fine differences in the behaviour of pretensioned and post-tensioned members are blurred by the simplifying assumptions used in routine design.

In simplified calculations the small additional stresses that develop in the tendon and the reinforcing steel during the application of external load are ignored. Nevertheless, simple estimates of these stress increments can be made by evaluating the stress increments in the adjacent concrete, and then multiplying them by the appropriate modular ratio, i.e. by E_p/E_c or E_s/E_c. A more accurate analysis can be made using the analysis in Appendix A.

EXAMPLE 4.2

USE OF EQUIVALENT LOAD METHOD TO CALCULATE STRESSES

The equivalent load method is used to calculate the stresses due to prestress and full load for the beam of Example 4.1.

SOLUTION

Stresses due to prestress alone

The uniformly distributed upward load exerted by the tendon along the span is obtained using Equation 4.10:

$$w_p = -8(1200 \times 0.25)/10^2 = -24 \text{ kN/m}$$

The slope of the tendon at each end is:

$$\theta_o = 4h/L = 4 \times 0.25/10 = 0.1 \text{ radians}$$

The downward forces at each end of the tendon are:

$$S = 0.1 \times 1200 = 120 \text{ kN}$$

The horizontal equivalent load at each end of the tendon is 1200 kN. As a statical check on the vertical forces, we note that $2S = w_p L = 240$ kN. The equivalent loads are shown in Figure 4.9(a).

At mid-span, the moment due to the equivalent loads is:

$$M = -120 \times 5 + 24 \times 5 \times 2.5 = -300 \text{ kNm}$$

The extreme fibre stresses due to this moment are:

$$\sigma_{cb} = -\sigma_{ca} = 300 \times 10^6 / 42.67 \times 10^6 = 7.03 \text{ MPa}$$

The equivalent axial forces produce a uniform compressive stress of:

$$\sigma_c = 1200 \times 10^3/(800 \times 400) = 3.75 \text{ MPa}$$

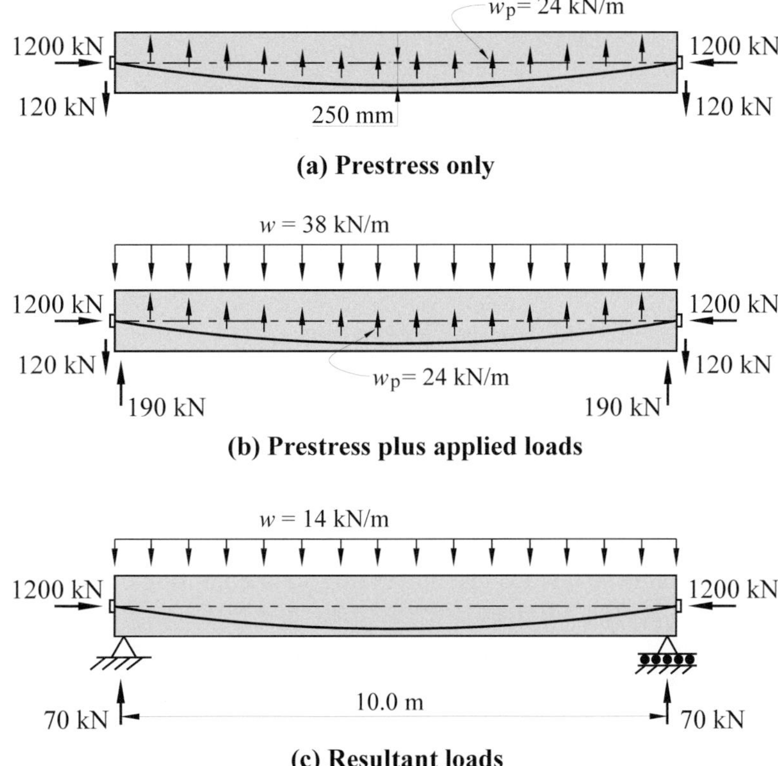

Figure 4.9 Equivalent loads, Example 4.2

At mid-span the total extreme fibre stresses due to prestress are thus:

$$\sigma_{ca} = +3.75 - 7.03 = -3.28 \text{ MPa (tension)}$$

$$\sigma_{cb} = +3.75 + 7.03 = +10.78 \text{ MPa (compression)}$$

Stresses due to prestress, self weight and live load

To determine the combined stresses we add the applied distributed load to the equivalent load to obtain $30 + 8 - 24 = 14$ kN/m . This nett load is shown in Figure 4.9(c) with the resultant reactions at each end of 70 kN. The bending moment at mid-span due to the nett load is:

$$M = 70 \times 5 - 14 \times 5 \times 2.5 = 175 \text{ kNm}$$

This moment induces extreme fibre stresses of:

$$\sigma_{ca} = -\sigma_{cb} = 175{\times}10^6 / 42.67{\times}10^6 = 4.1 \text{ MPa}$$

As before, the uniform compressive force due to the axial load is 3.75 MPa, and the final stresses are:

$$\sigma_{ca} = +3.75 + 4.10 = +7.85 \text{ MPa (compression)}$$

$$\sigma_{cb} = +3.75 - 4.10 = -0.35 \text{ MPa (tension)}$$

4.4 Load balancing

We have seen how prestress is used in the design of concrete beams to improve structural behaviour under working load conditions. By choosing a suitable cable profile (defined by the varying eccentricity $e(x)$ along the beam) together with the magnitude of the prestressing force P, it is possible to control deflections and crack widths for a specific design service load. The term **load balancing** refers to the choice of a cable profile and cable force to produce equivalent loads that are equal and opposite to a chosen external design load, so that a condition of zero deflection is achieved. In this condition, each cross-section along the beam is in a state of uniform compression, or balance, as indicated already in Figure 4.4(d). The external design load is said to be '*balanced*' by the prestress. Load balancing is a simple application of the equivalent load concept and provides a very convenient and useful preliminary design tool for choosing appropriate prestress details.

Consider for example the beam from Example 4.2 and Figure 4.9, but subjected now to a distributed external load, including self-weight, of just 24 kN/m, which is equal to the equivalent load w_p exerted by the tendon.

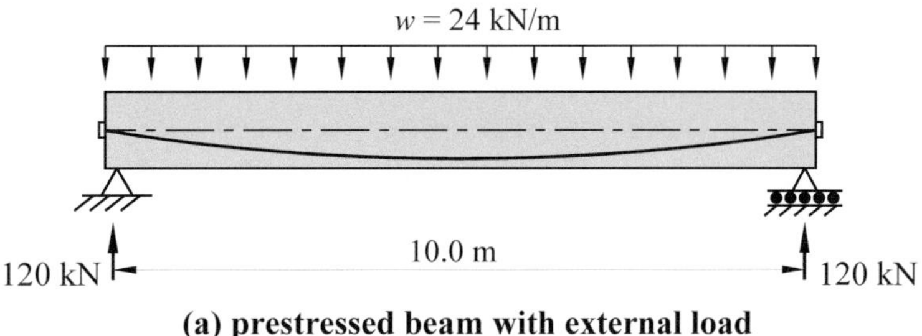

(a) prestressed beam with external load

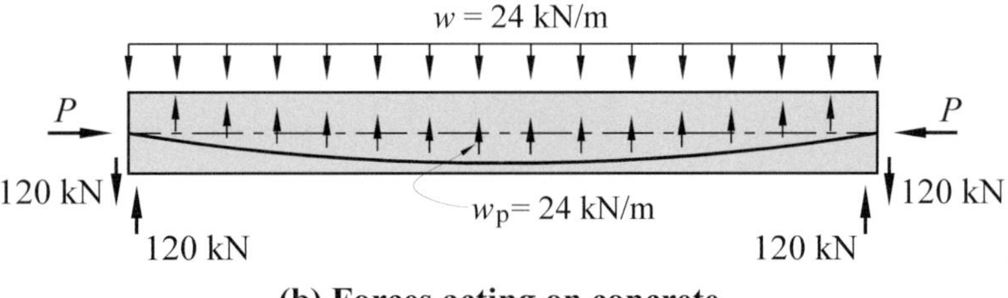

(b) Forces acting on concrete

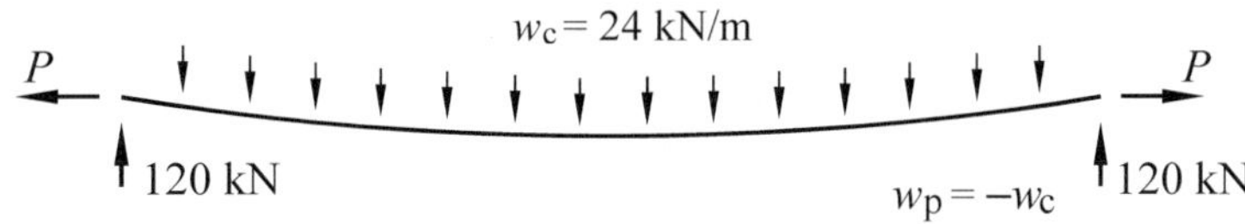

(c) Forces acting on the prestressing cable for balanced load

Figure 4.10 Load balancing

The beam with this load is shown in Figure 4.10(a), and as a free body diagram of the concrete, without the tendon, in Figure 4.10(b). At every section the external load w is balanced by the equivalent load w_p. The bending moment (and shear force) in the concrete due to w is equal and opposite to that produced by w_p. The concrete is in a state of uniform axial compression with the concrete stress $\sigma = P/A_g = 3.75$ MPa. The concrete undergoes purely axial contraction. There is no deflection and of course no cracking. The tendon, with its curved shape, is in a state of pure tension: it carries the applied vertical distributed load to the supports, as shown in the free body in Figure 4.10(c).

By choosing an appropriate design load to be balanced, that is to say, the load at which the deflection is to be zero, the designer can use a very simple calculation to determine appropriate values for the cable force and cable profile. The load that is chosen to be balanced might be the self-weight of the structure, the self-weight plus a proportion of the permanent load, or even the self-weight plus the total permanent load plus a proportion of the live load. The serviceability requirements for design are automatically satisfied for the specific design load that is balanced. However, it is usually necessary to check that the serviceability requirements are satisfied at other relevant design working load combinations, such as at transfer and at full design load.

EXAMPLE 4.3

USE OF LOAD BALANCING TO DETERMINE PRESTRESSING DETAILS

A post-tensioned one-way slab is 300 mm thick and is to carry a superimposed load of 12 kPa over a simply supported span of 10 metres. Determine the prestressing details to keep the slab level under its self-weight plus 4 kPa of live load. Take the total deferred losses as 18 per cent and the friction loss at mid-span as 5 per cent of the initial prestress (see Chapter 9 in regard to losses). Also, determine the load that causes decompression at the bottom fibres after the losses of prestress have occurred.

Other Data

For a one metre wide strip of slab:

$$A_g = 300\times10^3\ \text{mm}^2/\text{m}$$

$$I_g = 1000 \times 300^3/12 = 2.25\times10^9\ \text{mm}^4/\text{m}$$

$$Z = I_g/(0.5D) = 2.25\times10^9/(0.5 \times 300) = 15.0\times10^6\ \text{mm}^3/\text{m}$$

Self-weight load: $w_G = 0.3 \times 25 = 7.5$ kN/m

Load to be balanced: $w_p = 7.5 + 4 = 11.5$ kN/m

SOLUTION

Details of prestressing tendon

A parabolic tendon with constant force P and zero end eccentricities provides a uniformly distributed upward equivalent load. The maximum available eccentricity at mid-span is estimated to be 104 mm. Taking the sag as this value, $h = 104$ mm (0.104 metres), we calculate the required tendon force after all losses, P_e, using a rearrangement of Equation 4.10:

$$P_e = \frac{w_p L^2}{8h} = \frac{11.5 \times 10^2}{8 \times 0.104} = 1382 \text{ kN/m}$$

This is the cable force per metre width of slab at mid-span after losses. Allowing for friction and deferred losses we obtain the initial tendon jacking force at an end as:

$$P_i = \frac{1382}{0.95 \times 0.82} = 1774 \text{ kN/m}$$

In the denominator, 0.95 allows for the estimated friction loss between the jack and the section considered, while 0.82 accounts for the deferred losses.

A tendon consisting of four 12.7 mm diameter super grade strands stressed to 80 per cent of the minimum breaking stress gives a force of 589 kN. The required transverse spacing of these tendons would be:

$$s = 1000 \times 589 / 1774 = 330 \text{ mm}$$

Decompression moment

After losses, this prestressing tendon produces a uniformly distributed concrete stress under self-weight plus a live load of 4 kN/m. The stress is:

$$\sigma_c = 1382 \times 10^3 / 300 \times 10^3 = 4.61 \text{ MPa}$$

The additional moment that brings the extreme lower fibre to zero stress is therefore:

$$M = \sigma_c \times Z = 4.61 \times 15.0{\times}10^6 = 69.1 \text{ kNm/m}$$

This corresponds to an additional load of:

$$w = 8 \times 69.1 / 10^2 = 5.53 \text{ kN/m/m}$$

The decompression load and decompression moment are thus:

$$w_{dec} = 11.5 + 5.5 = 17.0 \text{ kPa}$$

$$M_{dec} = 17.0 \times 10^2 / 8 = 213 \text{ kNm/m}$$

4.5 Creep and shrinkage effects in beams

Figure 4.11(a) shows an unreinforced cross-section subjected to an initial prestressing force P_i. The initial concrete strains just after application of prestress, represented by line AB in Figure 4.11(a), are elastic and correspond to the stress distribution EF, also shown in Figure 4.11(a). The stresses and strains in the lower part of the section are compressive, but the eccentricity e of the prestress determines whether or not the top fibres are in compression or tension. The figure shows small tensile stresses and strains in the uppermost fibres, with the point of zero stress within the section at distance y_G above the centroidal axis.

The initial tensile strain in the tendon, just after the application of prestress, is ε_{pi}, which is represented in Figure 4.11(a) by line DQ, with origin at D. The initial elastic compressive concrete strain at this level, ε_{ci}, is represented by line CQ with origin at C. Both origins have been chosen such that any subsequent increments in strain at this level will occur equally in both the concrete and the tendon. This recognises that, with full bonding, any compressive strain increment in the concrete will occur together with a tensile strain decrement of equal magnitude in the adjacent prestressing steel.

As a result of creep and shrinkage, the compressive concrete strains increase markedly over time. The new state of strain in the concrete at some time t is represented by line A'B' in Figure 4.11(b).

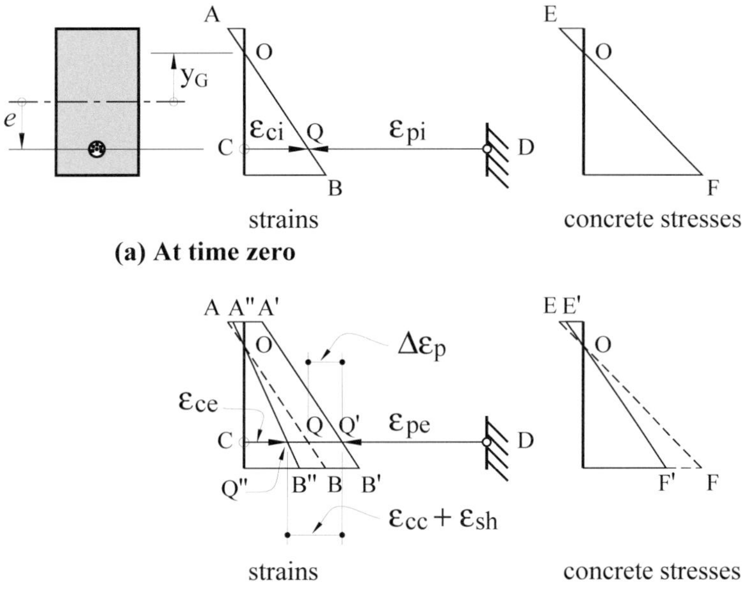

Figure 4.11 Effect of creep and shrinkage

The total concrete strain at any level now consists of elastic, creep and shrinkage components. The decrease in tensile strain in the prestressing steel, $\Delta\varepsilon_p$, is equal to the increase in compressive strain in the concrete at this level (which is $\Delta\varepsilon_{cp}$), and the new, reduced total tensile strain in the tendon is:

$$\varepsilon_{pe} = \varepsilon_{pi} - \Delta\varepsilon_p \tag{4.11}$$

The prestressing force has decreased from P_i to P_e, where:

$$P_e = P_i - \Delta\varepsilon_p E_p A_p \tag{4.12}$$

The terms E_p and A_p are, respectively, the elastic modulus and the cross-sectional area of the prestressing steel.

There is an equal loss of prestress in the concrete, and a proportional reduction in the concrete stress at all levels by the ratio P_e/P_i. In the new stress distribution shown in Figure 4.11(b), the point of zero stress remains at O. There has been a rotation of the stress line about O, from EF to E'F'. The elastic strain line likewise rotates about point O to the new elastic line A"B". At the tendon level the elastic strain in the concrete has reduced to CQ", and the inelastic strain is Q"Q'. The band between A"A' at the top fibre and B"B' at the bottom fibre (with Q"Q' at the tendon level) represents the total inelastic (creep and shrinkage) strain in the concrete. Although the total compressive strain in a typical fibre has increased markedly, there has been a nett decrease in the elastic component of strain and hence also a corresponding decrease in the compressive stress in the same fibre.

Any bonded reinforcing steel in the section undergoes the same strain increments as the adjacent concrete. In the lower fibres, tensile reinforcement is put into compression when the prestress is applied. As creep and shrinkage strains increase in the adjacent concrete, there is an equal increase in strain in the steel, and a corresponding elastic increase in compressive steel stress. In effect, prestress is progressively transferred from the concrete to the reinforcing steel, which results in a further loss of prestress in the concrete. It is important to note that the forces C and P, in the concrete and tendon, will now differ, sometimes considerably, because part of the compressive force is now being carried by the reinforcing steel. If compressive reinforcement is present in the upper fibres, a similar process takes place there. However, the initial compressive concrete stresses in the upper fibres due to prestress are small (and possibly even tensile), so that the effect is quantitatively much less significant.

If a sustained moment acts on the section together with the prestress, the initial stress distribution will be modified, with larger compressive stresses in the upper fibres and smaller compressive stresses in the lower fibres, which may even be tensile. The level of zero stress, point O in Figure 4.11, will now be on the lower side of the cross-section and may even be below the bottom fibre. The effects of creep and shrinkage will be substantial, as already discussed, although any compressive reinforcement will now play the important role that the tensile reinforcement previously played.

The shrinkage strains develop progressively in the concrete, irrespective of stress. Their effect is to transfer compressive prestress from the concrete to the steel and tendon, thereby increasing the loss of prestress in the concrete. Uniform shrinkage in a plain, unreinforced, concrete member would produce only axial shortening. However, restraint is provided by the unsymmetrically placed prestressing tendon and reinforcing steel, and this results in a strain gradient in the section, with curvature and overall deflection. This is called **shrinkage warping**. It can become significant in shallow members such as slabs and planks and is accentuated if there is differential shrinkage between the upper and lower surfaces of the member, for example because of differences in ambient conditions.

Shrinkage and creep together bring about large, long-term, prestress losses, especially when there is a large amount of reinforcing steel in the section. The overall increase in local deformations (and hence overall deflection) can be much larger than the initial elastic deformations, although the presence of reinforcing steel reduces the magnitude of the long-term deformations and deflection. Loss of prestress in the lower concrete fibres can reduce the cracking moment for the section significantly. This is discussed further in Chapter 5, in relation to the calculation of the cracking moment M_{cr}.

As already mentioned, Equation 4.1 provides a reasonably good estimate of the stresses due to prestress in a concrete section, even after losses have occurred, provided the reduced value of the compressive concrete force C is used, and not the tendon force P. Clearly, the long-term strains and deformations cannot be evaluated from an elastic analysis.

4.6 Analysis of creep and shrinkage effects

For design, quantitative methods are needed to evaluate the losses in prestress and the additional long-term deformations and deflections that result from creep and shrinkage of the concrete. To avoid complex analysis, simplified, approximate methods are sometimes used. For example, AS 3600 allows multiplying factors to be applied to the short-term deflections of reinforced concrete beams in order to estimate long-term deflections. Unfortunately, the Standard does not allow such simplifications in the case of prestressed concrete members. Indeed, AS 3600 requires that long-term deflections in prestressed members be evaluated from first principles, together with the design creep and shrinkage values φ_{cc} and ε_{cs} as defined in the Standard, even though the long-term deflections can be very small when load balancing is used for design.

Three methods for evaluating creep effects and shrinkage effects, of varying accuracy and complexity, are presented in Appendix B, together with a simple, order-of-magnitude, calculation procedure. The methods are based on first principles. The most appropriate method to use in any particular design case will depend on the accuracy required, the computational resources available, and the accuracy of the creep and shrinkage properties of the concrete that are available. These methods are only very briefly described here, but are fully explained in Appendix B.

Order-of-magnitude calculations: A simple, approximate procedure is explained in Section B.2 of Appendix B which is useful in preliminary design, and also for making order-of-magnitude checks on the results of more complex calculations and computer output. As this method will be used in the following examples in this chapter, the expressions for the long-term creep curvature and long-term deflection are given below in Equations 4.13 and 4.14. By assuming that free creep occurs in a section under the initial elastic concrete stresses, the long-term creep curvature κ_c^* is approximated by the total free creep curvature κ_{co}^*, which is obtained from the elastic curvature κ_o and the long-term creep value φ_o^*:

$$\kappa_c^* \approx \kappa_{co}^* = \varphi_o^* \kappa_o \tag{4.13}$$

Likewise the factor φ_o^* is used to obtain a rough estimate of the long-term creep deflection from the short-term deflection, Δ_e:

$$\Delta_c^* = \varphi_o^* \Delta_e \tag{4.14}$$

One-step analysis using the age-adjusted effective modulus: A one-step cross-section analysis is mathematically fairly complex, but can be used in detailed design calculations that are best undertaken with spreadsheets. Use of the age-adjusted effective modulus, $k^* E_c^*$ in lieu of the long-term elastic concrete modulus, E_c^*, allows approximately for changing concrete stress over time and improves accuracy. Section B.3 of Appendix B gives full details of this method.

Step-by-step numerical analysis: In a multi-step time analysis, the states of stress and strain and deformation are determined in a cross-section at a sequence of pre-chosen times. Section B.4 of Appendix B provides details of this method of analysis. Variations in concrete stress over time are allowed for in this analysis, and full information on conditions at intermediate times are determined, as well as at t^*. The accuracy of the results depends on the number of time steps used. The analysis is computationally intensive and is best undertaken by computer.

Approximate, simplified closed-form equations: Simple, approximate closed-form equations are presented in Section B.5 of Appendix B. These can be used to estimate the long-term loss of prestress and increase in curvature in a section due to creep and shrinkage. The results are less accurate than the step-by-step and one-step methods because approximations are introduced to derive simplified equations. The equations can be used in the design of members that are not deflection sensitive.

4.7 Deflections of uncracked beams

During its in-service life, a structural member is subjected typically to a sustained load (including self-weight) of nearly constant magnitude, but with intervening repeated short-term live loads that vary greatly in magnitude and even in configuration.

We consider the prestressed beam in Figure 4.12(a) that is subjected only to time-varying uniformly distributed loads. Prestressing is applied at time t_o, and the self-weight also acts from this time. The full dead load acts at time t_1. Live loads of short duration and of variable magnitude occur after t_1. The self-weight load is w_{SW}, the full dead load (including self-weight) is w_G, the live load is w_Q and the dead load plus full design live load is w_{GQ}. The variation of loading over time is shown in Figure 4.12(b). Although the prestress acting alone induces an upwards camber, the downwards self-weight also acts from t_o and the nett result will usually still be an upwards camber. This is shown in Figures 4.12(c) and (d). Of course, with a sufficiently low level of prestress, the nett deflection would be downward under self-weight, but this possibility is unusual and is not shown in the figure.

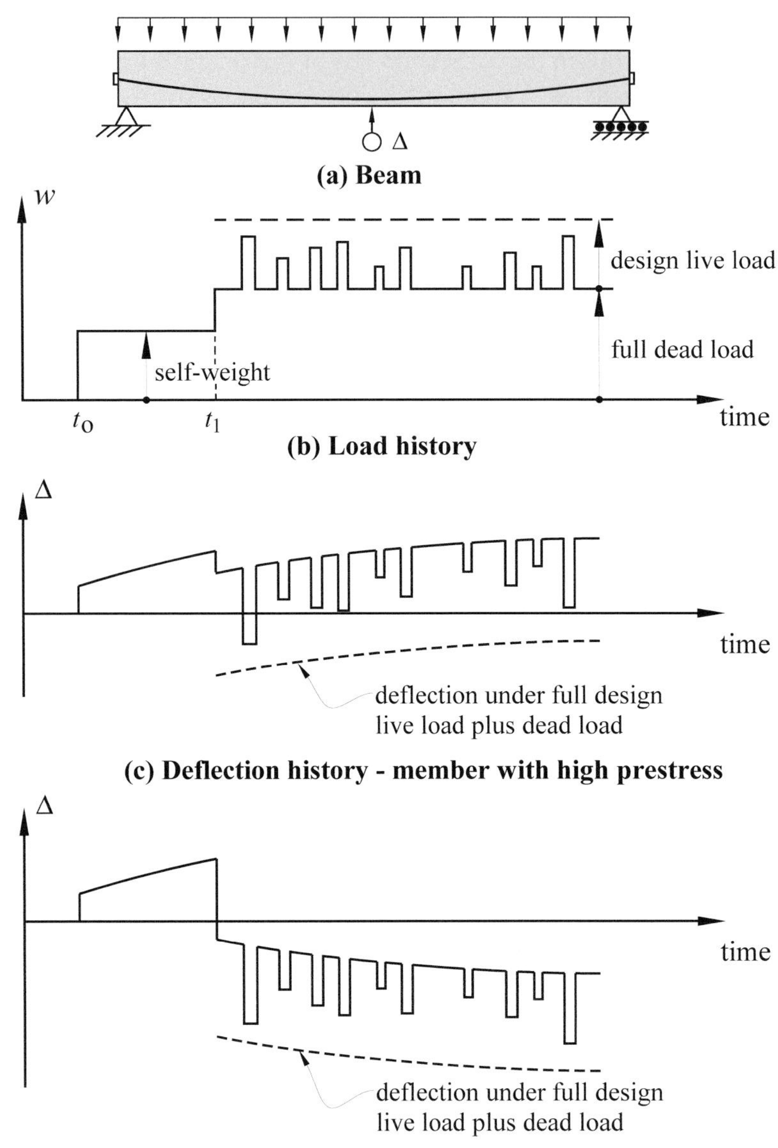

Figure 4.12 Deflection histories, as affected by creep

When the full dead load is applied at time t_1 there is a sudden downwards increment in deflection. If the prestress is fairly high, the nett result will still be an overall upwards deflection, or camber, as is shown in Figure 4.12(c), but, with a lower level of prestress, it will be downwards as in Figure 4.12(d).

Over time, the deflection changes due to creep and shrinkage. On the one hand, shrinkage will usually result in a small downward deflection increment, but this only becomes significant in slender members such as slabs and shallow beams with high span-to-depth ratios. On the other hand, creep will magnify the existing short-term deflections at times t_o and t_1. The trends of the full lines in Figures 4.12(c) and (d) indicate these changing deflections.

The short-term live load 'spikes' in Figure 4.12(b) produce downward short-term spikes in deflection. These deflection spikes are shown in Figures 4.12(c) and (d). The trends of the total deflections under the combined action of prestress, plus dead load and full design live load w_{GQ}, are indicated by the dashed lines. With higher prestress (Figure 4.12(c)), the total downward deflection actually decreases with time, because creep magnifies the nett initial upward deflection. With lower prestress, the total deflection in Figure 4.12(d) trends downwards.

If the prestressing force had been chosen to balance the full dead load, then the initial deflection at t_1 would be zero and there would be minimal change in deflection, if any, over time due to creep.

At any stage of loading, the deflection curve $y(x)$ for a flexural member is obtained by double integration of the curvature, taking due account of the boundary conditions:

$$y(x) = \iint \kappa(x)\,dx\,dx \tag{4.15}$$

This is a purely geometric relation and therefore applies irrespective of whether the material behaviour is elastic or inelastic.

4.7.1 Calculation of short-term deflections

The short-term curvature $\kappa(x)$ in the cross-section at distance x along the beam depends on the bending moment, $M(x)$. For short-term loading, the relation is linear prior to cracking and the areas of reinforcement and tendon in the section have little effect on the bending stiffness of the uncracked section. Short-term deflections are therefore calculated using the elastic bending stiffness of the plain concrete section, $E_c I_g$:

$$\kappa(x) = \frac{M(x))}{E_c I_g(x)} \tag{4.16}$$

so that:

$$y(x)) = \iint \frac{M(x))}{E_c I_g(x)} dx dx \tag{4.17}$$

These equations apply to both determinate and indeterminate beams. Standard formulae and coefficients are available that cover a wide range of practical cases and in practice these can be used in lieu of Equation 4.17. Appendix D of this book summarises useful information for deflection calculations.

Initial deflections due to prestress and dead load

The value of E_c increases with the age of the concrete, so that different values may need to be used when, for example, the prestress is applied at an early age and the dead load is applied in increments at later ages. Example 4.6 below takes into account variations in E_c. It is convenient to evaluate the initial elastic deflections due to dead load and prestress separately, and then add them together. The equivalent load concept can be used to calculate the initial deflections due to prestress.

In Figure 4.13(a) the cable eccentricity in the beam varies parabolically from e_1 at each end to a maximum of e_2 at mid-span. The equivalent loads, shown in Figure 4.13(b), consist of a negative (upwards) uniformly distributed load, downward vertical end reactions, horizontal axial end forces (which cause pure compression in the concrete) and negative end moments, Pe_1.

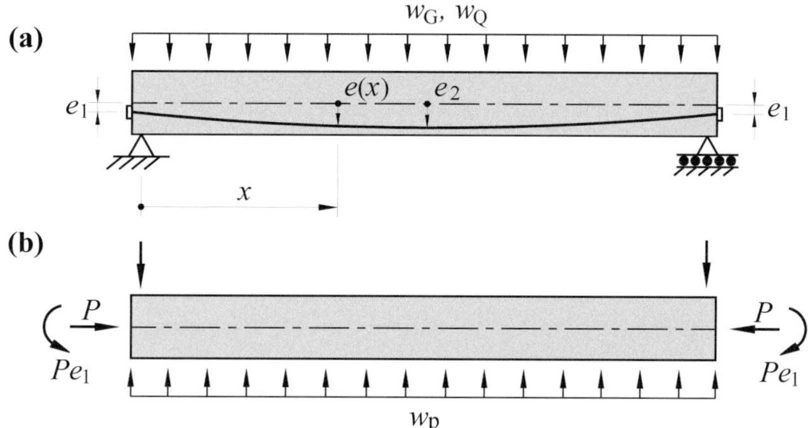

Figure 4.13 Deflection calculations: non-zero end eccentricities

The upward equivalent load:

$$w_{\mathrm{p}} = 8P \cdot (e_2 - e_1)/L^2 \tag{4.18}$$

produces an upward deflection at mid-span which, from Appendix D (Table D.1, item (c)), is:

$$\Delta_1 = \frac{5}{384} \cdot \frac{1}{E_{\mathrm{c}} I_{\mathrm{g}}} \cdot 8P(e_2 - e_1)L^2 \tag{4.19}$$

Friction loss causes variations in the prestressing force along the beam, but these are usually ignored in deflection calculations, with the value of P taken at mid-span. The constant end moments Pe_1 produce an upward deflection (Appendix D) of:

$$\Delta_2 = \frac{1}{8} \cdot \frac{1}{E_{\mathrm{c}} I_{\mathrm{g}}} \cdot Pe_1 \cdot L^2 \tag{4.20}$$

The total upward deflection at mid-span due to prestress is:

$$\Delta_P = \Delta_1 + \Delta_2 = \frac{PL^2}{E_c I_g}\left(\frac{e_1}{8} + \frac{5}{384}8(e_2 - e_1)\right) \quad (4.21)$$

The dead load produces a downward deflection:

$$\Delta_G = \frac{5}{384} \cdot \frac{w_G L^4}{E_c I_g} \quad (4.22)$$

If the total dead load acts from the time of prestressing, then the same value of E_c will be used in Equations 4.21 and 4.22. The total deflection due to prestress and dead load is obtained by adding the two components:

$$\Delta_{PG} = -\Delta_P + \Delta_G \quad (4.23)$$

If, furthermore, the prestressing details have been chosen on the basis of load balancing, then Δ_{PG} can be calculated in one operation as in Example 4.2. If the member is statically determinate, the deflection due to prestress can also be calculated from the bending moment produced in each section by the prestressing force P and the local cable eccentricity $e(x)$. Assuming, as before, a constant cable force, we have $M_P(x) = Pe(x)$, and the deflection due to prestress alone is obtained by substituting this expression for moment into Equation 4.17. The reader can confirm that the total camber (upward deflection) at mid-span, calculated in this manner, is that given by Equation 4.21.

Short-term deflections due to live load

In an uncracked beam, deflection increments due to live load w_Q can be calculated on the assumption of elastic behaviour, using deflection formulae if available, but with E_c chosen for the age at load application. For the beam in Figure 4.13, the downward deflection due to live load is:

$$\Delta_Q = \frac{5}{384} \cdot \frac{w_Q L^4}{E_c I_g} \quad (4.24)$$

A useful rearrangement of this equation can be obtained in terms of the curvature κ_Q that corresponds to the maximum moment $M_{Q.max}$ at mid-span.

With $\kappa_Q = M_{Q.max}/(E_c I_g)$ and $M_{Q.max} = w_Q L^2/8$, we obtain:

$$\Delta_Q = \frac{5}{48} \cdot \kappa_Q L^2 \tag{4.25}$$

This is a purely geometric equation and can be used also to evaluate inelastic long-term deflections provided the inelastic curvatures are distributed along the beam span in approximately the way that the elastically calculated moments are distributed.

4.7.2 Calculation of long-term creep deflections

According to AS 3600, the shrinkage and creep components of the long-term deflection must be determined separately, using first principles and the design shrinkage strain, ε_{cs} (also written here as ε^*_{sh}), and the design creep coefficient, φ_{cc}. A first-principles calculation of deflection is made by evaluating the curvatures at key points along the beam and then integrating them.

Evaluation of creep curvature

In choosing an appropriate method to evaluate creep curvature, it is important to consider the accuracy that is needed together with the reliability of the available creep data. Detailed computations are not justified unless based on reliable creep data. The methods of creep analysis listed previously in Section 4.6, and explained in detail in Appendix B, can be used to evaluate the creep curvature.

Creep deflection calculations

Creep deflections can always be obtained by integrating the creep curvatures at key cross-sections using Equation 4.15. Approximate expressions such as Equation 4.25 can alternatively be used when applicable.

For an elastic, uncracked member supporting a uniformly distributed load on span L, the mid-span deflection can be determined using the approximation:

$$\Delta = \frac{L^2}{96}[\kappa_L + 10\kappa_M + \kappa_R] \tag{4.26}$$

where κ_M, κ_L and κ_R are the curvatures at the middle, left and right ends, respectively. If the moment at a section is positive, then the curvature is also positive, and vice versa. In a simply supported beam the curvatures at the supports are zero and the above expression reduces to:

$$\Delta = \frac{\kappa_M L^2}{9.6} \tag{4.27}$$

Such geometric expressions can be used to evaluate creep deflections approximately from creep curvatures in key sections.

4.7.3 Calculation of long-term shrinkage deflections

The long-term shrinkage-induced curvature in a prestressed cross-section, κ_{sh}^*, depends on the geometrical properties of the section, the eccentricities of the restraining steel and tendon, and the final free shrinkage strain, which is denoted in AS 3600 as ε_{cs}, and here as ε_{sh}^*.

Methods are described in Appendix B for evaluating the long-term shrinkage curvature κ_{sh}^*, as well as the creep curvature. The step-by-step method (Section B.3) is always applicable, but a simpler alternative is to use the one-step age-adjusted effective modulus method (Section B.4). Even simpler but more approximate calculations can be made using the closed form equations for shrinkage presented in Section B.5 of Appendix B.

In calculating shrinkage deflection, the number of sections that need to be considered depends on how the reinforcement varies along the beam. In simple situations the curvature at three sections, at mid-span and at the quarter points, or the end points, will usually be sufficient.

If there is uniform shrinkage curvature along a simply supported member, with $\kappa_L = \kappa_M = \kappa_R$, Equation 4.26 reduces to:

$$\Delta = \frac{1}{8}\kappa_M L^2 \tag{4.28}$$

Equations 4.26 to 4.28 will often be sufficient for routine long-term deflection calculations.

EXAMPLE 4.4

DEFLECTION CALCULATIONS FOR AN UNCRACKED BEAM

Calculate the short-term and long-term deflections for the beam in Example 4.1, which is simply supported on a 10 metre span and has a parabolic tendon with an eccentricity at mid-span of $e = 250$ mm and zero at the ends. The initial cable force is 1200 kN, the live load is $w_Q = 30$ kN/m and the self-weight load is $w_G = 8$ kN/m. At mid-span, $M_G = 100$ kNm, $M_Q = 375$ kNm, and $M_p = -1200 \times 0.25 = -300$ kNm. The equivalent load due to prestress is:

$$w_p = -8(300)/10^2 = -24 \text{ kN/m}$$

Other data

$$E_c = 30{\times}10^3 \text{ MPa} ;\ \varphi_{cc} = \varphi_o^* = 3.0 ;\ \varepsilon_{cs} = 600{\times}10^{-6} ;$$

$$A_g = 320{\times}10^3 \text{ mm}^2 ;\ I_g = 17.07{\times}10^9 \text{ mm}^4 ;\ E_c^* = 33{\times}10^3 \text{ MPa}$$

SOLUTION

Initial elastic deflection and deformation

For a simply supported beam with uniformly distributed load, the elastic short-term deflection is:

$$\Delta = \frac{5}{384} \cdot \frac{wL^4}{EI}$$

The short-term deflection due to prestress and self-weight is:

$$\Delta_{PG} = \frac{5}{384} \frac{(8 - 24)10000^4}{30{\times}10^3 \times 17.07 \times 10^9} = -4.1 \text{ mm}$$

where the negative sign indicates an upward camber.

The short-term deflection due to the live load $w_Q = 30$ kN/m is:

$$\Delta_Q = -4.1 \times 30/(8 - 24) = 7.7 \text{ mm downwards}$$

The nett short-term deflection is thus:

$$\Delta_{el} = \Delta_Q + \Delta_{pG} = 7.7 - 4.1 = 3.6 \text{ mm downwards}$$

In estimating long-term deflections below, we will need the value of the short-term curvature at mid-span, $\kappa_o = M/EI$. With a nett load of $(8 - 24) = -16$ kN/m, the nett moment is:

$$M = -16 \times 10^2/8 = -200 \text{ kNm}$$

and we obtain:

$$\kappa_o = \frac{-200 \times 10^6}{17.07 \times 10^9 \times 30 \times 10^3} = -0.391 \times 10^{-6} \text{ mm}^{-1}$$

Creep deflections

Note: The following calculations use methods that are explained in Appendix B. It is recommended that the reader review the relevant parts of Appendix B, and Examples B.2, B.3, B.7, B.8 and B.9, before continuing with this example.

Approximate estimate: A first rough estimate of the creep curvature at mid-span, and the creep deflection, is obtained by ignoring the presence of the tendon. From a free creep calculation using Equations 4.13 and 4.14:

$$\kappa^*_{co} = \varphi^*_o \kappa_o = -3.0 \times 0.391 \times 10^{-6} = -1.170 \times 10^{-6} \text{ mm}^{-1}$$

$$\Delta^*_c = \varphi^*_o \Delta_{el} = 3.0 \times (-4.1) = -12.3 \text{ mm}$$

Age-adjusted effective modulus calculation: A more accurate calculation of the long-term curvature at mid-span can be obtained using a one-step analysis and an age-adjusted effective modulus of concrete with $k^* = 0.295$ and $k^* E^*_c = 9735\,\text{MPa}$. The calculation follows the procedure described in Sections B.3.1 and B.3.2, and as used in Example B.1 of Appendix B. The detailed calculations (omitted here) give the elastic curvature correction as $\Delta\kappa^*_c = 169.4 \times 10^{-9} \text{ mm}^{-1}$, and from this we obtain:

$$\kappa^*_c = \kappa^*_{co} + \Delta\kappa^*_c = (-1.170 + 0.169) \times 10^{-6} = -1.001 \times 10^{-6} \text{ mm}^{-1}$$

To estimate the creep deflection we take:

$$\Delta_c^* = \frac{\kappa_c^*}{\kappa_o} \times \Delta_{PG} = \frac{(-1.001)}{(-0.391)} \times (-4.1) = -10.5 \text{ mm}$$

Approximate closed-form equation: For a simpler but less accurate calculation, we can use the closed form Equations B.49, B.53 and B.54 in Section B.5.1 of Appendix B to obtain:

$$\Delta\kappa_c^*(A_p) = 168 \times 10^{-9} \text{ mm}^{-1}$$

From Equation B.55 we obtain:

$$\kappa_c^* = (-1.170 + 0.168) \times 10^{-6} = -1.002 \times 10^{-6} \text{ mm}^{-1}$$

and hence:

$$\Delta_c^* = \frac{(-1.002)}{(-0.391)} \times (-4.1) = -10.5 \text{ mm}$$

As is to be expected, for a section without reinforcement, this is the same value as that obtained from the age-adjusted effective modulus calculation.

Shrinkage deflection

We use the simplified formulae in Section B.6 of Appendix B to evaluate the parameters γ_3 and γ_4 (Equations B.77 and B.79) and then the shrinkage curvature κ_{sh}^* (Equation B.78). The values of the non-dimensional parameters are: $\gamma_3 = 0.189$ and $\gamma_4 = 1.014$.

With $\varepsilon_{cs} = \varepsilon_{sh}^* = 0.0006$, the long-term shrinkage curvature is calculated as:

$$\kappa_{sh}^* = 0.189 \times 1.014 \times 0.0006/800 = 0.144 \times 10^{-6} \text{ mm}^{-1}$$

An estimate of the final shrinkage deflection is:

$$\Delta_{sh}^* = \frac{\kappa_{sh}^*}{\kappa_o} \times \Delta_{PG} = \frac{(0.144)}{(-0.391)} \times (-4.1) = 1.5 \text{ mm}$$

Total deflection

The total long-term deflection is obtained by summing the components due to the applied load, prestress, creep and shrinkage:

$$\Delta = \Delta_{el} + \Delta_c^* + \Delta_{sh}^* = 3.6 - 10.5 + 1.5 = -5.4 \text{ mm (upwards)}$$

Discussion

This example illustrates an important point that will be observed repeatedly by designers. That is, although the absolute values of deflections can be quite small (especially if the design is based on load balancing) considerable computational effort is usually needed in order to satisfy the AS 3600 requirement that deflections be evaluated from first principles.

EXAMPLE 4.5

DEFLECTIONS CALCULATIONS FOR A T-BEAM

A beam with the cross-section shown in Figure 4.14 is simply supported over a span of 9.5 metres. The eccentricity of the cable varies parabolically from zero at each end to 390 mm at mid-span.

The superimposed dead load, excluding self-weight, is 18 kN/m, the live load is $w_Q = 40$ kN/m and the short-term and long-term live load factors are $\psi_s = 0.7$ and $\psi_l = 0.2$, respectively. Calculate the total final deflection if the effective prestressing force in the cable, including losses, is to be 1745 kN for a prestress area of $A_p = 1660$ mm^2. It will be assumed that the loads are applied over a short time period, so that a single value of E_c is used.

Other data:

$$A_g = 440\times10^3 \text{ mm}^2\text{ ; } I_g = 22.07\times10^9 \text{ mm}^4\text{ ; } d_g = 276 \text{ mm}$$

$$E_c = 31.6\times10^3 \text{ MPa ; } E_p = 200\times10^3 \text{ MPa ; } \varphi_o^* = 2.0\text{ ;}$$

$$\varepsilon_{cs} = 600\times10^{-6}$$

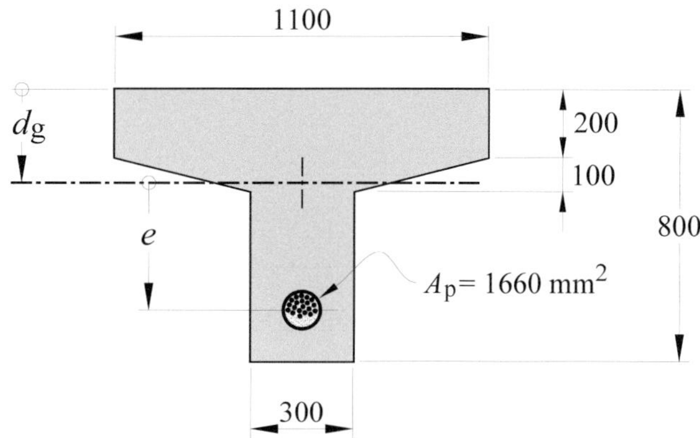

Figure 4.14 cross-section of T–beam for Example 4.5

SOLUTION

Loads

Self weight: $w_{SW} = 25 \times 440{\times}10^{3} \times 10^{-6} = 11.0$ kN/m

Total dead load: $w_G = 11.0 + 18.0 = 29.0$ kN/m

Equivalent load due to prestress (Equation 4.10):

$$w_p = -8(1745 \times 390{\times}10^{-3})/9.5^2 = -60.3 \text{ kN/m (upwards)}$$

Long-term live load: $w_{Ql} = \psi_l\, w_{Ql} = 0.2 \times 40 = 8.0$ kN/m

Short-term live load: $w_{Qs} = \psi_s\, w_{Qs} = 0.7 \times 40 = 28.0$ kN/m

Initial elastic deflections

With only the dead load, prestress and long-term component of the live load acting, the resultant long-term load is:

$$w_l = 29.0 - 60.3 + 8.0 = -23.3 \text{ kN/m (upwards)}$$

The moment at mid-span is:

$$M_G = \frac{w_1 L^2}{8} = -\frac{23.3 \times 9.5^2}{8} = -262.9 \text{ kNm}$$

The corresponding elastic curvature is:

$$\kappa_o = M_G/(E_c I_g) = -0.378 \times 10^{-6} \text{mm}^{-1}$$

The short-term component of deflection due to this sustained load is:

$$\Delta_{el} = \frac{5}{384} \cdot \frac{(-23.3) \cdot 9500^4}{31.6{\times}10^3 \times 22.07{\times}10^9} = -3.5 \text{ mm (upwards)}$$

The additional elastic deflection due to the difference in the short-term and long-term live load components is a deflection of:

$$\Delta_{Qs} - \Delta_{Ql} = \frac{(0.7 - 0.2)\ 40}{23.3} \times 3.5 = 3.0 \text{ mm (downwards)}$$

Creep deflection

Preliminary estimate: A rough, preliminary estimate of the long-term creep deflection is made using the free creep curvature κ^*_{co}. From Equation 4.14 we obtain:

$$\Delta^*_c = \varphi^*_o \Delta_{el} = 2.0 \times (-3.5) = -7.0 \text{ mm (upwards)}$$

This will be a conservative over-estimate. It is to be added to the initial upwards camber from the elastic component of the deflection.

One-step analysis with age-adjusted effective modulus: A more accurate value is obtained from a one-step analysis as described in Appendix B in Section B.3. From detailed calculations not included here, but which follow the same calculation steps as in Example B.2, the creep curvature correction is $\Delta\kappa^*_c = 0.990{\times}10^{-6}$ mm^{-1}. With $\kappa^*_{co} = -1.938{\times}10^{-6}$ mm^{-1} we obtain:

$$\kappa^*_c = (-1.938 + 0.990) \times 10^{-6} = -0.946 \text{ mm}^{-1}$$

The creep deflection is estimated as:

$$\Delta_c^* = \frac{\kappa_c^*}{\kappa_o} \times \Delta_o = \frac{-0.946}{-0.698} \times -3.5 = -4.7 \text{ mm (upwards)}$$

Shrinkage deflection

To estimate the shrinkage deflection we determine the long-term shrinkage curvature using the simplified equations in Section B.5.5. Using a shrinkage-only calculation with Equations B.77, B.78 and B.79, we obtain:

$$\gamma_3 = 0.1185; \ \gamma_4 = 4.499; \ \kappa_{sh}^* = 0.400 \times 10^{-6} \text{ mm}^{-1}$$

The shrinkage deflection is then estimated to be:

$$\Delta_{sh}^* = \frac{\kappa_{sh}^*}{\kappa_o} \times \Delta_{el} = \frac{0.400}{-0.698} \times -3.5 = 2.0 \text{ mm (positive and downwards)}$$

Final deflections

The final deflection due to the sustained loads, including the permanent part of the live load, is:

$$\Delta_{sus}^* = \Delta_{el} + \Delta_{co}^* + \Delta_{sh}^* = -3.5 - 7.0 + 2.0 = -8.5 \text{ mm (upwards)}$$

and the total final deflection is:

$$\Delta_{tot}^* = \Delta_{sus}^* + (\Delta_{Qs} - \Delta_{Ql}) = -8.5 + 3.0 = -5.5 \text{ mm (upwards)}$$

In this example the deflections are quite small. This is to be expected when the prestress is selected to carry a large portion of the permanent loads.

EXAMPLE 4.6

DEFLECTIONS OF A BEAM WITH CANTILEVERED ENDS

The beam ABCDE shown in Figure 4.15 is simply supported at B and D, with cantilevered ends AB and DE. The span length BCD is $b = 16.6$ metres, while the length of each cantilever AB and DE is $a = 5.83$ metres. The cross-section of the member is constant, 400 mm wide by 800 mm deep. In the preliminary design, a piece-wise parabolic cable has been chosen, with zero eccentricity and zero slope at the ends A and E, a negative (upward) eccentricity of 135 mm at B and D, and a positive eccentricity of 140 mm at the mid-span point C. Although the kinks at B and D will be smoothed out in the final design, the simplified cable profile in Figure 4.15 will be used for the preliminary deflection calculations.

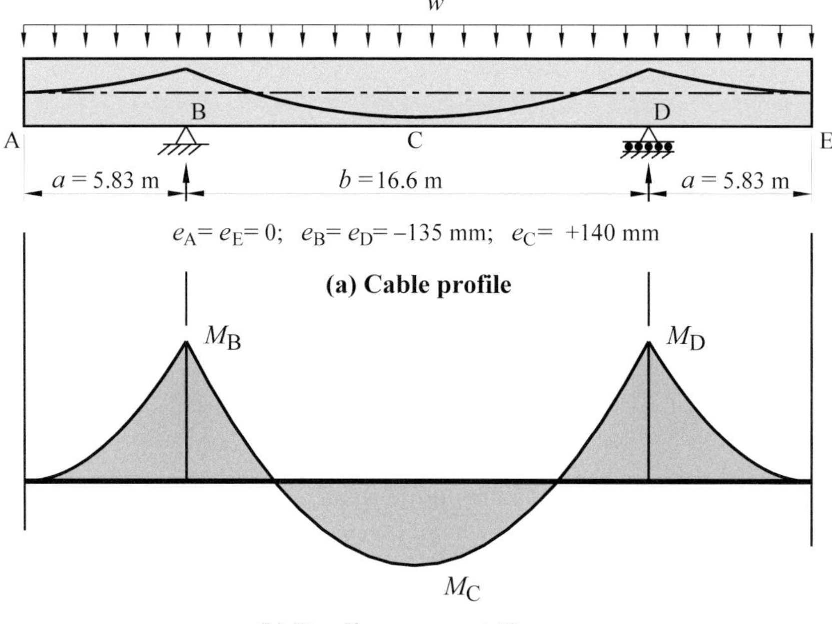

Figure 4.15 Cantilevered beam - Example 4.6

The beam is to be prestressed at 14 days with a cable force of 2000 kN, when it will also carry its own self-weight. The area and the elastic modulus of the cable are, respectively, $A_p = 2100 \text{ mm}^2$ and $E_p = 195\times10^3$ MPa. For preliminary deflection calculations, the cable force will be assumed to be constant throughout the length AE, ignoring variations due to friction.

At 28 days, a permanent dead load of 12 kN/m will be applied, and from 90 days the beam will be subject to a live load of 22 kN/m with $\psi_l = 0.3$ and $\psi_s = 0.7$.

The concrete strength is to be 50 MPa at 28 days. The estimated values of the elastic moduli at ages of 14, 28 and 90 days and long-term are as follows:

$$E_c(14) = 35\ 000 \text{ MPa}; \quad E_c(28) = 39\ 000 \text{ MPa}$$

$$E_c(90) = E_c^* = 42\ 000 \text{ MPa.}$$

The creep coefficients $\varphi_c(t, t_o)$ are:

$$\varphi_c(28,14) = 0.36; \quad \varphi_c(90,14) = 0.90; \quad \varphi_c(90,28) = 0.82; \quad \varphi_o^* = 1.71$$

The shrinkage strains are assessed as:

$$\varepsilon_{sh}(14) = 150\times10^{-6}; \ \varepsilon_{sh}(28) = 300\times10^{-6}; \ \varepsilon_{sh}(90) = 500\times10^{-6};$$

$$\varepsilon_{sh}^* = 600\times10^{-6}$$

The deflections are to be calculated for the relevant load conditions.

Other data:

$$A_g = 320\times10^3 \text{ mm}^2; \ I_g = 17.07\times10^9 \text{ mm}^4$$

$$Z_{bot} = Z_{top} = 42.67\times10^6 \text{ mm}^3$$

Solution

Dead and prestress loads

Self weight load: $w_{SW} = 25 \times 0.32 = 8.0$ kN/m

Total dead load: $w_G = 12.0 + 8.0 = 20.0$ kN/m

On span BCD, the upward equivalent load due to prestress is obtained from Equation 4.18 as:

$$w_p = -\frac{8Ph}{b^2} = -\frac{8 \times 2000 \times (0.135 + 0.140)}{16.6^2} = -16.0 \text{ kN/m}$$

The equivalent loads on the cantilevers have the same value:

$$w_p = -2P\frac{h}{a^2} = -16 \text{ kN/m}$$

Stress check for cracking

We first calculate the stresses under full load conditions to check the assumption that the beam is uncracked. Under full load the stresses consist of a uniform compressive stress of:

$$P/A_g = 2000{\times}10^3/320{\times}10^3 = 6.25 \text{ MPa}$$

plus the bending stress due to the moment produced by a uniformly distributed load of:

$$w = w_G + \psi_s w_Q + w_p = 20.0 + 0.7 \times 22 - 16.0 = +19.4 \text{ kN/m}$$

At B: $M_B = -wa^2/2 = -19.4 \times 5.83^2/2 = -329.7$ kNm

The extreme fibre concrete stresses at B due to the moment are:

$$\sigma_{\text{bot, top}} = \pm M/Z = \pm 329.7{\times}10^6/42.67{\times}10^6 = \pm 7.73 \text{ MPa}$$

The stress in the extreme tensile (bottom) fibre is thus:

$$\sigma_{\text{bot}} = 6.25 - 7.73 = -1.48 \text{ MPa (tensile)}$$

which is sufficiently low that cracking does not occur (see Chapter 5 for full details on the calculation of tensile stress at cracking).

At C: $M_{\text{C}} = w\left(\frac{b^2}{8} - \frac{a^2}{2}\right) = 19.4\left(\frac{16.6^2}{8} - \frac{5.83^2}{2}\right) = 338.5 \text{ kNm}$

The extreme fibre concrete stresses at C due to the moment are:

$$\sigma_{\text{bot, top}} = \pm M/Z = \pm 338.5\times10^6/42.67\times10^6 = \pm 7.93 \text{ MPa}$$

The stress in the extreme tensile (bottom) fibre is thus:

$$\sigma_{\text{bot}} = 6.25 - 7.93 = -1.68 \text{ MPa (tensile)}$$

which, again, is sufficiently low that cracking does not occur.

Elastic analysis of cantilevered beam

In the following calculations to determine deflections we will consider the member at various times and, hence, with varying values of E_{c} and with different load levels. In all cases, however, the same uniformly distributed load acts on the beam and the cantilevers. It is therefore convenient to carry out just one elastic analysis for a specific load level and a particular elastic modulus, to determine the deflections in the structural system. Deflections at other load levels and for other values of E_{c} can then be determined by a simple linear pro rata calculation. A computer analysis for $w = 10$ kN/m and using the elastic modulus at 14 days was carried out. The stiffness EI was determined using the following values:

$$E_{\text{c}}(14) = 35.0\times10^3 \text{ MPa}; \; I_{\text{g}} = 17.07\times10^9 \text{ mm}^4$$

Values obtained for the deflections at mid-span C and at the cantilever ends A and E are:

At C: $\Delta_{Gp} = 6.75 \text{ mm (downwards)}$

At A and E: $\Delta_{Gp} = 2.42 \text{ mm (upwards)}$

The authors recommend that all computer calculations be checked, either with an order of magnitude estimate or by an approximate calculation. Formulae for standard load and support conditions are readily available for this purpose. In the present case, deflections can be obtained using several formulae from Appendix D. For the cantilevered beam in Figure 4.15 we consider two load cases, as shown in Figure 4.16. In the first case only the beam span is loaded; in the second case only the two cantilevers are loaded. By adding the deflections obtained separately for load cases (1) and (2) we obtain the deflections for the total loading. The mid-span deflection for load case (1) is obtained using the formula given in Load Case (c) of Appendix D. Substitution of values gives:

$$\Delta_{C1} = \frac{5}{384} \cdot \frac{wb^4}{EI} = \frac{5}{384} \cdot \frac{10 \times 16600^4}{35 \times 10^3 \times 17.07 \times 10^9} = 16.55 \text{ mm}$$

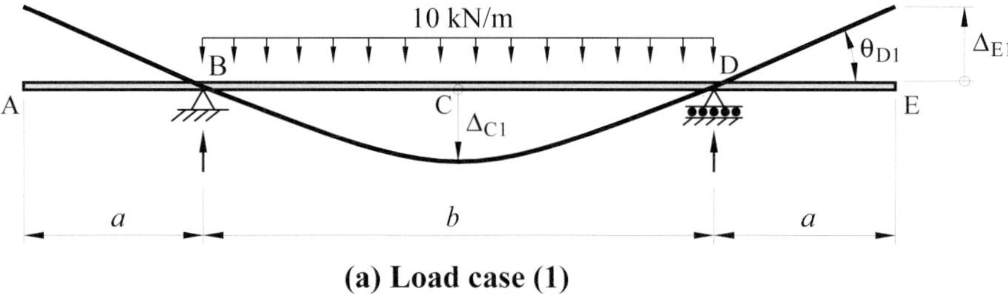

(a) Load case (1)

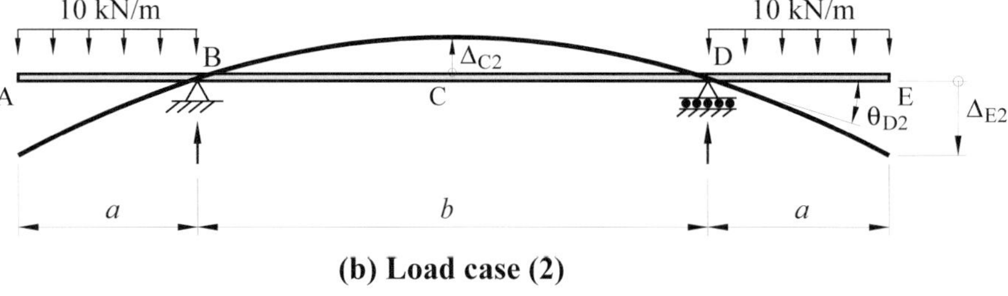

(b) Load case (2)

Figure 4.16 Load cases (1) and (2) for deflection calculation check

The slope at D for this loading is also evaluated from the formula given for Case (c) in Appendix D, and this gives:

$$\theta_{D1} = \frac{1}{24} \cdot \frac{wb^3}{EI} = \frac{1}{24} \cdot \frac{10 \times 16600^3}{35{\times}10^3 \times 17.07{\times}10^9} = 3.190{\times}10^{-3} \text{ rads}$$

The end deflection at E for load case (1) is $\Delta_{E1} = \theta_{D1} \times a$, where $a =$ 5830 mm is the span distance DE. The upwards deflection at E for load case (1) is thus $\Delta_{E1} = 3.190{\times}10^{-3} \times 5830 = -18.60 \text{ mm}$ upwards.

For load case (2) we note that the internal moment at B and D is $M_o = wa^2/2$ = 169.9 kNm. To obtain the upwards mid-span deflection at C for this load case we simply consider span BCD with two equal negative end moments of M_o acting at B and D. Case (b) in Appendix D deals with a span under equal end moments and substituting in the formula gives:

$$\Delta_{C2} = \frac{1}{8} \cdot \frac{M_o b^2}{EI} = \frac{1}{8} \cdot \frac{(-169.9{\times}10^6) \times 16600^2}{35{\times}10^3 \times 17.07{\times}10^9} = -9.80 \text{ mm}$$

The upwards deflection at C is thus $\Delta_C = +16.55 - 9.80 = 6.75 \text{ mm}$.

For load case (2) the slope of the beam at D is also obtained from the formula from Case (b) in Appendix D. This gives:

$$\theta_{D2} = \frac{1}{2} \cdot \frac{M_o b}{EI} = \frac{1}{2} \cdot \frac{(-169.9{\times}10^6) \times 16600}{35{\times}10^3 \times 17.07{\times}10^9} = -2.360{\times}10^{-3} \text{ rads}$$

For load case (2) the end deflection Δ_{E2} is the sum of Δ_{E2a} and Δ_{E2b}, where Δ_{E2a} is the component of the deflection due to the slope at D for the loadings of case (2):

$$\Delta_{E2a} = \theta_{D2} \times a = 2.360{\times}10^{-3} \times 5830 = 13.76 \text{ mm (downwards)}$$

and Δ_{E2b} is the deflection of the cantilever DE with end D treated as fixed. From Case (e) in Appendix D we obtain:

$$\Delta_{E2b} = \frac{1}{8} \cdot \frac{wb^4}{EI} = \frac{1}{8} \cdot \frac{10 \times 5830^4}{35 \times 10^3 \times 17.07 \times 10^9} = 2.42 \text{ mm (downwards)}$$

The resultant deflection at E is thus:

$$\Delta_E = -18.60 + 13.76 + 2.42 = -2.42 \text{ mm (upwards).}$$

The hand calculations in this simple case match the computer calculations. In more complex situations, approximate hand calculations can be made using simplifications and approximations. Such approximate values provide an important check on computer output.

Deflections at $t = 14$ days

At 14 days the equivalent load w_p on span BCD and cantilevers AB and DE is –16 kN/m. The resultant load is therefore $w = -16 + 8 = -8$ kN/m (upwards). The mid-span deflection of span BCE is upwards. At this time shrinkage is very low and can be ignored. The creep deflection is zero.

An elastic beam analysis of the member with $E_c(14) = 35.0 \times 10^3$ MPa, $I_g = 17.07 \times 10^9$ mm^4 and $w = -8$ kN/m gives:

At C: $\Delta_{Gp} = -5.4$ mm (upwards)

At A and E: $\Delta_{Gp} = 1.9$ mm (downwards)

Deflections at $t = 28$ days

At 28 days, the full dead load is applied. In addition to the elastic deflections there will be some small deflection components due to shrinkage and creep under the previously applied loads and prestress.

The additional dead load is $w_{G1} = 12$ kN/m and the additional deflections due to this are calculated with $E_c(28) = 39.0 \times 10^3$ MPa as:

At C: $\Delta_{G1} = +7.3$ mm (downwards)

At A and E: $\Delta_{G1} = -2.6$ mm (upwards)

A conservative estimate of the creep deflections at $t = 28$ days for the load of w_{Gp} first applied at $t_o = 14$ days can be obtained using Equation 4.14 (from the free creep analysis described in Section B.2 of Appendix B):

$$\textbf{At C: } \Delta_{cGp}(28, 14) = 0.36 \times (-5.4) = -1.9 \text{ mm (upwards)}$$

At A and E:

$$\Delta_{cGp}(28, 14) = 0.36 \times (+1.9) = +0.7 \text{ mm (downwards)}$$

For a simplified calculation of the shrinkage deflections, we shall use the 28 day reduced elastic modulus and Equation B.78 (Section B.5.5 of Appendix B) to calculate the warping curvature:

$$n_p = E_p/E_c = 195{\times}10^3/39.0{\times}10^3 = 5.00$$

$$p_p = A_p/A_g = 2100/320{\times}10^3 = 0.0066$$

Equations B.77 and B.79 are used to evaluate γ_3 and γ_4:

$$\gamma_3 = \frac{1}{1 + \dfrac{1}{5.00 \times 0.0066} + \dfrac{140^2 \times 320{\times}10^3}{17.07{\times}10^9}} = 0.032$$

$$\gamma_4 = \frac{320{\times}10^3 \times 140 \times 800}{17.07{\times}10^9} = 2.10$$

From Equation B.78 the final shrinkage curvature is:

$$\kappa_{sh}(28) = 0.032 \times 2.10 \times 300{\times}10^{-6}/800 = 25.2{\times}10^{-9} \text{ mm}^{-1}$$

A conservative over-estimate of the shrinkage deflection at mid-span point C can be obtained by assuming the central shrinkage curvature extends over the full span. This gives:

$$\textbf{At C: } \Delta_{sh}(28) = 25.2{\times}10^{-9} \times 16\,600^2/8 = +0.9 \text{ mm}$$

At A and E: $\Delta_{sh}(28) \approx 0$ and is ignored.

The total deflections at 28 days are:

At C:
$$\begin{aligned}\Delta(28) &= \Delta_{Gp}(14) + \Delta_{G1}(28) + \Delta_{cGp}(28, 14) + \Delta_{sh}(28) \\ &= -5.4 + 7.3 - 1.9 + 0.9 \\ &= 0.9 \text{ mm (downwards)}\end{aligned}$$

At A and E:

$$\begin{aligned}\Delta(28) &= \Delta_{Gp}(14) + \Delta_{G1}(28) + \Delta_{cGp}(28, 14) + \Delta_{sh}(28) \\ &= +1.9 - 2.6 + 0.7 + 0.0 \\ &= 0.0 \text{ mm}\end{aligned}$$

Deflections at *t* = 90 days

At 90 days, the shrinkage induced deflections and the creep deflections due to the loads applied at 14 and 28 days will all have increased. An additional elastic live-load deflection will also occur due to live load.

The elastic deflections at 90 days due to the long-term live load $w_{Ql} = 0.3 \times 22 = 6.6$ kN/m and $E_c(90) = 42.0 \times 10^3$ MPa are:

At C: $\Delta_{Ql} = +3.7$ mm (downwards)

At A and E: $\Delta_{Ql} = -1.3$ mm (upwards)

The short-term deflections are due to the live load:

$$w_{Qs} = \psi_s w_Q = 0.7 \times 22 = 15.4 \text{ kN/m}$$

For $E_c(90) = 42.0 \times 10^3$ MPa, the resulting elastic deflections are:

At C: $\Delta_{Qs} = +8.7$ mm (downwards)

At A and E: $\Delta_{Qs} = -3.1$ mm (upwards)

The creep deflections at $t = 90$ days for the load of w_{Gp} first applied at $t_o = 14$ days are conservatively estimated, as before, using the free creep curvature:

At C: $\Delta_{cGp}(90, 14) = 0.90 \times (-5.4) = -4.9$ mm (upwards)

At A and E:

$$\Delta_{cGp}(90, 14) = 0.90 \times (+1.9) = +1.7 \text{ mm (downwards)}$$

The creep deflections at $t = 90$ days for the load of w_{G1} first applied at $t_o = 28$ days are:

At C: $\Delta_{cG1}(90, 28) = 0.82 \times (-7.3) = -6.0$ mm (upwards)

At A and E:

$$\Delta_{cG1}(90, 28) = 0.82 \times (+2.6) = +2.1 \text{ mm (downwards)}$$

The shrinkage deflection at point C at 90 days is obtained as before, but using a uniform shrinkage strain of 0.0005. From Equations B.77, B.78 and B.79 we obtain:

$$\kappa_{sh}(90) = 0.032 \times 2.10 \times 500 \times 10^{-6} / 800 = 42.0 \times 10^{-9} \text{ mm}^{-1}$$

The final shrinkage deflection is determined as:

At C: $\Delta_{sh}(90) = 42.0 \times 10^{-9} \times 16\,600^2 / 8 = +1.4$ mm

At A and E: $\Delta_{sh}(90) \approx 0$ and can be ignored.

The total deflections at 90 days are:

At C:
$$\begin{aligned}\Delta(90) &= \Delta_{Gp}(14) + \Delta_{G1}(28) + \Delta_{Qs}(90) + \Delta_{cGp}(90, 14) \\ &\quad + \Delta_{cG1}(90, 28) + \Delta_{sh}(28) \\ &= -5.4 + 7.3 + 8.7 - 4.9 - 6.0 + 1.4 \\ &= +1.1 \text{ mm (downwards)}\end{aligned}$$

At A and E:

$$\begin{aligned}\Delta(90) &= \Delta_{Gp}(14) + \Delta_{G1}(28) + \Delta_{Qs}(90) + \Delta_{cGp}(90, 14) \\ &\quad + \Delta_{cG1}(90, 28) + \Delta_{sh}(28) \\ &= 1.9 - 2.6 - 3.1 + 1.7 + 2.1 + 0.0 \\ &= 0.0 \text{ mm}\end{aligned}$$

Final Deflections (t = 30 years):

At C: The total elastic deflections are:

$$\Delta_{el} = \Delta_{Gp} + \Delta_{G1} + \Delta_{Qs} = -5.4 + 7.3 + 8.7 = 10.6 \text{ mm}$$

For calculating the final creep, the sustained component of the elastic deflection is:

$$\Delta_{el.sus} = \Delta_{Gp} + \Delta_{G1} + \Delta_{Ql} = -5.4 + 7.3 + 2.0 = 3.9 \text{ mm}$$

and the creep deflection is calculated approximately as:

$$\Delta_c^* = \varphi_o^* \times \Delta_{el.sus} = 1.71 \times 3.9 = 6.7 \text{ mm}$$

The final shrinkage curvature for a section at mid-span is calculated from Equation B.78. In evaluating γ_3 (Equation B.77) note that the age-adjusted effective modulus is being used. With $k^* = 0.3$ in Equation B.77 we obtain $\gamma_3 = 0.09$ and with $\gamma_4 = 2.10$:

$$\kappa_{sh}^* = 0.09 \times 2.10 \times 600{\times}10^{-6}/800 = 142{\times}10^{-9} \text{ mm}^{-1}$$

The final shrinkage deflection is determined as:

$$\Delta_{sh}^* = 142{\times}10^{-9} \times 16\,600^2/8 = 4.9 \text{ mm}$$

The final deflection at Section C is then:

$$\Delta_C = \Delta_{el.sus} + \Delta_c^* + \Delta_{sh}^* = 10.6 + 6.7 + 4.9 = 22.2 \text{ mm}$$

At A and E: The total elastic deflections are:

$$\Delta_{el} = \Delta_{Gp} + \Delta_{G1} + \Delta_{Qs} = 1.9 - 2.6 - 3.1 = -3.8 \text{ mm}$$

For calculating the final creep, the sustained component of the elastic deflection is:

$$\Delta_{el.sus} = \Delta_{Gp} + \Delta_{G1} + \Delta_{Ql} = 1.9 - 2.6 - 0.7 = -1.4 \text{ mm}$$

and the creep deflection is then calculated conservatively as:

$$\Delta_c^* = \varphi_o^* \times \Delta_{el.sus} = 1.71 \times (-1.4) = 2.4 \text{ mm}$$

The final deflection at A and E is then:

$$\Delta_E = \Delta_{el.sus} + \Delta_c^* + \Delta_{sh}^* = -3.8 + 2.4 + 0 = -1.4 \text{ mm}$$

Discussion

In the above calculations, the progressive loss of prestress due to creep and shrinkage has been ignored, so that the upward deflection caused by prestress and creep has been slightly overestimated. However, the deferred losses are only in the order of 20 per cent, which is well within the order of accuracy of the value of the creep coefficient. If considered desirable, account may be taken of the progressive loss of prestress by estimating an increment of equivalent load at several times, to correspond to the decrement in prestress, and calculating the corresponding deflection increments. For example, estimating a total deferred loss of 20 per cent, and estimating that at 28 days the loss is eight per cent or 200 kN, we then introduce a permanent downward load applied at 28 days of $200 \times 20/2500 = 1.6$ kN/m. The effect of this load increment is calculated as for the external load of 12 kN/m, also applied at 28 days. Similarly, load increments at several later time increments can be introduced to account for the time effect of loss in prestress.

4.8 References

AS 3600–2018, *Concrete Structures*, Standards Australia, Sydney, Australia.

Foster, S J, Kilpatrick, A E and Warner, R F, 2021, *Reinforced Concrete Basics*, 3E, (3rd Edn) Pearson, Melbourne, Australia., 589 pp.

CHAPTER 5

Flexural behaviour in the post-cracking range

The AS 3600 method for calculating M_{cr} is explained, and the flexural behaviour of a typical cracked prestressed member is then followed qualitatively, from cracking through the service load range and into overload and high overload. The effects of creep and shrinkage in cracked members are also described. Procedures for evaluating stresses, strains, deformations and deflections in cracked members under short-term and long-term loading are presented. These are used in Chapters 10 and 11 in the design of determinate and indeterminate beams, and in Chapter 12 in the design of slabs.

5.1 Cracking moment

Flexural cracking occurs in a cross-section in a prestressed beam when the tensile stress in the extreme concrete fibre, due to applied load, exceeds the existing compressive prestress by an amount equal to the flexural tensile strength of the concrete. However, there is a progressive and significant decrease in the cracking moment, M_{cr} over time. This is caused by progressive decreases in the extreme fibre prestress, due to creep and shrinkage.

The decrease in compressive prestress in the extreme fibre occurs in part because the reinforcement in the section restrains the compressive creep and shrinkage strains as they develop in the concrete, which leads to a transfer of some compressive force from the concrete to the steel. AS 3600 specifically requires that restrained shrinkage be allowed for in the calculation of M_{cr}.

The tensile stress induced in the extreme fibre by restrained shrinkage is designated as σ_{cs}, and the expression for this stress, according to the Standard, is:

$$\sigma_{cs} = E_s \varepsilon^*_{cs} (2.5p_w - 0.8p_{cw})/(1 + 50p_w) \tag{5.1}$$

Here, ε^*_{cs} is the final design shrinkage strain for the concrete (which was discussed in Chapter 2) and p_w is the total area of reinforcing steel and prestressing steel, A_{st} and A_p, in the lower tensile region of the web, as a proportion of the concrete area:

$$p_w = (A_{st} + A_p)/(b_w d) \tag{5.2}$$

where d is the depth to the resultant tensile force, and b_w is the width of the web. In a rectangular beam, $b_w = b$. The compressive reinforcement ratio, p_{cw}, is also included in Equation 5.1. It allows for the fact that the restraint provided by any top fibre compressive reinforcement will induce compressive stress in the bottom fibres, and so increase the cracking moment. The ratio p_{cw} is defined as:

$$p_{cw} = A_{sc}/(b_w d) \tag{5.3}$$

where A_{sc} is the area of compressive reinforcement. In AS 3600, the cracking moment is expressed in terms of the section modulus Z as:

$$M_{cr} = Z(f'_{ct.f} - \sigma_{cs} + P_e/A_g) + P_e e \tag{5.4}$$

The term $f'_{ct.f}$ in Equation 5.4 is the characteristic flexural tensile strength of the concrete, which is best determined by test, but without test data it can be evaluated approximately as:

$$f'_{ct.f} = 0.6\sqrt{f'_c} \tag{5.5}$$

The contribution of the steel and tendon to the stiffness of the uncracked section is small, and is ignored when the section modulus Z is evaluated.

Although AS 3600 attributes the stress σ_{cs} to restrained shrinkage, the steel restrains both the shrinkage and creep strains. As the expression for σ_{cs} in Equation 5.1 is empirical, it is an overall allowance for the restraint effect.

Equation 5.4 gives an estimate of when cracking occurs in the extreme fibre of the concrete. In fact, internal cracking may occur much earlier. Grout is often introduced under pressure into the ducts of post-tensioned members shortly after prestressing. As the grout is not itself prestressed, it is prone to local cracking long before the full service load is applied. However, such cracking has a minor effect on subsequent flexural behaviour.

EXAMPLE 5.1

CALCULATION OF CRACKING MOMENT

Calculate the cracking moment and the cracking load for the beam in Example 4.1. There is no reinforcement in the section.

Other Data

$$f_c' = 35 \text{ MPa}; \; f'_{ct.f} = 0.6\sqrt{35} = 3.55 \text{ MPa};$$

$$A_g = 320 \times 10^3 \text{ mm}^2; \; I_g = 17.07 \times 10^9 \text{ mm}^4;$$

$$Z_{bot} = Z_{top} = I_g/(0.5D) = 17.07 \times 10^9/400 = 42.67 \times 10^6 \text{ mm}^3$$

SOLUTION

Residual shrinkage-induced stress

For the mid-span section we have: $p_w = \dfrac{1000}{400 \times 650} = 0.00385; p_{cw} = 0$

The residual stress, calculated from Equation 5.1, is:

$$\sigma_{cs} = \frac{200 \times 10^3 \times 565 \times 10^{-6} \times 2.5 \times 0.00385}{1 + 50 \times 0.00385} = 0.91 \text{ MPa}$$

Cracking moment

The average prestress in the section is:

$$\frac{P_e}{A_g} = \frac{1200\times10^3}{320\times10^3} = 3.75 \text{ MPa}$$

The cracking moment is obtained using Equation 5.4:

$$\begin{aligned} M_{cr} &= Z(f'_{ct.f} - \sigma_{cs} + P_e/A_g) + P_e e \\ &= 42.67\times10^6(3.55 - 0.91 + 3.75) + 1200\times10^3 \times 250 \\ &= 273.7\times10^6 + 300.0\times10^6 \\ &= 574\times10^6 \text{ Nmm} \quad (574 \text{ kNm}) \end{aligned}$$

Cracking load

The uniformly distributed load that causes cracking is:

$$w_{cr} = \frac{8M_{cr}}{L^2} = \frac{8\times574}{10^2} = 45.9 \text{ kN/m}$$

From Example 4.1 the self weight $w_G = 8.0$ kN/m. The additional uniformly applied load that can be carried above the self weight, without cracking, is $45.9 - 8 = 37.9$ kN/m.

EXAMPLE 5.2

Cracking Moment for Prestressed Section with Reinforcement

Recalculate the cracking moment for a similar section to the one considered in Example 5.1, except that in addition to the tendon it contains tensile reinforcing steel of 4-N24 bars for which $A_{st} = 1800$ mm^2.

SOLUTION

Using a slightly larger effective depth of $d = 700$ mm to allow for the equivalent depth of the combined steel area (tendon and reinforcing steel), we have:

$$p_w = \frac{1000 + 1800}{400 \times 700} = 0.0100\,; \; p_{cw} = 0$$

The shrinkage-induced tensile stress is calculated from Equation 5.1:

$$\sigma_{cs} = \frac{200\times10^3 \times 565\times10^{-6} \times 2.5 \times 0.0100}{1 + 50 \times 0.0100} = 1.88 \text{ MPa}$$

The cracking moment is (Equation 5.4):

$$\begin{aligned} M_{cr} &= Z(f'_{ct.f} - \sigma_{cs} + P_e/A_g) + P_e e \\ &= 42.67\times10^6(3.55 - 1.88 + 3.75) + 1200\times10^3 \times 250 \\ &= 231.3\times10^6 + 300.0\times10^6 \\ &= 531\times10^6 \text{ Nmm} \quad (531 \text{ kNm}) \end{aligned}$$

The cracking moment has been reduced because of the added shrinkage restraint provided by the tensile reinforcement.

5.2 Post-cracking flexural behaviour

We now examine the post-cracking flexural behaviour in the region of the beam surrounding the section of maximum moment where the first crack has formed.

5.2.1 Transition to fully cracked condition

At M_{cr} when the crack forms, there is a sudden loss of tensile stress in the extreme concrete tensile fibres resulting in a temporary, localised loss of stability. Under **load control** conditions, that is to say where the load on the member is being progressively increased, there is a sudden jump in curvature and a slight increase in deflection. The crack must rise immediately to a height above the tensile steel so that a new equilibrium condition is achieved, whereby the tensile force T is carried jointly by the tendon and the tensile reinforcement below the crack, and an equal compressive force C is carried by the intact concrete above the crack. The lever arm between C and T is z and the moment in the section is $M = M_{cr} = Tz = Cz$. In Figure 5.1, curvature in the section is plotted against moment. The jump in curvature at M_{cr} is shown as a short horizontal dashed line. There is also a sudden loss of flexural stiffness when the crack appears, so that the slope of the moment-curvature line for $M > M_{cr}$ is less than for $M < M_{cr}$. For comparison purposes, Figure 5.1 shows the moment-curvature relation for an un-prestressed, reinforced beam.

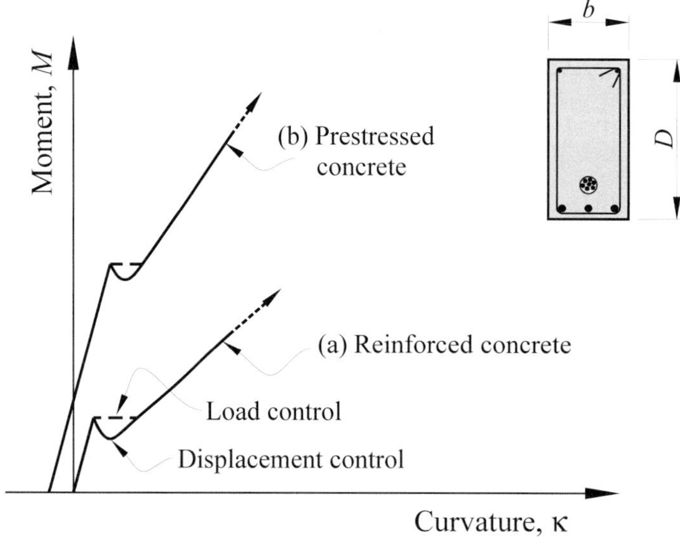

Figure 5.1 Transitional behaviour following cracking

In the case of **displacement control**, that is to say where small increments of displacement are being added, rather than small increments of load, there is a slight fall in moment at cracking, and the new equilibrium configuration occurs at a moment slightly less than M_{cr}. The moment further decreases with increasing displacement and curvature, but then increases again to reach M_{cr}

at an increased curvature. The moment-curvature relationship then follows the same path as that for load-control. The case of deformation control is indicated in Figure 5.1 by a curved full line.

The eccentric prestress in the section induces an initial negative curvature at $M = 0$, which is also shown in Figure 5.1. The external positive moment first has to overcome this initial negative curvature before it produces positive curvature. Other effects of the prestress are to increase the cracking moment substantially, and to reduce the magnitude of the 'jump' in the moment-curvature relationship at M_{cr}.

5.2.2 Post-cracking behaviour

The prestress also affects the post-cracking behaviour in the working load range. In a reinforced section without prestress, a fully developed crack forms at M_{cr}, which immediately extends up to a level high in the section. The neutral axis is just above the tip of the crack, and does not change as M increases, provided the member is not overloaded. This is shown in Figure 5.2(a). As explained in Chapter 4, the tensile force T in the steel and the compressive force C in the concrete above the neutral axis both increase proportionally as M increases, but the lever arm z remains constant. For equilibrium we have $M = T_s z = C_c z$.

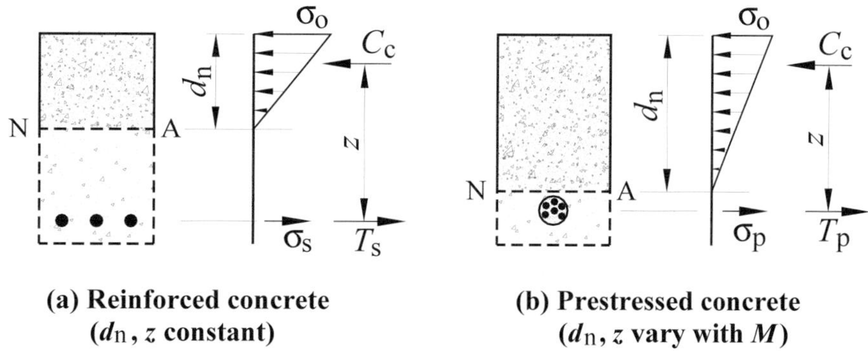

(a) Reinforced concrete (d_n, z constant)

(b) Prestressed concrete (d_n, z vary with M)

Figure 5.2 Post-cracking stress distribution

In contrast to that of a reinforced concrete section, the neutral axis in a cracked prestressed section initially rises only to a level slightly above the tensile steel (or tendon) but then continues to rise as M increases above M_{cr}. This is indicated in Figure 5.2(b). The internal lever arm z and the forces C and T all increase progressively as M increases, although for equilibrium we still have $M = T_p z = C_c z$.

Provided the prestressing steel and the reinforcement (if present) are well bonded to the surrounding concrete, the first crack remains narrow. With increasing moment, additional cracks appear progressively on either side of the initial crack, and the cracked region extends outwards from the peak moment region towards the supports. The height of the crack at a cross-section depends on the amount of prestress in the section as well as on the moment M.

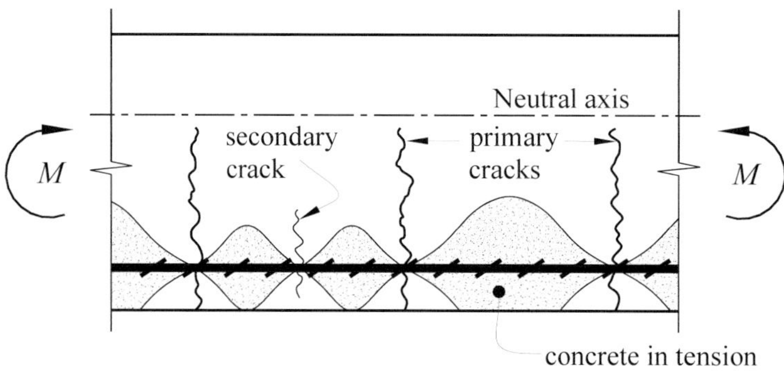

Figure 5.3 Flexural cracks and tension stiffening regions

The visible cracks on the side faces and soffit of a beam are called **primary cracks**. Initially they are rather irregularly spaced and extend up almost to the (slightly varying) height of the neutral axis (Figure 5.3). In the region between adjacent primary cracks, tensile stress is carried by the intact concrete bonded to the reinforcing steel and prestressing steel. This reduces the tensile stress and strain in the adjacent steel, and so reduces the average deformation as compared with the maximum local deformation at the cracked section. This effect is known as **tension stiffening**. With increasing moment, internal **secondary cracks** form progressively between the primary cracks, thus decreasing further the volume of concrete contributing to tension stiffening and so the tension stiffening itself.

Small internal **inclined microcracks**, called Goto-cracks (after Goto, 1971), also form around the steel bars at the ribs, further reducing the tension stiffening. By the time yielding of the tensile steel occurs, little tension stiffening remains. The bond between prestressing steel and the adjacent concrete is usually less effective than for the reinforcing steel, so that tension stiffening is more significant in a prestressed member when reinforcing steel is present.

When a cracked member is subjected to sustained moment, creep occurs in the compressive concrete above the neutral axis. This leads to a progressive increase in the compressive strains and hence in the local deformations and the overall deflection. If compressive reinforcement is present, it provides some restraint to the creep, and so attenuates the inelastic compressive strains in the top fibres and the overall deflection. Some tensile creep will occur in the concrete in the tension stiffening zone, but the additional local deformations here are mainly due to progressive cracking. The changes in the stresses in a section under sustained moment are complex, but not usually relevant for practical design, except for the prestress loss. The long-term increases in curvature and overall deflection are of course very important and need to be considered in the serviceability design calculations.

5.3 Elastic analysis for a rectangular cracked section

The analysis of post-cracking flexural behaviour is complex, and becomes more complex if the section is non-rectangular and the steel and prestressing tendons are distributed throughout the section. To simplify our presentation, we will proceed in several steps. We will first analyse the elastic behaviour of a rectangular prestressed member containing a single layer of tensile reinforcing steel and a prestressing tendon. The effects of creep and shrinkage will then be discussed. A general trial-and-error approach is given in Section 5.5 to analyse complex, non-rectangular cracked members that have several layers of tensile and compressive steel and tendons. This is needed for the analysis of the T- and I-sections that are used in bridge design. Non-linear behaviour at overload is briefly discussed in Section 5.6.

In the analysis of a cracked rectangular section at working load we assume that the compressive concrete and the reinforcing steel and tendons all behave elastically, and that the tensile concrete carries no stress. In Figure 5.4 the tensile steel area A_{st} is at depth d_{st} and the prestressing steel area A_p is at depth d_p. The applied moment M is larger than M_{cr}, and so large that the neutral axis lies above both the steel and tendon. In this elastic analysis, the neutral axes of strain and stress coincide and are both at depth d_n. This is not so when inelastic strains are taken into account.

In the cracked section, the tensile forces in the prestressing steel and tensile reinforcement are T_p and T_s, respectively. The compressive force in the concrete, C, acts at depth d_c below the top fibre. For horizontal force equilibrium:

$$C = T_s + T_p \tag{5.6}$$

The bending moment in the section is obtained by taking moments of these forces about the top fibre:

$$M = T_s d_{st} + T_p d_p - C d_c \tag{5.7}$$

The compressive stress in the concrete is linearly distributed, so that:

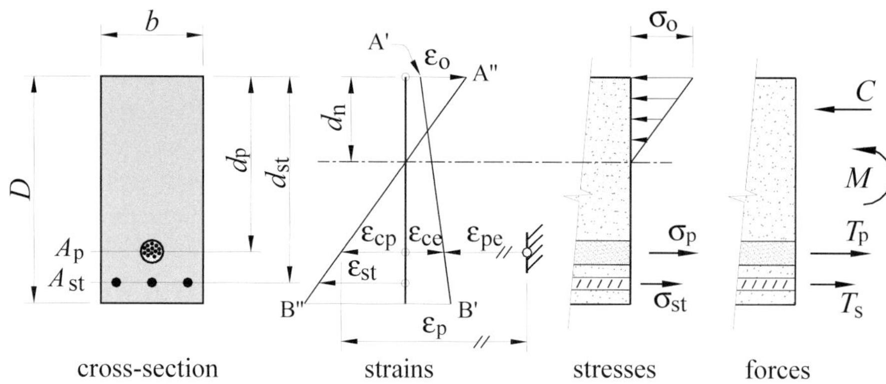

Figure 5.4 Elastic cracked section analysis

$$C = 0.5\sigma_o b d_n \tag{5.8}$$

where σ_o is the extreme top fibre stress. The line of action of C is at depth:

$$d_c = d_n/3 \tag{5.9}$$

With the assumptions of linear strain distribution and no slip (perfect bond), the strains in all fibres in the section can be expressed in terms of the compressive strain ε_o in the extreme top fibre and the neutral axis depth d_n. From Figure 5.4 we see that the tensile strains in the concrete at the levels of the tensile reinforcement and prestressing steel are, respectively:

$$\varepsilon_{cst} = \varepsilon_o \cdot \frac{d_{st} - d_n}{d_n} \tag{5.10}$$

$$\varepsilon_{cp} = \varepsilon_o \cdot \frac{d_p - d_n}{d_n} \tag{5.11}$$

With perfect bonding, the tensile strain in the concrete at the level of the reinforcing steel is equal to the tensile strain in the reinforcing steel, $\varepsilon_{st} = \varepsilon_{cst}$, but at the level of the prestressing steel, the tensile strain in the concrete, ε_{cp}, is much smaller than the total strain ε_p in the prestressing steel. From Figure 5.4 we see that the difference is made up of the initial compressive concrete strain due to the prestress, ε_{ce}, plus the initial strain in the tendon due to prestress just prior to loading, ε_{pe}. The initial tendon strain is:

$$\varepsilon_{pe} = \frac{P_e}{A_p E_p} \tag{5.12}$$

Although we are ignoring inelastic strains in this analysis, it is important to use the best estimate of the prestressing force with losses included, P_e, to determine ε_{pe}. The compressive strain in the concrete at the tendon level due to prestress, ε_{ce}, is obtained from the elastic stress, as described in Chapter 4:

$$\varepsilon_{ce} = \frac{P_e}{E_c}\left[\frac{1}{A_g} + \frac{e^2}{I_g}\right] \tag{5.13}$$

In Figure 5.4, the elastic strains in the concrete due to prestress, and just prior to loading, are represented by line A'B' with the strains ε_{ce} and ε_{pe} shown. The origins for measuring ε_{ce} and ε_{pe} in the figure have been chosen using the convention previously introduced in Chapter 4 in Figure 4.11 and discussed in Section 4.5.

As the external moment increases to M, the concrete strain diagram in Figure 5.4 rotates from A'B' to A"B". The concrete at the tendon level goes from compression into tension, with a tensile strain increment of $\varepsilon_{ce} + \varepsilon_{cp}$. The prestressing steel is bonded to the concrete and the tensile strain increment in the tendon is the same as in the adjacent concrete. The total strain in the prestressing steel at moment M is:

$$\varepsilon_p = \varepsilon_{pe} + \varepsilon_{ce} + \varepsilon_o \cdot \frac{d_p - d_n}{d_n} \tag{5.14}$$

For elastic behaviour we have:

$$T_s = A_{st}E_s\varepsilon_{st} \tag{5.15}$$

$$T_p = A_pE_p\varepsilon_p \tag{5.16}$$

The curvature in the section at moment M is:

$$\kappa = \varepsilon_o / d_n \tag{5.17}$$

These equations can be used, if necessary in iterative calculations, to evaluate the stresses, strains, internal forces in the section and the moment M that correspond to a chosen neutral axis depth, d_n. Moment-curvature and stress-moment relations in the working load range can thus be produced using these equations.

To determine the moment-curvature relationship for a cross-section, the following steps are undertaken:

1. Choose a progressively decreasing sequence of values of d_n that begin at a value of d_p.
2. For a chosen d_n value, find the corresponding top fibre strain ε_o, if necessary by trial and error, that satisfies the requirement of force

equilibrium, as represented by Equation 5.6. An equilibrium check is made for a trial value of ε_o by calculating the forces $T_s + T_p$ from Equations 5.15 and 5.16 and comparing them with the value of C from Equation 5.8. Trial values of ε_o are tested and modified until Equation 5.6 is satisfied.

3. With the values of ε_o and d_n fixed, evaluate the stresses and strains in the section using Equations 5.10 to 5.14.
4. Using Equation 5.7, evaluate the moment M that corresponds to the initial value chosen for d_n and calculate the curvature from Equation 5.17.
5. Return to Step 2 and repeat the calculations for other values of d_n.

In the upper range of moments, when d_n becomes small, a check should be made to ensure that the calculated stresses in the steel and the tendon have not exceeded yield and that the maximum concrete stress is still in the linear range. When the limits of the linear analysis are reached, a non-linear calculation is required. Such an analysis is outlined below in Section 5.6.

In the simple case of a rectangular section, a closed-form algebraic expression can be derived for the top fibre strain ε_o for a known value of d_n. This expression can be used in Step 3, in lieu of a trial and error search. To derive the expression for ε_o we express the forces C, T_s and T_p in terms of strains, using Equations 5.10 to 5.16. Substituting these expressions into the equilibrium equation (Equation 5.6) and re-arranging, we obtain the expression:

$$\varepsilon_o = \frac{A_p E_p d_n(\varepsilon_{ce} + \varepsilon_{pe})}{0.5E_c b d_n^2 - E_s A_{st}(d_s - d_n) - E_p A_p(d_p - d_n)} \tag{5.18}$$

On the one hand, tension stiffening in the concrete between cracks has been ignored in this analysis, and so the average flexural deformation is over-estimated. On the other hand, the inelastic creep and shrinkage strains that are present in the region are also ignored, and in this sense the analysis tends to underestimate the real average flexural deformation in the region. These matters will be considered further when we come to calculate deflections.

THE BATTLE FOR PARTIAL PRESTRESSING

In the design of the first successful prestressed concrete bridges and structures by Freyssinet and others immediately after the Second World war, the philosophy was to provide sufficient prestressing force to eliminate cracking under service conditions. Very little ordinary reinforcing steel was used and the resulting structures were said to be 'fully prestressed'. These slender and elegant structures popularised prestressed concrete as a new structural material. Indeed, prestressed concrete and reinforced concrete came to be treated as distinct, different materials of construction with different design criteria and design standards.

The design principle of full prestressing was repeatedly questioned by both researchers and practitioners. As early as 1948, Guyon, one of the pioneers of prestressed concrete, speculated that a design with only half the amount of prestress needed for a fully prestressed design would be possible. Guyon argued that cracks that appear under a full design live load application would disappear when the load is removed.

Paul W Abeles used the term 'partial prestressing' to describe his proposal to mix tensioned prestressing steel with untensioned prestressing steel in concrete flexural members. Using laboratory tests he showed that the behaviour and strength of such members were satisfactory and he designed a number of pretensioned members with some wires unstressed in order to reduce the precompression. Roof girders and bridge beams were produced using this design concept. The ideas of Abeles were criticised harshly by Freyssinet. In 1991, Bruggeling quoted a remark of Freyssinet to Abeles as follows:

> *"The bolts connecting the wheels to the axles of racing cars are always turned as tightly as possible and not 'partly tight'. This must also be the case with prestressing."*

Already by the mid-1950s, however, Swiss design engineers, including Max Birkenmaier, a director of the prestressing company BBR, had found practical advantages in using normal reinforcement combined with stressed prestressing tendon in flexural members. Since the 1970s, the advantages of using lower levels of prestress have been increasingly accepted by design engineers. Partial prestressing has become the norm in prestressed concrete design throughout the world and the term 'partial prestressing' has become redundant.

EXAMPLE 5.3

ELASTIC CRACKED SECTION ANALYSIS

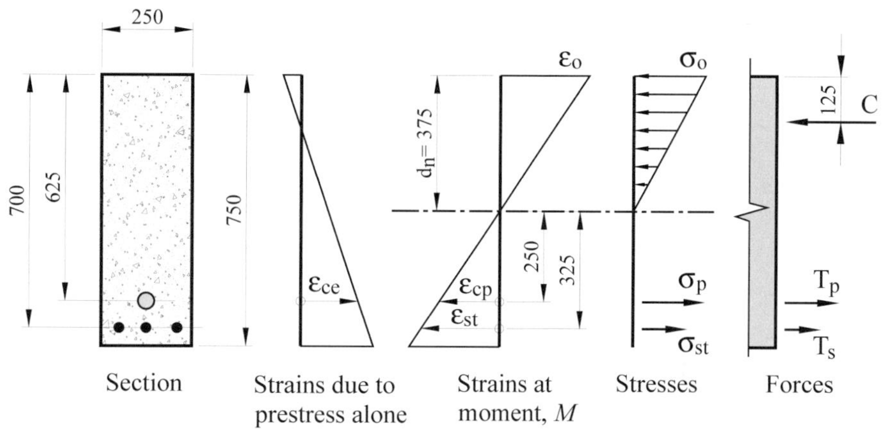

Figure 5.5 Cross-section for Example 5.3

The rectangular section shown in Figure 5.5 is 250 mm by 750 mm deep. It contains a prestressing tendon of area $A_p = 405\ \text{mm}^2$ and elastic modulus $E_p = 200\times10^3$ MPa at depth 625 mm and carries an effective prestressing force of $P_e = 445$ kN. The section has 3-N24 reinforcing bars at depth 700 mm with area $A_{st} = 1350\ \text{mm}^2$. An analysis is to be carried out for the section in the elastic range both before and after cracking. The strength of the concrete at the time of the analysis is $f_c' = 40$ MPa and its elastic modulus is $E_c = 32\,000$ MPa. The shrinkage strain is $\varepsilon_{sh}(t) = 600\times10^{-6}$.

Other data

$$A_g = 187.5\times10^3\ \text{mm}^2\,;\ I_g = 8.789\times10^9\ \text{mm}^4\,;\ y_b = y_t = 375\ \text{mm}\,;$$

$$Z_{bot} = Z_{top} = I_g/y_b = 8.789\times10^9/375 = 23.43\times10^6\ \text{mm}^3\,;$$

$$e = 625 - 375 = 250\ \text{mm}\,.$$

SOLUTION

Effective depth of uncracked section

As the section is elastic, the effective depth is calculated by taking the first moment of the forces in the tensile reinforcement and tendon about the extreme compressive fibre:

$$d = \frac{E_p A_p d_p + E_s A_{st} d_s}{E_p A_p + E_s A_{st}}$$

Taking $E_p = E_s$, we obtain:

$$d = \frac{405 \times 625 + 1350 \times 700}{405 + 1350} = 683 \text{ mm}$$

Initial conditions

The top and bottom fibre stresses due to prestress are:

$$\sigma_{cpa} = \frac{445{\times}10^3}{187.5{\times}10^3} - \frac{445{\times}10^3 \times 250 \times 375}{8.789{\times}10^9} = -2.37 \text{ MPa (tension)}$$

$$\sigma_{cpb} = \frac{445{\times}10^3}{187.5{\times}10^3} + \frac{445{\times}10^3 \times 250 \times 375}{8.789{\times}10^9} = +7.12 \text{ MPa (comp.)}$$

The corresponding extreme fibre strains are:

$$\varepsilon_{cpa} = -2.37/32000 = -74.02{\times}10^{-6} \text{ (tension)}$$

$$\varepsilon_{cpb} = 7.12/32000 = 222.5{\times}10^{-6} \text{ (compression)}$$

so that the initial curvature due to prestress is:

$$\kappa_{po} = (-74.02 - 222.5){\times}10^{-6}/750 = -395{\times}10^{-9} \text{ mm}^{-1}$$

Cracking moment

Taking:

$$f'_{\text{ct.f}} = 0.6\sqrt{f'_c} = 0.6\sqrt{40} = 3.79 \text{ MPa}$$

$$p_w = \frac{A_p + A_{st}}{b\,d} = \frac{405 + 1350}{250 \times 683} = 0.0103\;;\; p_{cw} = 0$$

the residual stress due to shrinkage is calculated from Equation 5.1 as:

$$\sigma_{cs} = \frac{200{\times}10^3 \times 600{\times}10^{-6} \times 2.5 \times 0.0103}{1 + 50 \times 0.0103} = 2.04 \text{ MPa}$$

and the cracking moment (Equation 5.4) is:

$$M_{cr} = Z(f'_{\text{ct.f}} - \sigma_{cs} + P/A_g) + Pe$$

$$= 23.43{\times}10^6 \left(3.79 - 2.04 + \frac{445{\times}10^3}{187.5{\times}10^3}\right) + 445{\times}10^3 \times 250$$

$$= 95.4{\times}10^6 + 111.3{\times}10^6$$

$$= 207{\times}10^6 \text{ Nmm} \quad (207 \text{ kNm})$$

The increment in elastic curvature induced by M_{cr} is:

$$\Delta\kappa_{cr} = \frac{M_{cr}}{E_c I_g} = \frac{207{\times}10^6}{32{\times}10^3 \times 8.789{\times}10^9} = 736{\times}10^{-9} \text{ mm}^{-1}$$

The total elastic curvature at cracking is thus:

$$\kappa_{po} + \Delta\kappa_{cr} = -395{\times}10^{-9} + 736{\times}10^{-6} = +341{\times}10^{-9} \text{ mm}^{-1}$$

Calculations for the cracked section

For the cracked section analysis we choose progressively decreasing values of d_n, starting with $d_n = d_p = 625$ mm. For each value of d_n we evaluate ε_o

using first principles and then the elastic stresses and strains and finally the moment in the section.

The effective prestress and the corresponding tensile strain in the prestressing steel are, respectively:

$$\sigma_{pe} = P_e/A_p = 445\times10^3/405 = 1100 \text{ MPa}$$

$$\varepsilon_{pe} = \sigma_{pe}/E_p = 1100/200\times10^3 = 5500\times10^{-6}$$

The corresponding strain in the concrete at the same level is (Equation 5.13):

$$\varepsilon_{ce} = \frac{P_e}{E_c}\cdot\left[\frac{1}{A_g}+\frac{e^2}{I_g}\right] = \frac{445\times10^3}{32\times10^3}\left[\frac{1}{187.5\times10^3}+\frac{250^2}{8.789\times10^9}\right]$$

$$= 173\times10^{-6}$$

To illustrate the principles of the procedure, detailed calculations are presented for one value, $d_n = 375$ mm. The rest are tabulated.

Calculation for $d_n = 375$ mm

The value of the top fibre strain is first determined from the condition of force equilibrium: $C = T_p + T_s$.

The force in the concrete is:

$$C = 0.5\sigma_o bd_n = 0.5\varepsilon_o\times32\times10^3\times250\times375 = 1.50\times10^9\varepsilon_o \text{ (N)}$$

The strain in the reinforcing steel is obtained from the geometry of the strain diagram, Figure 5.5:

$$\varepsilon_{st} = \varepsilon_o\times325/375 = 0.867\varepsilon_o$$

The tension in the reinforcing steel is:

$$T_s = \varepsilon_s E_s A_s = 0.867\varepsilon_o \times 200\times10^3 \times 1350 = 234\times10^6\varepsilon_o \text{ (N)}$$

With prestress alone acting ($M = 0$) the tensile strain in the prestressing steel is $\varepsilon_{pe} = 5500\times10^{-6}$ and the compressive strain in the concrete adjacent to the prestressing steel is $\varepsilon_{ce} = 173\times10^{-6}$.

Under the action of the moment M, the tensile prestressing strain increases by ε_{cp}, an increment equal to the change in strain in the adjacent concrete, and calculated from the strain diagram (Figure 5.5) as:

$$\varepsilon_{cp} = \varepsilon_o \times 250/375 = 0.667\varepsilon_o$$

The strain in the prestressing steel is thus (Figure 5.4):

$$\begin{aligned}\varepsilon_p = \varepsilon_{pe} + \varepsilon_{ce} + \varepsilon_{cp} &= 5500\times10^{-6} + 173\times10^{-6} + 0.667\varepsilon_o \\ &= 5673\times10^{-6} + 0.667\varepsilon_o\end{aligned}$$

and the prestressing force is:

$$\begin{aligned}T_p = \varepsilon_p E_p A_p &= (5673\times10^{-6} + 0.667\varepsilon_o) \times 200\times10^3 \times 405 \\ &= 54.03\times10^6\varepsilon_o + 459.5\times10^3 \text{ (N)}\end{aligned}$$

Substituting into the force equilibrium equation, $C - T_s - T_p = 0$, we obtain $\varepsilon_o = 379\times10^{-6}$.

Using this value the stresses and forces in the concrete, reinforcing steel and tendon are calculated as:

$\sigma_c = 12.1$ MPa; $C = 569$ kN

$\sigma_s = 65.7$ MPa; $T_s = 88.7$ kN

$\sigma_p = 1185$ MPa; $T_p = 480$ kN

The bending moment is calculated by taking moments about the top fibre:

$$M = T_p d_p + T_s d_s - C d_c$$
$$= 480 \times 0.625 + 88.7 \times 0.700 - 569 \times 0.125$$
$$= 291 \text{ kNm}$$

and the curvature is calculated from the slope of the strain diagram as:

$$\kappa = \varepsilon_o / d_n = 379 \times 10^{-6} / 375 = 1.011 \times 10^{-6} \text{ mm}^{-1}$$

Calculations for other values of d_n

Values of the moment and curvature for other values of d_n are presented in Table 5.1. Reviewing the results, we see that the last entry for $d_n = 250$ mm goes beyond the elastic range of the concrete and is, thus, unreliable.

TABLE 5.1 Moment-curvature points for various values of d_n.

d_n mm	ε_o x10^{-6}	σ_o MPa	C kN	σ_s MPa	T_s kN	σ_p MPa	T_p kN	M kNm	κ x10^{-6} mm^{-1}
550	217	6.24	468	6.95	16	1141	462	212	0.395
500	245	7.86	491	20	27	1147	465	227	0.491
450	284	9.09	511	32	43	1157	469	246	0.631
400	340	10.9	544	51	69	1173	475	273	0.850
375	379	12.1	569	66	89	1185	480	291	1.011
350	431	13.8	603	86	116	1202	487	315	1.231
325	503	16.1	654	116	157	1227	497	350	1.547
300	611	19.6	733	163	220	1267	513	401	2.036
275	793	25.4	872	245	331	1336	541	490	2.833
250	1171	37.5	1171	422	569	1486	602	677	4.683

5.4 The effect of prior creep and shrinkage

When we take account of prior creep and shrinkage strains in the cross-section, the neutral axes of strain and stress no longer coincide and this adds to the complexity of the analysis. A method is presented in Appendix C, with an example, for calculating stresses and curvatures with and without allowance for prior creep and shrinkage. In most cases the differences in calculated stresses are marginal and not of practical significance. However, the total curvature can be much larger than the elastic curvature, supporting the conclusion that prior creep and shrinkage are not of practical design importance in regard to the stresses and internal forces in a member at working load, or indeed for the strength limit state, but are of great importance when deformations and deflections and losses are to be determined.

5.5 Cracked section analysis: general trial-and-error method

If the section has an irregular shape, or if the steel is placed at several levels, as in Figure 5.6, an iterative trial and error calculation procedure is needed. An adaptation of the method in Section 5.3 can be used for sections that are symmetric about a vertical axis. A sequence of values for d_n is first chosen. For each d_n value, the corresponding value of ε_o is found by trial and error, such that equilibrium is satisfied for the longitudinal forces in the section.

In a non-rectangular section, the compressive concrete above the neutral axis is broken into component areas, with compressive forces C_1, C_2... C_i ..., so that, in checking a value for ε_o, we require:

$$\sum_i C_i = \sum_k T_{pk} + \sum_j T_{sj} \tag{5.19}$$

The tensile force in the k-th tendon is T_{pk}, while T_{sj} is the tensile force in the j-th reinforcing bar. The trial and error calculations for ε_o are continued until Equation 5.19 is satisfied. The moment in the section is:

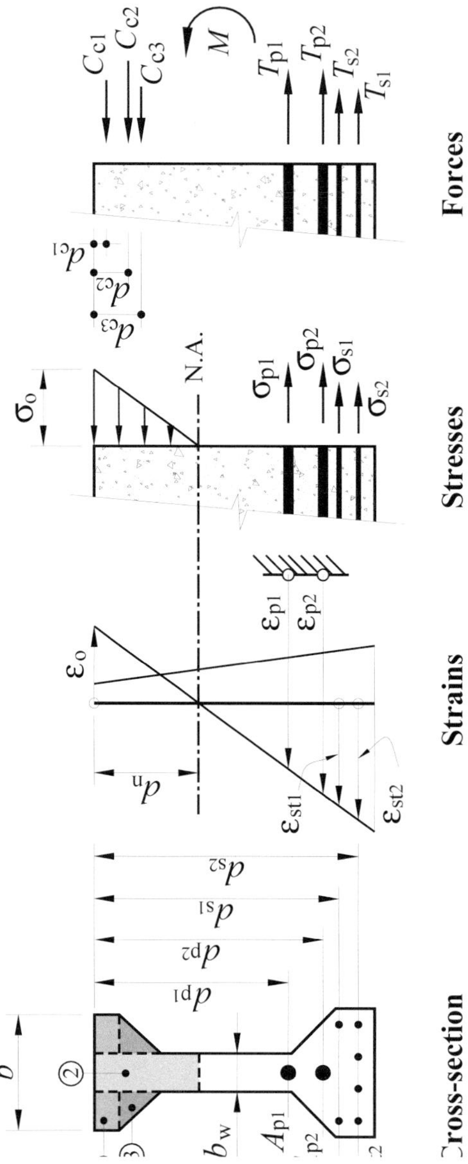

Figure 5.6 General section analysis

$$M = \sum_k T_{pk} \cdot d_{pk} + \sum_j T_{sj} \cdot d_{sj} - \sum_i C_i \cdot d_{ci} \tag{5.20}$$

where the d terms are all measured from the top fibre of the section to the line of action of the relevant force. If prior creep and shrinkage are to be allowed for, the procedure described in Appendix C can be used.

5.6 Non-linear analysis at high overload

As the moment increases into the overload range, the compressive stress block becomes increasingly non-linear and the extreme compressive fibre σ_o eventually approaches the strength of the concrete. Yielding of one or more groups of steel may also occur. The linear analysis clearly becomes increasingly inaccurate.

Flexural behaviour in the non-linear overload range of loading is not normally of interest in design, although a treatment of this range of behaviour is needed if full-range moment-curvature relations and stress-moment relations are required. The trial and error approach is also applicable to the high overload range of behaviour. For the rectangular section of Figure 5.4, Equations 5.6 and 5.7 express the equilibrium requirements at all stages of loading, and for the section at the crack the assumption of linear strain distribution, defined by ε_o and d_n, can be maintained, even at high overload. As in the case of linear behaviour, a unique set of values of ε_o and d_n exists for each moment M. As before, the calculation procedure consists of choosing a value for d_n and then finding by trial and error the corresponding value of ε_o in the top fibre, and hence the stresses and strains and moment M. However, appropriate non-linear stress-strain relationships for the concrete, reinforcing steel and tendons have to be used to obtain the stresses for each trial strain distribution. These have been discussed in Chapter 2. For tendons and reinforcing steel, an elastic-plastic idealisation or a bi-linear approximation will be adequate.

For a chosen value of d_n and a trial value of ε_o, the strains and stresses in the concrete are calculated, and hence the force C. Ignoring prior inelastic creep and shrinkage strains, the compressive strain and stress at depth y below the extreme fibre are, respectively:

$$\varepsilon(y) = \varepsilon_o \times (d_n - y)/d_n \tag{5.21}$$

$$\sigma(y) = fn[\varepsilon(y)] \tag{5.22}$$

where $fn[\varepsilon(y)]$ expresses stress as a function of strain. The force is:

$$C = b\int_0^{d_n} \sigma(y)\, dy \tag{5.23}$$

With C evaluated and with the steel stresses calculated directly from non-linear stress-strain relations, a check is made to see whether Equation 5.6 is satisfied; ε_o is then adjusted as necessary to satisfy the equilibrium requirement. The calculations to determine ε_o and hence the stresses, strains, moment M and curvature κ for a sequence of values of d_n can be carried out, even for sections of more complex shape, and with the steel placed in various layers. Alternatively, a layered section analysis may be undertaken (Warner and Lambert, 1974).

5.7 Moment-curvature and stress-moment relationships

Moment-curvature relations provide a good indication of the overload behaviour and ductility of flexural members. They can also be used to calculate deflections under overload conditions. Stress-moment relations are needed if the fatigue resistance of a member is to be investigated. In this section we look briefly at the moment-curvature and stress-moment relations for sections in a prestressed flexural member.

5.7.1 Moment-curvature relationships

In Figure 5.7 the complete moment-curvature relationship is shown for the cross-section considered in Example 5.3. The calculations were carried out by using Response 2000 (Bentz, 2000) that adopts the non-linear method of analysis described in Section 5.6. For comparison purposes, three points from the linear calculations in Example 5.3 are also plotted and are shown as points (c), (d) and (e). The non-linear analysis departs significantly from the simpler elastic analysis only after yielding of the reinforcing steel has occurred.

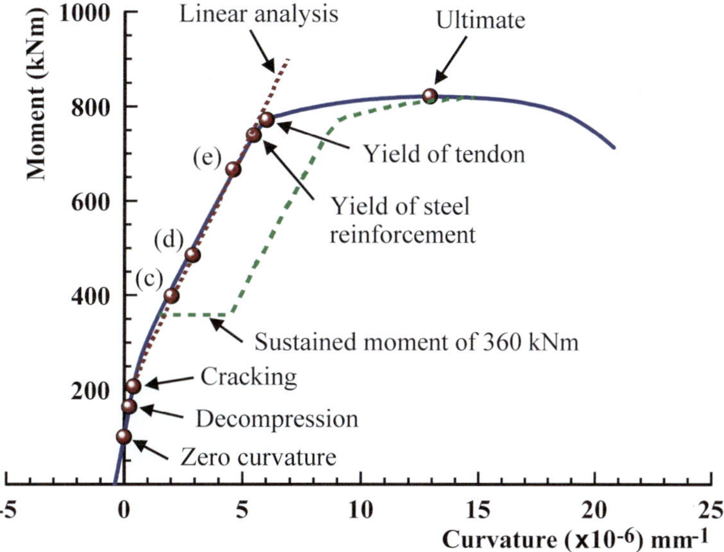

Figure 5.7 Moment-curvature relation

The broken line in Figure 5.7 shows the effect on behaviour that would occur if a constant sustained moment of 360 kNm were to act for an indefinitely long period of time. Although there is a large increase in curvature as a result of creep, the flexural stiffness of the beam under the final short-term loading to failure is almost unaffected. The final moment capacity of the section is likewise not perceptibly affected by the prior history of the sustained service loading. This behaviour is characteristic of both reinforced concrete and prestressed concrete members that do not contain an excessive amount of tensile steel.

To evaluate the peak strain (or stress) in a reinforcing bar or tendon, conditions need to be considered at the crack, where the concrete carries no tension. However, for deflection calculations, average curvatures should be taken over multiple cracks, with the tension stiffening effect allowed for. The portion of the M-κ relation most influenced by tension stiffening is that just after cracking.

Tension stiffening dissipates quickly after cracking and is most significant in members where the service load moment is not much greater than the cracking moment. In highly prestressed members that contain only small quantities of tensile steel, as is the case in Figure 5.7, the tension stiffening effect may not be very significant. However, in partially prestressed members that contain a relatively large proportion of reinforcing steel and also in slabs that contain relatively small quantities of tensile reinforcement, the flexural stiffness may be seriously underestimated by Equation 5.17 and a correction to the bending stiffness is needed when the deflections are calculated.

In the high moment range, the M-κ curve in Figure 5.7 is typical of an under-reinforced section. Two kinks typically occur, when the reinforcing steel and the prestressing steel yield. When the second yielding occurs, the curve flattens out. The relatively small range of the flat yield plateau between yield and M_u in Figure 5.7, with the curvature ratio κ_u/κ_y at about two, suggests that this member has limited ductility. With less tensile steel in the section the ratio would become much higher.

The shape of the moment-curvature relationship, and indeed the flexural behaviour generally, is very much influenced by the quantity of tensile steel in the section, both reinforced and prestressed, and also by the stress σ_{pe} in the prestressing steel, after losses are taken into account. An example of a family of M–κ curves for varying quantities of prestressing steel is given in Figure 5.8 for the section shown. It is seen that an increased quantity of tensile steel in the section results in an increased moment capacity, but also in reduced ductility. This is typical of both reinforced and prestressed concrete beams.

In a further example, the section of Figure 5.8 is analysed for the single case of a 12-strand tendon but with varying initial effective prestress. The results are plotted in Figure 5.9. As demonstrated by the example, for a given section with a fixed quantity of steel, an increase in the initial stress σ_{pe} in the prestressing steel increases both the initial negative curvature (which produces camber) and the cracking moment. A significant reduction in deformation at working moment occurs with increasing prestress, although there is no perceptible change in moment capacity or ductility.

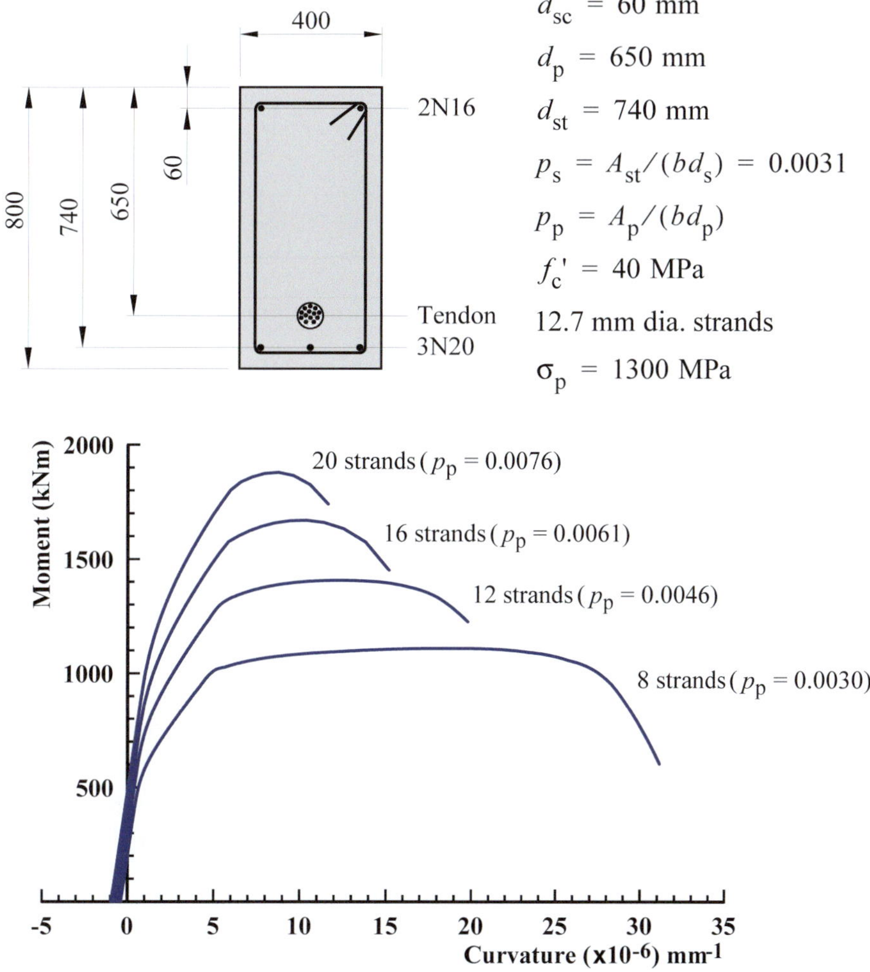

Figure 5.8 Moment-curvature relationships: effect of p_p

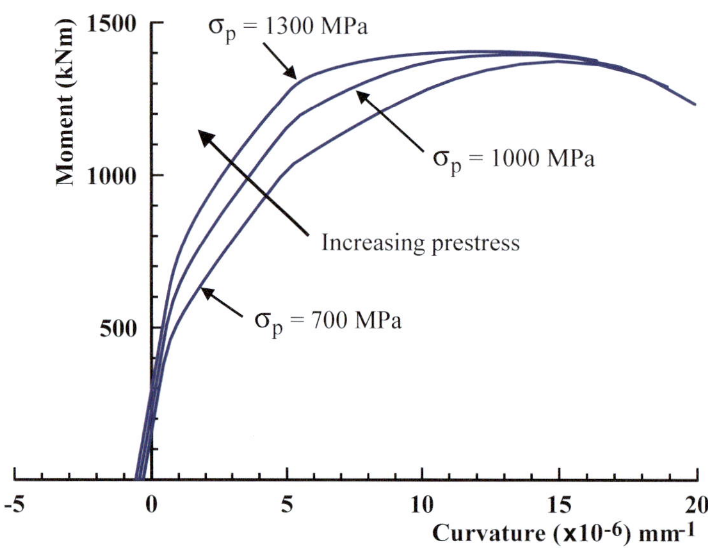

Figure 5.9 Moment-curvature relationships: effect of stress σ_p in the prestressing steel

5.7.2 Stress-moment relations

In Figure 5.10, stress-moment relations are presented for the cross-section in Figure 5.8 for the case of a 12-strand tendon with a prestress of σ_p = 1000 MPa. In the post-cracking range, stresses initially vary almost linearly with moment. The reinforcing steel yields at 500 MPa when M = 1110 kNm, while the prestressing steel yields just before the moment capacity of 1400 kNm is reached, and after the initial stages of crushing of the concrete. The compressive concrete stress reaches its peak value just prior to the ultimate moment M_u of about 1350 kNm and then decreases significantly. A small increase in stress occurs in the tendon in the pre-cracking range, but this is slight in comparison with the increases occurring in the post-cracking range. It should be noted that the stress-moment curves give no indication of the very large deformations that occur as the moment capacity of the section is approached. In practice, stresses only need to be calculated in the immediate post-cracking range, where the linear-elastic analysis described in Sections 5.3 gives adequate results.

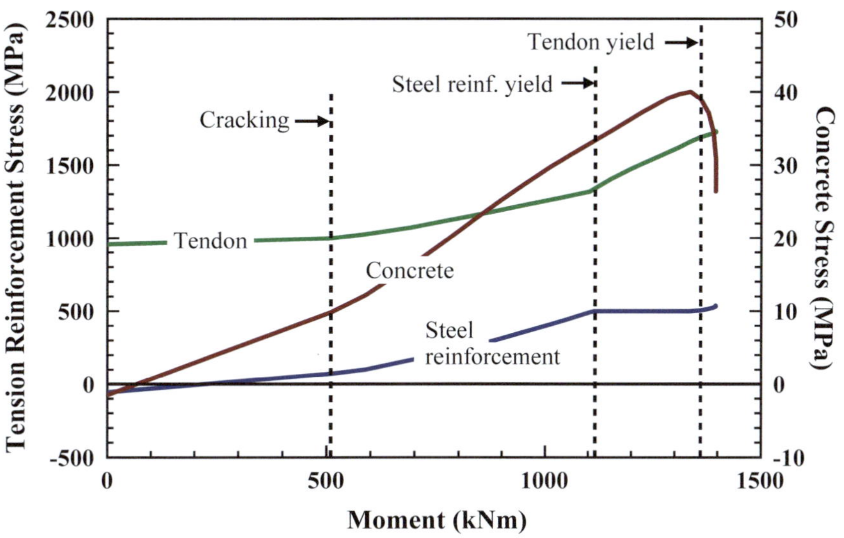

Figure 5.10 Stress-moment relationships

5.8 Deflection calculations

5.8.1 Introduction

We now consider methods for calculating deflections in cracked, statically determinate members. These will be used in Chapters 10 and 12 in the design of statically determinate beams and slabs. Deflections in indeterminate members are discussed in Chapter 11.

5.8.2 Short-term deflections

In the post-cracking range, the behaviour of a cross-section in a prestressed member is non-linear, even though the concrete acts elastically in compression. This is because of the complex conditions in the cracked concrete. The trial and error methods described in Sections 5.3 can be used to find the stresses and strains in a cracked section for a given moment $M(x)$ in the working load range. Although the relationship between moment and curvature at a cracked section is not in a closed form, it can be expressed as follows:

$$\kappa(x) = f[M(x)] \tag{5.24}$$

However, the cracked-section analysis ignores the complex behaviour in the concrete on either side of the crack and the tension stiffening effect, and so over-estimates the average curvature under short-term loading. Modifications can be made to the calculation procedures to take account of tension stiffening and so predict a more accurate average curvature (Kawano and Warner, 1975), but the calculations are complex and not well suited for design.

To provide a simple method of calculating deflections in cracked prestressed members, while also allowing for tension stiffening, Branson (1963) proposed an '*effective*' flexural stiffness, E_cI_{ef}, to be used in linear elastic deflection calculations. The previous Australian Standard, AS 3600-2009, followed this approach and used modifications of Branson's original equations in an elastic calculation procedure for deflections in reinforced concrete and prestressed concrete beams.

Short-term deflection calculations according to AS 3600

In the current Australian Standard, AS 3600–2018, two alternative methods are allowed for calculating deflections in both reinforced concrete and prestressed concrete beams. The first is by refined calculations that take account of cracking and tension stiffening and any other relevant effects. Such calculations require a first principles approach such as a segmental analysis.

The second method provided by the Standard is a simplified procedure that uses an effective flexural stiffness, E_cI_{ef}, together with elastic deflection formulae as appropriate. In AS 3600-2018, the value given for I_{ef} is similar to one proposed by Bischoff (2005), and is identical to the Eurocode equation for effective stiffness, albeit of a slightly different form. It is:

$$I_{ef} = \frac{I_{cr}}{1-\left(1-\frac{I_{cr}}{I}\right)\left(\frac{M_{cr.t}}{M_s^*}\right)^2} \leq I_{ef.max} \tag{5.25}$$

In applying Equation 5.25 to a simply supported beam, I_{ef} is evaluated at mid-span, and at the support for a cantilever. It is used in appropriate elastic deflection equations. The application to indeterminate beams is explained in Chapter 11.

The term M_s^* in Equation 5.25 is the maximum bending moment in the section, and corresponds to the short-term serviceability load or the construction load, as appropriate. The term $M_{cr.t}$ represents the cracking moment at the time of loading, which was previously given as M_{cr} in Equation 5.4. For convenience it is repeated here:

$$M_{cr.t} = Z(f'_{ct.f} - \sigma_{cs} + P_e/A_g) + P_e e \geq 0 \tag{5.26}$$

where Z is the section modulus of the uncracked section, and σ_{cs} is the shrinkage-induced residual tensile stress in the extreme concrete fibre. The equation for σ_{cs}, given earlier, is also repeated here:

$$\sigma_{cs} = \frac{2.5p_w - 0.8p_{cw}}{1 + 50p_w} E_s \varepsilon_{cs} \tag{5.27}$$

The strain ε_{cs} (or ε_{cs}^*) is the design shrinkage strain at the time at which the deflection is to be calculated.

In Equation 5.25, I is the second moment of area for the uncracked section. The upper limiting value, $I_{ef.max}$, is taken to be I. The second moment of area of the cracked section, $I_{cr,}$ is evaluated either using the modular ratio method, with the modular ratio of reinforcing steel $n = E_s / E_c$ and the tendon $n_p = E_p / E_c$, or from equilibrium and compatibility as outlined in Foster et al. (2021). The prestress in the tendon does not enter into the calculations in either method. Sample calculations of I_{cr} and I_{ef} are given in Examples 5.4 and 5.5, below.

Equation 5.25 is used by AS 3600 for both reinforced and prestressed members, but for it to be applicable to a cracked prestressed member, we need to replace the moment terms M_{cr} (or $M_{cr.t}$) and M_s^* by the following increments:

$$M'_{cr} = M_{cr} - M_o \tag{5.28}$$

$$M'_s = M_s^* - M_o \tag{5.29}$$

The reason for this is shown in Figure 5.11, where the moment M_o corresponds to zero curvature in the cross-section, which occurs at load w_o. Note that the initial deflection Δ_o due to prestress plus the load w_o is calculated for the uncracked condition, using the information provided in Chapter 4.

The total short-term deflection due to service load w_s^* is:

$$\Delta_s = \Delta_o + \Delta \tag{5.30}$$

where Δ_o is the deflection due to prestress plus the zero-curvature load w_o, and Δ is due to the load increment $(w_s^* - w_o)$. In many situations, Δ_o will be zero, or very close to zero, especially if the load-balancing approach to design has been used.

In design calculations, the total service load w_s^* will be made up of dead load plus the sustained component of the live load, $\psi_l Q$, plus the transient component of the live load, $\psi_s Q$. It is necessary to determine the deflection components separately. If the sustained component of the load is less than the cracking load, then the associated deflection due to prestress, G and $\psi_l Q$, $(\Delta_p + \Delta_{G+\psi_l Q})$, is calculated as in Chapter 4.

If the member is already cracked under the effect of the sustained loads, then $\Delta_p + \Delta_{G+\psi_l Q}$ is determined using the cracked section stiffness. If deflections for the first loading cycle are required, then $E_c I_{ef}$ has to be calculated separately for each load increment. For subsequent load cycles, the path follows the unloading-reloading line, as shown in Figure 5.11(a), and it is sufficient to evaluate $E_c I_{ef}$ for the maximum applied service moment and to use this stiffness to calculate all deflection increments in the post-cracking range.

Another method of evaluating $E_c I_{ef}$ is to determine the curvature κ_s that corresponds to the maximum moment M_s^* using the moment-curvature calculation method previously explained, and then to calculate the stiffness as $E_c I_{ef} = (M_s^* - M_o)/\kappa_s$. This is the effective stiffness of the beam and is used to calculate the short-term increment in deflection caused by the load increment $(w_s^* - w_o)$. Physically, this corresponds to using the reloading line (shown in Figure 5.11(a)) after a previous loading has been made up to the full working load. As already noted, the tension stiffening effect is not pronounced in prestressed concrete beams, and a simple moment-curvature calculation that ignores tension stiffening usually provides a reasonable value of I_{ef}.

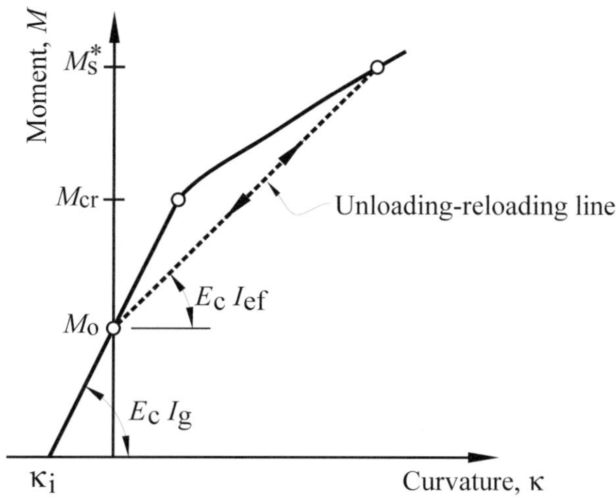

(a) Moment versus curvature

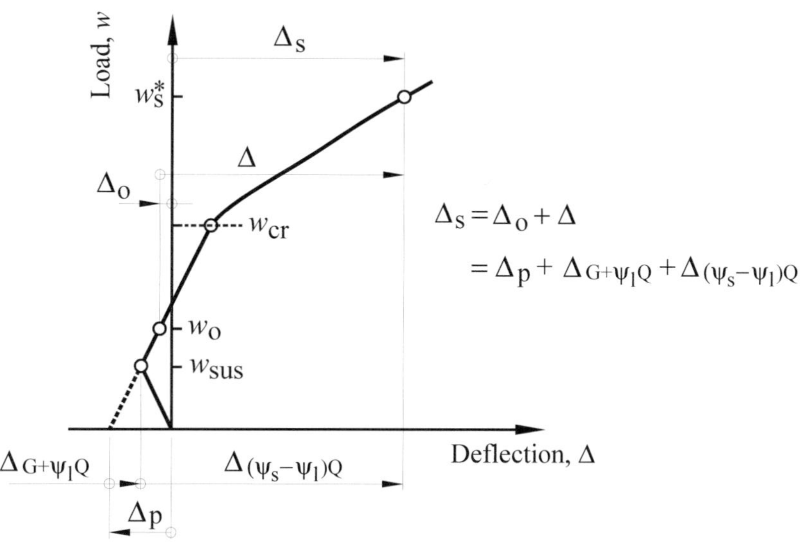

(b) Load versus deflection

Figure 5.11 Deflection of a cracked beam under short-term loading

5.8.3 Long-term deflections

In routine design, the sustained loads that act on prestressed flexural members will not usually be large enough to cause cracks to remain open. It is not often necessary, therefore, to calculate long-term deflections for cracked members, although the situation can arise. The following discussion is therefore brief.

As indicated in Figure 5.12, the long-term creep deflection Δ_c is a magnification of the initial deflection due to the sustained load w_{sus}. If the prestressed member is cracked under sustained load, the initial deflection will be positive (downwards) as in Figure 5.12(b). Shrinkage usually induces a further small downwards deflection. The total deflection, Δ_{tot}, is then the short term deflection plus the creep component Δ_c, plus the shrinkage deflection, Δ_{sh}.

Creep deflections for cracked members

In the peak moment region of a cracked member, creep in the compressive concrete above the neutral axis is the prime contributor to the creep curvature. Some creep will also develop in the tension stiffening regions of concrete on either side of the main cracks, but this is secondary. A simplified yet reasonable treatment of the creep deformations is shown in Figure 5.13, where ε_{cea} is the initial extreme fibre elastic concrete compressive strain due to prestress and the sustained moment, M_{sus}, and ε_{se} is the steel elastic tensile strain increment, not including prestress. As the result of creep over an extended period of time, the top fibre strain consists of the elastic component ε_{cea}, plus the creep component, ε_{cca}. An approximate expression for the creep strain is:

$$\varepsilon_{cca} = \varphi_{cc}\varepsilon_{cea}/\alpha \tag{5.31}$$

where α is a parameter that allows for the restraining effect on creep deformation of any nearby reinforcement in the section.

An estimate of the creep curvature is:

$$\kappa_c = \frac{\varepsilon_{cca}}{d} = \frac{\varphi_{cc}\varepsilon_{cea}}{\alpha d} \tag{5.32}$$

where d is the depth to the centroid of the tensile steel (which may be made up of both reinforcement and prestressing tendon). For a preliminary assessment

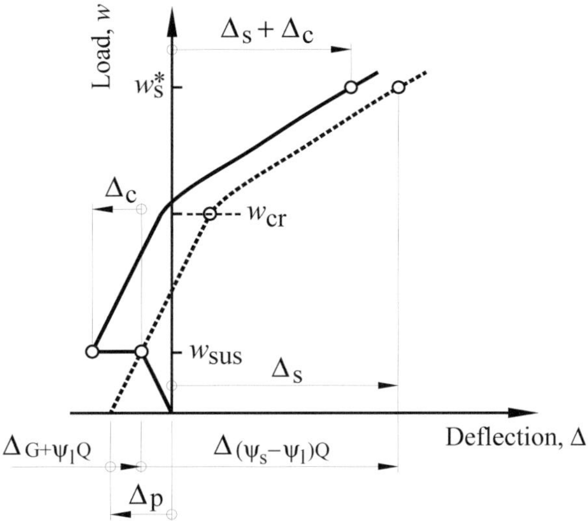

(a) Camber under sustained load

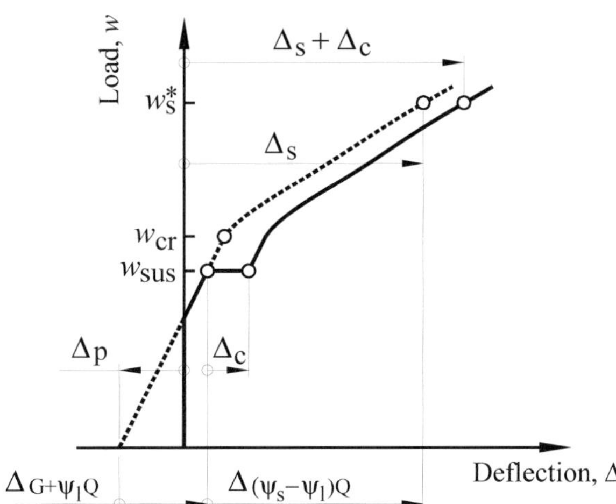

(b) Downward deflection under sustained load

Figure 5.12 Long-term deflections, cracked beam

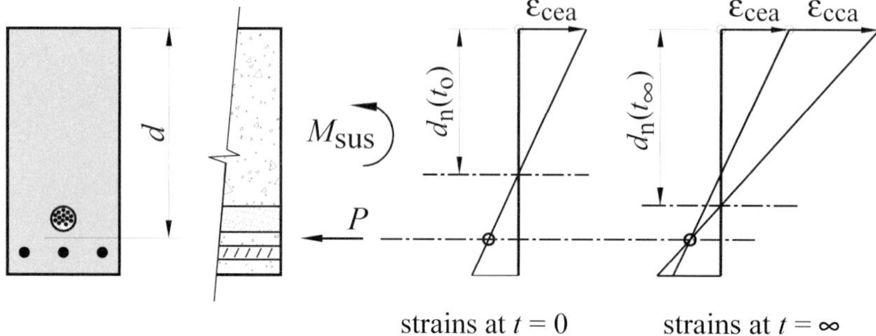

Figure 5.13 Simplified treatment of creep in a cracked section

of creep deflections, the model of Gilbert and Ranzi (2011) provides a simple approach from which further judgement can be made. They evaluate the parameter α using Equations 5.33 to 5.35, below. With κ_c evaluated, say at mid-span and at several intermediate sections, the creep deflection can be evaluated by numerical integration.

For cracked reinforced concrete sections:

$$\alpha = \alpha_1 = \frac{1}{2\sqrt{p_o}} \cdot \sqrt[3]{\frac{I_{cr}}{I_{ef}}} \cdot \left\{1 + (125\,p_o + 0.1)\left(\frac{A_{sc}}{A_{eq}}\right)^{1.2}\right\} \tag{5.33}$$

For uncracked reinforced or prestressed concrete sections:

$$\alpha = \alpha_2 = 1.0 + [45\,p_o - 900\,p_o^2] \cdot \left\{1 + \left(\frac{A_{sc}}{A_{eq}}\right)\right\} \tag{5.34}$$

For cracked partially prestressed concrete sections:

$$\alpha = \alpha_2 + (\alpha_1 - \alpha_2)\left(\frac{d_{n1}}{d_n}\right)^{2.4} \tag{5.35}$$

In this last equation, α_1 and α_2 are calculated as above, d_n is the neutral axis depth for the cracked section under the full service load, including the prestress, and d_{n1} is the neutral axis depth of an equivalent section treating the tendon as additional conventional reinforcement (i.e. ignoring the prestress). In the above equations, A_{sc} is the area of compressive reinforcement and $p_o = A_{eq}/(bd_o)$ is the reinforcement ratio, where d_o is the depth from the extreme compressive fibre to the centroid of the outermost layer of tensile reinforcement, and A_{eq} is the equivalent reinforcement area as given by:

$$A_{eq} = A_p + A_{st} \tag{5.36}$$

If deflection becomes a prime design consideration, then a first principles calculation may be required using a more rigorous approach, such as a layered section analysis, together with a division of the span into a number of segments, and breaking time into several steps and using the step-by-step analysis explained in Appendix B.

Shrinkage deflections for cracked members

If the sustained load is large enough to keep the cracks open, shrinkage will proceed primarily in the uncracked upper compressive fibres. The behaviour in the cracked tensile region below the neutral axis is quite complex, with compressive shrinkage strains developing in the uncracked tensile stiffening regions of concrete between adjacent primary cracks, and producing additional tensile concrete stress because of the restraint from the adjacent steel. This leads to additional cracking and further loss of tension stiffening.

A simple approximation for the design shrinkage curvature has been proposed by Gilbert and Ranzi (2011). They suggest that for cracked partially prestressed concrete the shrinkage curvature can be obtained from:

$$\kappa_{sh} = k_r \varepsilon_{sh}/D \tag{5.37}$$

where D is the overall depth of the section, ε_{sh} is the shrinkage strain and k_r is a factor that depends on the quantity of bonded reinforcement and its depth in the section.

For a cracked reinforced concrete section:

$$k_r = k_{r1} = 1.2\left(\frac{I_{cr}}{I_{ef}}\right)^{2/3}\left(1-\frac{A_{sc}}{2A_{eq}}\right)\left(\frac{D}{d_o}\right) \tag{5.38}$$

For an uncracked section, reinforced or prestressed:

$$k_r = k_{r2} = (100p_o - 2500p_o^2)\left(1 - \frac{A_{sc}}{A_{eq}}\right)^{4/3}\left(\frac{2d_o}{D} - 1\right) \quad \text{for } p_o \leq 0.01 \quad (5.39)$$

$$k_{r2} = (40p_o - 0.35)\left(1 - \frac{A_{sc}}{A_{eq}}\right)^{4/3}\left(\frac{2d_o}{D} - 1\right) \quad \text{for } p_o > 0.01 \quad (5.40)$$

For a cracked partially prestressed section:

$$k_r = k_{r2} + (k_{r1} - k_{r2})\left(\frac{d_{n1}}{d_n}\right) \quad (5.41)$$

where d_n, d_{n1}, A_{eq} and p_o are as previously defined beneath Equation 5.35.

For a more accurate calculation of the long-term shrinkage deflection for a cracked member, a step-by-step analysis of the cracked section would be required. Further information can be obtained from Gilbert and Ranzi (2011).

EXAMPLE 5.4

CALCULATION OF NEUTRAL AXIS DEPTH AND SECOND MOMENT OF AREA OF A PARTIALLY PRESTRESSED CONCRETE SECTION

The section shown in Figure 5.14 carries a factored service load moment of $M_s^* = 1231$ kNm. Determine the depth to the neutral axis for an effective prestressing force of $P_e = 1590$ kN and calculate the second moment of area of the fully cracked section, I_{cr}. The tendon is at a depth $d_p = 750$ mm and consists of 16–12.7 mm diameter strands with $A_p = 1580$ mm^2. The section also contains conventional reinforcement consisting of 3–N24 bars in tension ($A_{st} = 1350$ mm^2) at $d_{st} = 840$ mm and 8–N16 bars in compression ($A_{sc} = 1600$ mm^2) at $d_{sc} = 60$ mm.

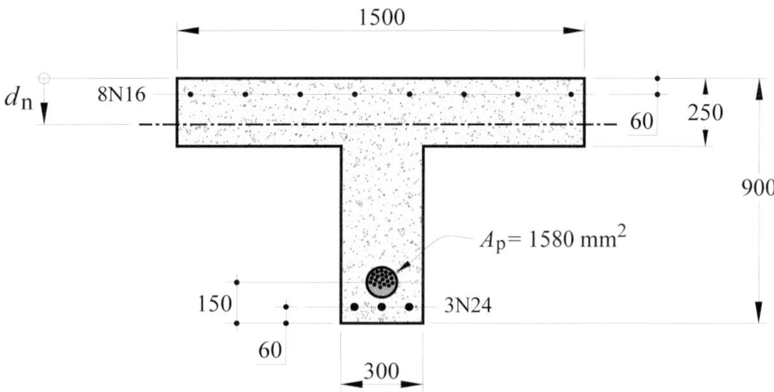

Figure 5.14 Section for Example 5.4

Other data

$$E_c = 30.0\times10^3 \text{ MPa} ; E_s = E_p = 200\times10^3 \text{ MPa}$$

$$n = n_p = E_s/E_c = 200\times10^3/30.0\times10^3 = 6.67$$

$$A_g = 570\times10^3 \text{ mm}^2 ; I_g = 34.80\times10^9 \text{ mm}^4 ; y_{top} = 279 \text{ mm} .$$

SOLUTION

Neutral axis depth

For the prestressed member, the depth of the neutral axis depends on the magnitudes of the prestress and the applied moment. In this case the system is non-linear and a simple modular ratio method is not applicable. Instead, equilibrium and compatibility are used to calculate d_n.

Using the approach outlined in Section 5.5 and assuming that $d_n \geq 250$ mm , we first calculate the tensile strain in the prestressing steel due to the prestress and the compressive strain in the concrete due to the prestress. Thus, from Equations 5.12 and 5.13:

$$\varepsilon_{pe} = \frac{P_e}{A_p E_p} = \frac{1590\times10^3}{1580 \times 200\times10^3} = 5032\times10^{-6}$$

$$\varepsilon_{ce} = \frac{P_e}{E_c}\left[\frac{1}{A_g} + \frac{e^2}{I_g}\right]$$

$$= \frac{1590\times10^3}{30\times10^3}\left[\frac{1}{570\times10^3} + \frac{(750-279)^2}{34.80\times10^9}\right] = 430.8\times10^{-6}$$

The forces in the steel and concrete are then:

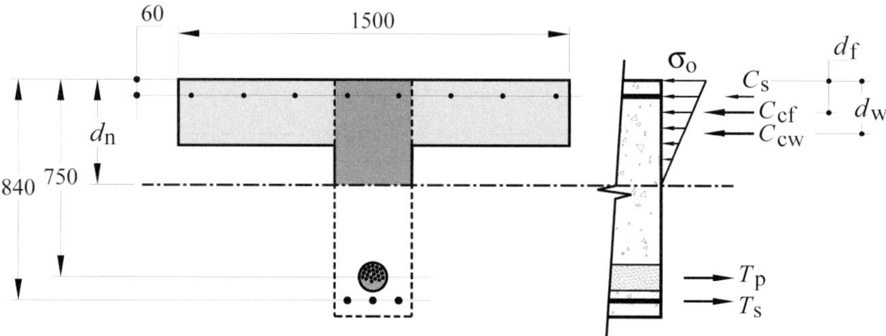

Compressive forces in the concrete and steel:

$$C_{cf} = 0.5(b-b_w)\,t\,(\sigma_o + \sigma_t)$$

$$= 0.5(b-b_w)\,t\,\varepsilon_o E_c\left(1 + \frac{d_n - t}{d_n}\right) = 4.50\times10^9\left(1 + \frac{d_n - 250}{d_n}\right)\varepsilon_o$$

$$C_{cw} = 0.5 b_w d_n \sigma_o = 0.5 b_w d_n \varepsilon_o E_c = 4.50\times10^6 d_n \varepsilon_o$$

$$C_s = A_{sc}\sigma_{sc} = A_{sc}\varepsilon_{sc}E_s = 320\times10^6\left(\frac{d_n - 60}{d_n}\right)\varepsilon_o$$

The tensile forces in the steel and tendon:

$$T_s = A_{st}\sigma_{st} = A_{st}\varepsilon_{st}E_s = 270\times10^6\left(\frac{840-d_n}{d_n}\right)\varepsilon_o$$

$$T_p = A_p\sigma_p = A_p\varepsilon_p E_p = 316\times10^6\left\{\varepsilon_{pe}+\varepsilon_{ce}+\left(\frac{750-d_n}{d_n}\right)\varepsilon_o\right\}$$

$$= 316\times10^6\left\{5463\times10^{-6}+\left(\frac{750-d_n}{d_n}\right)\varepsilon_o\right\}$$

From equilibrium of forces in the section we write:

$$\left\{4.50\times10^9\left(1+\frac{d_n-250}{d_n}\right)+4.50\times10^6 d_n+320\times10^6\left(\frac{d_n-60}{d_n}\right)\right.$$

$$\left.-270\times10^6\left(\frac{840-d_n}{d_n}\right)-316\times10^6\left(\frac{750-d_n}{d_n}\right)\right\}\varepsilon_o = 1.726\times10^6$$

The moment in the section is:

$$M_s^* = 750T_p + 840T_s - 60C_s - d_f C_{cf} - d_w C_{cw} = 1231\times10^6 \text{ Nmm}$$

where:

$$d_f = \left(\frac{3d_n-2t}{6d_n-3t}\right)\cdot t = \frac{3d_n-500}{6d_n-750}\times250 \text{ and } d_w = d_n/3$$

Solving these equilibrium and compatibility equations, we obtain:

$$\varepsilon_o = 243.9\times10^{-6}$$

$$T_p = 1809.3 \text{ kN};\ T_s = 87.3 \text{ kN}$$

$$C_s = 65.1 \text{ kN};\ C_{cf} = 1435.2 \text{ kN};\ C_{cw} = 396.3 \text{ kN}$$

and the depth to the neutral axis is calculated as $d_n = 361.1$ mm. This confirms our original assumption that $d_n \geq 250$ mm .

Cracked second moment of area

To determine the fully cracked second moment of area of the section, I_{cr}, we require the calculation of a 'fictitious' depth to the neutral axis including all reinforcing steel and tendons but as if the tendons carry no stress. For the purpose of the calculation, the tendon is unstressed and the modular ratio method may be used. The 'fictitious' neutral axis depth will be determined by taking the first moments of area. To begin, we assume $d_{n0} \leq 250$ mm :

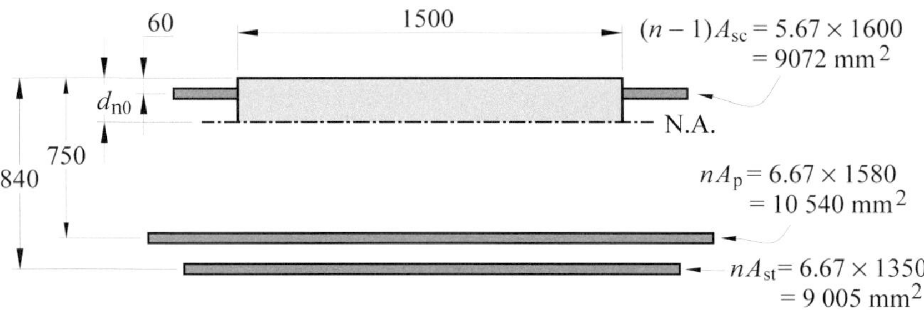

$$1500d_{n0}^2/2 + 9072(d_{n0} - 60)$$
$$= 10.54\times10^3(750 - d_{n0}) + 9005(840 - d_{n0})$$

and solving to give $d_{n0} = 128.3$ mm (< 250 mm OK).

With d_{n0} determined, the fully cracked second moment of area of the section, I_{cr}, is calculated by taking the second moment of area about the notional neutral axis:

$$I_{cr} = \frac{1500 \times 128.3^3}{3} + 9072(128.3 - 60)^2$$
$$+ 10.54\times10^3(750 - 128.3)^2$$
$$+ 9005(840 - 128.3)^2$$
$$= 9.733\times10^9 \text{ mm}^4$$

Discussion

In a prestressed concrete member, cracking is delayed by the presence of the prestress, as compared with an equivalent non-prestressed concrete member. After cracking, the increments of strain and curvature that occur with increments of moment depend on the material in the section; the pre-existing stress in the tendon is largely irrelevant to the incremental response. This is why the slope of the moment-curvature graph between cracking and yield, obtained from a cracked section analysis, as in Example 5.3, agrees well with that calculated from I_{cr} using areas only and ignoring prestress, despite the fact that the neutral axis depths are different.

EXAMPLE 5.5

DEFLECTION CALCULATIONS FOR A SIMPLY SUPPORTED BEAM

In this example we determine the deflection for the T-section beam of Example 5.4 The beam has a single span of 20 metres and is subjected to a live load of 14.8 kN/m, in addition to its self weight. Details of the section are given in Figure 5.13.

The tendon has a parabolic profile with zero eccentricity at its ends and has an eccentricity of 471 mm at the mid-span. The tendon is stressed with an initial prestress of P_i = 1990 kN (σ_{pi} = 1260 MPa) and after losses the effective prestress is P_e = 1590 kN (σ_{pe} = 1010 MPa).

Other data

$$f_c' = 40 \text{ MPa} ; \; f'_{ct.f} = 0.6\sqrt{40} = 3.79 \text{ MPa}$$

$$E_c = 30\times10^3 \text{ MPa} ; \; E_s = E_p = 200\times10^3 \text{ MPa}$$

$$\varphi_{cc} = \varphi_o^* = 2.4 ; \; \varepsilon_{cs} = 600\times10^{-6} ; \; \psi_s = 0.7 ; \; \psi_l = 0.0$$

$$A_g = 570\times10^3 \text{ mm}^2 ; \; I_g = 34.80\times10^9 \text{ mm}^4$$

$$y_{top} = 279 \text{ mm}; \; y_{bot} = 621 \text{ mm}$$

$$Z_{top} = I_g / y_{top} = 34.8\times10^9 / 279 = 124.7\times10^6 \text{ mm}^3$$

$$Z_{bot} = I_g / y_{bot} = 34.8\times10^9 / 621 = 56.04\times10^6 \text{ mm}^3$$

SOLUTION

Stresses in the mid-span section due to prestress

The average prestress acting on the section is:

$$P_e / A_g = 1590\times10^3 / 570\times10^3 = 2.79 \text{ MPa}$$

and the extreme fibre component bending stresses are:

$$P_e e / Z_{bot} = 1590\times10^3 \times 471 / 56.04\times10^6 = 13.36 \text{ MPa (comp.)}$$

$$P_e e / Z_{top} = 1590\times10^3 \times (-471) / 124.7\times10^6 = -6.00 \text{ MPa (tens.)}$$

Thus the extreme fibre stress due to the prestress only are:

$$\sigma_{pb} = 2.79 + 13.36 = 16.15 \text{ MPa (compressive)}$$

$$\sigma_{pt} = 2.79 - 6.00 = -3.21 \text{ MPa (tensile)}$$

for the bottom and top fibres, respectively.

Design loads and moments

The equivalent load for the prestressing cable is (Equation 4.10):

$$w_p = \frac{8P_e h}{L^2} = \frac{8 \times 1590\times10^3 (750 - 279)}{20\,000^2} = 15.0 \text{ N/mm (15.0 kN/m)}$$

and the moment due to the prestress is:

$$M_p = P_e e = 1590 \times 10^3 \times (750 - 279)$$
$$= 749 \times 10^6 \text{ Nmm (749 kNm)}$$

Taking the unit weight of prestressed concrete as 25 kN/m^3, the self-weight moment is:

$$w_G = 25 \times 570 \times 10^3 \times 10^{-6} = 14.25 \text{ kN/m}$$

$$M_G = 14.25 \times 20^2 / 8 = 712.5 \text{ kNm}$$

For the live load, $w_Q = 14.80$ kN/m and thus:

$$M_Q = 14.8 \times 20.0^2 / 8 = 740.0 \text{ kNm}$$

The factored serviceability design load and design moment are then:

$$w_s^* = w_G + \psi_s w_Q$$
$$= 14.25 + 0.7 \times 14.80 = 24.6 \text{ kN/m}$$

$$M_s^* = M_G + \psi_s M_Q$$
$$= 712.5 + 0.7 \times 740.0 = 1231 \text{ kNm}$$

Effective second moment of area

The effective depth is:

$$d = \frac{1350 \times 500 \times 840 + 1580 \times 1533 \times 750}{1350 \times 500 + 1580 \times 1533} = 770 \text{ mm}$$

For the mid-span section we have:

$$p_w = \frac{1350 + 1580}{300 \times 770} = 0.0127 \text{ and } p_{cw} = \frac{1600}{300 \times 770} = 0.00069$$

The residual stress, calculated from Equation 5.1, is:

$$\sigma_{cs} = \frac{200 \times 10^3 \times 600 \times 10^{-6}(2.5 \times 0.0127 - 0.8 \times 0.00069)}{1 + 50 \times 0.0127} = 1.92 \text{ MPa}$$

Using Equation 5.4, we calculate the cracking moment:

$$\begin{aligned} M_{cr} &= Z_{bot}(f'_{ct.f} - \sigma_{cs} + P/A_g) + Pe \\ &= 56.04\times10^6\left(3.79 - 1.92 + \frac{1590\times10^3}{570\times10^3}\right) + 1590\times10^3 \times (471) \\ &= 997\times10^6 \text{ Nmm} \quad (997 \text{ kNm}) \end{aligned}$$

From Equations 5.28 and 5.29, the increments in cracking and service moment are:

$$M'_{cr} = M_{cr} - M_o = 997 - 748.9 = 248 \text{ kNm}$$

$$M'_s = M^*_s - M_o = 1231 - 748.9 = 482 \text{ kNm}$$

For the long-term condition ($k = 0.5$) the effective second moment of area is (Equation 5.25):

$$I_{ef} = \frac{9.733\times10^9}{1 - \left(1 - \dfrac{9.733\times10^9}{34.80\times10^9}\right)\left(\dfrac{248\times10^6}{482\times10^6}\right)^2} = 15.46\times10^9 \text{ mm}^4$$

Short-term deflection

We obtain the short-term deflection as the sum of two terms, $\Delta_o + \Delta_1$, where Δ_o is the deflection due to the prestress plus a load w_o, w_o being the load that causes a moment $M_o = P_e e$ at mid-span. The second component Δ_1 is the deflection due to the load increment $(w^*_s - w_o)$.

As the load due to the prestress closely balances that due to the self-weight of the member, and both are uniformly distributed, $\Delta_o = 0$ and we calculate Δ_1 as:

$$\begin{aligned} \Delta_1 &= \frac{5}{384} \cdot \frac{(w^*_s - w_o)L^4}{E_c I_{ef}} \\ &= \frac{5 \times (24.6 - 15.0) \times 20\,000^4}{384 \times 30.0\times10^3 \times 15.46\times10^9} = +43 \text{ mm} \end{aligned}$$

The initial short term deflection is estimated to be $\Delta_s = 43$ mm downwards.

Deflection due to creep

The short-term deflection due to self-weight and prestress is required in order to estimate the creep deflection. The equivalent load from effective prestress, P_e, was 15.0 kN/m and that from the dead load and sustained component of the live load 14.25 kN/m. The nett load is therefore a small upward load of -0.75 kN/m, which acting on the uncracked section induces a camber of:

$$\Delta_1 = \frac{5}{384} \cdot \frac{(w_G - w_p)L^4}{E_c I_g}$$

$$= \frac{5 \times (-0.75) \times (20.0 \times 10^3)^4}{384 \times 30.0 \times 10^3 \times 34.8 \times 10^9} = -1.5 \text{ mm (upwards)}$$

Immediately after transfer, the initial prestress P_i causes an equivalent load of:

$$w_p = \frac{8P_i h}{L^2} = \frac{8 \times 1990 \times 10^3 (750 - 279)}{20\ 000^2} = 18.75 \text{ N/mm (18.75 kN/m)}$$

so that the nett load is an upward load of $14.25 - 18.75 = -4.5$ kN/m, inducing an initial camber of about -9.0 mm at the centre of the 20 metre span.

The creep deflection is significantly influenced by loss of prestress, and the rate of loss will decrease with time. In the calculations for the creep deflection, we shall take the deflection of the sustained portion of the load as the average: $\Delta_{sus} = (-1.5 - 9.0)/2 = -5.3$ mm.

To determine the creep deflection we shall adopt the simplified approach of Gilbert and Ranzi (2011) as given by Equations 5.31 to 5.35. We start by calculating the equivalent area of tensile reinforcement and the web reinforcement ratio as:

$$A_{eq} = 1580 + 1350 = 2930 \text{ mm}^2$$

$$p_o = \frac{2930}{1500 \times 840} = 0.0023$$

and terms α_1 and α_2 are determined from Equations 5.33 and 5.34 as:

$$\alpha_1 = \frac{1}{2\sqrt{0.0023}} \times \sqrt[3]{\frac{9.733\times10^9}{12.34\times10^9}} \times \left\{1 + (125 \times 0.0023 + 0.1)\left(\frac{1600}{2930}\right)^{1.2}\right\} = 11.44$$

$$\alpha_2 = 1.0 + [45 \times 0.0023 - 900 \times 0.0023^2] \times \left\{1 + \frac{1600}{2930}\right\} = 1.15$$

From Example 5.4, for the prestressed section with $M_s^* = 1231\ \text{kNm}$, the neutral axis depth was calculated as d_n =361.1 mm, while for the non-prestressed section d_{n1} = 128.3 mm. The parameter α is determined as (Equation 5.35):

$$\alpha = 1.15 + (11.44 - 1.15)\left(\frac{128.3}{361.1}\right)^{2.4} = 2.0$$

The creep deflection is:

$$\Delta_c = \varphi_o^* \Delta_{sus}/\alpha = 2.4 \times (-5.3)/2.0 = -6\ \text{mm (upwards)}$$

which is small relative to the length of the member.

Deflection due to shrinkage

For the determination of the shrinkage curvatures we shall adopt the simplified approach given by Equations 5.37 and 5.41. The equivalent area of reinforcement and the web reinforcement ratio as determined for creep and the terms k_{r1} and k_{r2} are evaluated from Equations 5.38 and 5.40 as:

$$k_{r1} = 1.2\left(\frac{9.733\times10^9}{12.34\times10^9}\right)^{2/3}\left(1 - \frac{1600}{2\times2930}\right)\left(\frac{900}{840}\right) = 0.798$$

$$k_{r2} = (100 \times 0.0023 - 2500 \times 0.0023^2)\left(1 - \frac{1600}{2930}\right)^{4/3}\left(\frac{2\times840}{900} - 1\right)$$

$$= 0.066$$

With $d_n = 361.1$ mm and $d_{n1} = 128.3$ mm, the curvature factor k_r is calculated as (Equation 5.41):

$$k_r = 0.066 + (0.798 - 0.066)\left(\frac{128.3}{361.1}\right) = 0.326$$

and the shrinkage curvature at the mid-span is estimated to be:

$$\kappa_M = k_r \varepsilon_{sh} / D = 0.326 \times 600 \times 10^{-6} / 900 = 217 \times 10^{-9} \text{ mm}^{-1}$$

For the simply supported span, assuming a constant shrinkage warping over the length of the member, the shrinkage deflection at mid-span is obtained by Equation 4.28 as:

$$\Delta = \frac{\kappa_M L^2}{8} = \frac{217 \times 10^{-9} \times 20\,000^2}{8} = 11 \text{ mm (downwards)}$$

Total deflection

The total deflection under full service load is estimated to be 43 – 6 + 11 = 48 mm; this represents a deflection of less than span/300 and will be satisfactory provided that live load vibration is not a critical issue.

Discussion

In this example, the loads due to the prestress and the dead load were reasonably balanced, so that the deflection is almost wholly due to the transient live loading where, often, a more tightly defined limit is needed.

In the creep deflection calculation, the results are sensitive to the input assumptions and the accuracy of the calculated elastic deflection, which can vary due to the assumed density of the concrete and the elastic modulus, as well as the choice of the final creep coefficient φ_0^*. The elastic modulus of the concrete can differ by as much as 30 percent of that predicted by the models of AS 3600, while the density of the concrete may vary by five or even ten percent. So, while the sum of the elastic and creep components of the deflection is a camber of – 5.3 – 6 = – 11 mm, it could be somewhat larger or even downwards. When creep deflections are a critical issue, then a more accurate method of assessment will be needed. In this example, the deflection is small and not of concern.

The deflection induced by shrinkage warping is a relatively small component of the total deflection and at 11 mm downwards it balances the small upwards creep deflection. In summary, the total time dependent deflection is estimated to be small with the critical deflection being that due to the transient component of the live loading.

5.9 Crack control

Excessive cracking does not usually pose a problem in the design of prestressed flexural members, but it is sometimes necessary to show that cracking is not a design problem. This is done using a cracked section analysis to evaluate either the stresses in the tendon or reinforcement, or the crack widths.

Flexural cracking in a prestressed member is deemed by AS 3600-2018 to be under control if the maximum tensile concrete stress in the region, calculated for the uncracked section, is not greater than $0.25\sqrt{f'_c}$. Even if this value is exceeded, flexural cracking is still considered to be under control provided reinforcement is present in the cracked region at a spacing of not more than 300 mm and that at least one of the following three requirements is satisfied:

(i) the tensile concrete stress calculated does not exceed $0.6\sqrt{f'_c}$; or

(ii) the calculated increment in stress in the steel near the tensile face does not exceed the relevant value given in Table 5.2; or

(iii) the calculated crack width, w, does not exceed a maximum characteristic crack width, w'_{max}, which is chosen by the designer and depends on the exposure of the cracked surface and its function.

Although AS 3600 limits the bar spacing to 300 mm, it does not specify a bar size. Clearly the best control is obtained using small diameter bars at close spacing and placed close to the concrete surface. When the load balancing method is used, there will normally be a need for tensile reinforcement in the section to achieve the required flexural strength, and this can often serve the dual purpose of crack control.

TABLE 5.2 Maximum steel stress increment for crack control

Bar diameter	Maximum increment in steel stress, MPa #
12 mm or smaller	350
16 mm	300
20 mm	260
24 mm	225
28 mm or greater	210
All bonded tendons	210

Note: # For crack widths of w'_{max} = 0.4 mm

Crack control by limiting the steel stress increment

A cracked section analysis, as described above in Sections 5.3 and 5.5, is carried out to evaluate the stress increments that occur in the reinforcing steel in the cracked region under investigation, to ensure that the limits in Table 5.2 are not exceeded. An example of the stress calculation procedure is given in Example 5.6.

Crack control by crack width calculation

According to AS 3600, the crack width calculations for a prestressed flexural member are carried out in the same way as for a reinforced beam, which is explained in the companion text by Foster et al. (2021). A brief explanation of the crack width calculation for prestressed concrete is given here. Although such calculations will rarely be required in the design of prestressed concrete members, Example 5.6 contains crack width calculations for crack control.

According to the Standard, the crack width, w, is calculated by:

$$w = s_{r,max}(\varepsilon_{sm} - \varepsilon_{cm}) \tag{5.42}$$

and must not exceed an appropriate characteristic maximum crack width, w'_{max}, which is chosen by the designer. Typical values of w'_{max} are 0.4 mm when only modest control of cracking is needed, and 0.3 mm and 0.2 mm when the control is to be moderate or strict, respectively.

In Equation 5.42, $s_{r,max}$ is the maximum final crack spacing, and $(\varepsilon_{sm} - \varepsilon_{cm})$ is the difference between the mean strain in the main tensile steel and the mean strain in the concrete.

The maximum final crack spacing is evaluated as:

$$s_{r,max} = 3.4c + 0.3k_1k_2\frac{d_b}{p_{eff}} \leq 1.3(D - kd) \tag{5.43}$$

where c is the cover to the longitudinal reinforcement, d_b is the bar diameter, and p_{eff} is the effective reinforcement ratio for the tension chord.

The tension chord is an effective volume of concrete that surrounds the tensile steel between adjacent cracks. The height and cross-sectional area of the tension chord have to be determined. AS 3600 adopts the CEB-FIP Model Code (1993) depth of the tension chord, $h_{c.eff}$, of:

$$h_{c.eff} = \text{smallest of} \begin{cases} 2.5y_c = 2.5[D - d] \\ [D - d_n]/3 \\ D/2 \end{cases} \tag{5.44}$$

where y_c is the distance from the extreme tensile fibre of the beam to the centroid of the tensile steel, as shown in Figure 5.15, d is the effective depth to the geometric centroid of the tensile steel (including tendons), and d_n is the neutral axis depth of the cracked section.

The effective area of the tension chord is $A_{c.eff}$ and is equal to the cross-sectional area defined by its height $h_{c.eff}$ and the width(s) of the section over this height. For a rectangular section, $A_{c.eff} = h_{c.eff}b$.

The ratio p_{eff} is $A_{st}/A_{c.eff}$. Equation 5.43 is only valid, according to AS 3600, provided the bonded reinforcement in the section has a spacing of not greater than $s_{r.max} = 5(c + 0.5\,d_b)$. When this spacing is exceeded, but not by much, a reasonable approach would be to increase the calculated crack width by the ratio ($s_b/s_{b.max}$) and ensure that this increased value does not exceed the limiting value w'_{max}.

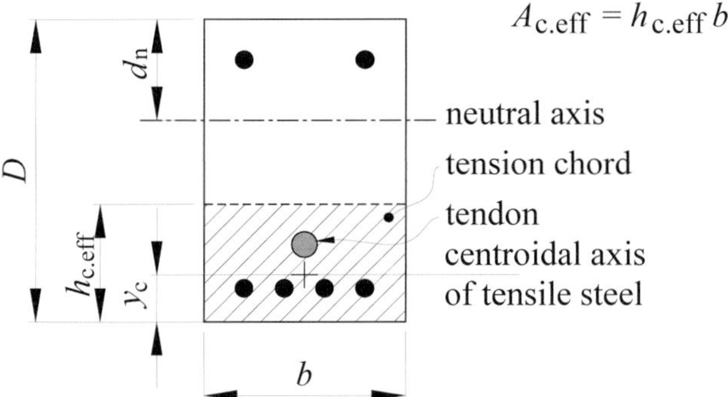

Figure 5.15 Tension chord in a concrete beam

The terms k_1 and k_2 in Equation 5.43 account respectively for the bond properties of the reinforcement, and the longitudinal strain distribution in the tension chord. AS 3600 values for k_1 are taken to be 0.8 for deformed bars and 1.6 for plain bars. Values for k_2 are 0.5 for a member in bending and 1.0 in pure tension.

In Equation 5.42 the term $(\varepsilon_{sm} - \varepsilon_{cm})$ is evaluated as:

$$\varepsilon_{sm} - \varepsilon_{cm} = \frac{\sigma_{scr}}{E_s} - 0.6\frac{f_{ct}}{E_s p_{eff}}(1 + n_e p_{eff}) + \varepsilon_{cs} \geq 0.6\frac{\sigma_{scr}}{E_s} \tag{5.45}$$

The term σ_{scr} is the stress in the tensile reinforcement, obtained from a cracked section analysis; f_{ct} is the mean axial tensile strength of the concrete at the time of cracking; p_{eff} is the reinforcement ratio for the tension chord, which for prestressed concrete the authors recommend be taken as $p_{eff} = (A_{st}+A_p)/A_{c.eff}$[1]; n_e is an effective modular ratio, $(1 + \phi_{cc})E_s/E_c$, which allows for creep; and ε_{cs} is the final long-term shrinkage strain.

1. AS 3600 does not include the area of the prestressing steel in the calculation of p_{eff}.

In addition to the above, the Standard requires that additional longitudinal reinforcement be placed in the side faces of beams that are more than 750 mm deep. This reinforcement is to consist of 12 mm bars spaced vertically at 200 mm centres, or 16 mm bars at 300 mm centres.

EXAMPLE 5.6

CRACK CONTROL CALCULATIONS

Crack control checks are to be carried out for the rectangular section of Example 5.3 (shown in Figure 5.5), for serviceability moments of 300 kNm and 400 kNm. A post-cracking elastic analysis has been carried out in Example 5.3 for this cross-section, for moments increasing from M_{cr} into the overload range. The results are contained in Table 5.1 and are used in the following crack control calculations. The final creep and shrinkage values to be used are: φ_{cc} (= φ_o^*) = 2.5 and ε_{cs} (= ε_{cs}^*) = 600×10^{-6}. Note that typical stress-moment relations, but for a different section, are shown in Figure 5.10.

SOLUTION

The cracking moment for the section was evaluated in Example 5.3 as M_{cr} = 207 kNm,. The moments of 300 kNm and 400 kNm are clearly well in excess of M_{cr}, and the tensile concrete stress will be greater than $0.25\sqrt{f_c'}$ and $0.6\sqrt{f_c'}$. However, the AS 3600 requirement that there is steel reinforcement close to the tensile face and at a spacing of less than 300 mm is satisfied, so that checks on crack control can be made, either by calculating the steel stress increment or the crack width. Both methods will be used here. As the moment values of 300 kNm and 400 kNm do not appear in Table 5.1, interpolation is used to evaluate key variables at these moments.

Other data

The effective depth to the geometric centroid of the tensile steel is:

$$d = \frac{1350 \times 700 + 405 \times 625}{1350 + 405} = 683 \text{ mm}$$

Stress data for moment $M = 300$ kNm

For d_n values of 375 mm and 350 mm in Table 5.1, the corresponding moments are 291 kNm and 315 kNm, respectively. We interpolate to evaluate d_n, σ_s and σ_p for $M = 300$ kNm as follows:

At $M = 291$ kNm:

$d_n = 375$ mm; $\sigma_c = 12.1$ MPa; $\sigma_s = 66$ MPa; and $\sigma_p = 1185$ MPa.

At $M = 315$ kNm:

$d_n = 350$ mm; $\sigma_c = 13.8$ MPa; $\sigma_s = 86$ MPa; and $\sigma_p = 1202$ MPa.

For $M = 300$ kNm, we obtain by interpolation:

$$d_n = 375 - \frac{25(300 - 291)}{315 - 291} = 366 \text{ mm}$$

and similarly, $\sigma_s = 74$ MPa; and $\sigma_p = 1191$ MPa.

Steel stress increment for $M = 300$ kNm

For N24 bars the allowable stress increment in the steel, from Table 5.2, is 225 MPa. Clearly, with $\sigma_s = 74$ MPa, this requirement is satisfied.

Check on crack width for $M = 300$ kNm

To calculate the crack width we first evaluate $s_{r,max}$ and $(\varepsilon_{sm} - \varepsilon_{cm})$. The depth of the tension chord, $h_{c,eff}$, is (Equation 5.44):

$$h_{c.eff} = \text{smallest of} \begin{cases} 2.5[750 - 683] = 168 \\ [750 - 366]/3 = 128 \\ D/2 = 375 \end{cases} = 128 \text{ mm}$$

and:

$$A_{c.eff} = 250 \times 128 = 32.0 \times 10^3 \text{ mm}^2$$

$$p_{eff} = (1350 + 405)/32.0 \times 10^3 = 0.055$$

With $c = (750 - 700 - 12) = 38$ mm, $k_1 = 0.8$ and $k_2 = 0.5$, the maximum crack spacing from Equation 5.43 is:

$$\begin{aligned} s_{r,max} &= 3.4 \times 38 + 0.3 \times 0.8 \times 0.5 \times 24/0.055 \\ &= 129.2 + 52.4 = 182 \text{ mm} \end{aligned}$$

With:

$$n_e = \frac{(1 + 2.5) \times 2{\times}10^5}{30{\times}10^3} = 23.3$$

$$f_{ct} = 1.4 f'_{ct} = 1.4 \times 0.36 \times \sqrt{40} = 3.19 \text{ MPa and } \varepsilon_{cs} = 600{\times}10^{-6}$$

we obtain the difference between the average steel strain and the average concrete strain in the tension chord as (Equation 5.45):

$$\begin{aligned} \varepsilon_{sm} - \varepsilon_{cm} &= \frac{74.0}{2{\times}10^5} - \frac{0.6 \times 3.19}{2{\times}10^5 \times 0.055}(1 + 23.3 \times 0.055) + 600{\times}10^{-6} \\ &= (370 - 397 + 600) \times 10^{-6} \\ &= 573{\times}10^{-6} \end{aligned}$$

which is greater than:

$$\frac{0.6\sigma_{sct}}{E_s} = 0.6 \times \frac{74}{200{\times}10^3} = 222{\times}10^{-6} \quad (\therefore \text{OK})$$

The calculated crack width is:

$$w = 182 \times 573{\times}10^{-6} = 0.10 \text{ mm}$$

and is comfortably less than the a characteristic maximum value of 0.2 mm, which applies when strong crack control is required.

As the depth of the section is not greater than 750 mm, it is not necessary to include side-face reinforcement.

Stress data for moment $M = 400$ kNm

Interpolating in Table 5.1 between moment values of 350 kNm and 401 kNm, we obtain for $M = 400$ kNm:

$d_n = 300.5$ mm; $\sigma_s = 162$ MPa; and $\sigma_p = 1266$ MPa.

Steel stress increment for $M = 400$ kNm

Again, $\sigma_s < 210$ MPa and so the control is satisfactory.

Check on crack width for $M = 400$ kNm

As before, the depth of the tension chord, $h_{c,eff} = 128$ mm, and we obtain $s_{r,max} = 198$ mm.

With the steel stress now at 162 MPa, the difference between the average steel strain and the average concrete strain in the tension chord becomes (Equation 5.45):

$$\begin{aligned}\varepsilon_{sm} - \varepsilon_{cm} &= \frac{162}{2\times10^5} - \frac{0.6 \times 2.27}{2\times10^5 \times 0.055}(1 + 23.3 \times 0.055) + 600\times10^{-6} \\ &= (810 - 282 + 600) \times 10^{-6} \\ &= 1128\times10^{-6}\end{aligned}$$

which is greater than $0.6\sigma_{sct}/E_s = 0.6 \times 810 = 486\times10^{-6}$ ($\therefore$ OK)

The calculated crack width is:

$$w = 182 \times 1128\times10^{-6} = 0.21 \text{ mm}$$

which is sufficient for a moderate to strong degree of crack control. It is observed that the crack increased by about 60 per cent as the moment increased by 100 kNm, from 300 kNm to 400 kNm, while still being in the working load range.

5.10 References

Bentz, E.C., 2000, *Sectional Analysis of Reinforced Concrete Members*, Doctoral Thesis, Graduate Department of Civil Engineering, University of Toronto, Toronto, Canada, 316 pp.

Bishchoff, P H, 2005, Reevaluation of deflection prediction for concrete beams reinforced with steel and FRP bars, Journal of Struct. Eng., ASCE, Vol. 131, No. 5, pp. 752-767.

Branson, D.E. 1963, *Instantaneous and Time Dependent Deflection of Simple and Continuous Reinforced Concrete Beams*, HPR Report No. 7, Alabama Highway Department., US Bureau of Public Roads.

Branson, D.E. and Trost, H. 1982, Application of the I-Effective Method in Calculating Deflections of Partially Prestressed Members, *PCI Journal*, Vol. 27, No. 5, pp. 62–77.

CEB-FIP Model Code 1990, 1993, *Comite Euro-International du Beton*, Thomas Telford, London.

Foster, S J, Kilpatrick, A E and Warner, R F, 2021, *Reinforced Concrete Basics*, 3E, (3rd Edn) Pearson, Melbourne, Australia., 589 pp.

Gilbert R.I. and Ranzi, G. 2011, *Time dependent behaviour of concrete structures*, Spon Press, Oxon, UK, 426 pp.

Goto, Y. 1971, Cracks Formed in Concrete Around Deformed Tension Bars, *ACI Journal Proceedings*, Vol. 68, No. 4, pp. 244-251.

Kawano, A. and Warner, R.F. 1975, *Nonlinear Analysis of Time Dependent Behaviour of Reinforced Concrete Frames*, The University of Adelaide, Research Report R-125, January, 41 pp.

Kilpatrick, A.E. and Gilbert, R.I. 2009, Prediction of Short-term Deflections in Reinforced Concrete One-way Continuous Members, *Concrete09, 24th Biennial Conference*, Concrete Institute of Australia, 17–19 September, Sydney, Paper 1b-3.

Warner, R.F. 1973, Simplified Model of Creep and Shrinkage Effects in Reinforced Concrete Flexural Members, *Civil Engineering Transactions*, Institution of Engineers, Australia, Vol. CE 1.5, Nos 1 and 2.

Warner, R.F. and Lambert, J.H. 1974, Moment-curvature-time Relations for Reinforced Concrete Beams, *IABSE Publications*, Zurich, Vol. 34, No. 1, 181–203.

CHAPTER 6

Flexural strength analysis

Methods for calculating the ultimate flexural strength of prestressed concrete beams are presented in this chapter. The simple case of a rectangular section is first dealt with, and other more complicated sections are then considered, including non-rectangular sections and sections with tensile steel reinforcement and tendons at various levels. A general trial-and-error method of analysis is described which can be used for complex cross-sections. The chapter concludes with a review of the AS 3600 requirements for the strength design of beams.

6.1 Overload behaviour and ultimate strength

The behaviour of a prestressed concrete flexural member at high overload and at failure is similar to that of a reinforced concrete member and the method of determining the moment capacity M_u follows a similar procedure to that for reinforced concrete. The assumptions concerning the concrete compressive stress block are the same in both cases.

We first consider the simple case of a rectangular section with tensile reinforcing steel at depth d_{st}, compressive steel at depth d_{sc}, and a tendon at depth d_p. At all stages of loading after cracking, the internal forces in the cracked section are as shown in Figure 6.1, with tensile forces T_s and T_p, and compressive forces C_c and C_s in the concrete and compressive steel above the neutral axis. As we have already seen in Chapter 5, the equilibrium requirements for a section with tensile reinforcement and a prestressing tendon, at all stages of loading including the ultimate condition when M_u acts, are:

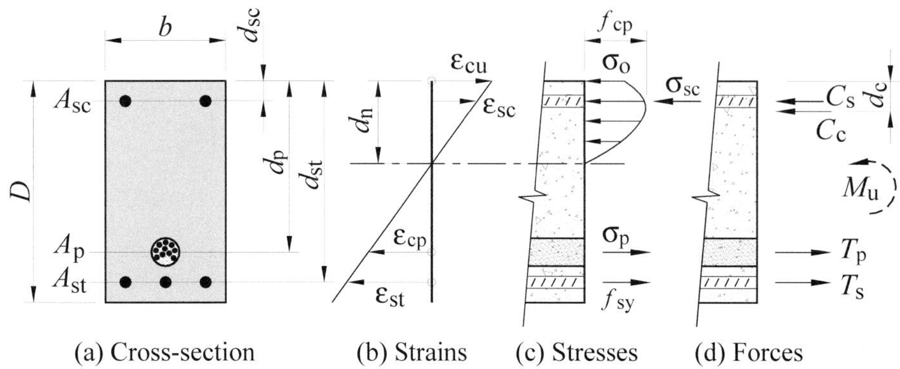

Figure 6.1 Conditions at M_u

$$C_c + C_s = T_s + T_p \tag{6.1}$$

$$M = T_s d_{st} + T_p d_p - C_c d_c - C_s d_{sc} \tag{6.2}$$

As the moment M increases into the high overload range, the stresses in the compressive concrete above the neutral axis become increasingly non-linear and begin to reflect the shape of the stress-strain curve for concrete under short-term loading. The top fibre concrete stress σ_o progressively increases until it reaches a maximum value σ_u, which is some proportion of the mean in situ strength of the concrete in uniaxial compression, f_{cmi}. With further increases in moment, the extreme fibre stress σ_o starts to decrease and the peak value σ_u moves downwards away from the top fibre. At this stage, the compressive steel will have yielded. The tensile stresses and strains increase in both the prestressing tendon and the reinforcing steel. Provided the member is under-reinforced the reinforcing steel and tendon both yield well before the peak moment M_u is reached. Even if the stress-strain curve for the prestressing steel has no sharply defined yield point, the tendon stress σ_p will exceed the 0.2 per cent proof stress before M_u is approached. The flexural deformations thus increase rapidly in the beam section in the final stage of loading. When M_u acts, the compressive concrete stress block is as shown in Figure 6.1(c).

In some prestressed concrete members, however, some of the tensile reinforcing steel or tendon may remain elastic at M_u, even when the section is under-reinforced. This can occur, for example, if the prestressing steel is placed at several different depths in the section as in Figure 6.2. Likewise, if the reinforcing steel is placed at various levels, the interior bars near mid-depth will remain elastic. Although in these situations the flexural deformations will still be large when M_u acts, the evaluation of M_u becomes more complicated, and, as we shall see shortly, will require a trial-and-error calculation procedure.

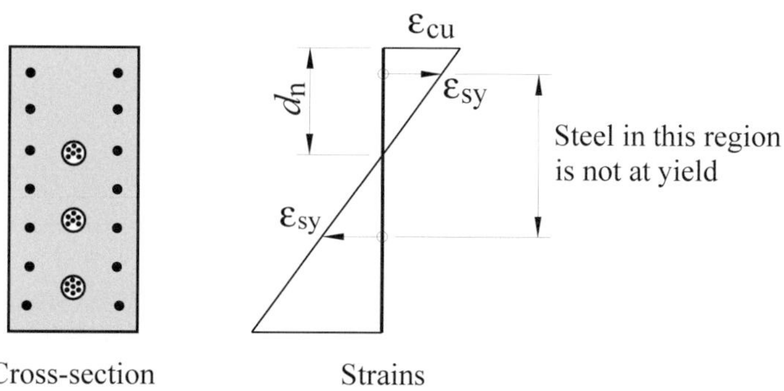

Figure 6.2 Non-yielding of portion of reinforcement at M_u

It is also possible, in sections where large quantities of reinforcing and prestressing steel are used, that the section is over-reinforced so that even with the relatively simple arrangement of steel shown in Figure 6.1(a), yield does not occur. In this case a different approach is required for calculating M_u.

As in Chapter 5, we will first analyse a rectangular beam in which the prestressing steel and reinforcing steel are concentrated at depths d_p and d_{st}. More general cases will then be dealt with in turn. However, we first need to discuss the assumptions made in flexural strength analysis and in particular the conditions of stress and strain that are assumed to exist in the compressive concrete above the neutral axis at the time when M_u acts.

6.2 Assumptions for ultimate strength analysis

Several assumptions are usually made to simplify the calculation of moment capacity. These relate to:

- the distribution of compressive stresses in the concrete above the neutral axis;
- the magnitude of the extreme fibre concrete strain;
- the distribution of strains throughout the depth of the section; and
- the stress-strain relations for the prestressing steel and reinforcing steel.

We discuss these in turn.

Rectangular Stress Block

At high overload, the compressive concrete stress block above the neutral axis shown in Figure 6.1(c) is similar in shape to the stress-strain curve for concrete in compression. Numerous studies have shown that the calculated moment capacity M_u is not very sensitive to the assumed shape of the stress block. The Australian Concrete Structures Standard, AS 3600, allows the use of the equivalent rectangular stress block shown in Figure 6.3. A uniform stress of $\alpha_2 f_c'$ is taken to act over an area A' above the neutral axis, A' being defined by the edges of the cross-section and a line drawn parallel to, but somewhat above, the neutral axis of stress at depth d_n. The area A' extends to a depth γd_n, where γ is less than unity. It should be noted that in our analysis of moment capacity the term d_n will always refer to the depth to the neutral axis of stress at the instant when M_u acts.

The following values for α_2 and γ are specified in AS 3600 as:

$$\gamma = 0.97 - 0.0025 f_c' \quad \text{within the limit } \gamma \geq 0.67 \tag{6.3}$$

$$\alpha_2 = 0.85 - 0.0015 f_c' \quad \text{within the limit } \alpha_2 \geq 0.67 \tag{6.4}$$

The value for α_2 given by Equation 6.4 is reduced by 5 per cent for circular sections, and 10 percent for sections where the width of the stress block reduces between the neutral axis and the extreme compressive fibre, such as occurs for a triangular shaped section with its apex on the compressive side.

The lower limit of 0.67, placed on the stress block parameters γ and α_2, only becomes relevant for very high strength concrete. Values of α_2 and γ for standard concrete grades are given in Table 6.1.

TABLE 6.1 Rectangular stress block parameters for various grades of concrete

f_c'	25	32	40	50	65	80	100	120
γ	0.91	0.89	0.87	0.85	0.81	0.77	0.72	0.67
α_2	0.81	0.80	0.79	0.78	0.75	0.73	0.70	0.67

The rectangular stress block values in Table 6.1 reflect the manner in which the shape of the stress-strain curve for concrete changes with compressive strength. They have been chosen so that the rectangular stress block gives very nearly the same line of action (at depth d_c) for a given value of C_c as would be obtained from a more accurate curvilinear stress block analysis.

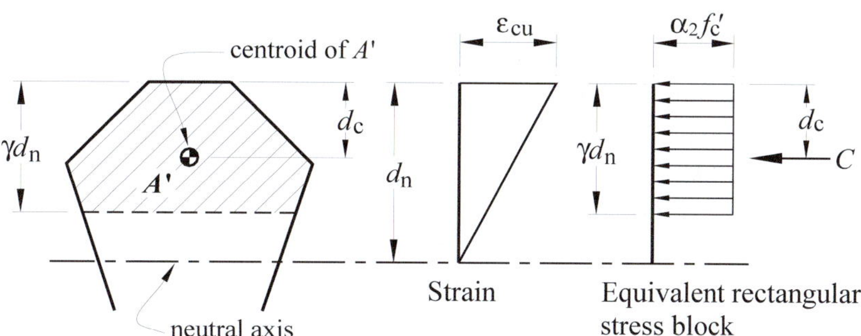

Figure 6.3 Equivalent rectangular compressive stress block

Using the rectangular stress block concept, we obtain the following expressions for the compressive force in the concrete, C_c, and its depth from the extreme fibre, d_c, in a beam with a rectangular cross-section (Figure 6.1):

$$C_c = \alpha_2 f_c' b \gamma d_n \tag{6.5}$$

$$d_c = 0.5 \gamma d_n \tag{6.6}$$

In the case of a non-rectangular cross-section (Figure 6.3), the rectangular stress block can lead to greater errors in the values of C_c and d_c than are obtained with a non-linear stress block. Nevertheless, the error is rarely significant when compared with errors introduced by other simplifying assumptions. The force C_c is taken to act at the centroid of area A' as defined above and its value is:

$$C_c = \alpha_2 f_c' A' \quad (6.7)$$

The rectangular stress block concept provides a simplified and rather idealised picture of the compressive concrete stresses above the neutral axis. It is commonly used in Australia and in the United States, whereas a slightly more complex and more realistic stress block is used in Europe and elsewhere.

Limiting Strain Criterion

In simple flexural strength theory it is assumed that the compressive strain in the concrete extreme fibre reaches the value ε_{cu} when the ultimate moment M_u acts. In AS 3600, the value for ε_{cu} is taken to be:

$$\varepsilon_{cu} = 0.003 \quad (6.8)$$

It should be noted that ε_{cu} represents stress-dependent strain: it does not include any inelastic creep and strain components that may exist in the extreme fibre at the time of loading.

The justification for the assumption that ε_{cu} is a constant is simply that, provided the section is reasonably ductile, the relationship between moment and extreme fibre strain is relatively flat in the vicinity of M_u. This means that the calculated moment capacity is not sensitive to the value chosen for ε_{cu}.

Linear distribution of strain in section

The concrete strains in the cross-section at ultimate moment are assumed to be linearly distributed as in Figure 6.1(b), so that strains at any level are completely defined by the top fibre strain ε_{cu}, together with the depth d_n to the neutral axis of strain. In simple strength calculations, the neutral axis of strain is also assumed to be at depth d_n from the extreme compressive fibre. In more accurate calculations where the effects of creep and shrinkage are considered, the neutral axis of strain, d_{ne}, must be distinguished from that for stress, d_n. Numerous calculations suggest, however, that for normal design no significant error in M_u will occur if creep and shrinkage strains are ignored, at least in the case of under-reinforced sections.

Although local irregularities develop at a cracked section as the result of slip of the steel relative to the concrete around the crack, the assumption of linear strain distribution does apply to average strains, and is sufficiently accurate for practical ultimate strength calculations. A no-slip condition is assumed to exist between the concrete and the bonded reinforcing steel and prestressing tendon.

Stress-strain relations for prestressing steel

In most practical situations the moment capacity M_u can be calculated with reasonable accuracy by replacing the actual stress-strain curve for the prestressing steel by an idealised elastic-plastic diagram as shown in Figure 2.1. It is reasonable to approximate the 'yield' stress of the prestressing steel as the 0.2 per cent offset stress, even though the latest, 2018 edition of AS 3600 specifies the more conservative value of the 0.1 per cent offset stress as a nominal measure of the yield stress. It also allows the yield stress to be assumed to be 0.82 times the characteristic minimum breaking stress, f_{pb}.

If flexural strength becomes a critical design consideration, a more accurate calculation can be obtained by using a bi-linear stress-strain relation for the prestressing steel, or by a formula fitted to test data. An example of such a formula is given in Devalapura and Tadros (1992). The calculation of M_u should then be based on trial strain distributions, as explained below in Section 6.7.

6.3 Rectangular section: calculation of ultimate moment

We first consider M_u for an under-reinforced rectangular beam section as in Figure 6.1, but without compressive steel reinforcement. The analysis also applies to a beam with a T-section as in Figure 6.6, provided the neutral axis depth d_n is not greater than the flange thickness t. We initially ignore the creep and shrinkage strains which exist in the section at the time of loading, and take ε_o to be equal to ε_{cu}, i.e. 0.003.

The tensile forces in the prestressing and reinforcing steels act at depths d_p and d_{st} respectively, and are assumed to be the yield values:

$$T_p = T_{py} = A_p f_{py} \quad (6.9)$$

$$T_s = T_{sy} = A_s f_{sy} \quad (6.10)$$

The compressive force C_c and its depth d_c are given by Equations 6.5 and 6.6.

To find the neutral axis depth d_n we equate tensile and compressive forces in the section, $C_c = T_p + T_s$, which gives:

$$d_n = \frac{T_p + T_s}{\alpha_2 f_c' \gamma b} \quad (6.11)$$

The moment capacity is obtained by taking moments about the top fibre:

$$M_u = T_p d_p + T_s d_s - C_c d_c \quad (6.12)$$

For any under-reinforced section, a good approximation to M_u can be obtained by assuming that the resultant compressive force acts at depth $0.15d_s$ below the top fibre, so that, by taking moments about the assumed line of action of the resultant compressive force:

$$M_u = T_p (d_p - 0.15d_s) + T_s (0.85d_s) \quad (6.13)$$

where T_{py} and T_{sy} are calculated from Equations 6.9 and 6.10, respectively. This approximate expression gives good results for under-reinforced sections because the value of M_u is not very sensitive to the location of the centroid of the compressive stress block.

A check must be made on the assumption of steel yield. This is done by calculating the strains in both the reinforcing steel and prestressing steel. The procedure is similar to the one used in Chapter 5 in the analysis of stresses in a cracked section. The convention used in Figure 6.4 to represent strains in the tendon and adjacent concrete is also the same as was used in Chapter 5.

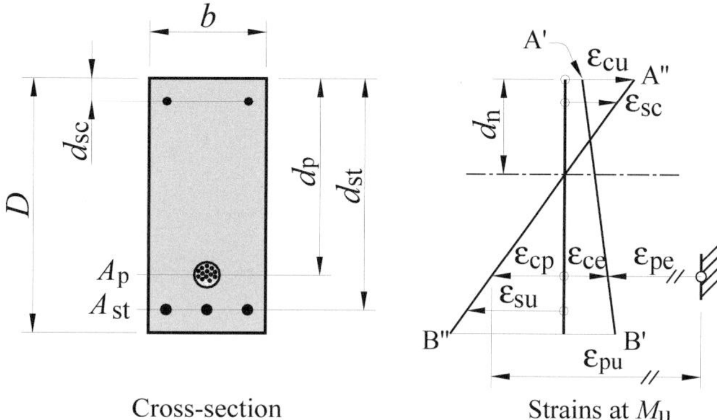

Figure 6.4 Strains at M_u

The strain in the reinforcing steel is equal to the strain in the adjacent concrete, but this is not true of the prestressing steel because of its initial prestrain. To determine the total strain in the prestressing steel at M_u, two steps are needed.

We first consider the strains due to prestress only, with no bending moment acting on the section and ignoring prior creep and shrinkage. The concrete strains at this stage are represented by line A'B' in Figure 6.4. At the level of the prestressing tendon, the compressive concrete strain is ε_{ce}, which is readily calculated from the section details:

$$\varepsilon_{ce} = \frac{\sigma_{ce}}{E_c} = \frac{1}{E_c}\left[\frac{P_e}{A_g} + \frac{P_e e^2}{I_g}\right] \tag{6.14}$$

At the same stage the tendon has a tensile strain $\varepsilon_{pe} = P_e/(A_p E_p)$.

We next consider conditions at M_u, remembering that after prestressing the tendon was bonded to the concrete, so that for any increment of strain that occurs in the concrete adjacent to the tendon, due to bending moment, the same increment of strain must also occur in the tendon.

In Figure 6.4, the concrete strain at M_u is represented by line A"B". At the tendon level, the concrete strain has changed from the compressive strain ε_{ce}

to a tensile strain ε_{cp}, so that the concrete has undergone a tensile strain increment equal to the absolute sum of ε_{ce} and ε_{cp}. The tensile strain in the prestressing steel has therefore also increased by this amount, and the total strain in the prestressing tendon is:

$$\varepsilon_{pu} = \varepsilon_{pe} + \varepsilon_{ce} + \varepsilon_{cu}(d_p - d_n)/d_n \tag{6.15}$$

The strain in the tensile steel is:

$$\varepsilon_{st} = \varepsilon_{cu}(d_{st} - d_n)/d_n \tag{6.16}$$

These strains must be greater than ε_{py} and ε_{sy} respectively, for the analysis to be applicable. If either yield strain is not reached, then the tensile force(s) is not given by Equations 6.9 and/or 6.10. The strain compatibility analysis, which is described below in Section 6.7, can then be used to determine M_u.

Allowance for Prior Creep and Shrinkage

Inelastic strains in the cross-section have been ignored in Figure 6.4 and in Equations 6.15 and 6.16. No serious error is to be expected from this simplification, particularly if the prestressing and reinforcing steels are at yield. If desired, account can be taken of creep and shrinkage strains that have developed in the section up to the time of loading using the method outlined in Appendix C.

EXAMPLE 6.1

Moment Capacity for a Rectangular Section

Calculate the moment capacity for the beam section shown in Figure 6.5, which is the same as that analysed in Example 5.3 to undertake an elastic cracked section analysis.

The section has 3-N24 reinforcing bars in the tensile region of area A_{st} = 1350 mm^2 and a prestressing tendon of area A_p = 405 mm^2. The effective prestress is σ_{pe} = 1100 MPa and the yield stress of the tendon is $f_{py} = 0.82 f_{pb}$ = 1533 MPa.

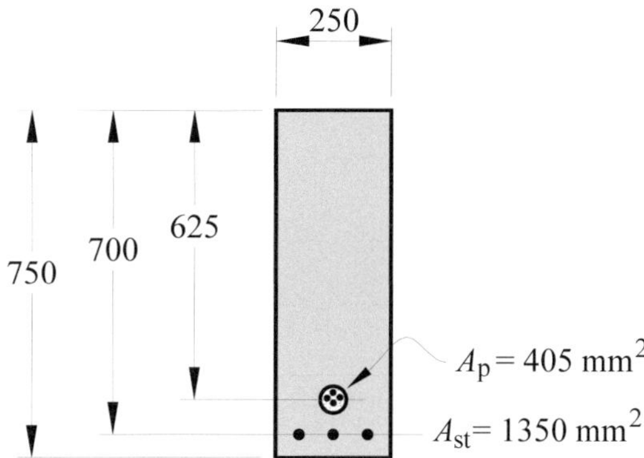

Figure 6.5 Cross-section for Example 6.1

Other data:

$$f_c' = 40 \text{ MPa} ; \alpha_2 = 0.79; \gamma = 0.87; \ E_p = E_s = 200\times10^3 \text{ MPa}$$

SOLUTION

Calculate neutral axis depth:

Assuming, subject to check, that both steels yield:

$$T_p = T_{py} = 405 \times 1533 = 621\times10^3 \text{ N}$$

$$T_s = T_{sy} = 1350 \times 500 = 675\times10^3 \text{ N}$$

Concrete force:

$$C = \alpha_2 f_c' b\gamma d_n = 0.79 \times 40 \times 250 \times 0.87 d_n = 6873 d_n$$

For equilibrium $\Sigma T = \Sigma C$ and solving for d_n gives:

$$d_n = \frac{(621 + 675)\times10^3}{6873} = 189 \text{ mm}$$

Check the steel yield assumptions:

From Example 5.3, $\varepsilon_{ce} = 0.00017$, $\varepsilon_{pe} = 0.00550$ and from the strain diagram:

$$\varepsilon_{cp} = 0.003\left(\frac{625 - 189}{189}\right) = 0.00692$$

The strain in the tendon is thus:

$$\varepsilon_{pu} = \varepsilon_{ce} + \varepsilon_{pe} + \varepsilon_{cp} = 0.00017 + 0.00550 + 0.00692 = 0.01259$$

and this is compared to the yield strain of the tendon of:

$$\varepsilon_{py} = f_{py}/E_p = 1533/200\times10^3 = 0.00767$$

As $\varepsilon_{pu} > \varepsilon_{py}$ our assumption of yielding of the tendon is correct.

The strain in the reinforcing steel is:

$$\varepsilon_{st} = \varepsilon_{cu}(d_{st} - d_n)/d_n = 0.003(700 - 189)/189 = 0.00811$$

and as $\varepsilon_{st} > \varepsilon_{sy} = 0.0025$, this steel is also at yield. Thus, as our assumption of both steels yielding is correct, our calculation for d_n is also correct.

Calculate the effective depth:

The effective depth is calculated by taking the first moment of the forces of the tensile steel and tendon about the extreme compressive fibre:

$$d = \frac{T_p \times d_p + T_s \times d_{st}}{T_p + T_s} = \frac{621 \times 625 + 675 \times 700}{621 + 675} = 664 \text{ mm}$$

Calculate the ultimate moment, M_u:

The force in the concrete is obtained from equilibrium as:

$$C = 6873 \times 189 = 1299\times10^3 \text{ N}$$

and a check on equilibrium shows that $C - T_p - T_s = 0$.

The centroid of C is located at a depth:

$$d_c = 0.5\gamma d_n = 0.5 \times 0.87 \times 187 = 81.3 \text{ mm}.$$

We calculate the moment capacity M_u by taking moments about the centroid of the tensile forces:

$$\begin{aligned} M_u = C(d - d_c) &= 1299 \times 10^3 (664 - 81.3) \\ &= 757 \times 10^6 \text{ Nmm (757 kNm)} \end{aligned}$$

Check section ductility:

We shall check the section against the AS 3600 ductility measure of $k_{uo} \le 0.36$, which is explained below in Section 6.9.2.

The neutral axis depth parameter at the ultimate state is thus:

$$k_{uo} = d_n / d = 189/664 = 0.28 < 0.36$$

and the section is therefore okay.

EXAMPLE 6.2

Approximate Estimate of Moment Capacity

In this example we obtain a quick estimate of the ultimate moment capacity M_u for the section of Example 6.1.

Solution

Assuming that both steels are beyond yield then by Equation 6.13:

$$\begin{aligned} M_u &= 621 \times 10^3 \times (625 - 0.15 \times 700) + 675 \times 10^3 \times 0.85 \times 700 \\ &= 724 \times 10^6 \text{ Nmm (724 kNm)} \end{aligned}$$

This value is within 5 per cent of that obtained in the previous example and is sufficiently accurate for preliminary design calculations, provided that it is later shown that the assumption that all steel has yielded is correct.

6.4 T- and I-sections: calculation of ultimate moment

The expressions for M_u in Equations 6.12 and 6.13 apply also to flanged sections provided that the stress block depth, γd_n is less than the flange thickness, t. If γd_n is greater than t, the non-uniform width of the compressive stress block must be considered. In the following we ignore the inelastic strains in the section, so that the neutral axes of stress and strain are assumed to be the same. The procedure for including the inelastic strains is described in Appendix C.

For the T-section shown in Figure 6.6, we divide the total compressive concrete force into a component C_f in the outstanding flanges of area $(b - b_w)\,t$, and a component C_w in the web, of width b_w:

$$C_f = \alpha_2 f_c' (b - b_w) t \qquad (6.17)$$

$$C_w = \alpha_2 f_c' b_w \gamma d_n \qquad (6.18)$$

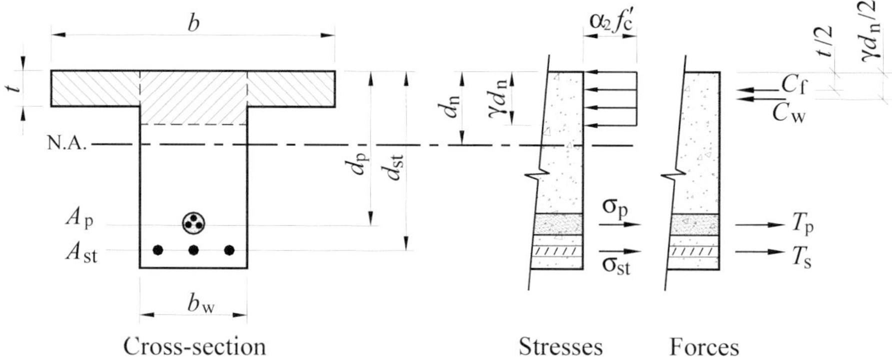

Figure 6.6 Flanged section, internal forces at M_u

For equilibrium:

$$C_f + C_w = T_p + T_s \tag{6.19}$$

Assuming all the steel to be at yield, so that Equations 6.9 and 6.10 apply, we obtain from Equations 6.17 to 6.19:

$$d_n = \frac{1}{\alpha_2 f_c' \gamma b_w} [A_p f_{py} + A_{st} f_{sy} - \alpha_2 f_c' (b - b_w) t] \tag{6.20}$$

and hence:

$$M_u = T_p d_p + T_s d_{st} - C_f \frac{t}{2} - C_w \frac{\gamma d_n}{2} \tag{6.21}$$

To check that the steels are in fact at yield, Equations 6.15 and 6.16 are used to calculate ε_{pu} and ε_{st}, respectively.

EXAMPLE 6.3

MOMENT CAPACITY OF A T-SECTION – NEUTRAL AXIS IN FLANGE

Determine M_u for the T-section beam shown in Figure 6.7. The section contains 8-N32 bars, giving $A_s = 6400$ mm^2, and $A_p = 2800$ mm^2 at the depths shown. The effective prestress in the tendon is $\sigma_{pe} = 1100$ MPa.

Other data:

$$f_c' = 40 \text{ MPa}; \ \alpha_2 = 0.79; \ \gamma = 0.87$$

$$E_p = E_s = 200 \times 10^3 \text{ MPa}; \ f_{py} = 1750 \text{ MPa}; \ f_{sy} = 500 \text{ MPa}$$

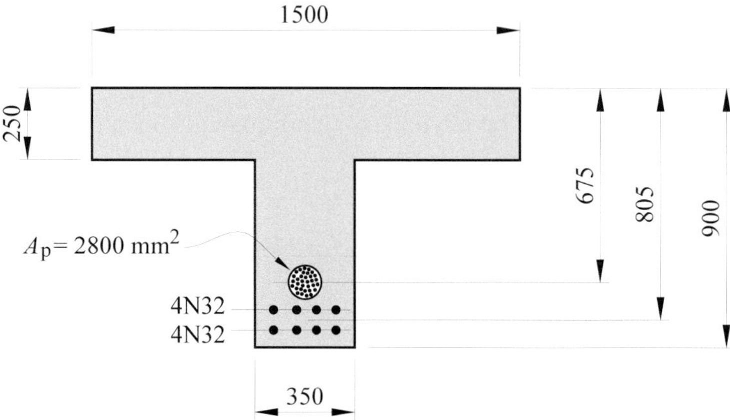

Figure 6.7 T-beam section for Example 6.3

SOLUTION

Depth to neutral axis:

We assume, subject to later check, that the neutral axis lies in the flange, and that both steels are at yield. The forces on the section are then:

$$T_p = 2800 \times 1750 = 4900 \times 10^3 \text{ N} \quad (4900 \text{ kN})$$

$$T_s = 6400 \times 500 = 3200 \times 10^3 \text{ N} \quad (3200 \text{ kN})$$

$$C = 0.79 \times 40 \times 1500 \times 0.87 d_n = 41.25 \times 10^3 d_n \text{ N}$$

Setting $C = T_p + T_s = 8100 \times 10^3$ N we obtain $d_n = 196$ mm and the depth of the stress block is then $\gamma d_n = 159$ mm, which is less than the depth of the flange. The strains in the concrete at the level of the prestressing steel is:

$$\varepsilon_{cp} = 0.003 \times \frac{675 - 196(206)}{196} = 0.00733$$

The initial strain in the prestressing steel is:

$$\varepsilon_{pe} = \sigma_{pe}/E_p = 1100/200\times10^3 = 0.00550$$

The total strain in the tendon at ultimate is thus:

$$\varepsilon_{pu} = 0.00733 + 0.00550 + \varepsilon_{ce} = 0.0128 + \varepsilon_{ce}$$

which exceeds the yield strain of $\varepsilon_{py} = f_{py}/E_p = 0.00875$.

The strain in the reinforcing steel is:

$$\varepsilon_{st} = 0.003 \times \frac{805 - 196}{196} = 0.00932$$

which is also well in excess of yield; thus our initial assumptions are valid.

Check section ductility:

$$d = \frac{T_p \times d_p + T_s \times d_{st}}{T_p + T_s} = \frac{4900 \times 675 + 3200 \times 805}{4900 + 3200} = 726 \text{ mm}$$

$$k_{uo} = d_n/d = 196/726 = 0.27 \quad (< 0.36 \therefore \text{OK})$$

Calculate M_u:

Taking moments about the top fibre:

$$M_u = T_p d_p + T_s d_{st} - C d_c$$

$$= 4900 \times 0.675 + 3200 \times 0.805 - 8100\,(0.5 \times 0.87 \times 0.196)$$

$$= 5193 \text{ kNm}$$

EXAMPLE 6.4

MOMENT CAPACITY OF A T-SECTION – NEUTRAL AXIS IN WEB

The moment capacity is to be determined for a T-section similar to that in Example 6.3, except that the flange depth is $t = 160$ mm.

SOLUTION

Depth to neutral axis:

We assume, subject to later check, that the neutral axis lies in the flange, and that both steels are at yield. The forces in the section are then:

$$T_p = 2800 \times 1750 = 4900{\times}10^3 \text{ N} \quad (4900 \text{ kN})$$

$$T_s = 6400 \times 500 = 3200{\times}10^3 \text{ N} \quad (3200 \text{ kN})$$

$$C = 0.79 \times 40 \times 1500 \times 0.87 d_n = 41.24{\times}10^3 d_n \text{ N}$$

From the previous example, it is determined that the stress block γd_n lies in the web. As previously, assuming both the bars and tendon to be at yield, T_p = 4900 kN and T_s = 3200 kN.

The compressive force in the outstanding flanges is:

$$C_f = 0.79 \times 40(1500 - 350) \times 160 = 5814{\times}10^3 \text{ N} \quad (5814 \text{ kN})$$

and the compressive force in the web is:

$$C_w = 0.79 \times 40 \times 350 \times 0.87 d_n = 9622 d_n \text{ N} \quad (9.622 d_n \text{ kN})$$

From equilibrium we have $C_w = T_p + T_s - C_f$, so that:

$$C_w = 4900 + 3200 - 5814 = 2286 = 9.622 d_n$$

Thus $d_n = 2286/9.622 = 238$mm ($\gamma d_n = 206$ mm).

A check of strains similar to that of the previous example shows that both steels are in fact at yield.

Check section ductility:

As for Example 6.3, the effective depth is $d = 726$ mm and:

$$k_{uo} = d_n/d = 238/726 = 0.33 \quad (< 0.36 \therefore \text{OK})$$

Calculate M_u:

The ultimate moment is then calculated by taking moments about the top fibre:

$$\begin{aligned} M_u &= 4900 \times 0.675 + 3200 \times 0.805 - 5814 \times 0.160/2 \\ &\qquad - 2286 \times (0.5 \times 0.87 \times 0.238) \\ &= 5182 \text{ kNm} \end{aligned}$$

Discussion:

In comparing the result for M_u of the section of this example with that of the previous one, the moment capacities are seen to be similar. This is because, while in this example some of the web below the flange is in compression, the bulk of the compressive force remains in the flange. The centroid of the forces is then largely unchanged from that of the previous example and, thus, the length of the internal lever arm is largely unchanged. As both the bars and tendon have yielded, the total tensile, and thus compressive, force is unchanged and indeed the moment capacities of the sections are almost equal.

This is the case for most under-reinforced sections, provided that all the tensile reinforcement (tendons and bars) are at or beyond yield. Using the approximate method of Equation 6.13, the moment capacity is estimated as $M_u \approx 4910$ kNm, just six per cent lower than the more accurate calculations.

6.5 Moment capacity with some steel not at yield

As already discussed in relation to Figure 6.2, tendons and reinforcing bars placed near the mid-depth of the cross-section will not reach yield when M_u acts. Various other situations arise that result in some steel not reaching yield. For example, pretensioned beams sometimes have some strands near the top of the section in order to control cracking in the upper fibres at the ends of the beam where there is little compressive stress due to external loading. Figure 6.8 below shows such a section. The top strands are bonded to compressive concrete and therefore their initial tensile stress decreases as the bending moment increases, and will not therefore be at yield at M_u.

Ultimate moment calculations in such situations should be undertaken on a case-by-case basis, as in Example 6.5 below, or by means of the general trial and error procedure which will be described in Section 6.7.

EXAMPLE 6.5

MOMENT CAPACITY FOR A SECTION WITH SOME STEEL NOT AT YIELD

Calculate M_u for the pretensioned section shown in Figure 6.8, which has prestressing strands in both the top and the bottom of the section. At the top there are two 12.7 mm strands ($A_{p1} = 200$ mm^2) and at the bottom five 15.2 mm strands ($A_{p2} = 716$ mm^2). The effective prestress σ_{pe} in all strands is 1000 MPa tension.

Other data:

$$f_c' = 40 \text{ MPa}; \ \alpha_2 = 0.79; \ \gamma = 0.87; \ E_c = 32 \times 10^3 \text{ MPa}$$

$$E_p = 200 \times 10^3 \text{ MPa}; \ f_{py} = 1700 \text{ MPa}$$

$$A_g = 180 \times 10^3 \text{ mm}^2; \ I_g = 5.4 \times 10^9 \text{ mm}^4$$

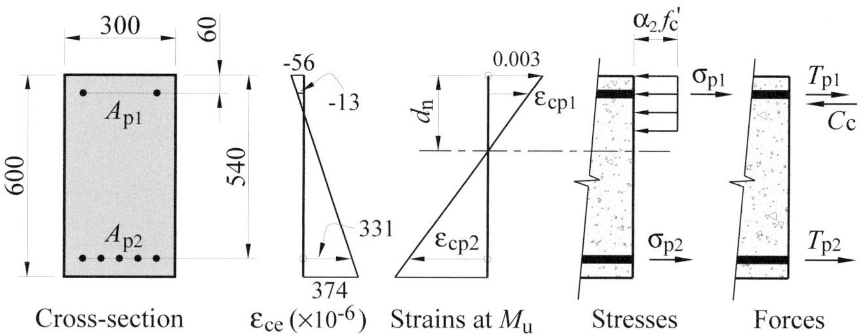

Figure 6.8 Pretensioned section for Example 6.5

SOLUTION

Strains due to prestress:

The top strands have a prestressing force P_{e1} = 200 kN, the bottom strands have P_{e2} = 716 kN. The resultant prestressing force is 916 kN acting at a distance d_{pr} from the top of the section, where:

$$d_{pr} = \frac{(200 \times 60) + (716 \times 540)}{916} = 435.2 \text{ mm}$$

The eccentricity of prestress is 600/2 – 435 = 135.2 mm, and the concrete stress and strain due to prestress at distance y from the centroidal axis are:

$$\sigma_{ce} = \frac{916 \times 10^3}{180 \times 10^3} - \left(\frac{916 \times 10^3 \times 135.2}{5.4 \times 10^9} \right) y = 5.09 - 0.02293 y \text{ MPa}$$

$$\varepsilon_{ce} = \frac{\sigma_{ce}}{E_c} = (159.1 - 0.7166 y) \times 10^{-6}$$

The concrete strains due to prestress at the two steel levels are:

Top: y = 240 mm; $\varepsilon_{ce1} = -13 \times 10^{-6}$ (tensile)

Bottom: $y = -240$ mm; $\varepsilon_{ce2} = 331\times10^{-6}$ (compressive)

Steel strains due to effective prestress are:

$$\varepsilon_{pe1} = \varepsilon_{pe2} = \sigma_{pe}/E_p = 1000/200\times10^3 = 0.00500 \quad \text{(tensile)}$$

Concrete strains at M_u:

From the strain diagram, strains at the two steel levels are:

Top: $$\varepsilon_{cp1} = 0.003\left(\frac{d_n - 60}{d_n}\right) \quad \text{(compressive)}$$

Bottom: $$\varepsilon_{cp2} = 0.003\left(\frac{540 - d_n}{d_n}\right) \quad \text{(tensile)}$$

The changes in concrete strain from $M = 0$ to $M = M_u$ are:

Top steel increment:

$$\Delta\varepsilon_1 = -13\times10^{-6} + 0.003\left(\frac{d_n - 60}{d_n}\right)$$

Bottom steel increment:

$$\Delta\varepsilon_2 = +331\times10^{-6} - 0.003\left(\frac{540 - d_n}{d_n}\right)$$

The total tensile strains at M_u are $\varepsilon_{pu} = \varepsilon_{pe} + \Delta\varepsilon = 0.0050 + \Delta\varepsilon$, giving:

Top steel: $$\varepsilon_{pu1} = 0.00499 - 0.003\left(\frac{d_n - 60}{d_n}\right)$$

Bottom steel: $$\varepsilon_{pu2} = 0.00533 + 0.003\left(\frac{540 - d_n}{d_n}\right)$$

Steel stresses at M_u:

It is reasonable to assume, subject to later check, that the bottom steel has yielded. However the tensile stress in the top steel decreases with bending moment since it is bonded to the compressive concrete and, therefore, remains in the elastic range at M_u.

$$\text{Top steel:} \quad \sigma_{pu1} = E_p \varepsilon_{pu1} = 998.0 - 600.0\left(\frac{d_n - 60}{d_n}\right) \text{ MPa}$$

$$T_{p1} = A_{p1}\sigma_{pu1} = 199.6{\times}10^3 - 120.0{\times}10^3\left(\frac{d_n - 60}{d_n}\right)$$

$$= 79.6{\times}10^3 + 7.20{\times}10^6/d_n \quad \text{N}$$

$$\text{Bot. steel:} \quad T_{p2} = A_{p2} f_{py} = 716 \times 1700 = 1217{\times}10^3 \text{ N}$$

$$\text{Concrete:} \quad C = \alpha_2 f_c' \gamma d_n b = 0.79 \times 40 \times 0.87 d_n \times 300$$

$$= 8248 d_n \text{ N}$$

Neutral axis location:

The neutral axis depth d_n is found using the equilibrium condition $C - T_{p1} - T_{p2} = 0$. Multiplying by d_n we have:

$$8248 d_n^2 - 79.6{\times}10^3 d_n - 7.20{\times}10^6 - 1217{\times}10^3 d_n = 0$$

$$d_n^2 - 157.2 d_n - 872.9 = 0$$

Solving the quadratic gives $d_n = 162.6$ mm.

Check steel yield assumptions:

The strain in the top strands is:

$$\varepsilon_{pu1} = 0.00499 - 0.003\left(\frac{162.6 - 60}{162.6}\right)$$

$$= 0.00310 < \varepsilon_{py} = f_{py}/E_p = 0.00850$$

and in the bottom strands:

$$\varepsilon_{pu2} = 0.00533 + 0.003\left(\frac{540 - 162.6}{162.6}\right) = 0.00123 > \varepsilon_{py}$$

Thus the strain assumptions were correct.

Check section ductility:

$$k_{uo} = d_n / d = 162.6/540 = 0.30 \quad (< 0.36 \therefore \text{OK})$$

Moment capacity:

The stress in the top strands is:

$$\sigma_{pu1} = \varepsilon_{pu1} E_p = 0.00310 \times 200 \times 10^3 = 620 \text{ MPa}$$

and the tension forces are:

$$T_{p1} = A_{p1}\sigma_{p1} = 200 \times 620 \times 10^{-3} = 124 \text{ kN} \quad \text{(tension)}$$

$$T_{p2} = 1217 \text{ kN} \quad \text{(tension)}$$

For the concrete:

$$C = 8248 \times 162.3 \times 10^{-3} = 1339 \text{ kN} \quad \text{(compression)}$$

A check on equilibrium shows that $\Sigma F = 1217 + 124 - 1341 = 0$ kN.

Taking moments about the top fibre gives:

$$M_u = 1217 \times 0.540 + 122 \times 0.060 - 1341\left(\frac{0.87 \times 0.1626}{2}\right)$$

$$= 570 \text{ kNm}$$

6.6 Effect of incomplete bond

In the methods of analysis considered so far, it has been assumed that full bond exists between the tendons and the surrounding concrete. Thus, any change in concrete strain at the steel level has been assumed to occur also in the tendon. The assumption of full bond is appropriate for pretensioned construction using strand or indented wires and also for properly grouted post-tensioned construction. However, it would be inappropriate in the case of ungrouted post-tensioned construction and also for beams with external unbonded prestressing tendons.

In the absence of adequate bond, slip occurs between tendon and concrete, and the strains in the prestressing strand tend to equalise over significant lengths of the beam, thus reducing the peak stress levels that would otherwise occur at cracks. The increment in stress in the prestressing steel does not then follow, even approximately, the distribution of bending moment along the beam.

In the extreme case of zero bond and zero friction depicted in Figure 6.9, the steel stress is constant over the full length of the beam. It would then be reasonable to assume that the stress in the prestressing steel remains at its initial value, ε_{pe}, even when M_u acts. In reality, considerable friction develops between the steel and the duct, so that some concentration of stress does occur near the crack.

In an unbonded member, the local increase in strain in the prestressing steel in a peak moment region due to externally applied load will be considerably less than the value ($\varepsilon_{ce} + \varepsilon_{cp}$) that applies to a bonded member at M_u. The additional strain for the unbonded steel may be written as:

$$\Delta\varepsilon_p = \alpha\, \varepsilon_{cp} \tag{6.22}$$

The value of the reduction factor α is less than 1.0, and depends on a wide range of variables, such as the cable profile along the beam, the shape of the bending moment diagram, and the amount of friction between tendon and duct. Accurate values for α are difficult to obtain. Even when careful tests are carried out, the values obtained for α cannot confidently be extrapolated to similar, but different, structural systems.

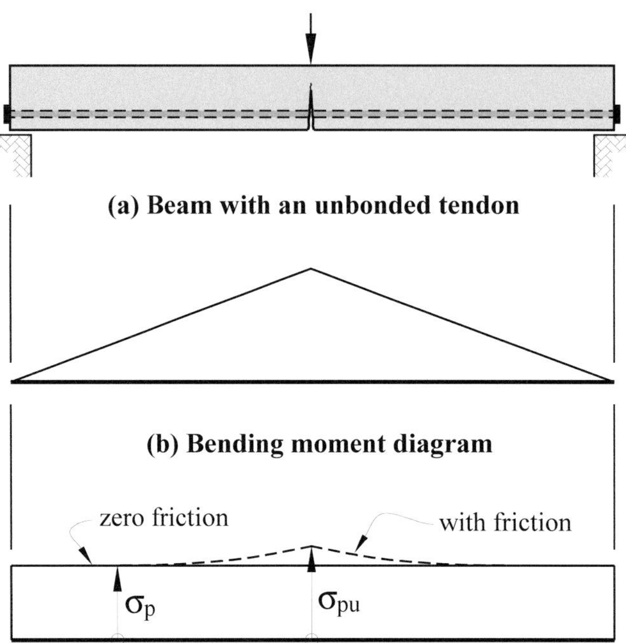

Figure 6.9 Beam with unbonded tendon

Various conservative approximations have therefore been introduced to allow the tendon stress at ultimate moment to be estimated. In AS 3600, the following expressions are given for the stress in unbonded tendons:

For a member with a span-to-depth ratio of up to 35:

$$\sigma_{pu} = \sigma_{pe} + 70 + \frac{f_c' \, b_{ef} \, d_p}{100 \, A_{pt}} \leq \sigma_{pe} + 400 \tag{6.23}$$

For a member with a span-to-depth ratio exceeding 35:

$$\sigma_{pu} = \sigma_{pe} + 70 + \frac{f_c' \, b_{ef} \, d_p}{300 \, A_{pt}} \leq \sigma_{pe} + 200 \tag{6.24}$$

Both values are also limited to $\sigma_{pu} \leq f_{py}$

The term A_{pt} refers to the area of the prestressing steel that is in tension under ultimate strength conditions and normally will be equal to A_p.

The provision of normal reinforcing steel in a member with unbonded prestressing tendons improves the cracking pattern at overload, and restrains the widening of the cracks. The ultimate moment for such a member may be estimated by assuming that the reinforcing steel is at yield and that the unbonded tendons are stressed to the level given by Equations 6.23 and 6.24.

6.7 General analysis by trial strain distributions

In the case of an irregular cross-section, or one with steel located at many different levels (Figure 6.2), it is not practical to derive equations for the ultimate moment. An iterative calculation procedure can be used.

We again ignore the inelastic section strains in the following, but these can be allowed for in the manner explained in Appendix C. A trial neutral axis depth d_n is first chosen. This, together with the extreme compressive fibre strain, $\varepsilon_{cu} = 0.003$, defines the strains throughout the section, so that the strains in all steel elements can be evaluated, and thereby the steel forces.

Considering the j-th reinforcing element at depth d_{sj}, we calculate its total tensile strain by application of Equation 6.16 as:

$$\varepsilon_{sj} = \varepsilon_u \cdot \frac{d_{sj} - d_n}{d_{sj}} \tag{6.25}$$

The stress σ_{sj} corresponding to e_{sj} is obtained from the stress-strain relationship for the steel. If $\varepsilon_{sj} > \varepsilon_{sy}$ then $\sigma_{sj} = f_{sy}$. If ε_{sj} is negative, then the stress is compressive. The force in the j-th reinforcing element, T_{sj}, of area A_{sj} is determined as:

$$T_{sj} = \sigma_{sj} A_{sj} \tag{6.26}$$

At the level d_{pk} of the k-th prestressing tendon, of area A_{pk}, the initial elastic compressive strain in the concrete is ε_{cek} and the total strain in the tendon is:

$$\varepsilon_{pk} = \varepsilon_{pek} + \varepsilon_{cek} + \varepsilon_{cu} \cdot \frac{d_{pk} - d_n}{d_n} \tag{6.27}$$

in which ε_{pek} is the initial strain in the tendon prior to the application of moment. The stress σ_{puk} is determined from the stress-strain relationship, and the force T_{pk} is thus:

$$T_{pk} = \sigma_{puk} A_{pk} \tag{6.28}$$

If the concrete compressive zone is rectangular, then C_c and d_c can be determined from the stress block (Equations 6.5 and 6.6). For a flanged section, compressive forces in the flange and web areas can be determined from Equations 6.17 and 6.18.

For more complex sections, such as the one shown in Figure 6.10, the concrete compressive area can be divided into a number of regions and using the force in the i-th region, C_i, and its corresponding depth d_{ci}. The requirement of horizontal equilibrium then becomes:

$$\sum C_i = \sum T_{sj} + \sum T_{pk} \tag{6.29}$$

This expression is used to check the assumed neutral axis position. If the resultant compressive force is greater than the tensile force then d_n must be decreased, and vice versa. Adjustments are made to d_n until Equation 6.29 is satisfied. The moment capacity is then determined by taking moments about the top fibre:

$$M_u = \sum T_{sj} d_{sj} + \sum T_{pk} d_{pk} - \sum C_i C d_{ci} \tag{6.30}$$

EXAMPLE 6.6

M_U CALCULATION BY TRIAL STRAIN DISTRIBUTIONS

Calculate the moment capacity for the I-section girder shown in Figure 6.10 where:

bars: A_{s1} =3720 mm^2; $A_{s2} = A_{s3} = 400$ mm^2; $A_{s4} = 1800$ mm^2

tendons: $A_{p1} = 1200$ mm^2; $A_{p2} = 500$ mm^2; $\sigma_{pe} = 1100$ MPa

concrete: $f_c' = 35$ MPa ($\gamma = 0.88$; $\alpha_2 = 0.80$)

Other Data:

$$E_c = 32\times10^3 \text{ MPa}; \; E_p = E_s = 200\times10^3 \text{ MPa};$$

$$f_{py} = 1750 \text{ MPa}; \; f_{sy} = 500 \text{ MPa}$$

$$I_g = 75.2\times10^9 \text{ mm}^4; \; A_g = 375\times10^3 \text{ mm}^2; \; y_t = y_b = 600 \text{ mm}$$

SOLUTION

Concrete stresses and strains due to prestress:

The forces in the tendons due to the effective prestress are:

$$P_{e1} = A_{p1}\sigma_{pe} = 1200 \times 1100\times10^{-3} = 1320 \text{ kN}$$

$$P_{e2} = A_{p2}\sigma_{pe} = 500 \times 1100\times10^{-3} = 550 \text{ kN}$$

Total: $P_e = 1870$ kN

and the effective depth of the equivalent prestressing force and its eccentricity are:

$$d_p = \frac{(1320 \times 1060) + (550 \times 800)}{1870} = 984 \text{ mm}$$

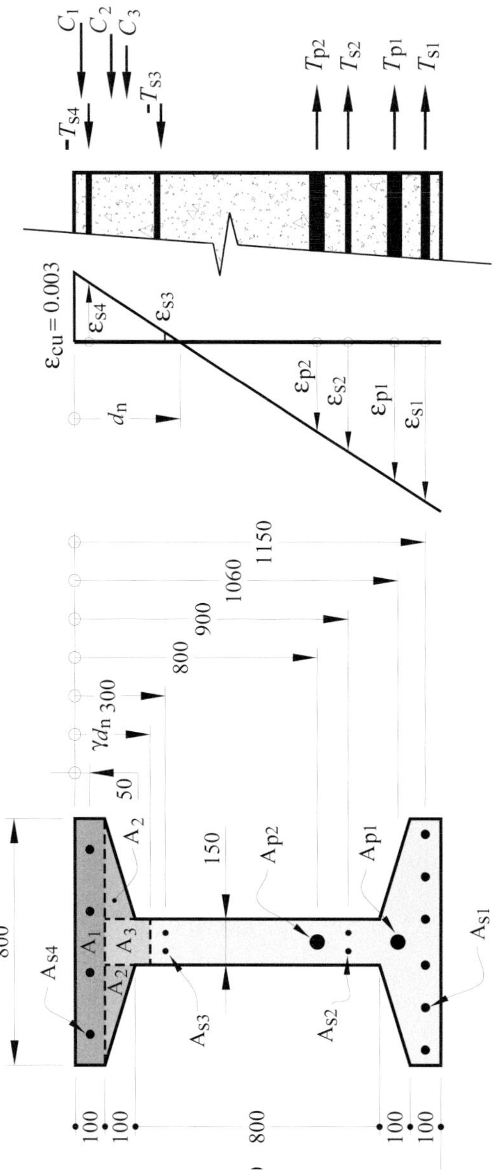

Figure 6.10 Cross-section for Example 6.6

$$e = d_\text{p} - y_t = 984 - 600 = 384 \text{ mm}$$

The steel strain ε_pe due to P_e is:

$$\varepsilon_\text{pe} = \sigma_\text{pe}/E_\text{p} = 1100/200\times10^3 = 0.0055$$

For multiple tendons in a section, the concrete stresses due to prestress P_e are calculated from:

$$\sigma_\text{ce} = \frac{P_\text{e}}{A_\text{g}} + \frac{P_\text{e}\times e\times(d_\text{p} - y_\text{t})}{I_\text{g}}$$

and, thus, at the level of A_p1 the concrete stress and associated strain are:

$$\sigma_\text{ce1} = \frac{1870\times10^3}{375\times10^3} + \frac{1870\times10^3\times384\times(1060-600)}{75.2\times10^9} = 9.38 \text{ MPa}$$

$$\varepsilon_\text{ce1} = \frac{9.38}{32.0\times10^3} = 0.00029$$

$$\varepsilon_\text{pe} + \varepsilon_\text{ce1} = 0.00550 + 0.00029 = 0.00579$$

At the level of A_p2:

$$\sigma_\text{ce1} = \frac{1870\times10^3}{375\times10^3} + \frac{1870\times10^3\times384\times(800-600)}{75.2\times10^9} = 6.90 \text{ MPa}$$

$$\varepsilon_\text{ce1} = \frac{6.90}{32.0\times10^3} = 0.00022$$

$$\varepsilon_\text{pe} + \varepsilon_\text{ce2} = 0.00550 + 0.00022 = 0.00572$$

Forces at M_u:

The concrete compressive area is divided into three parts as shown in Figure 6.10. We shall assume, subject to a later check, that the stress block depth γd_n exceeds 200 mm; the compressive forces are:

$$C_1 = 0.80 \times 35 \times 800 \times 100 = 2240 \times 10^3 \text{ N}$$

$$C_2 = 0.80 \times 35 \times 325 \times 100 = 910 \times 10^3 \text{ N}$$

$$C_3 = 0.80 \times 35 \times 150(\gamma d_n - 100) = 4400(0.88 d_n - 100) \text{ N}$$

In dealing with the reinforcing steel bars that lie above the neutral axis, i.e. the compressive bars, the forces are treated as negative (tension) in the following equations. The force at the level of A_{s1} is:

$$T_{s1} = A_{s1} \times \varepsilon_{cu} \left(\frac{d_{s1} - d_n}{d_n} \right) E_p$$

$$\text{... within the limits } -A_{s1} f_{sy} \leq T_{s1} \leq A_{s1} f_{sy}$$

$$= 2232 \times 10^3 \left(\frac{1150 - d_n}{d_n} \right) \text{ N}$$

$$\text{... within the limits } \quad -1860 \times 10^3 \leq T_{s1} \leq 1860 \times 10^3 \text{ N}$$

and similarly:

$$T_{s2} = 240 \times 10^3 (900 - d_n)/d_n \text{ N} : \; -200 \times 10^3 \leq T_{s2} \leq 200 \times 10^3 \text{ N}$$

$$T_{s3} = 240 \times 10^3 (300 - d_n)/d_n \text{ N} : \; -200 \times 10^3 \leq T_{s3} \leq 200 \times 10^3 \text{ N}$$

$$T_{s4} = 1080 \times 10^3 (50 - d_n)/d_n \text{ N} : \; -900 \times 10^3 \leq T_{s4} \leq 900 \times 10^3 \text{ N}$$

In the above equations for T_{s1} to T_{s4}, both the negative (compressive) force and positive (tensile) force limits are included in each case. For bars that lie below the neutral axis, only the tensile limit is of significance, whereas above the neutral axis only the negative limit is strictly needed. As no assumptions are made in the following analysis as to the location of the neutral axis, both limits are maintained in this example.

For the tendons, the forces are:

$$T_{p1} = A_{p1} \times \left(\varepsilon_{pe} + \varepsilon_{ce} + \varepsilon_{cu} \left(\frac{d_{p1} - d_n}{d_n} \right) \right) E_p$$

$$\text{... within the limits } -A_{p1} f_{py} \le T_{p1} \le A_{p1} f_{py}$$

$$= 1390 \times 10^3 + 720 \times 10^3 \left(\frac{1060 - d_n}{d_n} \right)$$

$$\text{... within the limits } -2100 \times 10^3 \le T_{p1} \le 2100 \times 10^3$$

$$T_{p2} = 572 \times 10^3 + 300 \times 10^3 \left(\frac{800 - d_n}{d_n} \right)$$

$$\text{... within the limits } -875 \times 10^3 \le T_{p1} \le 875 \times 10^3$$

Location of neutral axis by trial strain distributions:

Calculations can now be made by trial and error with trial values of d_n, or with the aid of computational tools such as computer software or a spreadsheet. The solution is converged when the sum of the forces in the section is zero (that is: $\Sigma T = \Sigma C$).

Using this approach, it is determined that the depth to the neutral axis is d_n = 319.5 mm, and the assumption of $d_n \ge 200$ mm is satisfied. The forces on the section are calculated as:

bars: $T_{s1} = 1860$ kN; $T_{s2} = 200$ kN; $T_{s3} = -27$ kN; $T_{s4} = -900$ kN

tendons: $T_{p1} = 2100$ kN; $T_{p2} = 875$ kN

concrete: $C_1 = 2240$ kN; $C_2 = 910$ kN; $C_3 = 761$ kN

Calculation of M_u:

Taking moments about the extreme compressive fibre, the moment is:

$$M_u = (1860 \times 1.15) + (200 \times 0.90) - (27 \times 0.30) - (900 \times 0.05) + (2100 \times 1.06) + (875 \times 0.80) - (2240 \times 0.05) - (910 \times 0.133) - 761[0.10 + (0.88 \times 0.3195 - 0.10)/2]$$

$$= 4814 \text{ kNm}$$

Check on section ductility:

We check the section against the AS 3600 ductility measure of $k_{uo} \le 0.36$, which is explained below in Section 6.9.2 of this chapter. The effective depth, *d*, is defined as the depth from the extreme compressive fibre to the centroid of the forces in the bars and tendons that lie on the tensile side of the neutral axis (compressive reinforcement and tendons that lie on the compressive side of the neutral axis are not included in the calculation). Thus:

$$d = \frac{\sum_i T_{si} d_{si} + \sum_j T_{pj} d_{pj}}{\sum_i T_{si} + \sum_j T_{pj}}$$

$$= \frac{1860 \times 1150 + 200 \times 900 + 2100 \times 1060 + 875 \times 800}{1860 + 200 + 2100 + 875}$$

$$= 1042 \text{ mm}$$

The neutral axis depth parameter is then $k_{uo} = d_n/d = 319.5/1042 = 0.31$ and, thus, the member ductility is okay.

6.8 Stress in bonded tendons at ultimate

As discussed in Section 6.2, it is common to represent the stress-strain curve for prestressing steel by an elastic-plastic approximation, especially for the purposes of preliminary design. AS 3600 suggests, conservatively, that the yield stress in such an approximation be taken as either the 0.1 per cent proof stress, or $0.82f_{pb}$ for strands.

In the final calculation of the moment capacity M_u, a more accurate value for σ_{pu} may be needed. Reference to the stress-strain curve for prestressing steel given in Figure 2.1 shows that the steel stress may increase substantially above the yield stress f_{py}. The stress increment above f_{py} increases as the ductility of the section increases, and this should be reflected in the calculation for M_u.

This is best done by a strain compatibility analysis as outlined earlier in this chapter. The relationship between steel stresses and strains can be entered manually by reading values from the σ-ε graph, in which case an iterative approach becomes necessary. Alternatively a computer analysis can be used with the stress-strain relationship for the tendons taken as bi-linear, or by a formula such as that proposed by Develapura and Tadros (1992).

In lieu of such an approach, the Standard permits the use of an approximate, conservative, semi-empirical expression to evaluate the stress σ_{pu} in bonded tendons. In this approach the stress in the tendons at ultimate is:

$$\sigma_{pu} = f_{pb}\left[1 - \frac{k_1 k_2}{\gamma}\right] \tag{6.31}$$

where:

$k_1 = 0.40$, unless $f_{py}/f_{pb} > 0.9$, in which case the value is $k_1 = 0.28$

$$k_2 = \frac{A_{pt} f_{pb} + (A_{st} - A_{sc}) f_{sy}}{b_{ef} d_p f_c'}$$

and f_{pb} is the tensile strength of the tendon, f_{py} is its yield stress and A_{pt} is the cross-sectional area of the tendons in the zone that is tensile at the ultimate condition, b_{ef} is the effective width of the flange and d_p is distance from the extreme compressive fibre to the centroid of the tendons that are in the tensile zone at ultimate.

If compressive reinforcement is taken into account in calculating k_2, then k_2 cannot be taken less than 0.17 and d_{sc} cannot be greater than $0.15d_p$.

The value of k_2 in Equation 6.31 indirectly reflects the relationship between the amount of tensile steel in the cross-section and the neutral axis depth d_n at M_u. As the steel area decreases, d_n also decreases, and the concrete strain at the tendon level, ε_{cp}, increases (see Figure 6.4). This leads to an increase in the total steel strain ε_{pu} and, consequently, in the steel stress at ultimate moment, σ_{pu}.

EXAMPLE 6.7

Calculation of Ultimate Moment using AS 3600 Semi-Empirical Expressions for Steel Stress

For the section of Example 6.1, shown in Figure 6.5, estimate the ultimate moment capacity using the semi-empirical method for determining σ_{pu}. The breaking stress of the tendon is f_{pb} = 1870 MPa and the yield stress is taken as $0.82f_{pb}$.

Solution

The parameter k_2 used in Equation 6.31 is calculated as:

$$k_2 = \frac{405 \times 1879 + 1350 \times 500}{250 \times 625 \times 40} = 0.230$$

The steel stress σ_{pu} is obtained from Equation 6.31, and since $f_{py}/f_{pb} < 0.9$, $k_1 = 0.4$:

$$\sigma_{pu} = 1870\left(1 - \frac{0.4 \times 0.230}{0.77}\right) = 1647 \text{ MPa}$$

Tensile force in prestressing steel:

$$T_p = A_p\sigma_{pu} = 405 \times 1647 = 667{\times}10^3 \text{ N}$$

Tensile force in reinforcing steel:

$$T_s = A_{st}\,f_{sy} = 1350 \times 500 = 675{\times}10^3 \text{ N}$$

Concrete force:

$$C = 0.79 \times 40 \times 250 \times 0.87d_n = 6873d_n$$

and thus from $\Sigma C = \Sigma T$:

$$d_n = \frac{(667 + 675){\times}10^3}{6873} = 195.2 \text{ mm}$$

and the compression force is $C = 1342$ kN (and a check on a summation of the tension forces shows this to be correct).

The effective depth of the section is as calculated in Example 6.1 and $d = 662$ mm; the neutral axis depth parameter is thus:

$$k_{uo} = 195.2/662 = 0.29$$

and is less than the limit of 0.36 set by AS 3600.

Taking the moments about the extreme compressive fibre:

$$M_u = 667 \times 0.625 + 675 \times 0.700 - 1342 \times 0.87 \times 0.1952/2$$

$$= 775 \text{ kNm}$$

This compares with $M_u = 757$ kNm calculated in Example 6.1 using the more conservative estimate of taking $T_p = A_p f_{py}$ at ultimate.

6.9 Design considerations

6.9.1 Design strength in bending

The strength design criteria in AS 3600 have been discussed briefly in Chapter 3, and are summarised in Equation 3.1. As we have seen, when applied to the specific case of design for flexural strength the left hand term R_d in Equation 3.1 becomes the **design strength in bending**, M_d, and the right hand side becomes the **design ultimate moment**, M^*. The design criterion requires M_d to be not less than M^*.

To obtain the design strength in bending, M_d, it is necessary to reduce the moment capacity, M_u, by a capacity reduction factor ϕ. The design criterion for bending becomes:

$$\phi M_u \geq M^* \tag{6.32}$$

This design requirement has to be satisfied at all sections of a flexural member. For normal under-reinforced members failing in flexure, $\phi = 0.85$.

The design ultimate moment M^* is obtained from a linear analysis of the structural system when subjected to the most adverse combination of factored loads for strength design. The factored load combinations to be considered in design are specified in the loading standard, AS/NZS 1170.0 and have been summarised in Chapter 3. For example, with M_G and M_Q representing the moments due to dead load and live load respectively, the combinations for dead and live load to be considered are:

$$1.35\, M_G \tag{6.33}$$

$$1.2\, M_G + 1.5\, M_Q \tag{6.34}$$

$$1.2\, M_G + 1.5\, \psi_1\, M_Q \tag{6.35}$$

The parameter ψ_1 in Equation 6.35 varies typically between 0.4 and 0.6 but its maximum value is 1.2. AS/NZS 1170.0 provides detailed design information on load combinations and load factors.

6.9.2 Ductility Requirements in AS 3600

Although the bending strength requirement, Equation 6.32, can be satisfied by a wide range of section sizes with various amounts of reinforcing steel and prestressing steel quantities, AS 3600 indirectly but strongly encourages the use of ductile cross-sections, that is to say, sections that will undergo large deformations before the moment capacity M_u is reached. In AS 3600, the capacity reduction factor ϕ is expressed as a function of the neutral axis parameter $k_{uo} = d_n / d$ at ultimate moment, $\phi = 1.24 - k_{uo}(13/12)$. However, ϕ cannot be less than 0.65 nor greater than 0.85. Provided k_{uo} satisfies the following requirement:

$$k_{uo} \leq 0.36 \qquad (6.36)$$

ϕ takes on the maximum value of 0.85, as the section is considered to be ductile. In a section such as the one in Figure 6.1, d is as follows:

$$d = \frac{T_p d_p + T_s d_{st}}{(T_p + T_s)} \qquad (6.37)$$

Severe restrictions apply to sections that do not satisfy Equation 6.36 and such sections should be avoided in design.

The limit on k_{uo} of 0.36 is an indirect and approximate, but convenient, means of maintaining reasonable ductility in flexure. This is characterised by a moment-curvature relationship with a long, flat, plateau in the vicinity of the ultimate moment. It can be seen from Figure 5.8 that as more and more tensile steel is introduced into a beam section, the plateau becomes shorter and shorter. The limit of 0.36 on k_{uo} indirectly restricts the amount of tensile steel in any section.

The ductility limit, $k_{uo} = 0.36$, is not applicable in complex situations. For example it does not cover a section that contains non-prestressed tendons, or tendons prestressed to a low value. The criterion is only valid where normal levels of prestress are used, for example when the effective prestress is not less than about 500 MPa below the yield stress of the tendon.

Although AS 3600 does not specifically prohibit the use of sections with $k_{uo} > 0.36$, such sections incur reduced capacity reduction factors and an area of compressive steel of at least 0.01 times the area of the concrete in compression must be included.

6.9.3 Adding reinforcement to increase moment capacity

Frequently in the design of flexural members, the quantity of prestressing steel will first be determined on the basis of the serviceability limit states, for example using the load balancing concept. It then remains to determine an appropriate amount of reinforcing steel to place in critical sections so that Equation 6.32 is satisfied.

Assuming that both the reinforcing steel and the prestressing steel will be at yield, we can make a simple estimate of the required area of reinforcing steel using the approximate expression for M_u given in Equation 6.13. Substituting for M_u in Equation 6.32 and rearranging, we obtain

$$A_{st} = \frac{1}{0.85 f_{sy} d_{st}} \left[\frac{M^*}{\phi} - A_p f_{py} (d_p - 0.15 d_{st}) \right] \tag{6.38}$$

This expression can be used to estimate the steel area, but a proper ultimate strength calculation must also be carried out to ensure that Equations 6.32 and 6.36 are in fact satisfied.

EXAMPLE 6.8

ADDITIONAL REINFORCING STEEL TO INCREASE M_U

A post-tensioned beam has a rectangular cross-section with a width of 400 mm and an overall depth of 950 mm (Figure 6.11). It is simply supported on a 20 m span and carries a service live load of 15 kN/m. At mid-span the effective prestress, determined from serviceability requirements, is 972 kN at an eccentricity of 340 mm ($d_p = 815$ mm). The prestressing force is provided by a tendon with nine 12.7 mm diameter strands with an area of 900 mm^2.

Determine how much reinforcing steel is needed at the mid-span to provide the required ultimate moment capacity.

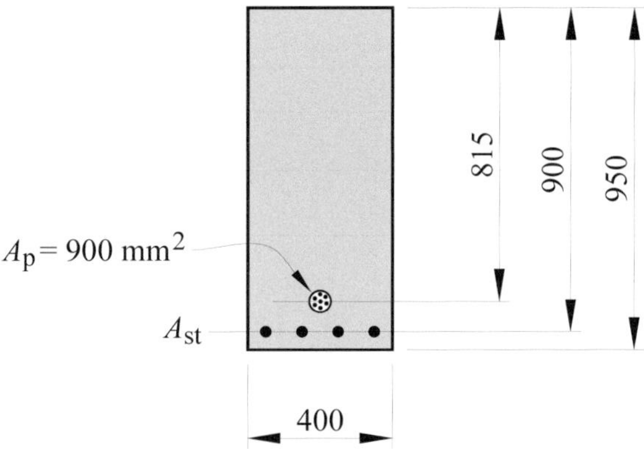

Figure 6.11 Cross-section for Example 6.8

Other Data:

$$f'_c = 40 \text{ MPa}; \ \alpha_2 = 0.79; \ \gamma = 0.87$$

$$f_{sy} = 500 \text{ MPa}; \ f_{py} = 1750 \text{ MPa}; \ \sigma_{pe} = 1100 \text{ MPa}$$

SOLUTION

The self-weight of the beam is $0.4 \times 0.95 \times 25 = 9.5$ kN/m, which produces a mid-span moment of:

$$M_G = 9.5 \times 20^2/8 = 475 \text{ kNm}$$

The live load moment is $15 \times 20^2/8 = 750$ kNm and so the design ultimate moment is:

$$M^* = 1.2 \times 475 + 1.5 \times 750 = 1695 \text{ kNm}$$

The required moment capacity M_u is $M^*/\phi = 1695/0.85 = 1994$ kNm.

A preliminary estimate of A_{st} is made using Equation 6.38 with d_{st} = 900 mm:

$$A_{st} = \frac{1994{\times}10^6 - 900 \times 1750\,(815 - 0.15 \times 900)}{0.85 \times 500 \times 900} = 2413 \text{ mm}^2$$

With a trial choice of 5-N28 bars, with A_{st} = 3100 mm^2, we now check for adequacy. Assuming yield in both steels:

$$T_p = 900 \times 1750 = 1575{\times}10^3 \text{ N}$$

$$T_s = 500 \times 3100 = 1550{\times}10^3 \text{ N}$$

$$C = 0.79 \times 40 \times 400 \times 0.87 d_n = 10997 d_n \text{ N}$$

Hence $10.997 d_n = 1575 + 1550 = 3125$ and d_n = 284 mm.

A check shows that k_{uo} is less than 0.36 and both steels are at yield. Taking moments about the top fibre:

$$M_u = 1575 \times 0.815 + 1550 \times 0.900 - 3125 \times (0.87 \times 0.284/2)$$

$$= 2293 \text{ kNm}$$

As $\phi M_u = 0.85 \times 2293 = 1949 \text{ kNm} \geq M^* = 1695 \text{ kNm}$, the section is adequate.

6.9.4 Minimum moment capacity

A sudden, non-ductile failure can occur in members that contain very small quantities of tensile steel. In such cases, failure occurs immediately the concrete cracks if the steel area is so small that it is unable to accept the tensile force previously carried by the concrete. For this reason, AS 3600 requires that the moment capacity of the section M_u must be at least twenty per cent higher than the cracking moment M_{cr}.

6.9.5 Section capacity at transfer

At transfer an eccentric prestress induces axial compression and bending moment in a beam cross-section. If the prestressing force is too high it is possible for failure to occur during the prestressing operation by crushing of the concrete in the lower fibres surrounding the cable. A design check must be made to ensure that this cannot occur. According to AS 3600, the check is made using a value of $\phi = 0.6$ on the capacity and a load factor of 1.15 on the prestress. As the self-weight moment normally opposes the moment due to prestress, a load factor of 0.9 is specified for the moment due to self-weight and any other dead load acting at transfer.

In lieu of a strength calculation, the strength requirement is deemed to be satisfied if the maximum compressive stress in the concrete at transfer does not exceed $0.5 f'_{cp}$ when the stress distribution is rectangular, or $0.6 f'_{cp}$ if, as is normal, the stress distribution is triangular.

Conditions at ultimate in a critical cross-section are shown in Figure 6.12.

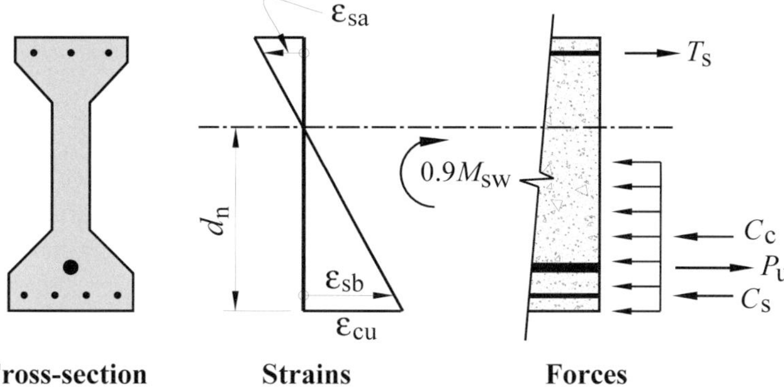

Figure 6.12 Conditions at transfer

We seek to calculate the magnitude of P_u, the cable force that will cause failure by concrete crushing in the bottom of the section. At failure, under combined bending and compression, we adopt the usual simplifying assumption of the equivalent rectangular stress block, and the limiting strain criterion of $\varepsilon_{cu} = 0.003$.

Using the linear strain diagram, the strains and, hence the stresses and forces in the top and bottom reinforcement can be expressed in terms of an unknown neutral axis depth d_n. The concrete force C_c is also written as a function of d_n. The two unknowns, P_u and d_n, are then evaluated from the two equations of equilibrium: $\Sigma H = 0$, giving $P_u = C_c + C_s - T_s$, and $\Sigma M = 0$. If moments are taken about the cable location, the value of d_n is obtained directly. The calculation procedure is illustrated for a simple section in Example 6.9.

Some notes on details of the calculation of strength at transfer follow:

Multiple cables: For sections containing two or more prestressing cables, separate calculations can be made to find P_u for each cable in turn. The procedure follows that illustrated in Example 6.9, with the prestressing force in any previously stressed cable appearing as a constant force in the calculation. The value of P_u for cable 2 (i.e. the cable stressed second) will be the same as that calculated for cable 1, less the magnitude of the prestressing force in cable 1, and so on.

However, for the purpose of checking whether the section satisfies AS 3600 requirements, the calculation should be made as though all cables were stressed together.

Effect of duct voids: In Example 6.9, the presence of the void created by the duct is ignored in calculating the concrete force C_c. This force is therefore overstated by an amount equal to $0.80 f_{cp} \pi (d_d^2/4)$, where d_d is the duct diameter, and the calculated capacity at transfer is overstated by the same amount. If, in Example 6.9, the diameter of the duct were 100 mm, the error is $0.80 \times 32 \times \pi \times 100^2/4 = 204$ kN, which would reduce the calculated value of P_u from 6705 kN to 6501 kN.

Simplified calculations: If the calculations are to be done manually, they may be simplified by ignoring either or both of (i) any conventional reinforcement in the section, (ii) the self-weight moment. If on this simplified, and conservative, calculation, the resulting value of P_u satisfies strength requirements, no further calculations need be made.

EXAMPLE 6.9

SECTION STRENGTH AT TRANSFER

A simply supported post-tensioned beam spans 25 m and has a rectangular cross-section 1200 mm deep and 350 mm wide. The strength of the concrete at transfer is $f'_{cp} = 32$ MPa. Locations of the reinforcing and prestressing steel at mid-span are shown in Figure 6.13. The top reinforcement consists of two 2-N28 bars, $A_{sa} = 1240$ mm^2. The bottom reinforcement comprises 3-N28 bars, $A_{sb} = 1860$ mm^2.

It is required to calculate the magnitude of the prestressing force, P_u, at which failure would occur by crushing of the concrete around the cable. Only the self-weight of the member acts at transfer.

Other data:

$$E_s = 200 \times 10^3 \text{ MPa}; \; f_{sy} = 500 \text{ MPa}; \; \varepsilon_{sy} = 0.0025$$

SOLUTION

Design conditions at time of transfer:

Self-weight: $w_G = 0.35 \times 1.2 \times 25 = 10.5$ kN/m

Mid-span moment: $M_G = 10.5 \times 25^2/8 = 820$ kNm

Factored moment: $0.9M_G = 738$ kNm $(738 \times 10^6$ Nmm$)$

Forces acting on the section:

We assume, subject to later check, that the top reinforcement has not yielded and the bottom steel has yielded. The strain in the top steel is then:

$$\varepsilon_{sa} = 0.003 \times (1150 - d_n)/d_n$$

and the tension force is:

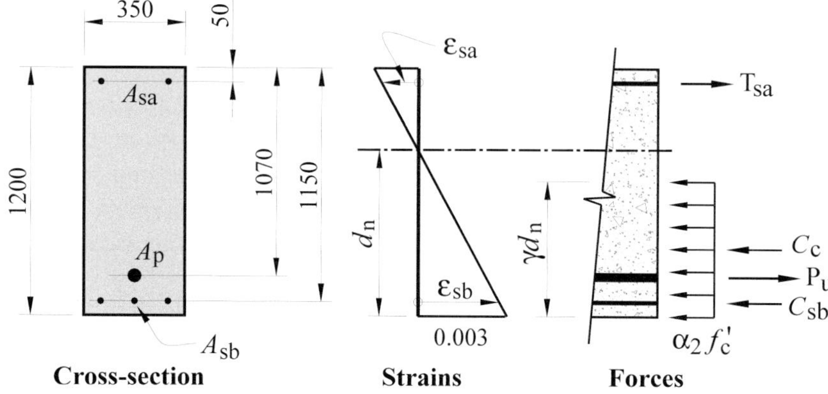

Figure 6.13 Section for Example 6.8

$$T_{sa} = E_s \varepsilon_s A_{sa} = 200\times10^3 \times 0.003 \times \frac{1150 - d_n}{d_n} \times 1240$$

$$= 744\times10^3 \left(\frac{1150 - d_n}{d_n}\right) \text{ N}$$

Assuming, subject to checking, that the bottom steel is beyond yield:

$$C_{sb} = A_{sb} f_{sy} = 1860 \times 500 = 930 \times10^3 \text{ N}$$

At transfer $f_{cp}' = 32$ MPa and thus $\alpha_2 = 0.80$ and $\gamma = 0.89$, the compressive stress resultant is then:

$$C_c = 0.80 \times 32 \times 350 \times 0.89 d_n = 7974 d_n \text{ N}$$

Neutral axis depth:

The neutral axis depth can be determined by taking moments about the cable location:

$$\Sigma M = T_{\text{sa}} \times 1020 + C_{\text{sb}} \times 80 - C_{\text{c}}\left(\frac{0.89d_{\text{n}}}{2} - 130\right) + 0.9M_{\text{G}} = 0$$

Substituting the values for T_{sa}, C_{sb}, C_{c} and M_{G}, we obtain after simplifying:

$$d_{\text{n}}^3 - 292d_{\text{n}}^2 - 26.9\times10^3 d_{\text{n}} - 246.1\times10^6 = 0$$

and solving gives $d_{\text{n}} = 724$ mm.

Checking the strain assumptions:

top steel: $\varepsilon_{\text{sa}} = 0.003 \times (1150 - 724)/724 = 0.00177 < \varepsilon_{\text{sy}}$

bot. steel: $\varepsilon_{\text{sb}} = 0.003 \times (724 - 50)/724 = 0.00280 > \varepsilon_{\text{sy}}$

Thus the assumed strain states for the calculation of d_{n} were correct.

Capacity at transfer:

The forces are:

Top steel: $T_{\text{sa}} = 0.00177 \times 200\times10^3 \times 1240 \times 10^{-3}$
$= 439$ kN (tension)

Bot. steel: $C_{\text{sb}} = 930$ kN (compression)

Concrete: $C_{\text{c}} = 7974 \times 724 \times 10^{-3} = 5773$ kN (compression)

For equilibrium, $\Sigma H = 0$ and thus:

$$P_{\text{u}} = C_{\text{c}} + C_{\text{sb}} - T_{\text{sa}} = 6264 \text{ kN}$$

Therefore, $P_{\text{u}} = 6264$ kN is the capacity at transfer for this section, i.e. the prestressing force that would lead to failure by concrete crushing in the bottom of the section. AS 3600 specifies a capacity reduction factor ϕ of 0.6 for P_{u}, and a load factor of 1.15 for the prestressing force. Thus, the maximum permissible force in this cable at transfer is:

$$P \leq 0.6 \times 6264/1.15 = 3268 \text{ kN}$$

6.10 References

AS 3600–2018, *Concrete Structures*, Standards Australia.

AS/NZ 1170.0 2002, *Structural design actions, Part 0: General principles*, Standards Australia.

Devalapura, R K, and Tadros, M K, Critical Assessment of ACI 318 Eq. (18-3) for Prestressing Steel Stress at Ultimate Flexure, *ACI Structural Journal*, Vol. 89, No. 5, September/October 1992, pp. 538-546.

CHAPTER 7

Shear and torsion

Although shear and torsion have little effect on the working load behaviour of most prestressed concrete beams, these actions can lead to premature and sudden failure. The aim in design, therefore, is to provide sufficient reinforcement to ensure that shear failure and torsion failure do not occur before the flexural capacity of the member is reached. In 2017 and 2018, the models for shear and torsion in AS 5100 and AS 3600 were considerably revised from those in previous editions. The new provisions for shear are derived from the simplified modified compression field theory of Bentz et al. (2006) and are based on physical (mechanical) models that are understandable and take account of the most important influences observed in tests (such as the dependence of shear strength on web strains and the size of the member). The design provisions in AS 5100.5–2017 and AS 3600–2018 are an adaptation of the *fib* Model Code 2010 rules, and have also been influenced by the 2014 AASHTO LRFD Bridge Design Specifications and the Canadian Standard CSA A23.3–14. In this chapter we look at the overload behaviour of prestressed beams in shear and torsion and appropriate code design procedures and introduce the new approaches.

7.1 Shear and torsion in prestressed concrete

In previous chapters of this book we have concentrated on the flexural behaviour of prestressed concrete members at service load and at failure. While shear and torsion also have to be considered in design, these actions have little effect on the service load behaviour of most members. Design procedures for shear and torsion are therefore focused on introducing transverse reinforcement, and possibly additional longitudinal reinforcement, in regions of high shear and torsion to ensure that the member reaches its full flexural capacity. With adequate transverse and longitudinal reinforcement present, truss-like action effectively develops in the member to prevent premature failure in shear or torsion.

The overload behaviour of prestressed concrete beams in shear and torsion is generally similar to that of reinforced concrete beams. The main effect of the prestress is to increase the load at which vertical and inclined cracks first appear. The simplified design procedures for shear and torsion in AS 3600 are therefore similar to those used for reinforced concrete, which are explained in detail in the companion text *Reinforced Concrete Basics*. In this Chapter we shall first consider the effects of shear on the overload behaviour of prestressed members and then look at appropriate design methods. The effects of torsion are then discussed together with appropriate design methods.

7.2 Overload behaviour in shear and bending

7.2.1 Inclined cracking in beams

In a flexural member subjected to transverse loading, shear force and bending moment occur together throughout the span. The shear force V at any cross-section is statically related to the rate of change of the bending moment M:

$$V = -\frac{dM}{dx} \tag{7.1}$$

When discussing the effects of shear in a region at overload, it is therefore necessary also to consider the presence of the bending moment.

Figure 7.1(a) shows three types of cracking that can be observed in a statically determinate prestressed concrete beam at overload and, in Figure 7.1(b) the same types of cracks are shown in a continuous beam. Region BC of the beam in Figure 7.1(a) has high moment but negligible shear, and **flexural cracks** develop at the extreme tensile fibres when the tensile strength of the concrete is exceeded. These then extend vertically into the section and are denoted as Type 1 in Figure 7.1.

In Figure 7.1(a), the shear force is constant in the shear spans AB and CD, but the bending moment is large at cross-sections close to the load points B and C. Vertical flexural cracks form here in the tensile face of the concrete, but with increasing load they become inclined because of the presence of the shear force. These cracks, denoted as Type 2, are often referred to as **flexure-shear** cracks.

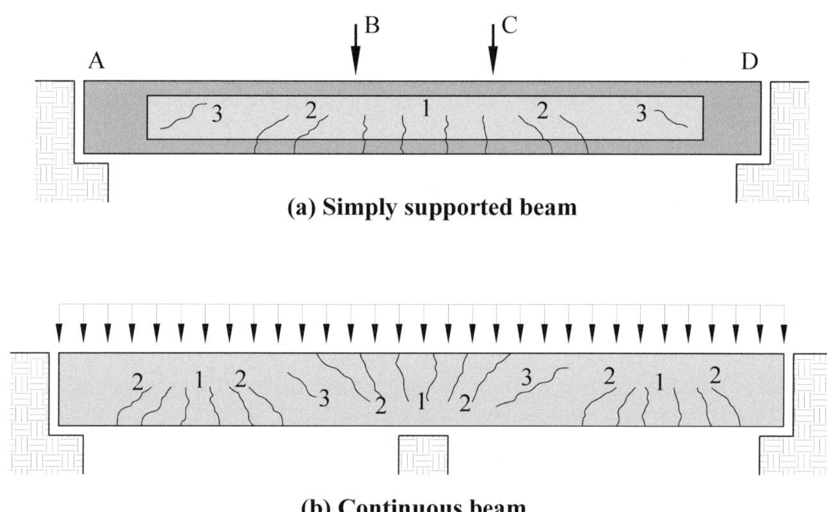

Figure 7.1 Types of cracking

Near the end supports A and D in Figure 7.1(a), the shear force is large but the bending moment is small. Inclined cracks will appear here in the web if the principal tensile stress reaches the tensile strength of the concrete. Such cracks, labelled Type 3 in Figure 7.1, appear in regions remote from any pre-existing flexural cracks. They are referred to as **diagonal tension** cracks or **web-shear** cracks and are most likely to develop in beams with thin webs.

The three types of cracks also form in regions of continuous members, as indicated in Figure 7.1(b). Note that the Type 1 flexural cracks form in the top fibres in the peak negative moment regions above the interior support, as well as in the bottom fibres in the positive high moment regions. The Type 2 flexure-shear cracks also form in both the negative moment and positive moment regions adjacent to the flexural cracks. In the interior regions of the continuous beam, Type 3 cracks may form in low-moment regions between adjacent positive moment Type 2 cracks and negative moment Type 2 cracks.

Laboratory tests indicate that beams without shear reinforcement are liable to fail at loads that are not significantly higher than the load at which the inclined cracks develop. The post-cracking behaviour of such beams is difficult to predict and shear failure can be sudden and catastrophic, especially if the prestress in the beam is high.

7.2.2 Load carrying mechanisms in the post-cracking range

The presence of properly anchored transverse stirrup reinforcement allows various mechanisms to develop, whereby shear force is safely transmitted through regions where inclined cracks have formed, so that the member can carry loads well above the inclined cracking load. In such beams some shear force can be carried by the intact concrete in the compressive stress block above the extremity of the inclined crack. Some shear transfer also occurs by bearing and friction between the adjacent jagged faces of the inclined cracks. This effect has come to be known as **aggregate interlock**. **Dowel action** may also occur where the prestressing tendons and longitudinal reinforcing bars cross the crack. While resistance to sliding on the crack is observed as the major component for beams with shear span-to-depth ratios greater than about 2.5, the tensile strength of the concrete plays some role, as do dowel action and shear on the uncracked concrete in the compressive zone. Research by Muttoni and Fernández Ruiz (2019) typically put the component of the shear force resisted by sliding along a critical shear crack as about 70 per cent of the total, with concrete tensile stresses, dowel action and shear on the uncracked concrete making up the balance. For members with shear span-to-depth ratios less than 2.5, arching (i.e., the strut-and-tie effect) plays a significant part in the transfer of shear forces to supports.

The transverse shear reinforcement crossing the inclined cracks allows **truss-like action** to develop in the member where, for the case of a sagging moment, the top region of the beam is in compression and the bottom in tension. The transverse stirrups are in tension and the web concrete carries diagonal compressive stress.

In a region where the tendon is inclined, the prestressing force P has a vertical component P_V. This will usually oppose the shear force induced by the external load.

Although it is difficult to assess the relative importance of the different load carrying mechanisms in shear, it is clear that the major contribution comes from the stirrup reinforcement. Aggregate interlock diminishes as the load increases and the cracks widen. Also, dowel action is destroyed if horizontal cracking occurs at the level of the longitudinal steel at high overload.

For routine design, the combined effects of aggregate interlock, dowel action and shear in the compressive concrete above the tip of the crack are lumped together in AS 3600 and represented by V_{uc}. This is added to a truss-action term, V_{us} to represent the total shear capacity V_u of the region of the member:

$$V_u = V_{uc} + V_{us} + \{P_v\} \tag{7.2}$$

In Equation 7.2, the vertical component of prestress, P_V, is included with the sectional capacity terms. We shall see later in this chapter that P_V may be treated as either increasing (or decreasing) the section capacity or as a load on the member. In Amendment 2 of AS 3600–2018, the vertical component of prestress is treated as a load, which is changed from earlier editions of the Standard.

7.2.3 Modes of shear failure

At high overload, the inclined cracks extend deeply into the top compressive region. Failure can occur either by (a) yielding of the stirrups, with subsequent crushing of the concrete in the flexural compressive zone, or (b) crushing of the compressive strut that forms in the web, prior to yielding of either the transverse (web) reinforcement or longitudinal steel (or prestress tendon).

If in case (a) crushing of the concrete in the flexural compressive zone occurs before the full moment capacity of the beam is achieved, the failure is sometimes called **shear-compression failure**. If failure arises after yielding of the stirrups and combined with loss of bond along the longitudinal reinforcement, due to the formation of a splitting crack, this is called **shear-tension** failure. Where failure occurs with yielding of the web steel followed by yielding of the longitudinal steel, but before the full flexural demand at the critical section for flexure is achieved, and prior to crushing of the diagonal struts, failure may be described as **shear-flexure**.

In case (b), in a beam containing a large quantity of stirrup steel, it is possible for the inclined concrete struts to fail by crushing, prior to steel yield. This is referred to as **web crushing** or **diagonal-compression failure**.

7.3 Web reinforcement behaviour in the post-cracking range

7.3.1 Simple truss model

The contribution of truss-like action to the shear capacity of a section or region can be estimated using models of varying accuracy and complexity. In the original truss analogy for reinforced concrete, developed more than a century ago, the behaviour of the tensile stirrups and concrete in diagonal compression was likened to vertical tension members and diagonal compression members in a simple pinned truss. This approach is still used as the basis for the current AS 3600 shear design clauses. In the case of prestressed concrete members, the effect of the prestressing tendon on truss action is ignored in the determination of the truss depth, as indicated in Figure 7.2(b). In Figure 7.2(b) truss-like action is provided by the vertical tensile ties (the stirrups), the inclined compressive struts (the compressive concrete between inclined cracks), the top compression chord (the compressive top-fibre concrete) and the bottom tension chord (the tensile reinforcement).

At a typical lower joint, Figure 7.2(c), V_S is the tensile force carried by a stirrup unit, D is the diagonal force in the inclined compressive concrete and T is the total tensile force in the stringer to the right of the joint. On the left, the tensile force has reduced to $T - \Delta T$, where ΔT is the horizontal component of D. The stirrup force is equal to the vertical component of D. If θ_V is the angle of inclination of the inclined strut, we have:

$$V_S = D\sin\theta_V + P_V \tag{7.3}$$

$$\Delta T = D\cos\theta_V \tag{7.4}$$

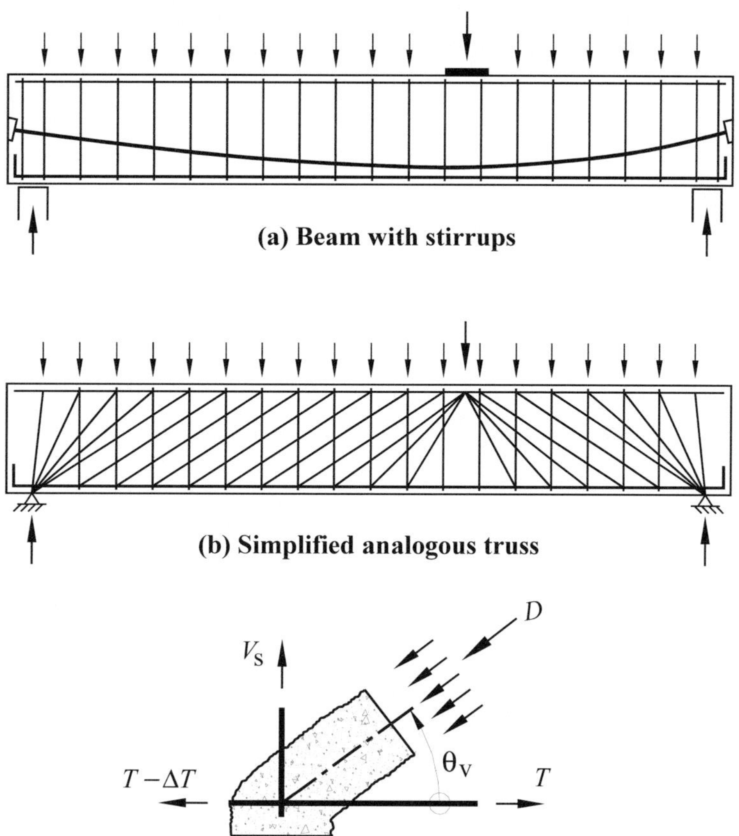

(c) Joint forces: lower joint in analogous truss

Figure 7.2 Simple truss model

The number of stirrups n crossing an inclined crack depends on the angle θ_V, and the total shear force carried by the truss is nV_S. The maximum shear force carried by the stirrups, V_{us}, occurs when the stirrups are at yield.

This truss model is obtained through considerable idealisation and simplification. For example, if the stirrups are widely spaced, a diagonal strut may not form exactly on the line between the extremities of adjacent stirrups as is suggested in Figure 7.2(b), and quite complex stress patterns may then be induced to allow an effective load carrying mechanism to develop. On the other hand, with closely spaced stirrups the internal forces tend to be more uniformly spread throughout the web, which then tends to act like a shear panel.

The truss model captures the broad aspects of web behaviour and provides a simple means for estimating the stirrup forces to be used in design. The model also suggests the ways in which shear failure can be expected to occur, by yielding of the stirrups, by crushing of the diagonal concrete struts and even by joint failure, that is to say by slip of inadequately anchored stirrups.

Nevertheless, there are various inadequacies that have to be considered. For example, the components of the truss are assumed to meet at perfect joints of zero size, thus ignoring the complex state of stress in the concrete in the joint regions. The simple truss model also ignores any contribution of the tendon to truss action. This is reasonable in situations when shear is not a significant design problem, as for example in a band beam in a floor system. In such situations the only contribution is the vertical component of the tendon force, P_V. If desired, the tendon can be included in the truss, as is shown in Figure 7.3 below. In this case the forces in the longitudinal reinforcement (bars and tendons) are accurately included in the model, together with that of the transverse reinforcement (stirrups or ties).

These and other inadequacies in the simple truss approach have led to the development of more sophisticated models, mostly based on the concepts of struts, ties, nodes (or joints) and stress-fields. Such models are developments of, and variations on, the truss concept (Muttoni et al, 1997).

7.3.2 Stress-field model

A better representation of the load carrying mechanism in the web of a prestressed concrete member is obtained using a series of stress-fields, as shown in Figure 7.3(a). In the middle region of the beam a diagonal compressive stress-field, carried by the concrete, is intersected by a transverse tensile stress-field, provided by the stirrups. These stress-fields are equilibrated in the flexural tensile region both by the tensile reinforcement and the prestress cable, and in the flexural compressive region by the compressive concrete.

An example of such a transfer of the diagonal compressive stress-field into forces in the stirrups is shown in Figure 7.3(b), where the diagonal stresses acting in the concrete are translated into forces D_1 and D_2, equilibrated by the cable and reinforcing bars, respectively, T_p and T_s are the forces in the cable and bar and ΔT_p and ΔT_s are the changes in these forces and V_s is the force carried by the stirrup. It is readily seen from the figure that, firstly, both the cable and tensile reinforcing steel contribute in the force transfer, secondly this transfer may occur at different levels in the section and, finally, any transverse component of the prestressing force adds to, or in the case of negative bending subtracts from, the shear capacity of the section.

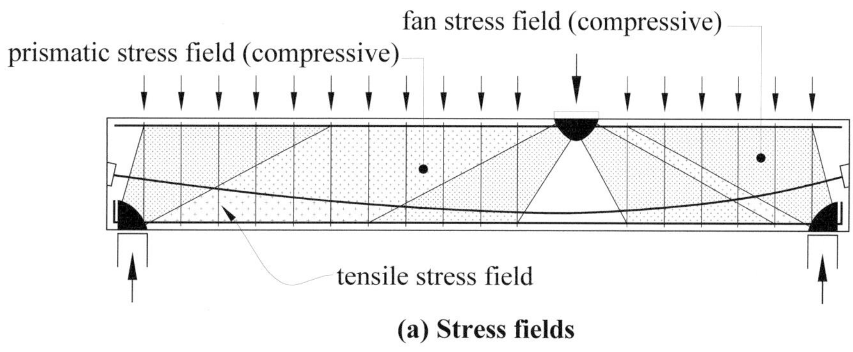

(a) Stress fields

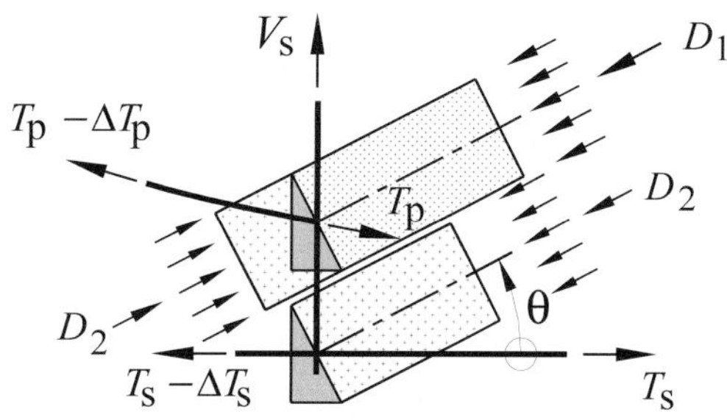

(b) Transfer of concrete compressive stresses to longitudinal and transverse reinforcement and prestress tendon

Figure 7.3 Internal load (stress) path for a partially prestressed concrete beam

Figure 7.3 also shows fan-shaped stress-fields at the ends of the beam, and triangular nodal (joint) regions. An explanation of the basics of strut-and-tie theory and stress-field theory is given in the companion text to this book (Foster et al., 2021).

Although we will use a simpler approach than the stress-field model for routine design calculations, the principles of the load carrying mechanism must not be overlooked, particularly when it comes to the detailing of the member. For more on stress-fields and their relationship with strut-and-tie modelling, see Section 3.7.2 of Chapter 3.

7.4 Effect of prestress on behaviour in shear

As already noted, the vertical component of the prestressing force has to be taken into account in determining the shear capacity of a member. However, prestressing also improves the overload behaviour in shear by delaying inclined cracking and, hence, indirectly increasing the strength in shear.

In most prestressed members, the tendon has a curved or draped profile and there are both horizontal and vertical components of the prestress acting in a section. We consider the effect of each component of prestress in turn on the behaviour in shear of a statically determinate member.

Effect of horizontal component of *P*

The horizontal component of the prestressing force delays the formation of both flexural (Type 1) and inclined (Types 2 and 3) cracks. In the case of web-shear cracking (Type 3), the horizontal prestress induces a compressive stress at the centroidal axis, which reduces the principal tensile stress there. Nevertheless, Type 3 cracking occurs more frequently in prestressed than in reinforced members. There are two reasons for this. Firstly, prestressed sections usually have thinner webs, and therefore higher shear stresses, than reinforced sections. Secondly, horizontal prestress inhibits flexural cracking in the bottom fibres, so that Type 3 cracks tend to occur in lieu of Type 2 cracks. The horizontal component of the prestressing force nevertheless increases both the inclined cracking load and the load capacity of the member in shear.

Effect of vertical component P_V

If a prestressing tendon is inclined to the horizontal at an angle α, as at section A–A in Figure 7.4, the vertical component is $P \sin \alpha$ or, if α is small, $P\alpha$ and this results in a shear force acting in the concrete. If the section is in positive bending (i.e. under a sagging moment), this force is usually in the opposite direction to the shear obtained from the externally applied forces. However, if the section being considered is in negative bending (i.e. under a hogging moment), this force is likely to be in the same direction as the shear obtained from the externally applied forces, thus increasing the shear demand on the section.

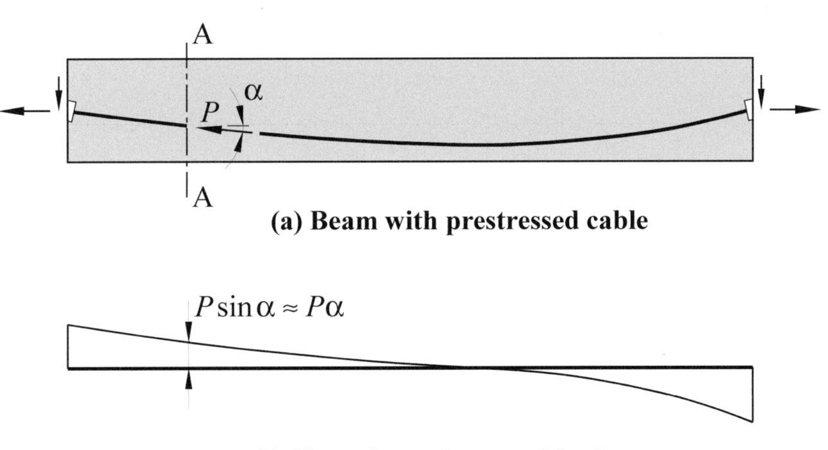

Figure 7.4 Vertical component of prestress

With increasing bending moment, the cable force P increases slightly, and so too does its vertical component. However, the regions where the cable slope α is large are usually regions of low bending moment, so that the change in P is least marked there. It is therefore usual to ignore any increase in prestressing force with load and to assume that the vertical component of prestress (P_V) at all stages of loading is:

$$P_V = P_e\alpha \tag{7.5}$$

where P_e is the effective prestress at the time of loading.

Shear in indeterminate prestressed structures (hyperstatic shears)

The introduction of prestress into a statically indeterminate member may induce additional support reactions, called **hyperstatic reactions**, which in turn cause **secondary moments** and **secondary shear forces** in the member. These secondary actions may be significant and are discussed in some detail in Chapter 11. They must be taken into account when the load at inclined cracking is calculated. In a region that is critical for shear design, the secondary shear force may either add to or subtract from the shear force produced by an external load. The effect of the secondary actions (both shear and bending moment) on load capacity depends on the degree of ductility inherent in the member. In essence, the load capacity of a ductile structure is unaffected by the secondary actions. However, they must be carefully considered in the design of members with limited ductility, and particularly in design for shear, as shear failure is often non-ductile. This matter is discussed further in Chapter 11. The present discussion is restricted to statically determinate members.

7.5 Web-shear cracking load for prestressed members

7.5.1 Introduction

Depending on the magnitude of the shear force and bending moment at a section, cracking in webs can occur due either to flexure-shear, where diagonal shear cracks are a continuation of initially formed flexural cracks, or by cracking that initiates in the web, i.e. web-shear cracks. Under certain conditions, web-shear cracks may impact on serviceability. This can be particularly important in members with thin webs. In this section the shear force at the formation of a web-shear crack is considered.

7.5.2 Web-shear cracking load

The vertical component of the cable force, P_V, is taken into account in evaluating the load to produce web-shear cracking. The slope of the cable is usually near its maximum value in regions close to the supports, which are prone to web-shear cracking, so that the contribution of the vertical component of the cable force can be quite significant. The shear force in the concrete for web-shear cracking is given by:

$$V_{cr} = V_t + P_v \qquad (7.6)$$

where V_t is the shear force in the concrete at which the maximum principal tensile stress reaches the tensile strength of the concrete f'_{ct}. Although the term P_v is added to V_t in Equation 7.6, there can be design situations where it opposes V_t.

As the web-shear crack forms in a region previously uncracked in flexure, V_t is evaluated by equating the principal tensile stress at a critical point in the web to the effective tensile strength of the concrete. The tensile strength is taken as the characteristic uniaxial tensile strength, which may be approximated as $f'_{ct} = 0.36\sqrt{f'_c}$.

In determining the principal stress, account must be taken of the shear stress caused by V_t as well as the axial stress due to the axial component of the prestressing force P and the bending moment M. Figure 7.5 shows the distribution of compressive stress and shear stress in a typical thin-web member. The direct stress due to prestressing force P and moment M varies with distance y from the centroidal axis. With y and e taken as positive downward, the tensile stress in the concrete in the direction of the longitudinal axis of the member, σ_{cx}, is:

$$\sigma_{cx} = \frac{My}{I} - \frac{P}{A} - \frac{Pey}{I} \qquad (7.7)$$

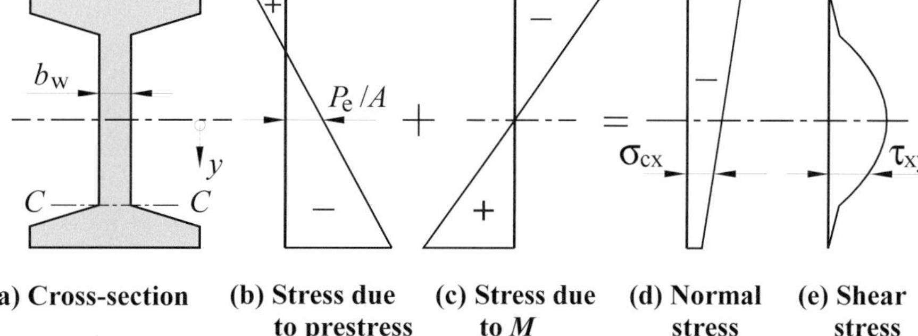

Figure 7.5 Distribution of direct and shear stresses in an uncracked section

The shear stress τ_{xy} also varies with y and simple beam theory gives us the following expression:

$$\tau_{xy} = V_t Q/(I b_v) \tag{7.8}$$

where V_t is the shear force carried by the concrete at the cross-section, b_v is the effective width of the web at the section being considered (discussed below), Q is the first moment of area about the centroidal axis of that part of the cross-sectional area that lies above (or below) the level at which τ_{xy} is being calculated and I is the second moment of area of the gross section taken about its centroidal axis.

The full width of the web is b_w. If, however, prestressing ducts are in the vicinity of a critical section in shear, the effective width of the web should be reduced. An effective web width is used, which is obtained by subtracting one-half of the diameter of any duct (d_d) that lies in any plane passing through the web. The effective width of the web (b_v) is thus:

$$b_v = b_w - k\Sigma d_d \tag{7.9}$$

where Σd_d is the sum of the diameters of the ducts that lie in a horizontal plane across the web and $k = 0.5$ and 0.8 for grouted steel and plastic ducts, respectively, and $k = 1.2$ for ungrouted ducts.

This reduced web thickness is used in calculating the sectional shear stress as given by Equation 7.8, and also when checks are made for web crushing, as in Equation 7.22 below.

Figure 7.6(a) shows the stresses acting on a small element of concrete in the web of a prestressed beam and orientated so that the x-direction aligns with the longitudinal axis of the member and the y-direction is transverse to this axis. In this element, σ_{cx} is the stress in the longitudinal direction of the beam and shown here as a negative, or compressive, stress; τ_{xy} is the shear stress on the element; and σ_{c1} and σ_{c2} are the major (tensile) and minor (compressive) principal stress directions, respectively. In our beam it is also noted that the normal stress in the transverse direction is zero, that is $\sigma_{cy} = 0$.

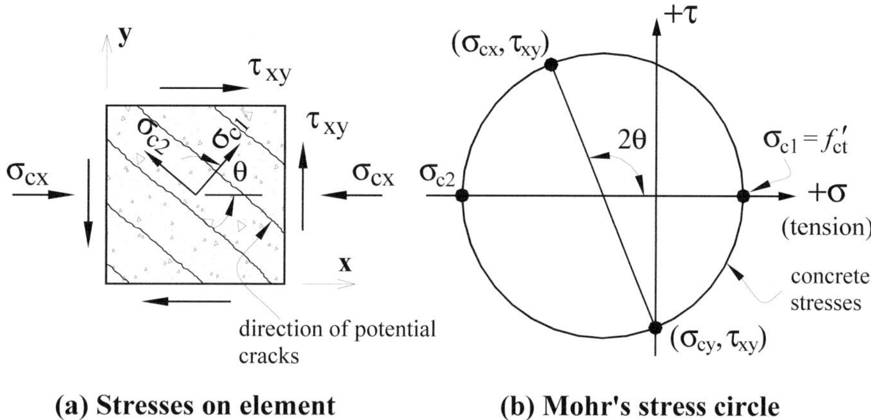

(a) Stresses on element **(b) Mohr's stress circle**

Figure 7.6 Mohr's circle construction for principal stresses

The corresponding Mohr's circle in Figure 7.6(b) can be used to express the major (tensile) principal stress in the concrete σ_{c1} as a function of the normal stresses σ_{cx} and τ_{xy} and is[1]:

$$\sigma_{c1} = \sqrt{(0.5\sigma_{cx})^2 + \tau_{xy}^{\ 2}} + 0.5\sigma_{cx} \tag{7.10}$$

noting that if σ_{cx} is compressive, it is entered into Equation 7.10 as a negative value.

The shear force V_t is evaluated by setting σ_{c1} equal to f'_{ct} and substituting Equations 7.7 and 7.8 into Equation 7.10.

The shear stress, which increases the value of σ_{c1}, is a maximum at the centroidal axis, whereas the compressive stress due to prestress and moment, which tends to reduce the value of σ_{c1}, normally decreases with distance y below the centroidal axis. It is therefore not obvious whether σ_{c1} will be a

1. **Sign convention:** Throughout the text we normally treat compressive stresses as positive, as we are dealing with prestressed concrete. However, for the Mohr's circle calculations undertaken to determine web-shear strength, the convention in mechanics of solids is to treat tension as positive.

maximum at the centroidal axis or somewhere else. It is normally sufficient to carry out the calculation at two points, one at the centroidal axis (where the direct stress due to bending is zero) and the other at the interface of web and tensile flange, shown as section *C-C* in Figure 7.5.

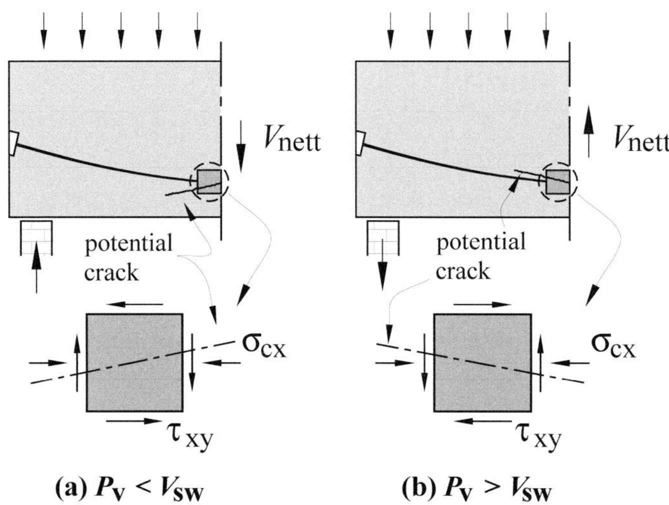

Figure 7.7 Effect of P_v on direction of principal tension and potential crack

At the web-flange junction, the direct stress σ_{cx} depends on the moment acting and this, in turn, is related to the applied shear force. A trial and error calculation is therefore necessary. A suitable procedure is to evaluate V_t for the stress conditions at the centroidal axis, and then use this shear force and its corresponding moment to calculate the principal tensile stress σ_{c1} at the web-flange intersection for this load condition. If σ_{c1} is less than the tensile strength f'_{ct} then V_t has been correctly determined. If $\sigma_{c1} > f'_{ct}$, then V_t has been overestimated and must be reduced. For the next trial, V_t may be reduced by the ratio of the tensile concrete strength to σ_{c1} and the process repeated.

In post-tensioned beams, the diameter of the cable ducts can significantly reduce the web thickness and special consideration may also need to be given to principal tensile stresses adjacent to ducts. Conditions in such weakened regions may also be critical at transfer, particularly if the vertical component P_v is larger than the shear force applied by the self-weight, V_{sw}. In this situation, the nett shear acting on the concrete is upward, so that the slope of the potential principal tensile crack is more or less aligned with the slope of the duct. This is shown in Figure 7.7(b). Splitting may then occur along the cable line.

EXAMPLE 7.1

CALCULATE THE LOAD ON A POST-TENSIONED BEAM IN SHEAR FOR A WEB-SHEAR CRACK TO FORM

A post-tensioned beam with the cross-section shown in Figure 7.8 is to span 25 metres between simple supports and carry a uniform live load of 24 kN/m. The beam is prestressed at each end by a parabolic cable with eccentricity of 524 mm at mid-span and zero at the ends. The cable comprises 22 strands of 12.7 mm diameter with $A_p = 2170$ mm^2, in a 112 mm diameter duct. Owing to friction the prestressing force varies in an approximately linear manner from 2435 kN at each end to 2300 kN at mid-span.

The beam also has longitudinal reinforcement consisting of 5-N28 bars ($A_{st} = 3100$ mm^2) at depth $d_{st} = 1160$ mm, throughout the span. For the longitudinal reinforcing steel and the stirrups we take $f_{sy} = 500$ MPa and $f_{sy.f} =$ 250 MPa (Grade R). The concrete strength is $f_c' = 40$ MPa.

Calculate the shear corresponding to the formation of a significant web-shear crack at a distance of $x = 1.0$ metres from the support.

Other data:

$$A_g = 488.4\times10^3 \text{ mm}^2;\; I_g = 83.8\times10^9 \text{ mm}^4$$

$$y_t = 552 \text{ mm};\; y_b = 668 \text{ mm}$$

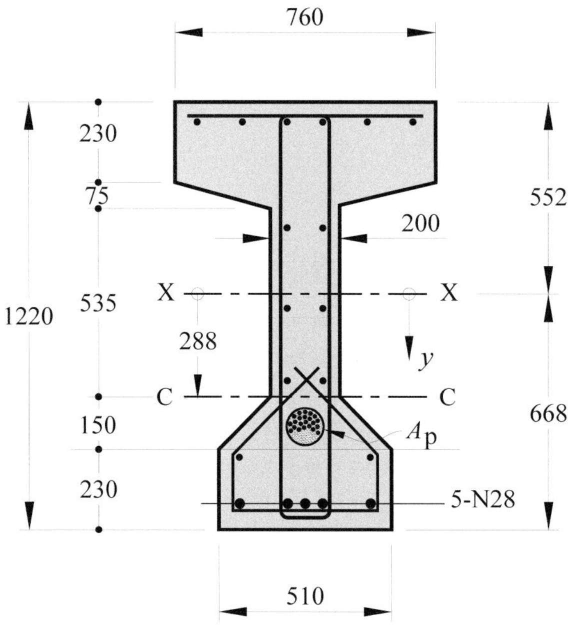

Figure 7.8 Cross-section for Examples 7.1, 7.2 and 7.3

SOLUTION

Self-weight:

$$w_G = A_g \times \gamma_c = 488.4 \times 10^3 \times 25 \times 10^{-6} = 12.2 \text{ kN/m}$$

Maximum service load:

Considering the service load factors $\psi_s = \psi_l = 1.0$, the maximum service load, w_S, that is applied to the girder is:

$$w_s = 12.2 + 24.0 = 36.2 \text{ kN/m}$$

For this load, the moment and shear at distance of x metres from the support are:

$$V = 452.5 - 36.2x \text{ kN}$$

$$M = 452.5x - 18.1x^2 \text{ kNm}$$

At x = 1.0 metres:

At this section the cable force is 2424 kN, the eccentricity (Equation 4.8) is e = 80.5 mm and the slope of the cable (Equation 4.9) is α = 0.077 radians.

We take the effective web width, after grouting, as:

$$b_v = 200 - 112/2 = 144 \text{ mm}$$

At this section the design stress resultants are:

$$V = 452.5 - 36.2 \times 1.0 = 416 \text{ kN}$$

$$M = 452.5 \times 1.0 - 18.1 \times 1.0^2 = 434 \text{ kNm}$$

Sign convention: As per footnote 1, we normally treat compressive stresses as positive, as we are dealing with prestressed concrete. However, for the Mohr's circle calculations, the convention in mechanics of solids is to treat tension as positive. To avoid any potential confusion we shall specify in the calculation results the state of the stress, tensile or compressive, in addition to its sign.

Web-shear cracking: The shear carried by the concrete at web-shear cracking, V_t, is that shear for which the principal tensile stress at some point in the web reaches the critical value of:

$$f_{ct} = 0.36\sqrt{f_c'} = 0.36\sqrt{40} = 2.28 \text{ MPa (tensile)}$$

We calculate first for conditions at the centroidal axis X–X (y = 0). At this level, the horizontal normal stress is:

$$\sigma_{cx} = \frac{P_e}{A_g} = \frac{-2424\times10^3}{488.4\times10^3} = -4.96 \text{ MPa (compressive)}$$

From Equation 7.8 the shear stress is: $\tau_{xy} = (V_t Q)/(I b_v)$, where Q is the first moment of area above (or below) the centroidal axis:

$$\begin{aligned} Q_{X-X} &= 760 \times 230 \times \left(552 - \frac{230}{2}\right) \\ &\quad + 75 \times \left(\frac{760-200}{2}\right) \times \left(552 - 230 - \frac{75}{3}\right) \\ &\quad + 200 \times (552 - 230) \times \left(\frac{552-230}{2}\right) \\ &= 92.99 \times 10^6 \ \text{mm}^3 \end{aligned}$$

The shear stress in the web is then:

$$\tau_{xy} = \frac{V_t \times 92.99 \times 10^6}{83.8 \times 10^9 \times 144} = 7.71 \times 10^{-6} V_t \ \text{MPa}$$

Taking $\sigma_{c1} = f_{ct} = 2.28$ MPa and evaluating Equation 7.10 we obtain:

$$\begin{aligned} \sigma_{c1} &= \sqrt{(0.5\sigma_{cx})^2 + \tau_{xy}^2} + 0.5\sigma_{cx} \\ &= \sqrt{(0.5(-4.96))^2 + (7.71 \times 10^{-6} V_t)^2} + 0.5(-4.96) \\ &= 2.28 \ \text{MPa (tensile)} \end{aligned}$$

and solving gives $V_t = 527 \times 10^3$ N (527 kN).

We now check conditions at the intersection of the web and bottom flange at the section C–C ($y = 288$ mm) for a shear force of $V_t = 527$ kN. The first moment of area below the axis is:

$$Q_{C-C} = 510 \times 230 \times \left(668 - \frac{230}{2}\right)$$
$$+ 150 \times \left(\frac{510-200}{2}\right) \times \left(668 - 230 - \frac{150}{3}\right)$$
$$+ 150 \times 200 \times \left(668 - 230 - \frac{150}{2}\right)$$
$$= 84.78{\times}10^6 \text{ mm}^3$$

and the shear stress is:

$$\tau_{xy} = \frac{527{\times}10^3 \times 84.78{\times}10^6}{83.8{\times}10^9 \times 144} = 3.70 \text{ MPa}$$

The bending moment corresponding to $V = 527$ kN may be obtained by factoring the moment for the design load relative to the shear for that moment; that is:

$$M = \frac{527}{416} \times 434 = 550 \text{ kNm}$$

and the normal stress due to this moment and the effective prestress at $y = +288$ mm on the section is determined from Equation 7.7 as:

$$\sigma_{cx} = \frac{My}{I_g} - \frac{P_e}{A_g} - \frac{P_e e y}{I_g}$$
$$= \frac{550{\times}10^6 \times 288}{83.8{\times}10^9} - \frac{2424{\times}10^3}{488.4{\times}10^3} - \frac{2424{\times}10^3 \times 80.5 \times 288}{83.8{\times}10^9}$$
$$= 1.890 - 4.963 - 0.671 = -3.74 \text{ MPa (compressive)}$$

Substituting $\tau_{xy} = 3.70$ MPa and $\sigma_{cx} = -3.74$ MPa into Equation 7.10:

$$\sigma_{c1} = \sqrt{(0.5\sigma_{cx})^2 + \tau_{xy}^{\ 2}} + 0.5\sigma_{cx}$$

$$= \sqrt{(0.5(-3.74))^2 + 3.70^2} + 0.5(-3.74)$$

$$= 2.28 \text{ MPa (tensile)}$$

Thus the major principal stress at the centroidal axis and the junction of the flange and web are equally critical and, thus, $V_t = 527$ kN.

The vertical component of the cable force is (Equation 7.5):

$$P_v = P_e \times \alpha = 2424 \times 0.077 = 187 \text{ kN}.$$

and by Equation 7.6, the shear at cracking is:

$$V_{cr} = 527 + 187 = 714 \text{ kN}$$

It is calculated that shear corresponding to web-shear cracking is well above that at service, 416 kN, and thus web-shear cracking under service conditions would not be expected. At ultimate at $x = 1.0$ metres, $V^* = 582$ kN and, similarly, web-shear cracking is not controlling for this section.

7.6 Strength in shear

7.6.1 Introduction

Both in recent research and in modern codes of practice, there has been a move to the use of physical (and more rational) models for evaluating the shear strength of concrete structural members. In particular, standards such as CSA A23.3 (2014), AASHTO (2015) AS 5100.5 (2017), AS 3600 (2018) and others, as well as the 2010 *fib* Model Code (2013), have based their treatment of shear on the modified compression field theory (MCFT).

Two distinct modes of failure have to be considered when evaluating the strength in shear of a region of a concrete member. These are (a) shear failure following yielding of the shear reinforcement, and (b) inclined crushing of the web concrete before yielding of the shear reinforcement can occur.

In the case of shear failure by inclined crushing of the concrete in the web, the strength is evaluated from a consideration of the crushing of the diagonal concrete struts, taking into account the effect of the transverse concrete strains. The calculation of shear strength for this mode of failure is dealt with in Section 7.6.4 following.

If failure occurs after stirrup yielding, the shear strength is obtained using Equation 7.2, that is to say, by adding the contributions of the concrete, V_{uc}, the shear reinforcement, V_{us}, and the added shear resistance provided by any inclined tendon in the region, P_v. A simple estimate of P_v is usually taken to be the vertical component of the inclined tendon force at the time of loading (i.e. taking account of losses) but prior to the application of load. The vertical component P_v of the inclined tendon force P usually opposes the applied shear, as for example in the case of a post-tensioned simply supported beam, but there are situations where P_v can act in the same direction as the applied shear and so decreases shear strength.

The contribution of the shear reinforcement, V_{us}, is usually determined on the assumption of truss-like action, although more sophisticated approaches are based on strut-and-tie analysis or stress-field theory. The evaluation of V_{us} is considered in Section 7.6.3 of this chapter.

Until recently, AS 3600 and other national codes of practice took a pragmatic approach to evaluating V_{uc}. The shear at inclined cracking (the smaller of the shear at diagonal web-cracking and the shear at flexure-shear cracking) was used as an acceptable estimate of V_{uc}. In the current edition of AS 3600, however, a new approach has been introduced, which is based on models developed from research that more closely follow observed overload behaviour, and account is taken of new variables such as the longitudinal strain in the web at failure, ε_x. In Section 7.6.2 following, a procedure for evaluating V_{uc} is presented which is based specifically on the AS 3600–2018 design provisions. While this approach will be broadly similar to those proposed by current researchers, the numerical data are aligned specifically to AS 3600. Strength calculations are often needed in the evaluation of existing buildings: the approach presented in Section 7.6.2 conforms with the relevant provisions of the current Australian Standard.

7.6.2 Contribution of concrete to shear strength, V_{uc}

In AS 3600, and other standards, the concrete component of the strength is given as:

$$V_{uc} = k_v \sqrt{f_c'}\, b_v d_v \qquad (7.11)$$

The term k_v is a parameter that determines the capacity of the web to resist the stresses that provide the concrete contribution to the shear strength, including the size effect and aggregate interlocking, d_v is the internal moment lever arm at the critical section for shear and b_v is the effective width of the web (Equation 7.9).

The term $\sqrt{f_c'}$ appears in Equation 7.11 because shear strength is related to frictional resistance along the surface of the critical shear crack, which is one of the variables influencing V_{uc}. The shear strength on a crack is usually assumed to vary with the square root of f_c'. For normal strength concrete, shear cracks generally follow a path around aggregate particles, providing this frictional resistance to sliding along the crack. However, as concrete strength increases, cracks tend to pass more through the aggregates, reducing the frictional strength. The strength of the concrete where this transition occurs depends on the strength of aggregate used, together with the strength of the cementitious matrix bond to these aggregates. An upper limit of 65 MPa is placed on the concrete's strength in this equation and, thus, $\sqrt{f_c'}$ is limited to 8.1 MPa, which is rounded to 8 MPa in the Standard.

At any critical shear section it is generally assumed in the design model that the longitudinal reinforcement is not at yield. That is, the strength of the section is not limited by its flexural capacity, but fails before yielding of the longitudinal reinforcement, with an inclined critical shear crack forming first. Thus, the depth to the centroid of the tensile force from the extreme compressive fibre is taken as approximately equal to the depth of the geometric centroid of the reinforcement in the tensile half of the section (which is approximately equal to the centroid of the tensile force for the case of the reinforcement not being at yield).

For several layers of longitudinal tensile reinforcement the effective depth may be taken as:

$$d = \frac{\sum_{i=1}^{n} A_{sti} d_{si} + \sum_{j=1}^{m} A_{pj} d_{pj}}{\sum_{i=1}^{n} A_{sti} + \sum_{j=1}^{m} A_{pj}} \tag{7.12}$$

where n and m are the number of bar and tendon layers, respectively, located within the tensile half of the cross-section; A_{sti} and A_{pj} are the areas of reinforcing steel and prestressing steel for layer i and j, respectively; and d_{si} and d_{pj} are the distances from the extreme compressive fibre to the bars and tendons located at layers i and j, respectively.

The internal lever arm, d_v, for this section is then taken as:

$$d_v = 0.9d\text{, but not taken as less than } 0.72D \tag{7.13}$$

For reinforced concrete members, the Standard offers two accuracy levels for the determination of k_v; however, the lower level model, referred to as the '*simplified method*', is not allowed for prestressed concrete and, thus, is not described here. The higher level model, the '*general method*', provides two equations for calculating the concrete contribution to the shear stress, depending on whether a minimum quantity of transverse reinforcement, $A_{sv.min}$ (Equation 7.16), is provided, or not. For an area of shear reinforcement of less than the minimum (i.e. $A_{sv} < A_{sv.min}$), the value of k_v is taken as (Sigrist et al., 2013):

$$k_v = \left[\frac{0.4}{1 + 1500\varepsilon_x}\right]\left[\frac{1300}{1000 + k_{dg} d_v}\right] \tag{7.14}$$

whereas, for $A_{sv} \geq A_{sv.min}$:

$$k_v = \left[\frac{0.4}{1 + 1500\varepsilon_x}\right] \tag{7.15}$$

For AS 3600, in Equation 7.14, for $f'_c \leq 65$ MPa and normal weight concrete:

$$k_{dg} = \left[\frac{32}{16 + d_g}\right] \text{ but not less than } 0.8$$

where d_g is the size of the maximum aggregate particles. For concrete compressive strengths greater than 65 MPa, and for light weight concrete, $k_{dg} = 2.0$.

Research shows a minimum level of transverse shear reinforcement is needed if depth (or size) effect in beams subject to a shear failure is to be avoided (Collins and Kuchma, 1999), and this minimum is generally accepted as:

$$A_{sv,min} = 0.08 b_v s \sqrt{f'_c} / f_{sy.f} \qquad (7.16)$$

where $f_{sy.f}$ is the yield strength of the stirrups provided.

In members with less than the minimum specified web reinforcement, shear strength is strongly influenced by two factors, the depth effect and the strain effect. Both, together with the spacing of the cracks, influence the width of the shear crack and thus the amount of sliding (aggregate interlock) force that it can resist – the wider the crack, the less force can be resisted. In Equation 7.14, the left hand bracketed term represents the strain effect, and the right the depth effect. For members without shear reinforcement, the spacing between cracks is about 0.7 to 1.0 times the depth of the member; for members with minimum shear reinforcement, and more, $k_{dg}d_v$ is approximately 300 mm, and Equation 7.14 becomes Equation 7.15. While the depth effect is not influential in the capacity of members with more than minimum web reinforcement, the longitudinal strain influence remains. Although it is not immediately obvious from the design equations, the strain term ε_x is related to the width of the shear crack, and it is the crack width that indeed controls the shear strength.

In lieu of more accurate calculations, the mid-depth strain parameter ε_x in the concrete can be determined by (Bentz and Collins, 2006):

$$\varepsilon_x = \frac{\dfrac{|\alpha M|}{d_v} + |\alpha V| - \gamma P_v + 0.5\alpha N - A_{pt} f_{po}}{2(E_s A_{st} + E_p A_{pt})} \qquad (7.17)$$

... within the limits $0 \le \varepsilon_x \le 3.0 \times 10^{-3}$

where αM and αV are the moment and shear acting on the section, respectively, and γP_v is the vertical component of the prestress at the section. For analysis of strength of a section, $\alpha = \gamma = 1.0$. For design, as discussed in Section 7.7, α and γ are load factors.

Should the calculated value of ε_x be less than zero, ε_x may be taken as zero or recalculated to account for force in the compressed concrete in the tension region. In this case:

$$\varepsilon_x = \frac{\dfrac{|\alpha M|}{d_v} + |\alpha V| - \gamma P_v + 0.5\alpha N - A_{pt} f_{po}}{2(E_s A_{st} + E_p A_{pt} + E_c A_{ct})} \qquad (7.18)$$

$$\text{... within the limits } -0.2\times10^{-3} \leq \varepsilon_x \leq 0$$

where A_{ct}, is the area of the section taken between the extreme tensile fibre and its mid-depth ($D/2$). For a symmetric section $A_{ct} = A_g/2$.

In Equations 7.17 and 7.18:

- The moment must be at least $\alpha M \geq (\alpha V - \gamma P_v)\, d_v$
- N is taken as positive for tension and negative for compression
- A_{st} and A_{pt} are the areas of reinforcing bars and prestressing tendons, respectively, in the half depth of the section containing the flexural tension zone (in the zone defined by a depth of $D/2$ from the extreme tensile fibre).
- f_{po} is the stress in the prestressed tendon at decompression of the adjacent concrete. According to AS 3600, f_{po} may be determined by calculation or taken as equal $0.5f_{pb}$, where f_{pb} is the tendon's characteristic breaking stress.

It is worthy of mention that Equations 7.17 and 7.18 contains a discontinuity when considering the area of prestressing steel that lies close to mid-depth of the cross-section, either on the compressive side or the tensile side. This can lead to significant differences in the calculated shear strength with only small changes to depths of prestressing steel.

When the prestressing steel lies close to the mid-depth of the section on the compressive side, ignoring this steel can lead to an overly conservative approximation in the calculation of ε_x and, thus, the concrete's contribution to the section's shear capacity. This can be important in cases, for example, such as prestressed slabs, where only a small area of, or no, conventional reinforcement is used on the tensile side.

In such cases, a more accurate procedure is needed for the calculation of the strain in the section at its mid-height. This can be undertaken using first principles, with consideration of bending, axial force, including the effect of prestress, and shear contributions. Alternatively, the authors recommend using the following modified version of the *fib* Model Code 2010 (2013) for calculation of ε_x:

$$\varepsilon_x = \frac{\frac{|\alpha M|}{d_v} + |\alpha V| - \gamma P_v + 0.5\alpha N - A_{pt} f_{po}}{2\left(E_s A_{st} \times \frac{d_s}{d} + E_p A_{pt} \times \frac{d_p}{d}\right)} \qquad (7.19)$$

$$\ldots \text{ within the limits } 0 \le \varepsilon_x \le 3.0 \times 10^{-3}$$

where d_s is the depth from the extreme compressive fibre to the geometric centroid of the bar reinforcement located in the tensile half-depth, and d_p is the depth to the geometric centroid of the prestressing steel from the extreme compressive fibre, and need not be located in the tensile half-depth.

7.6.3 Contribution of shear reinforcement to shear strength, V_{us}

In Figure 7.9 the assumptions of the truss model are illustrated and the forces crossing an inclined crack are shown. If the stirrups are at yield, the shear force carried across the crack is $nA_{sv}f_{sy.f}$, where n is the number of stirrups crossing the crack and A_{sv} is the area of one stirrup unit. If the crack is inclined at angle θ_v to the horizontal, its horizontal projection is approximately $d_1 = d_v \cot \theta_v$; thus, with vertical stirrups at spacing s, we have $n = d_1/s$, where d_1 is the length of the failure surface in the direction of the member and s is the spacing of the stirrups. The contribution of the stirrup reinforcement to the shear strength at the section of the member being considered is then:

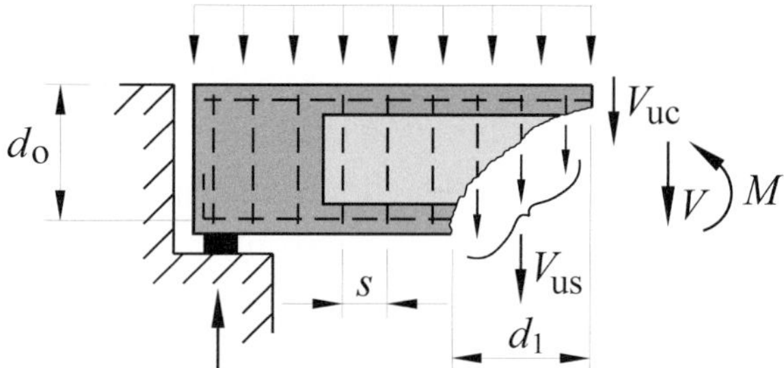

Figure 7.9 Shear forces at failure

$$V_{us} = f_{sy.f} \frac{A_{sv}}{s} d_v \cot \theta_v \tag{7.20}$$

where A_{sv} and s are the cross-sectional area of shear reinforcement and the spacing of the reinforcement, respectively, $f_{sy.f}$ is the yield strength of the web reinforcing steel and θ_v is the angle of the compressive strut relative to the longitudinal axis of the member, respectively.

The angle of inclination of the concrete compression strut to the longitudinal axis of the member (θ_v) is:

$$\theta_v = 29 + 7000\varepsilon_x \tag{7.21}$$

7.6.4 Shear strength due to web crushing

Failure can occur prior to yielding of the stirrup reinforcement (the fitments) due to inclined crushing of the concrete in the web. This is known as web crushing failure. In this case, the strength of the section is limited by the strength of the diagonally cracked concrete; it places an upper limit on the shear capacity of a member, which cannot be increased by adding additional stirrups.

In AS 3600 the upper limit on shear strength is denoted as $V_{u.max}$ and is calculated using the expression:

$$V_{u.max} = 0.55\left[0.9 f_c' b_v d_v \left(\frac{\cot(\theta_v) + \cot(\alpha_v)}{1 + \cot^2(\theta_v)}\right)\right]; \text{ or} \tag{7.22}$$

$$V_{u.max} = 0.55\left[0.9 f'_{cp} b_v d_v \left(\frac{\cot(\theta_v) + \cot(\alpha_v)}{1 + \cot^2(\theta_v)}\right)\right] \tag{7.23}$$

... at transfer

in which b_v is the reduced web width allowing for the presence of grouted or ungrouted ducts in the section as given by Equation 7.9. In the case of stirrups placed normal to the longitudinal axis, that is $\alpha_v = 90$ degrees, the web crushing limits may also be expressed as:

$$V_{u.max} = 0.55[0.9 f'_c b_v d_v \cos\theta_v \sin\theta_v]\text{; or} \tag{7.24}$$

$$V_{u.max} = 0.55[0.9 f'_{cp} b_v d_v \cos\theta_v \sin\theta_v] \text{ at transfer} \tag{7.25}$$

7.6.5 Shear strength at a specified cross-section

To analyse the shear strength of a member at a particular cross-section, in addition to knowledge of materials and sectional properties at the section considered, details of full-member loading arrangements and combinations are needed. It may be that several loading combinations require investigation to determine the most critical. The following approach is suggested for determining the shear strength of a member at a given cross-section[2]:

1. For the loading arrangement being studied, calculate the ratios of M/V and N/V; this may be done by analysing the member for loads that are in proportion to those of the case considered. For example, for a single span member under a uniform load, the member may be analysed for a unit uniform load and the moment and shear force, and hence M/V, at the section being investigated extracted from the analysis results.

2. The procedure described here will lead to slightly different ultimate shear strengths than when calculated from AS 3600. This is because the Standard uses values of ε_x corresponding to factored actions for determination of V_u. This provides a slightly lower estimate for ε_x than for the ultimate condition and, thus, slightly higher capacities. To distinguish the shear capacity at ultimate calculated through this analysis procedure, the strengths of the concrete and web reinforcement components are denoted as $\overline{V}_{uc}$ and $\overline{V}_{us}$, respectively, and their sum as $\overline{V}_u$.

2. Estimate an initial value for the mid-height strain ε_x and, with this value, calculate k_v from Equations 7.14 or 7.15, as appropriate, and θ_v, from Equation 7.21.
3. Determine V_{uc} from Equation 7.11, V_{us} from Equation 7.20, and $V_u + P_v$ from $V_u + P_v = V_{us} + V_{uc} + P_v$.
4. Calculate our estimate for M_u from $M_u = (V_u + P_v) \times M / V$.
5. With $\alpha = \gamma = 1.0$, determine a new value of ε_x from Equations 7.17 to 7.19, as appropriate.
6. Compare the initial value for ε_x and our newly calculated one. If not within the desired tolerance, make a new estimate for ε_x and repeat from Step 2. For positive values of ε_x, usually, taking one-third of the initial value plus two-thirds of our calculated value provides for speedy convergence. For negative values, adopting the value of ε_x determined at the end of the trial for the following estimate will usually provide for quick convergence.
7. Once converged and V_u is determined, check to ensure that the sectional capacity is not limited by its web crushing strength (Equations 7.22 to Equations 7.25).

To obtain the ultimate strength predicted by applying the equations of AS 3600, the same procedure may be used but with the mid-height strain parameter taken as:

$$\varepsilon_x = \frac{(\phi V_u + \gamma_p P_v) \cdot \dfrac{M}{V d_v} + \phi V_u + 0.5\phi N_u - A_{pt} f_{po}}{2(E_s A_{st} + E_p A_{pt})} \tag{7.26}$$

... within the limits $0 \le \varepsilon_x \le 3.0 \times 10^{-3}$; or

$$\varepsilon_x = \frac{(\phi V_u + \gamma_p P_v) \cdot \dfrac{M}{V d_v} + \phi(V_u) + 0.5\phi N_u - A_{pt} f_{po}}{2(E_s A_{st} + E_p A_{pt} + E_c A_{ct})} \tag{7.27}$$

... within the limits $-0.2 \times 10^{-3} \le \varepsilon_x \le 0$

where $V_u = V_{uc} + V_{us}$ and $\phi = 0.70$ or 0.75, as appropriate, and $\gamma_p = 0.9$.

For a reinforced concrete member, a value of 40 per cent of the yield strain of the longitudinal tensile reinforcement usually provides a reasonable starting estimate for ε_x, and for a fully prestressed member an initial value of $\varepsilon_x = 0$ is suggested. For a partially prestressed member, a value for ε_x of between 10 and 20 percent of the longitudinal tensile reinforcement yield strain is suggested. Examples of the above procedure are presented below and in the companion text *Reinforced Concrete Basics* (Foster et al., 2021).

EXAMPLE 7.2

Analysis of Shear Strength at a Section

The post-tensioned beam of Example 7.1, with the cross-section shown in Figure 7.8, is to carry a uniformly applied load along its full length. As before, the beam is prestressed at each end by a parabolic cable with eccentricity of 524 mm at mid-span and zero at the ends. The cable comprises 22 strands of 12.7 mm diameter with $A_p = 2170$ mm^2, in a 112 mm diameter duct. Owing to friction the prestressing force varies in an approximately linear manner from 2435 kN at each end to 2300 kN at mid-span. The beam also has longitudinal reinforcement consisting of 5-N28 bars ($A_{st} = 3100$ mm^2) at depth $d_{st} = 1160$ mm, throughout the span

If the member contains 2-leg R10 stirrups spaced at 270 mm centres throughout, calculate the shear strength of a section located at distance of $x = 1.0$ metres from the support.

Other data:

$f_c' = 40$ MPa; $f_{sy} = 500$ MPa; $f_{sy.f} = 250$ MPa; $f_{pb} = 1870$ MPa

$b_v = 144$ mm; $d_v = 878$ mm; $P_v = 187$ kN; $E_p = 195$ GPa

(note: b_v, d_v and P_v are calculated in Example 7.3).

SOLUTION

Bending moment to shear force ratio:

Consider an applied uniform load of 1.0 kN/m along the length of the girder. The moment and shear force at distance of $x = 1.0$ metres from the support are 12 kNm and 11.5 kN, respectively. The bending moment to shear force ratio is thus:

$$M/V = 12.0/11.5 = 1.044 \text{ metres}$$

As the section is heavily prestressed we shall assume an initial estimate of the mid-height strain as $\varepsilon_X = 0$. A check on the supplied web reinforcement shows that more than minimum is provided (see Example 7.3) and, thus, from Equations 7.15 and 7.11:

$$k_v = \left[\frac{0.4}{1 + 1500 \times 0.0}\right] = 0.40$$

$$\bar{V}_{uc} = 0.40\sqrt{40} \times 144 \times 878 = 320 \times 10^3 \text{ N} \quad (320 \text{ kN})$$

From Equations 7.21 and 7.20:

$$\theta_v = 29 + 7000 \times 0.0 = 29.0^o$$

$$\bar{V}_{us} = 250 \times \frac{160}{270} \times 878 \times \cot 29.0 = 235 \times 10^3 \quad (235 \text{ kN})$$

and thus:

$$\bar{V}_u + P_v = 320 + 235 + 187 = 742 \text{ kN}$$

$$\bar{M}_u = 742 \times 1.044 = 775 \text{ kNm}$$

From Equation 7.17 (with $\alpha = \gamma = 1.$):

$$\varepsilon_x = \frac{\dfrac{775 \times 10^6}{878} + 742{\times}10^3 - 187{\times}10^3 - 2170 \times 0.5 \times 1870}{2(200{\times}10^3 \times 3100 + 195{\times}10^3 \times 2170)}$$

$$= -591.2{\times}10^3/(2 \times 1.043{\times}10^9) = -283{\times}10^{-6}$$

$$\text{... within } 0 \le \varepsilon_x \le 3.0 \times 10^{-3}$$

As the calculated value of ε_x is less than zero, we shall recalculate it using Equation 7.18:

$$A_{ct} = \frac{1220}{2} \times 200 + 230(510 - 200) + \frac{150}{2}(510 - 200)$$

$$= 216.6{\times}10^3 \text{ mm}^2$$

$$\varepsilon_x = \frac{-591.2{\times}10^3}{2 \times (1.043{\times}10^9 + 32.8{\times}10^3 \times 216.6{\times}10^3)} = -36.3{\times}10^{-6}$$

$$\text{... and is greater than } -0.2{\times}10^{-3} \quad \therefore \text{OK}$$

We see that our calculated value of $\varepsilon_x = -36.3{\times}10^{-6}$ is reasonably close to that of our initial estimate of $\varepsilon_x = 0$. Nevertheless, we shall undertake a further trial with $\varepsilon_x = -36.3{\times}10^{-6}$:

By Equations 7.15 and 7.21:

$$k_v = \frac{0.4}{1 + 1500 \times (-36.3{\times}10^{-6})} = 0.423$$

$$\theta_v = 29 + 7000 \times (-36.3{\times}10^{-6}) = 28.7^{\circ}$$

and by Equations 7.11 and 7.20:

$$\bar{V}_{uc} = 0.423\sqrt{40} \times 144 \times 878 = 338 \text{ kN}$$

$$\bar{V}_{us} = 250 \times \frac{160}{270} \times 878 \times \cot 28.7 = 238 \times 10^3 \quad (238 \text{ kN})$$

Thus:

$$\bar{V}_u + P_v = 338 + 238 + 187 = 763 \text{ kN}$$

$$\bar{M}_u = 763 \times 1.044 = 797 \text{ kNm}$$

and our new estimate for ε_x is (Equation 7.18):

$$\varepsilon_x = \frac{\dfrac{797 \times 10^6}{878} + 763 \times 10^3 - 187 \times 10^3 - 2170 \times 0.5 \times 1870}{2 \times (1.043 \times 10^9 + 32.8 \times 10^3 \times 216.6 \times 10^3)}$$

$$= -545 \times 10^3 / 16.30 \times 10^9 = -33.4 \times 10^{-6}$$

With further iterations we find that $\varepsilon_x = -33.8 \times 10^{-6}$, and the shear strength of the member at $x = 1.0$ m is determined as $\bar{V}_u = 761$ kN. A check on web crushing shows that it does not govern (see Example 7.3).

It is worthy of note that the shear strength is a function of the ratio M/V at that section, and other load cases may prove more critical.

In this example, using the procedure of AS 3600, with $\alpha = \phi = 0.75$ and $\gamma = \gamma_p = 0.9$, $\varepsilon_x = -53.3 \times 10^{-6}$, and the capacity of the section is evaluated as $V_u = 773$ kN (including the vertical component of prestress, P_v). The treatment of design is discussed in the following section.

7.7 Design for shear according to AS 3600

7.7.1 Introduction

In design, prestress is sometimes treated as a load term, and sometimes as a resistance term. This has already been discussed in Chapter 4 in relation to load balancing. When it is treated as a load, a load factor is usually applied to it, while a capacity reduction factor is applied when it is treated as a resistance term. In the original 2018 edition of the current Australian Standard, P_v was treated as a resistance term and was added to the right hand side of Equation 7.28. However as mentioned in Section 7.2.2, in Amendment 2 to the Standard, this was changed, and P_v is now treated as a load term. This change was implemented to treat safety factors more consistently, reducing its influence when positively impacting shear strength and increasing when negatively impacting. In this section, design requirements specific to AS 3600 are outlined and a recommended design procedure provided.

7.7.2 Design requirements

As described in the previous section, in prestressed concrete members the total shear is resisted by components of concrete (V_{uc}) and of the reinforcing steel (V_{us}). The design equation is:

$$\phi V_u = \phi(V_{uc} + V_{us}) \geq V^* - \gamma_p P_v \tag{7.28}$$

where V_u is the ultimate shear capacity of the section, V^* is the factored design shear, ϕ is the capacity reduction factor (0.70 or 0.75, as appropriate), P_v is the vertical component of the prestress at the cross-section being considered (with a positive value reducing the shear on the section and a negative value increasing it) and γ_p is a load factor equal to 0.9 when P_v is positive and 1.15 when negative.

The design of beams for shear in AS 3600 is carried out using Equations 7.28, 7.11 and 7.20, with $\gamma = \gamma_p$ and the terms αM, αV, and αN replaced with the factored design stress resultants M^*, V^* and N^*, respectively. The capacity reduction factor ϕ for shear has the value 0.75 where at least minimum Class N transverse shear reinforcement is provided and failure is not by web crushing; otherwise, it is equal to 0.7 (Table 3.1). It is to be noted that the most adverse combination of strength design loads giving the highest value V^* need not be the most critical for design, as a member's capacity in shear also depends also on the actions of bending and axial force. For example, a loading scenario providing a higher value for M^* (or N^*), and lower V^*, may prove to be the critical condition.

The critical section for design, also referred to as the '*control section*', is dependent on the combination of the shear force and bending moment acting at each section of a member. In many cases the control section will occur near a support. However, if decisive, other control sections may also require consideration.

For cases where the member is:

- directly supported, i.e. where a diagonal compressive strut frames into the support (the member is not suspended); and
- no significant concentrated load or changes in distributed load occur on the member between the support face and the distance d_v from it,

the strength of the member closer than a distance of d_v from the support is deemed to be satisfied provided at least the minimum quantity and spacing of the transverse shear reinforcement determined at d_v from the support is continued, unchanged, to the support face. This recognises the beneficial effect of arch action in the end region.

When a concentrated load acts close to a support, part of this load may be transferred directly to the support by strut action. This effect is especially important when the concentrated loads are large in comparison to the distributed loads. In such cases the design of the region is best undertaken using the strut-and-tie method which is dealt with in Chapter 3 and in the companion text *Reinforced Concrete Basics* (Foster et al., 2021).

Several practical limits have to be imposed on the shear design process. For example, a maximum limit on the shear that can be carried by a beam section, $V_{u.max}$, as already introduced in Equations 7.22 and 7.23, is specified by AS 3600. Others include:

- Where web reinforcement is required, a minimum quantity, $A_{sv.min}$, must be provided. This is to avoid sudden shear failure during the formation of the inclined cracks as load is applied to the member, which is particularly important in deeper members, and to mitigate the size effect. The expression for $A_{sv.min}$ in given in Equation 7.16.

- As explained in Section 7.6.2, the angle θ_v of the inclined stress field is not fixed but, rather, may be treated as a design variable through adopting a higher value for ε_x than the minimums given by Equations 7.17 or 7.18, but not greater than +0.003. The values of k_v and θ_v are calculated with the designer's selected value for ε_x. Higher values than the minimum may be desirable, for example, where anchorage demands on longitudinal reinforcement are high and detailing is thus difficult.

  Although the requirement $\varepsilon_x \leq 0.003$ limits the stress field angle to a maximum of 50° degrees, the strut-and-tie approach allows the use of a variable strut angle of up to 60°, with a concrete component to the shear strength taken as zero (i.e. $V_{uc} = 0$).

Other design requirements also apply, some of which are outlined in Section 7.7.5. Based on the truss model, the shear reinforcement required at any cross-section must be continued for a distance $d_v \cot(\theta_v)$ in the direction of decreasing shear. In support regions where forces are applied to the beam in such a way as to produce vertical compressive stress in the beam, the critical section is taken to be at d_v from the face of the support, with the shear reinforcement requirements at this section being carried through to the support.

7.7.3 Additional longitudinal reinforcement required for shear, with or without torsion

In this section the additional reinforcement in the longitudinal direction resulting from the influence of shear is discussed (see Figure 7.3). The longitudinal force, and thus reinforcement, is also a result of torsion, which is discussed in Section 7.8. Each of the shear and torsional components of this force are outlined here, with a total area of additional longitudinal reinforcement determined to provide for both.

As mentioned, in addition to transverse forces for shear, forces are induced in the longitudinal direction, above that calculated for flexure, and must also be accounted for; this is due to the diagonal compression. As seen in Figure 7.10, the additional longitudinal tension force at the control section considered is equal to the longitudinal component of the of diagonal force F_d.

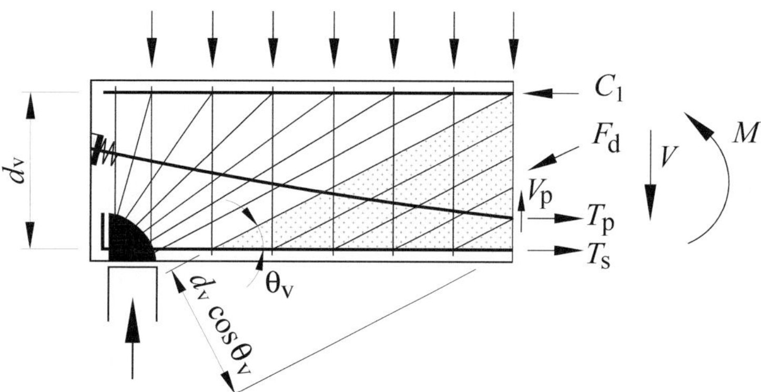

Figure 7.10 Forces acting on cross-section

For detailing, this force is split into two equal components, one placed at the centroid of the flexural compressive forces, and the other at the level of the flexural tensile forces. The additional force on each side (tensile and compressive) is:

$$\Delta F_{tds} = \min[0.5(V^* - \gamma_p P_v + \phi V_{uc}),\ V^* - \gamma_p P_v] \cot\theta_v \tag{7.29}$$

and a suitable area of tensile reinforcement (above that for flexure) is to be provided to meet the demand. In Equation 7.29, ϕ is that for shear.

On the tensile side of the neutral axis, the design tension force at a section is the sum of the flexural, axial and shear components, and is:

$$F_{td} = \gamma_{td}\left(\frac{M^*}{z} + \frac{N^*}{2}\right) + \Delta F_{td} \tag{7.30}$$

where M^* and N^* are the factored design moment and axial force (tension positive), respectively, ΔF_{td} is the additional force that results from the diagonal compression due to shear and torsion and z is the internal lever arm between the internal flexural tension and flexural compression forces from bending.

In Equation 7.30 the term γ_{td} is introduced; this term does not appear in the originally published edition of AS 3600–2018, but a variation of it did in the torsion rules of AS 3600–2009. The factor accounts for uncertainties where stress resultants that occur in opposite directions are summed; there is a quantifiable probability that the terms providing compression are over-predicted while, at the same time, the terms leading to tension are under-predicted, leaving a theoretical nett compression and an actual nett tension. In this book we recommend $\gamma_{td} = 1.0$ where the resultant of the sum of the applied bending and axial compression is a positive force, and $\gamma_{td} = 0.85$ where negative.

It will be seen in the treatment of torsion in later sections of this chapter that additional tensile longitudinal steel may also be needed when torsion acts. The additional tensile force for shear and torsion is:

$$\Delta F_{td} = \Delta F_{tds} + \Delta F_{tdt} \tag{7.31}$$

where the shear component ΔF_{tds} is given in Equation 7.29 and the torsion component ΔF_{tdt} by Equation 7.51, below.

The required area of longitudinal tension reinforcement is:

$$A_{st} \geq F_{td}/(\phi f_s) \tag{7.32}$$

where f_s is the maximum available stress in the developed bars (for fully developed bars, $f_s = f_{sy}$) and ϕ taken as that for members in combined bending and applied axial force (see Table 3.1).

On the flexural compressive side, the design force is:

$$F_{cd} = \gamma_{td}\left(\frac{-M^*}{z} + \frac{N^*}{2}\right) + \Delta F_{td} \tag{7.33}$$

and the required area of longitudinal compressive reinforcement is:

$$A_{sc} \geq F_{cd}/(\phi f_s) \tag{7.34}$$

7.7.4 Design of shear reinforcement

According to AS 3600, at least the minimum transverse shear steel for a concrete member is to be provided if the overall depth of the section (D) is equal to or greater than 750 mm, or where:

$$V^* - \gamma_p P_v > k_s \phi V_{uc} \tag{7.35}$$

$$k_s = (1000 - D)/700 \quad \text{... within the limits } 0.5 \le k_s \le 1.0 \tag{7.36}$$

For shear design, a number of cross-sections along the beam are considered in turn, commencing at the critical section of maximum shear. The shear steel requirements, calculated at any particular section, are carried through to the next section. Although the web-shear steel might theoretically be varied continuously along the beam, practical considerations dictate that it should only be changed at reasonably large intervals, depending on the shape of the shear force diagram.

The following steps can be used to determine the stirrup requirement at any particular section:

1. Determine the design bending moment M^* and shear force V^* for the section being considered.
2. Calculate the minimum allowed mid-height strain parameter ε_x from Equations 7.17 or 7.18, as appropriate[3]. Choose a value of ε_x between that calculated and the maximum allowed value of +0.003.
3. With the chosen value of ε_x, calculate the shear V_{uc} resisted by the concrete from Equation 7.11.
4. Calculate θ_v from Equation 7.21. By adjusting the value of ε_x, the angle θ_v can be taken as a design variable between a minimum value determined with the lowest possible ε_x and 50°.
5. Calculate $V_{u.max}$ from Equation 7.22 or 7.24. If $V^* - \gamma_p P_v > \phi V_{u.max}$ it is necessary to increase the dimensions of the section.
6. Recalculate the shear V_{uc} resisted by the concrete from Equation 7.11. If $V^* \le k_s \phi V_{uc}$, $A_{sv.min}$ is calculated from

3. In iterating to calculate ε_x, in lightly prestressed members an initial estimate of $\varepsilon_x = 0.0004$ is generally reasonable, whereas for fully prestressed members a value of zero is normally a good starting assumption.

Equation 7.16 and then no further design is necessary except if $D > 750$ mm, in which case $A_{sv.min}$ is provided.

7. If $V^* > k_s \phi V_{uc}$ or if $D > 750$ mm, calculate $A_{sv.min}$ from Equation 7.16.
8. Calculate A_{sv} / s from Equation 7.20 in which $\phi V_{us} = V^* - \phi V_{uc}$. Select the stirrup size and, thus, determine the stirrup spacing s.
9. Determine the longitudinal reinforcement (Equations 7.32 and 7.34) and detail the member.

In steps (2) and (4) above, it is mentioned that the designer may select a value of ε_x, and thus θ_v, that falls between the minimum determined through the equations and a value of +0.003. For most cases, the minimum value of ε_x (and θ_v) will likely be adopted. However, where detailing of longitudinal reinforcement dictates, for example in providing sufficient anchorage near beam ends, higher values may be desirable. Adopting a higher value for θ_v lessens the demand on the longitudinal reinforcement, while increasing it on the transverse reinforcement, thus requiring a higher quantity of stirrups. An example is provided in the companion text *Reinforced Concrete Basics* (Foster et al., 2021).

Where a higher strut angle (θ_v) than the minimum is to be used, a certain degree of ductility is needed to allow the structural system to rebalance. That is, for the angle to rotate to its design value, yielding, or slipping, of the longitudinal reinforcement is needed, allowing yielding of the transverse reinforcement and, thus, full shear capacity to be ascertained. In this case, the Standard requires that Class N reinforcement be used for both the transverse and longitudinal reinforcement.

For convenience in construction it is usual to use one stirrup size (A_{sv} constant) and vary the spacing s, as and if necessary, to suit the requirements in different regions of the beam.

According to AS 3600, the spacing s should not be more than $0.5D$ or 300 mm, whichever is less. The reason behind such limits on spacing is to ensure that at least one stirrup is crossed by the failure surface at its mid-height.

For shear reinforcement to be effective, it must be properly anchored at each end. For this reason U-stirrups are preferable to single-leg stirrups. Standard hooks are desirable in the compression zone. A problem arises in regions of negative bending. If a U-stirrup has the ends upward, these ends are in the tension zone and may not be adequately anchored. Closed hoops are somewhat better. In beams where the shear is particularly severe it is suggested that special precautions be taken with the end anchorage of the stirrups.

7.7.5 Design Details

Various design details need to be observed in order to ensure that the stirrup reinforcement functions adequately:

> **Anchorage:** All stirrups should be properly anchored on either side of any potential inclined crack. Deformed bars should always be used for stirrups to help ensure good anchorage and control of the widths of potential inclined cracks. Closed (rather than open) stirrup configurations are preferable, even though they are less convenient for construction purposes.
>
> **Stirrup Spacing:** In beams not greater that 1.2 metre depth, the maximum spacing of stirrups in the longitudinal direction is specified in AS 3600 as the lesser of 300 mm and $0.5D$. For sections 1.2 metres and greater, the spacing of the stirrups can not exceed 600 mm. In the transverse direction, the spacing of the stirrup legs can not exceed 600 mm.
>
> **Additional Longitudinal Steel in End Regions:** For stirrups to be effective, they must cross the potential inclined crack and meet up with longitudinal steel on the tensile face of the beam to ensure truss-like action. Where the eccentricity of the prestressing tendons is reduced in the end region of a beam, or if tendons in a pretensioned member are debonded, the stirrups will not be effective in resisting shear unless additional bottom reinforcement is provided, as shown in Figure 7.11, to resist the horizontal components of the inclined strut forces in the concrete.

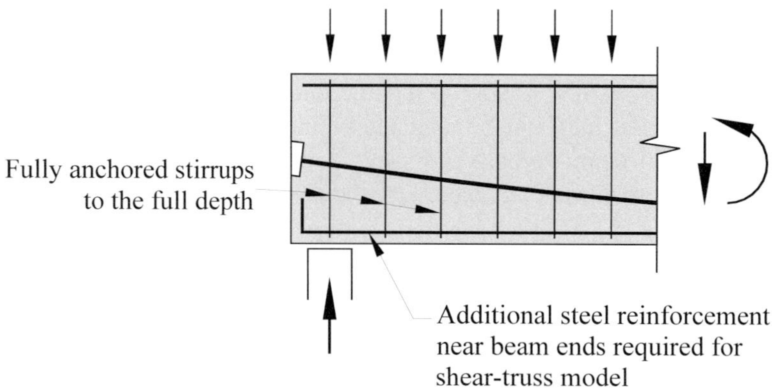

Figure 7.11 Steel needed in end regions to ensure truss action

Critical Sections for Shear: Where loads act on the top surface and reactions are provided under the beam, the critical section for shear may be taken to be no nearer than a distance d_v from the face of the support. In such cases the stirrup spacing s, calculated at d_v from the support, may be extended to the support. However, where the diagonal compression strut does not enter the support directly, as in the case of hanging supports, such arching directly into the supports cannot occur and all sections up to the face, or line, of the support must be considered.

EXAMPLE 7.3

Design of Post-tensioned Beam for Shear

Design the shear reinforcement for the beam given in Example 7.1. For the stirrups use Grade R for which $f_{sy.f}$ = 250 MPa. The tendon consists of 22 strands of 12.7 mm diameter and from Table 2.1 f_{pb} =1870 MPa and E_p = 195 GPa. The cross-section is shown in Figure 7.8.

SOLUTION

Design Load:

The beam has to carry a design load, w^*, of:

$$w^* = 1.2 \times 12.2 + 1.5 \times 24.0 = 50.6 \text{ kN/m}$$

For this factored load, the moment and shear at distance of x metres from the support are:

$$V^* = 633 - 50.6x \text{ kN};$$

$$M^* = 633x - 25.3x^2 \text{ kNm}$$

The force in the cable and slope at distance x from the support:

The prestressing force and slope in the cable distance x (in metres) from the support are:

$$P_e = 2435 - 10.8x \text{ kN} \quad (\text{for } x \leq 12.5 \text{ m})$$

The slope is calculated from Equation 4.9 as:

$$\alpha(x) = \frac{4 \times 524 \times 10^{-3}}{25.0}\left(1 - \frac{2x}{25.0}\right) = 0.0838\left(1 - \frac{x}{12.5}\right) \text{ radians}$$

Depth of tendon at distance x (in metres) from support:

From Equation 4.8, the eccentricity of the cable is given by:

$$e(x) = 4 \times 524 \times 10^{-3}\left[\frac{x}{25} - \left(\frac{x}{25}\right)^2\right]$$

$$= \frac{x}{11.93} - \frac{x^2}{298.2} \quad \text{(metres)}$$

Shear design at a control section at distance d_v from the support:

The internal lever arm for shear is determined from Equation 7.12 as:

$$d_v = 0.9\left[\frac{3100 \times 1160 + 2170 \times [552 + e(x = d_v)]}{3100 + 2170}\right] \geq 0.72 \times 1220$$

As d_v appears on both sides of the above equation, we solve by trial and error and find for $x = d_v$, $d_v = \max(845, 878)$ mm: thus $d_v = 878$ mm (noting that the limit of $0.72D$ governs).

Therefore, the first critical section for shear that we shall check is located at distance $d_v = 878$ mm from the support. At this section the cable force is 2426 kN, the eccentricity is $e = 71.0$ mm and the slope of the cable is $\alpha = 0.0779$ radians.

We take the effective web width, after grouting, as:

$$b_v = 200 - (0.5 \times (112)) = 144 \text{ mm}$$

The vertical component of the cable force is (Equation 7.5):

$$P_v = P_e \times \alpha = 2426 \times 0.0779 = 189 \text{ kN}$$

At this section the design stress resultants are:

$$V^* = 633 - 50.6 \times 0.878 = 589 \text{ kN}$$

$$M^* = 633 \times 0.878 - 25.3 \times 0.878^2 = 536 \text{ kNm}$$

Calculate concrete contribution to shear capacity:

From Equation 7.15:

$$k_v = \left[\frac{0.4}{1 + 1500\varepsilon_x}\right]$$

From Equation 7.17:

$$\varepsilon_x = \frac{|M^*/d_v + V^*| - \gamma_p P_v + 0.5N^* - A_{pt} f_{po}}{2(E_s A_{st} + E_p A_{pt})}$$

... within $0 \le \varepsilon_x \le 3.0 \times 10^{-3}$

and taking f_{po} as $0.5 f_{pb}$:

$$\varepsilon_x = \frac{\dfrac{536 \times 10^6}{878} + 589\times10^3 - 0.9 \times 189\times10^3 - 2170 \times 0.5 \times 1870}{2(200\times10^3 \times 3100 + 195\times10^3 \times 2170)}$$

$$= -999.6\times10^3/(2 \times 1.043\times10^9) = -479\times10^{-6}$$

As the calculated value of ε_x is less than zero, we shall recalculate it using Equation 7.18:

$$A_{ct} = \frac{1220}{2} \times 200 + 230(510 - 200) + \frac{150}{2}(510 - 200)$$

$$= 216.6\times10^3 \text{ mm}^2$$

$$\varepsilon_x = \frac{-999.6\times10^3}{2 \times (1.043\times10^9 + 32.8\times10^3 \times 216.6\times10^3)} = -61.3\times10^{-6}$$

... and is greater than -0.2×10^{-3} $\therefore$OK

By Equations 7.15 and 7.21:

$$k_v = \frac{0.4}{1 + 1500 \times (-61.3\times10^{-6})} = 0.441$$

$$\theta_v = 29 + 7000 \times (-61.3\times10^{-6}) = 28.6^{\circ}$$

and (Equation 7.11):

$$V_{uc} = 0.441\sqrt{40} \times 144 \times 878 = 352.6 \text{ kN}$$

$$\phi V_{uc} = 0.75 \times 352.6 = 264 \text{ kN}$$

Design for shear reinforcement:

The minimum shear that is required to taken by the stirrups is:

$$V_{us} \geq \frac{1}{\phi}[V^* - (\phi V_{uc} + \gamma_p P_v)] = \frac{1}{0.75}[589 - (264 + 0.9 \times 189)]$$

$$= 207 \text{ kN}$$

The area of shear reinforcement per spacing is (Equation 7.20):

$$A_{sv}/s = V_{us}/(f_{sy.f}\, d_v \cot\theta_v)$$

$$A_{sv}/s = 207 \times 10^3/(250 \times 878 \times \cot 28.6)$$

$$= 0.514$$

Minimum web reinforcement: As the section depth is greater than 750 mm, at least a minimum quantity of web reinforcement must be supplied. From Equation 7.16:

$$\frac{A_{sv.min}}{s} = 0.08\sqrt{40} \times 144/250 = 0.291$$

which is less than that required for strength. Thus $A_{sv}/s \geq 0.514$. With two-leg R10 ligatures giving $A_{sv} = 2 \times 80 = 160$ mm^2, the calculated shear reinforcement spacing is:

$$s \leq 160/0.514 = 311 \text{ mm}$$

This is greater than the maximum spacing allowed in AS 3600 of $s \leq 300$ mm. Therefore we can adopt a stirrup spacing of 300 mm.

Check web crushing by Equations 7.22 or 7.24 ($\alpha_v = 90^o$):

$$V_{u.max} = 0.55[0.9 f_c' b_v d_v \cos\theta_v \sin\theta_v]$$

$$V_{u.max} = 0.55[0.9 \times 40 \times 144 \times 878 \times \sin 28.6 \times \cos 28.6]$$

$$= 1052 \text{ kN}$$

$$\phi V_{u.max} = 0.70 \times 1052 = 736 \text{ kN} > V^* - \gamma_p P_v = 419 \text{ kN}$$

and therefore web crushing is not critical.

Thus for our design we will adopt **2–leg R10 stirrups at 300 mm** spacing.

Shear design at $x = 3.5$ metres

We consider the section at $x = 3.5$ m, where the effective force in the tendon is $P_e = 2397$ kN, its slope is $\alpha = 0.0603$ radians and the eccentricity is $e = 252$ mm.

The design shear force and bending moment at $x = 3.5$ m are:

$$V^* = 633 - 50.6 \times 3.5 = 456 \text{ kN}$$

$$M^* = 633 \times 3.5 - 25.32 \times 3.5^2 = 1906 \text{ kNm}$$

The vertical component of the prestress is:

$$P_v = 2397 \times 0.0603 = 145 \text{ kN}$$

$$b_v = 144 \text{ mm}$$

Calculate concrete contribution to shear capacity:

The effective depth d at the section considered is:

$$d = \frac{3100 \times 1160 + 2170 \times [552 + 252]}{3100 + 2170} = 1013 \text{ mm}$$

and the internal lever arm is:

d_v = maximum $(0.72D, 0.9d) = (878, 912)$. Thus, $d_v = 912$ mm.

From Equation 7.17:

$$\varepsilon_x = \frac{\dfrac{1906 \times 10^6}{912} + 456{\times}10^3 - 0.9 \times 145{\times}10^3 - 2170 \times 0.5 \times 1870}{2(200{\times}10^3 \times 3100 + 195{\times}10^3 \times 2170)}$$

$$= 386{\times}10^3/(2 \times 1.043{\times}10^9) = 185{\times}10^{-6}$$

By Equations 7.15 and 7.21:

$$k_v = \frac{0.4}{1 + 1500 \times 185{\times}10^{-6}} = 0.313$$

$$\theta_v = 29 + 7000 \times 185{\times}10^{-6} = 30.3^{\circ}$$

and from Equation 7.11:

$$V_{uc} = 0.313\sqrt{40} \times 144 \times 912 = 260 \text{ kN}$$

$$\phi V_{uc} = 0.75 \times 260 = 195 \text{ kN}$$

Design for shear reinforcement:

The minimum shear that is required to be taken by the stirrups is:

$$V_{us} \geq \frac{1}{\phi}[V^* - (\phi V_{uc} + \gamma_p P_v)] = \frac{1}{0.75}[456 - (195 + 0.9 \times 145)]$$

$$= 174 \text{ kN}$$

The area of shear reinforcement per spacing is (Equation 7.20):

$$A_{sv}/s = V_{us}/(f_{sy.f}\, d_v \cot\theta_v)$$

$$A_{sv}/s = 174{\times}10^3/(250 \times 912 \times \cot 30.3)$$

$$= 0.446$$

which is greater than the minimum reinforcement, calculated above for an effective web width of $b_v = 144$ mm as $A_{sv.min}/s = 0.291$.

Thus $A_{sv}/s \geq 0.446$. With $A_{sv} = 160$ mm^2, the calculated shear reinforcement spacing is:

$$s \leq 160/0.446 = 359 \text{ mm}$$

which is greater than the minimum allowed by the Standard of 300 mm.

Check web crushing by Equations 7.22 or 7.24 ($\alpha_v = 90^o$):

$$V_{u.max} = 0.55[0.9 f_c' b_v d_v \cos\theta_v \sin\theta_v]$$

$$V_{u.max} = 0.55[0.9 \times 40 \times 144 \times 912 \times \sin 30.3 \times \cos 30.3]$$

$$= 1133 \text{ kN}$$

$$\phi V_{u.max} = 0.70 \times 1133 = 793 \text{ kN} > V^* - \gamma_p P_v = 326 \text{ kN}$$

and therefore web crushing is not critical.

Thus for our design we will adopt **2–leg R10 stirrups at 300 mm** spacing.

Design for shear at other sections

The results of calculations for sections spaced along the beam are presented in Table 7.1. We provide R10 stirrups at 290 mm spacing throughout the beam.

To allow for possible non-uniformity in the load distribution on the span, in Table 7.1 a minimum design shear has been assumed for all sections of two-sevenths of the maximum of the design shear at the ends. The authors strongly recommend that such minimum values should always be applied to sections close to the mid-span.

TABLE 7.1 Shear design calculation results

x m	P_e kN	e mm	α rad	$\gamma_p P_v$ kN	d_v mm	ε_x ($\times 10^{-6}$)	ϕV_{uc}# kN	ϕV_{us}❄ kN	V^* kN
1.0	2424	80	0.0771	168	878	–57	262	152	582
2.0	2413	154	0.0704	153	878	–20	247	132	532
4.0	2392	282	0.0570	123	923	278	178	129	430
6.0	2370	382	0.0436	93	960	580	195	41	329
8.0	2349	456	0.0301	64	988	776	173	–	228
10.0	2327	503	0.0168	35	1005	907	162	–	181♣
12.5	2300	524	0.0	0	1012	979	156	25⌖	181♣

Notes: # $b_v = 144$ mm for $e \leq 344$ mm and $b_v = 200$ mm for $e > 344$ mm;
❄ the minimum ϕV_{us} needed is calculated as equal to $V^* - (\phi V_{uc} + \gamma_p P_v)$
♣ taken as 2/7 of the design shear at the supports; and
⌖ required to satisfy the imposed minimum design shear requirement

7.8 Analysis and design for torsion

7.8.1 Introduction

Torsion occurs in almost all structural members as a result of eccentricity of loading, out-of-straightness, and other secondary effects. Nevertheless, it often has little effect on working load behaviour and strength and can be ignored in the design of many concrete members. In some prestressed concrete members, such as bridge girders, spandrel beams in buildings and cranked and curved beams, torsion may become a major design consideration. If this is the case, then it is desirable to avoid torsionally inefficient sections such as I- and T-shapes and the design should be based on a hollow box section or solid sections of squarish shape.

As with shear, torsion usually has a minor effect on working load behaviour and becomes important only in the strength design of the member. This said, under in-service conditions, the torsional-shear stress resultants should be kept below the cracking threshold for any section where torsion, in combination with shear, is critical.

In contrast with shear, however, torsion is not statically linked with bending. It can occur in a member irrespective of whether or not bending is present. Large torsional moments can, for example, be applied to the ends of a member which otherwise carries only its self-weight, with negligible bending and shear. Nevertheless, torsion usually occurs together with significant bending moment and shear force, and the overload behaviour and load capacity depend on the relative magnitudes of the internal actions M, V and T.

Methods used for the design of reinforced concrete and prestressed concrete members in torsion assume that the overload behaviour, after inclined cracking, is like that of a three-dimensional truss. The truss concept was first applied to concrete members in torsion by Rausch (1929) and provides the basis for the current AS 3600 requirements for torsion design. The strut-and-tie method of analysis and design was developed as an extension of the truss concept in Europe in the 1970s (Thürlimann, 1976). It provides an alternative approach to torsion and shear design, which is allowed by AS 3600. A full treatment of strut-and-tie concepts is presented in the companion text, *Reinforced Concrete Basics* (Foster, et al., 2021), with examples for its application to combined bending, shear and torsion. We only consider the simple truss approach here.

The truss method of analysis is best suited to concrete girders with a hollow box section that is subjected predominantly to torsion. However, it can be adapted to deal with solid sections of rectangular and prismatic shapes.

The truss concept of overload behaviour in pure torsion is illustrated in Figure 7.12(a), where the thin walled concrete member contains longitudinal reinforcement in the corners and transverse closed stirrups. At overload, after the cracking torsion has been exceeded, parallel inclined cracks develop in each wall and spiral around the member. The compressive concrete between adjacent cracks acts like inclined parallel '*struts*' in a space frame, while the stirrups act like transverse '*ties*' and the reinforcement in the corners acts as tensile '*stringers*'. The equivalent truss is shown in Figure 7.12(b). The applied torsion induces inclined compressive forces D in the struts, tensile forces S in the ties, and tensile forces F in the corner stringers.

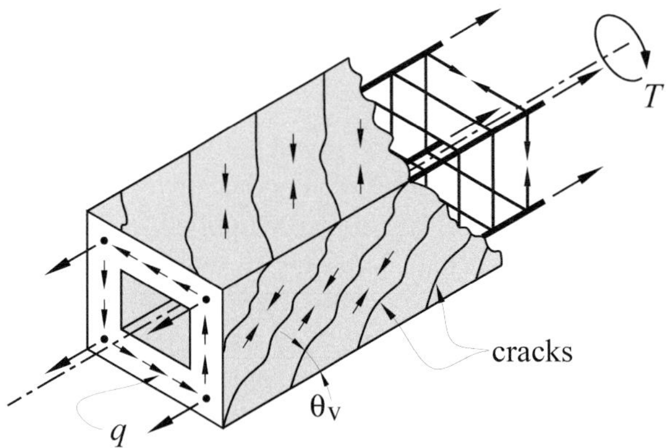

(a) Torsional cracks and shear flow

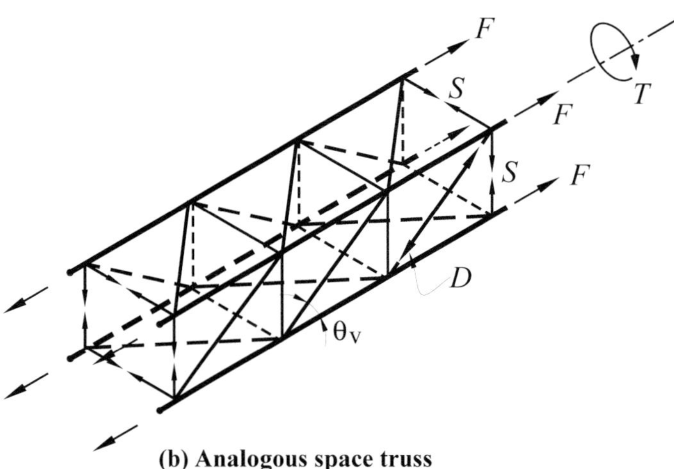

(b) Analogous space truss

Figure 7.12 Space truss model for torsion

Of course, the space-truss representation of the internal forces is highly idealised, just as the shear truss is an idealisation of beam behaviour in the presence of shear. To use the truss model for the design of a rectangular beam in torsion, various idealisations and approximations have to be made to fit the real member to the idealised truss. For example, appropriate values have to be chosen for the effective widths of the '*walls*'. Likewise, an effective area of corner reinforcement has to be chosen to represent each stringer. The choice is not obvious when the longitudinal tensile and compressive steel is uniformly distributed transversely across the section. Such assumptions and simplifications are incorporated in the AS 3600 design provisions for torsion for both reinforced concrete and prestressed concrete beams.

In principle, the original truss theory, together with its modern formulation in terms of strut-and-tie concepts, provides a lower bound, and hence safe, approach to design provided there is adequate ductility in the member.

7.8.2 Behaviour under load

On first application of forces on a beam that produce a pure torsional response, the member initially behaves linear-elastically. In the linear stage, $\theta_V = 45^o$. When cracking occurs in beams with torsional hoop reinforcement, a truss mechanism develops (see Figure 7.12), the tensile stress in the concrete reduces, whilst the stresses in the longitudinal and hoop reinforcement increase and diagonal compression occurs in the diagonal struts. Provided the struts do not fail, the direction of the cracks remain relatively stable and the forces in both the longitudinal and hoop steel steadily increase. In this stage, the strut angle remains at 45°. When either the longitudinal or hoop reinforcement yield, the internal forces continuously rebalance to support the applied load, until either the other reinforcement yields or the struts crush. In this stage θ_V either increases or decreases from 45°, depending on which of the reinforcements, longitudinal or hoop, is yielding. If the longitudinal bars yield before the hoop steel, $\theta_V < 45^o$, otherwise $\theta_V > 45^o$. If both yield simultaneously $\theta_V = 45^o$. The final stage occurs when the steel yields in the other direction or the concrete crushes. At this point, the directions of the principal stress stop rotating, and the ultimate capacity is realised.

At the ultimate condition, cracking is extensive, spiralling around the section, and may pass through both the flexural compressive and flexural tensile regions of the beam. The cover concrete will have spalled extensively from the section.

The strength of the section is affected by a number of factors, primarily:

- the relative magnitude of the torsion–moment and torsion–shear ratios;
- the ratio of the forces in the flexural longitudinal reinforcement to those in the closed stirrup (hoop) reinforcement; and
- the depth to width ratio (D/b) of the member.

7.8.3 Equilibrium torsion and compatibility torsion

In the design of members subjected to torsion, it is important to distinguish between **equilibrium torsion** and **compatibility torsion**. In Figure 7.13(a), torsion is induced in the simply supported bridge girder by loads that are eccentric to the longitudinal axis. Torsion is inevitably present, irrespective of the torsional stiffness of the member, because of the requirements of static equilibrium. In such situations, the torsion is described as equilibrium torsion.

On the other hand, torsion may be induced in a component of an indeterminate structure by imposed deformations. For example at point B in Figure 7.13(b) the end rotation of flexural member DB induces a torsional rotation of equal magnitude in section B of beam ABC, about its longitudinal axis. Provided ABC has torsional stiffness and provided also that the supports at A and C are torsionally stiff, torsion is induced throughout ABC as a result of loading on DB. In such situations, the torsion is not present because of the requirements of static equilibrium, but because of the requirement for compatibility of deformations throughout the structural system. This type of torsion is referred to as compatibility torsion.

In contrast with equilibrium torsion, the magnitude of the compatibility torsion depends on the torsional stiffness of the member and the relative stiffnesses of the other components in the indeterminate structure. At high overload and just prior to collapse, significant redistribution of the internal actions occur in a concrete structure if it is ductile. Generally speaking, the torsional stiffness of a concrete member decreases sharply as soon as inclined cracks form. This can lead to a sharp decrease in the compatibility torsion present at high overload. Different approaches are therefore adopted in designing for compatibility torsion and equilibrium torsion, as we shall discuss shortly.

In order to evaluate the design torsion, T^*, and the design shear, V^*, in critical regions of an indeterminate member, AS 3600 requires the elastic uncracked stiffness to be used in the structural analysis. This is to avoid excessive redistribution of loads, which requires extensive cracking. Ideally, under service loads the member should be uncracked and an example of proportioning of a member for working conditions is given in *Reinforced Concrete Basics* (Foster et al., 2021). The design of members for torsion for the strength limit condition follows the load path, and the forces obtained therefrom, as described above.

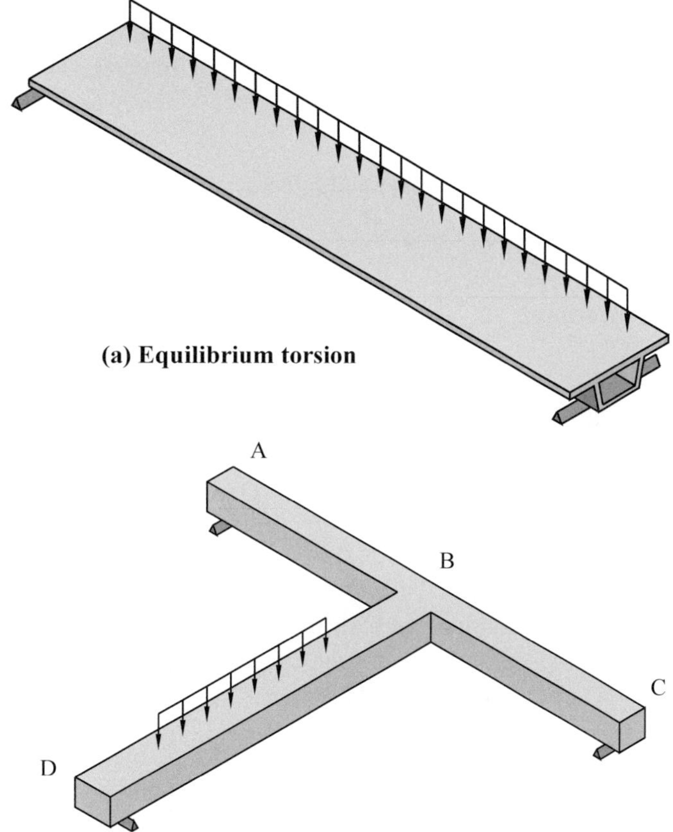

Figure 7.13 Equilibrium torsion and compatibility torsion

7.8.4 Design considerations for working loads

From a consideration of elasticity theory, the cracking torque for a concrete beam can be taken to be:

$$T_{cr} = 0.33 J_t \sqrt{f_c'} \times \sqrt{1 + \frac{\sigma_{cp}}{0.33\sqrt{f_c'}}} \qquad (7.37)$$

where J_t is the torsional section modulus (described below) and σ_{cp} is the average effective prestress in the section, i.e. $\sigma_{cp} = P_e/A_g$. The second term under the square root sign allows approximately for the effect of prestress on the torsional cracking strength of the section. The torsional modulus J_t for some common sections is shown in Figure 7.14. For flanged sections (T, L or I), J_t is calculated separately for each component rectangle and then obtained by their summation, as shown in the figure.

The torsional section modulus J_t is calculated as:

For solid sections: $J_t = A_{cp}^2/u_c$

For box sections: $J_t = \min(A_{cp}^2/u_c, 2A_o t_w)$

where u_c is the outside perimeter of the concrete section and A_{cp} is the area enclosed by this perimeter, A_o is the area enclosed by the shear flow path and t_w is the minimum wall thickness of the box. Examples for the calculation of J_t are shown in Figure 7.14.

7.8.5 Space truss model for pure torsion

The model adopted by AS 3600 for torsion design is a space truss, which is essentially a three-dimensional version of the model used in the design for shear. The model is discussed here for the case of pure torsion, and in the following section for combined torsion and shear.

The space truss, shown in Figure 7.12(b), consists of inclined compression members of concrete acting near the surface of the beam and inclined at an angle θ_v. These compressive struts are in equilibrium with the transverse (hoop) reinforcement and longitudinal reinforcement, both of which are in tension.

The shear flow q in a vertical cross-section (Figure 7.12(a)) is made up of the in-plane components of the inclined strut forces in each side. The units of q are N/mm. Taking moments about A in Figure 7.15 gives:

$$T = qx_o \times y_o + qy_o \times x_o \tag{7.38}$$

where x_o and y_o are the distances between the centres of the shear flow (Figure 7.15) and correspond to the dimensions of the median line shown in Figure 7.14.

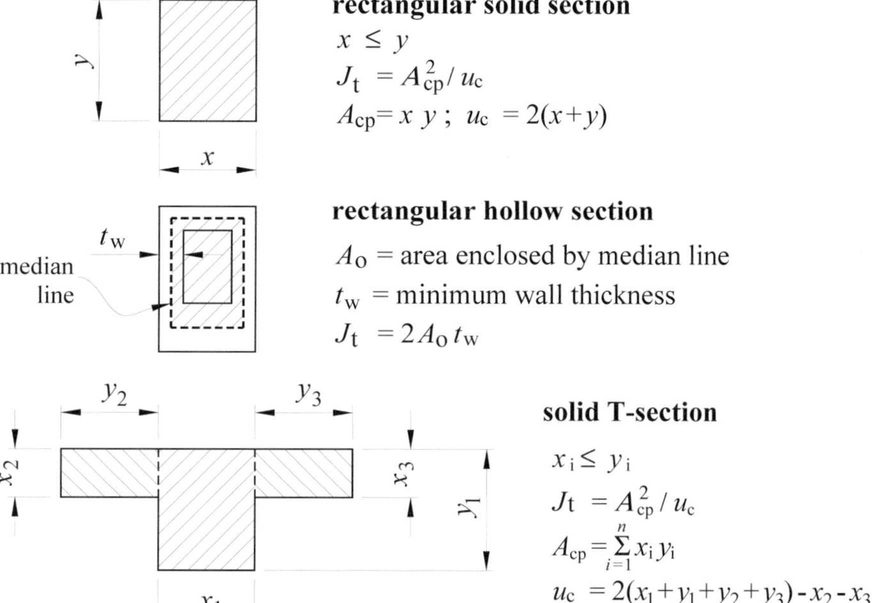

Figure 7.14 Calculation of torsional section modulus

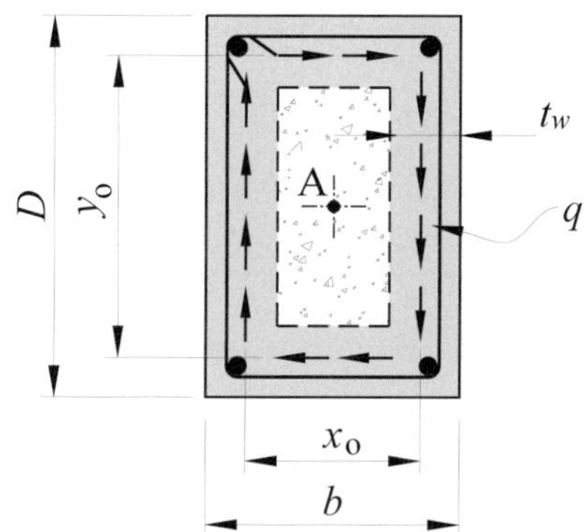

Figure 7.15 Torsion acting on a cross-section

Defining $A_o = x_o y_o$ as the area enclosed by the centre-line of the shear flow, then Equation 7.38 gives the shear flow as:

$$q = T/(2A_o) \tag{7.39}$$

For solid sections, the effective wall thickness of the tube, t_w, may be taken as (Collins and Mitchell, 1980):

$$t_w = 0.75A_{cp}/u_c \tag{7.40}$$

where A_{cp} and u_c are defined in Figure 7.14 for rectangular solid and box sections. After cracking, any flange outstands should be ignored.

Capacity limited by yielding of the reinforcing steel

With the longitudinal steel at yield, equilibrium of the longitudinal forces in the section gives:

$$A_s f_{sy} = q\cot\theta_v\ [2(x_o + y_o)] \tag{7.41}$$

where A_s is the total longitudinal steel area required for pure torsion and f_{sy} is its yield strength. Taking the perimeter of the centre-line of the shear flow (Figure 7.15) as $u_o = 2(x_o + y_o)$:

$$A_s f_{sy} = qu_o\cot\theta_v \tag{7.42}$$

Similarly, with the steel in the stirrups at yield, we have:

$$A_{sw} f_{sy.f} = qs/(\cot\theta_v) \tag{7.43}$$

where A_{sw} is the area of one leg of the closed fitment and s is the longitudinal spacing of the fitments.

Substituting Equation 7.39 into Equation 7.43 and arranging gives:

$$T_{us} = f_{sy.f}\,\frac{A_{sw}}{s}\,2A_o\cot\theta_v \tag{7.44}$$

where T_{us} is the ultimate strength of the section when subjected to pure torsion. Lastly, by equating the shear flow q in Equations 7.42 and 7.43, the total longitudinal reinforcement is obtained from:

$$A_s f_{sy} = f_{sy.f} \frac{A_{sw}}{s} u_o \cot^2 \theta_v \tag{7.45}$$

Note that one-half of the longitudinal reinforcement is to be provided in the flexural compressive region with the other half in the flexural tensile region.

Web crushing failure

To ensure that the section remains under-reinforced, and ductile, it is desirable that the member capacity be limited by the properties of the steel stirrups, and not the concrete struts. Thus, a limit on the stress in the concrete struts is required to avoid a compressive failure, referred to as web crushing.

AS 3600 allows several methods for design for torsion, with the two most common being sectional analysis and strut-and-tie-modelling. An example of the use of strut-and-tie modelling for the analysis and design for torsion is provided in the companion text *Reinforced Concrete Basics* (Foster et al., 2021) and is not presented further here.

The shear forces produced in the side walls from torsion may be determined from Equation 7.39 as (see Figure 7.15):

$$V_t^* = \pm q y_o = \pm \frac{T^* y_o}{2A_o} \tag{7.46}$$

noting that the resultant forces in the side walls are in opposite directions. In Equation 7.46, T^* is the factored design torsion at the section being considered.

In AS 3600, the distance between the centrelines of the shear flow in the *y*-direction is taken as $y_o \approx d_v$ and the area enclosed by the shear flow path $A_o \approx 0.85 A_{oh}$, where A_{oh} is the area bounded by the centreline of the external closed fitments (i.e. of the hoop reinforcement). With these approximations, the force in each web due to the torsional stress resultant is:

$$V_t^* = \pm q y_o = \pm \frac{T^* d_v}{1.7 A_{oh}} \tag{7.47}$$

The shear stress in the critical web is the sum of the shear stresses due to the applied actions (shear force and torsion) and is:

For box sections: $$\tau_w = \frac{V^* - \gamma_p P_v}{b_v d_v} + \frac{V_t^*}{t_w d_v} \le \frac{\phi V_{u.max}}{b_v d_v}\text{; and} \quad (7.48)$$

For solid sections: $$\tau_w = \sqrt{\left[\frac{V^* - \gamma_p P_v}{b_v d_v}\right]^2 + \left[\frac{T^* u_h}{1.7 A_{oh}^2}\right]^2} \le \frac{\phi V_{u.max}}{b_v\, d_v} \quad (7.49)$$

where u_h is the perimeter of the centreline of the hoop reinforcement and t_w is the width of a single side wall of a box section, but not to be taken as greater than A_{oh}/u_h (i.e. $t_w \le A_{oh}/u_h$). The maximum allowable force in the web, $V_{u.max}$, is determined from Equation 7.22, or Equation 7.23 at transfer.

Strength limit state requirement

To proportion the steel reinforcement to meet the strength limit requirement of torsion we need to ensure that:

$$\phi T_{us} \ge T^* \quad (7.50)$$

where T_{us} is the torsional capacity of the section, T^* is the factored design torsional moment and $\phi = 0.75$ is the capacity reduction factor for torsion (Table 3.1).

Detailing to AS 3600

The Australian Standard outlines a number of detailing requirements to ensure that a member in torsion performs satisfactorily in both the service and strength limit states. Reinforcement is to be provided such that:

- the torsional section capacity is not less than $0.25T_{uc}$; and
- a minimum area of fitments is provided equal to, or greater than, that needed for the minimum shear reinforcement requirements and these stirrups are fully closed and anchored such that they can develop full yield.

In addition to the above, to ensure that a minimum number of bars cross a torsional crack, the Standard requires that the spacing of the fully closed stirrups (fitments) does not exceed the lesser of $0.12u_h$ and 300 mm.

7.8.6 Combined torsion, bending and shear

The case of pure torsion was dealt with in the previous section. However, torsion rarely acts alone, but rather in combination with bending and shear. These can produce additional forces in each of the longitudinal and transverse directions, as outlined below.

Combining forces in the longitudinal direction

When bending, torsion and shear act together, the forces in the longitudinal direction are additive. That is, in the flexural tensile zone the forces due to torsion, flexure and shear are combined and the area of steel calculated accordingly. The additional force due to torsion in each of the flexural tensile and flexural compressive regions is obtained by substituting Equations 7.39 and Equations 7.41 into 7.45 and found as:

$$\Delta F_{tdt} = 0.5T^* \frac{u_o}{2A_o} \cot\theta_v \qquad (7.51)$$

Once ΔF_{tdt} is found, it is added to the component due to shear, ΔF_{tds}, to give ΔF_{td} (Equation 7.31). On the tensile side, this is then added to the force due to bending (Equations 7.30) and an area of steel reinforcement provided to carry the resultant (Equations 7.32). On the compressive side, the tension force ΔF_{td} (containing both the shear and torsion components) is subtracted from the compression force that arises due to flexure (Equation 7.33), and the reinforcement determined by Equation 7.34.

If the results of Equations 7.32 or 7.34 are less than the minimum longitudinal torsional reinforcement (or if Equation 7.33 is negative), then minimum reinforcement should be provided.

Combining forces in the transverse direction

As in the previous case, the reinforcement requirements obtained from the separate models for shear and torsion can also be added together when shear and torsion act together. This follows from the lower bound theorem of plasticity because the proportioning is carried out, using each model in turn, for the chosen equilibrating load path. However, the ductility of the member must be adequate to ensure that the load paths can be realised. It should be noted that the forces (and stresses) due to shear and torsion will be additive on one vertical side but will be subtractive, or opposing, on the other vertical side (Figure 7.16).

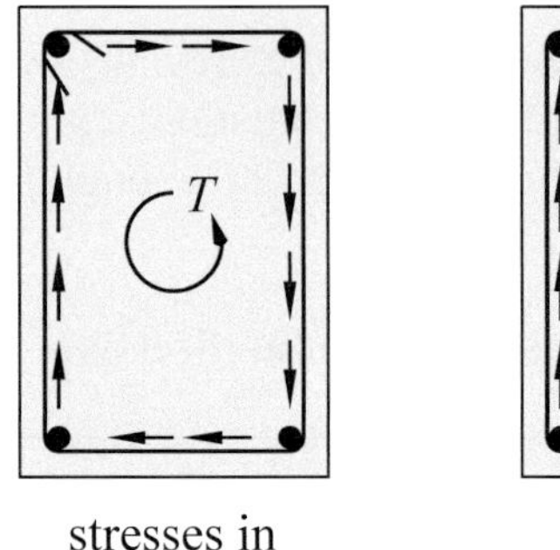

stresses in torsional stirrups

stresses in shear stirrups

Figure 7.16 Stresses in stirrups due to torsion and shear

When a member in combined torsion and shear is close to failure it is cracked extensively. The shear-carrying mechanisms in the concrete such as aggregate interlock and dowel action, which were discussed in Section 7.2.2, are much less effective in this case because of the wide cracking. When shear is combined with torsion, the mid-height strain parameter ε_x becomes:

$$\varepsilon_x = \frac{\dfrac{|M^*|}{d_v} + \sqrt{(|V^*| - \gamma_p P_v)^2 + \left[\dfrac{0.9T^* u_h}{2A_o}\right]^2} + 0.5N^* - A_{pt} f_{po}}{2(E_s A_{st} + E_p A_{pt})} \qquad (7.52)$$

$$\text{... within the limits } 0 \le \varepsilon_x \le 3.0 \times 10^{-3}$$

Should the calculated value of ε_x be less than zero, ε_x may be taken as zero or recalculated to account for the force in the compressed concrete in the tension region. In this case:

$$\varepsilon_x = \frac{\dfrac{|M^*|}{d_v} + \sqrt{(|V^*| - \gamma_p P_v)^2 + \left[\dfrac{0.9T^* u_h}{2A_o}\right]^2} + 0.5N^* - A_{pt} f_{po}}{2(E_s A_{st} + E_p A_{pt} + E_c A_{ct})} \qquad (7.53)$$

$$\text{... within the limits } -0.2\times10^{-3} \le \varepsilon_x \le 0$$

where A_{ct} is the area of the section taken between the extreme tensile fibre and its mid-depth ($D/2$). For a symmetric section $A_{ct} = A_g/2$.

From Equations 7.52 and 7.53:

- The design moment must be at least:

$$M^* \geq d_v \sqrt{(|V^*| - \gamma_p P_v)^2 + \left[\frac{0.9T^* u_h}{2A_o}\right]^2} \tag{7.54}$$

- N^* is taken as positive for tension and negative for compression
- A_{st} is the area of reinforcing bars located in the half depth of the section containing the flexural tension zone (in the zone defined by a depth of $D/2$ from the extreme tensile fibre).

The angle of the strut is calculated from Equation 7.21.

As for the case of combined bending and torsion, the design models for shear and torsion can be detailed independently. The torsional web reinforcement must be located near the outer faces of the section, while the shear reinforcement may be distributed through the section.

7.8.7 Design for torsion: AS 3600 method

The following steps can be used to determine the requirement at any particular section for strength:

1. Determine the design bending moment M^*, shear force V^* and torsion T^* acting on the section being considered.
2. Design for flexure at the critical section for bending, as outlined in Chapter 6 of this book, and determine details of longitudinal reinforcement at the shear control section under consideration.
3. Calculate the minimum allowed mid-height strain parameter ε_x from either Equations 7.52 or 7.53, as appropriate. Choose a value of ε_x between that calculated and the maximum value allowed of +0.003.
4. With the chosen value of ε_x, calculate the shear V_{uc} resisted by the concrete from Equation 7.11.

5. Calculate θ_v from Equation 7.21. By adjusting the value of ε_x, the angle θ_v can be taken as a design variable between a minimum value determined with the lowest possible ε_x and 50°. For members with high torsional moments and low shear, an angle close to 45° will require the lowest demand on ductility from cracking to ultimate. Where the shear demand is high, a solution closer to the minimum value is generally preferable.
6. Calculate the shear stress in the critical wall, τ_w, and ensure that it does not exceed the allowable factored ultimate stress.
7. Calculate A_{sv} / s from Equation 7.20 in which:

   $$\phi V_{us} = V^* - (\phi V_{uc} + \gamma_p P_v)$$

   and calculate the longitudinal reinforcement needed to sustain the shear force.
8. With the selected value of θ_v, calculate the transverse and longitudinal reinforcement required to sustain the torsional moments.
9. Add the transverse reinforcement components of shear and torsion and detail the transverse reinforcement. The spacing of the hoop reinforcement, s, can be determined from:

   $$1/s \geq (A_{sv}/s)/A_{sv} + (A_{sw}/s)/A_{sw}$$
10. Add the longitudinal reinforcement components of bending, shear and torsion and detail the longitudinal reinforcement.
11. Check minimum detailing requirements and detail the section.

The torsional web reinforcement must be located near the outer faces of the section, while the shear reinforcement may be distributed through the section. This may be accomplished with a single closed tie or with several overlapping closed ties as shown in Figure 7.17.

Many standards, including AS 3600, apply lower threshold limits at which if the torsional demands on a section are sufficiently small, such that there is little risk of torsional-shear cracking of the concrete, the torsion is deemed to be carried by the concrete section alone. In such cases the provision of torsional reinforcement, by way of closed fitments and additional longitudinal reinforcement, is unnecessary.

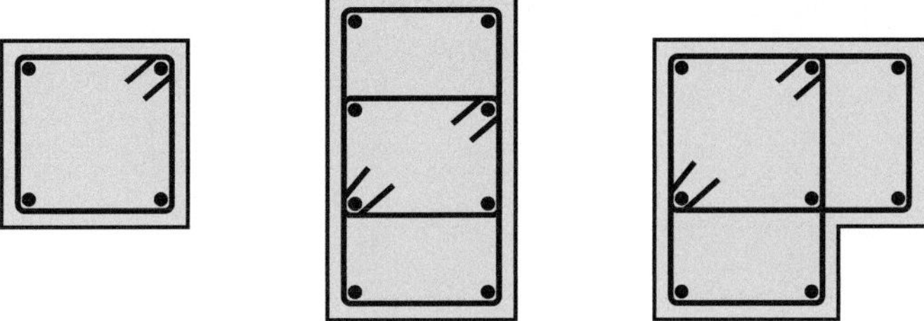

Figure 7.17 Closed ties for torsion

7.8.8 Threshold considerations for provision of torsional reinforcement

According to AS 3600 it is necessary to consider torsion in a member if:

$$\frac{T^*}{\phi T_{cr}} \geq 0.25 \tag{7.55}$$

where T_{cr} is determined from Equation 7.37. When $T^*/\phi T_{cr}$ is less than 0.25, torsional effects can be neglected. This is reasonable as torsional effects below this threshold have only a small influence on the capacity of the cross-section for shear and bending.

EXAMPLE 7.4

DESIGN OF PRETENSIONED GIRDER FOR BENDING, SHEAR AND TORSION

The pretensioned concrete beam shown in Figure 7.18 has a rectangular cross-section 400 mm wide and 550 mm deep. At a particular section the beam must resist the following factored stress resultants:

$M^* = 180$ kNm; $V^* = 240$ kN; $T^* = 60$ kNm

At the section considered, the member is reinforced with 4-N24 bars at the bottom (flexural tensile side), with an area of $A_{st} = 1800$ mm^2 and 2-N20 bars at the top (flexural compressive side), with an area of $A_{sc} = 620$ mm^2. The longitudinal reinforcement is fully developed at the section considered.

The effective prestressing force is 500 kN at an eccentricity of 100 mm and is supplied by five horizontal 12.7 mm super grade strands, with a total area of 493 mm^2. The breaking stress of the tendon is f_{pb} = 1870 MPa and the yield stress is taken as $0.82f_{pb}$ = 1533 MPa.

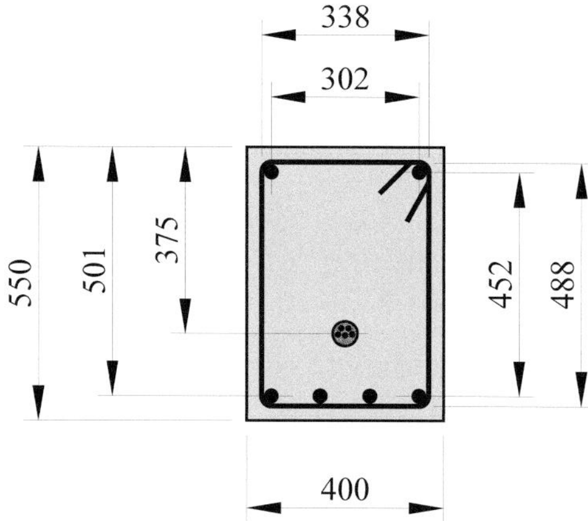

Figure 7.18 Cross-section for Example 7.4

Check that the longitudinal reinforcement for the section is adequate to carry the forces resulting from the combined bending-shear-torsion actions, and design the transverse reinforcement.

Other data:

f'_c = 40 MPa ; $\gamma = 0.87$; $\alpha_2 = 0.79$

$f_{sy} = f_{sy.f} = 500$ MPa

Clear cover to the stirrups = 25 mm; $P_v = 0$ (no drape)

SOLUTION

Dimensional properties

Assuming N12 stirrups:

$$D = 550 \text{ mm};\ b_v = 400 \text{ mm}$$

$$x = 550 - 2 \times (25 + 12/2) = 488 \text{ mm}$$

$$y = 400 - 2 \times (25 + 12/2) = 338 \text{ mm}$$

$$u_c = 2(550 + 400) = 1900 \text{ mm}$$

$$u_h = 2(488 + 338) = 1652 \text{ mm}$$

$$u_o = 0.92 \times 1652 = 1520 \text{ mm}$$

$$A_{cp} = A_g = 550 \times 400 = 220{\times}10^3 \text{ mm}^2$$

$$A_{oh} = 488 \times 338 = 165{\times}10^3 \text{ mm}^2$$

$$A_o = 0.85 \times 165{\times}10^3 = 140{\times}10^3 \text{ mm}^2$$

Consideration of flexure

Assuming both steels have yielded, and ignoring the compressive reinforcement, we calculate the depth from the extreme compressive fibre to the neutral axis as (Equation 6.11):

$$d_n = \frac{T_s + T_p}{\alpha_2 f_c' \gamma b} = \frac{1800 \times 500 + 493 \times 1533}{0.79 \times 40 \times 0.87 \times 400} = 151 \text{ mm}$$

$$\gamma d_n = 0.87 \times 151 = 131 \text{ mm}$$

The effective depth of the section is determined as:

$$d = \frac{1800 \times 500 \times 501 + 493 \times 1533 \times 375}{1800 \times 500 + 493 \times 1533} = 443 \text{ mm}$$

and the internal lever arm for flexure is:

$$z = 443 - 131/2 = 378 \text{ mm}$$

The tension force needed to carry the flexural stress resultant is thus:

$$T_b^* = M^*/z = (180\times10^6/378)\times10^{-3} = 476 \text{ kN}$$

Consideration of ductility

Assuming that all of the tensile steel is at yield, the depth to the neutral axis is 151 mm, and:

$$k_{uo} = d_n/d_o = 151/501 = 0.30 \le 0.36 \quad \therefore \text{OK}$$

Consideration of shear

Shear carried by concrete (Equation 7.11)

There are four 24 mm diameter longitudinal bars and five 12.7 mm diameter strands that are located in the tensile half-depth and both are fully developed at the section being investigated.

At the shear control section we assume that the stresses in the longitudinal reinforcement and prestress are less than yield for shear failure to occur. Thus, the effective depth for shear is calculated as (Equation 7.12):

$$d = \frac{1800\times501 + 493\times375}{1800+493} = 474 \text{ mm}$$

It is worthy of note that the effective depth taken for shear failure is different to that due to bending. This is due to the different assumptions of longitudinal steel yielding.

By Equation 7.13 we calculate the internal lever arm at the shear control section as:

$$d_v = \max[0.9\times474, 0.72\times550] = 427 \text{ mm}$$

With $P_v = 0$, we calculate (Equation 7.54):

$$d_v \sqrt{|V^*|^2 + \left[\frac{0.9T^* u_h}{2A_o}\right]^2} =$$

$$427 \times \sqrt{(240\times10^3)^2 + \left[\frac{0.9 \times 60\times10^6 \times 1652}{2 \times 140\times10^3}\right]^2} = 170\times10^6 \text{ Nmm}$$

which is less than M^*; thus the mid-height strain parameter is (Equation 7.52):

$$\varepsilon_x = \frac{\dfrac{180\times10^6}{427} + \sqrt{(240\times10^3)^2 + \left[\dfrac{0.9 \times 60\times10^6 \times 1652}{2 \times 140\times10^3}\right]^2}}{2 \times (200\times10^3 \times 1800 + 195\times10^3 \times 493)}$$

$$= 899\times10^{-6} \leq 3000\times10^{-6} \quad \text{OK}$$

A value of ε_x may be chosen between 899×10^{-6} and 0.003. We select $\varepsilon_x = 899\times10^{-6}$.

By Equation 7.15, the concrete parameter k_v is calculated as:

$$k_v = \frac{0.4}{1 + 1500 \times 899\times10^{-6}} = 0.170$$

and from Equation 7.11:

$$V_{uc} = 0.170\sqrt{40} \times 400 \times 427 = 184\times10^3 \text{ N (184 kN)}$$

$$\phi V_{uc} = 0.75 \times 184 = 138 \text{ kN}$$

Angle of compression strut

The angle of inclination of the concrete compression strut to the longitudinal axis of the member is calculated by Equation 7.21:

$$\theta_v = 29 + 7000 \times 899 \times 10^{-6} = 35.3^o$$

Check web crushing (Equation 7.24)

$$V_{u.max} = 0.55 \times 0.9 \times 40 \times 400 \times 427 \times \sin(35.3) \times \cos(35.3) \times 10^{-3}$$
$$= 1595 \text{ kN}$$

and the maximum allowable side wall stress is:

$$\tau_{u.max} = \frac{1595 \times 10^3}{400 \times 427} = 9.34 \text{ MPa, and}$$

$$\phi\tau_{u.max} = 0.7 \times 9.34 = 6.54 \text{ MPa}$$

The stress in the critical side wall is (Equation 7.49):

$$\tau_w = \sqrt{\left(\frac{240 \times 10^3}{400 \times 427}\right)^2 + \left(\frac{60 \times 10^6 \times 1652}{1.7 \times (165 \times 10^3)^2}\right)^2} = 2.56 \text{ MPa}$$

As $\tau_w \le \phi\tau_{u.max}$, web crushing is not critical.

Shear carried by reinforcement

The component shear force to be taken by stirrups is:

$$\phi V_{us} \ge V^* - (\phi V_{uc} + \gamma_p P_v) = 240 - (138 + 0) = 102 \text{ kN}$$

With $\phi = 0.75$, the ratio of the total stirrup area to stirrup spacing is found by rearranging Equation 7.20, and is:

$$\frac{A_{sv}}{s} \ge \frac{102 \times 10^3}{0.75 \times 500 \times 427 \times \cot(35.3)} = 0.451 \quad \text{mm}^2/\text{mm}$$

The additional design force in the longitudinal bars for the diagonal compression is (Equation 7.29):

$$\Delta F_{tds} = \min[0.5(240 + 138), 240]\cot 35.3^o = 267 \text{ kN}$$

Consideration of torsion

From Figure 7.14:

$$J_t = \frac{A_{cp}^2}{u_c} = \frac{(220\times10^3)^2}{1900} = 25.5\times10^6$$

The average effective prestress at the centroidal axis is:

$$\sigma_{cp} = P_e/A_g = 500\times10^3/220\times10^3 = 2.27 \text{ MPa}$$

and from Equation 7.37 we determine the torque required to produce cracking as:

$$T_{cr} = 0.33 J_t\sqrt{f_c'}\times\sqrt{1+\frac{\sigma_{cp}}{0.33\sqrt{f_c'}}}$$

$$= 0.33\times25.5\times10^6\times\sqrt{40}\times\sqrt{1+\frac{2.27}{0.33\sqrt{40}}}\times10^{-6} = 76.9 \text{ kNm}$$

As $0.25\phi T_{cr} = 0.25\times0.75\times76.9 = 14.4$ kNm is less than $T^* = 60$ kNm, the effect of torsional moments on the section must be considered.

Setting $\phi T_{us} = T^*$, the area of torsional web reinforcement per stirrup spacing (rearranging Equation 7.44) is:

$$\frac{A_{sw}}{s} = \frac{T^*}{\phi}\frac{1}{f_{sy.f}\,2A_o\cot\theta_v}$$

$$= \frac{60\times10^6}{0.75}\times\frac{1}{500\times2\times140\times10^3\times\cot35.3^{\circ}} = 0.405 \text{ mm}^2/\text{mm}$$

Longitudinal torsional steel in the flexural tensile zone (Equation 7.51):

$$\Delta F_{tdt} = 0.5\times60\times10^6\times\frac{1520}{2\times140\times10^3}\times\cot35.3^{\circ}\times10^{-3} = 230 \text{ kN}$$

Final details

Longitudinal reinforcement

The total force in the bottom reinforcement, tensile side, is obtained by combining the bending, shear and torsional components as given in Equations 7.30 and 7.31 (with $\gamma_{td} = 1.0$):

$$F_{td} = \gamma_{td} T_b^* + \Delta F_{tds} + \Delta F_{tdt} = 1.0 \times 476 + 267 + 230 = 973 \text{ kN}$$

noting that T_b^* is the tension force due to bending, calculated above.

The total force supplied by the tensile steel is:

$$A_{st} f_{sy} + A_{pt} \sigma_{pu} = (1800 \times 500 + 493 \times 1533) \times 10^{-3} = 1656 \text{ kN}$$

and:

$$\phi(A_{st} f_{sy} + A_{pt} \sigma_{pu}) = 0.85 \times 1656 = 1408 \text{ kN} \geq F_{td} \quad \therefore \text{OK}$$

As the design force on the tensile side supplied, $F_{td} = 973$ kN, is not greater than the factored capacity of 1408 kN, the longitudinal tensile reinforcement is sufficient.

In the flexural compressive zone (with $\gamma_{td} = 0.85$ for the flexural compressive side):

$$\begin{aligned} F_{cd} &= -\gamma_{td} T_b^* + \Delta F_{tds} + \Delta F_{tdt} \\ &= -0.85 \times 476 + 267 + 230 = 92 \text{ kN} \end{aligned}$$

The total force supplied by the compressive steel is:

$$A_{sc} f_{sy} + A_{pt} \sigma_{pu} = 620 \times 500 \times 10^{-3} = 310 \text{ kN}$$

and:

$$\phi A_{sc} f_{sy} = 0.85 \times 310 = 264 \text{ kN} \geq F_{cd} = 92 \text{ kN} \quad \therefore \text{OK}$$

Web reinforcement

For the web reinforcement, we shall try 2-leg N12 closed stirrups that have a bar area of 110 mm^2. As two legs are available for shear, $A_{sv} = 2 \times 110 = 220$ mm^2, and for torsion one leg is available giving $A_{sw} = 110$ mm^2. The spacing of the closed stirrups is calculated from:

$$\frac{1}{s} \geq \frac{A_{sv}/s}{A_{sv}} + \frac{A_{sw}/s}{A_{sw}} = \frac{0.451}{220} + \frac{0.405}{110} = 0.00573$$

which gives $s \leq 175$ mm, and is less than the maximum allowable spacing of $0.12u_h = 198$ mm and 300 mm. We try 2-leg N12 hoop reinforcement at 170 mm spacing, giving $A_{sv}/s = 220/170 = 1.29$.

Checking the minimum reinforcement (Equation 7.16):

$$\frac{A_{sv.min}}{s} = 0.08 \times 400 \times \frac{\sqrt{40}}{500} = 0.405 < 1.29 \qquad \therefore \text{OK}$$

Thus, we adopt **2-leg N12 closed stirrups at 170 mm spacing**.

7.9 References

AASHTO, 2017, LRFD Bridge Design Specifications, 8th Edition, Section 5, Concrete Structures.

AS 3600, 2009, *Concrete Structures*, Standards Australia, Sydney, Australia.

AS 3600, 2018, *Concrete Structures*, Standards Australia, Sydney, Australia.

AS 5100.5, 2017, Bridge Design, Part 5: Concrete, Standards Australia.

Bentz, E.C., Vecchio, F.J., and Collins, M.P., 2006, Simplified Modified Compression Field Theory for Calculating Shear Strength of Reinforced Concrete Elements, ACI Structural Journal, Vol. 103, No. 4, pp. 614-624.

Bentz, E.C, and Collins, M.P, 2006, Development of the 2004 Canadian Standards Association (CSA) A23.3 shear provisions for reinforced concrete, *Canadian Journal of Civil Engineering,* Vol. 33, pp. 521-534.

Collins MP, Mitchell D., 1980, Shear and Torsion Design of Prestressed and Non-Prestressed Concrete Beams. *PCI Journal,* Vol. 25, No. 5, pp. 32-100.

CSA A23.3, 2014, Design of concrete structures, CSA Group, Toronto, Canada.

Collins, M.P., and Kuchma, D., 1999, How Safe Are Our Large, Lightly Reinforced Concrete Beams, Slabs, and Footings?, *ACI Structural Journal*, Vol. 96, No. 4, pp. 482-491.

fib Model Code 2010, 2013, *fib Model Code for Concrete Structures 2010*, Ernst and Sohn, 402 pp.

Foster, S. J., Kilpatrick, A. E. and Warner, R. F., 2021, *Reinforced Concrete Basics*, 3rd Ed., Pearson, Australia, 587 pp.

Muttoni, A., and Fernández Ruiz, M., 2019, From experimental evidence to mechanical modeling and design expressions: The Critical Shear Crack Theory for shear design, *Structural Concrete*, Vol. 20, Issue 4, pp. 1464-1480.

Muttoni, A., Schwartz, J. and Thürlimann, B., 1997, Design of concrete structures with stress fields, Birkhäuser Verlag, Basel, 143 pp.

Sigrist, V., Bentz, E., Fernändez Ruiz, M., Foster, S. and Muttoni, A., 2013, Background to the Model Code 2010 Shear Provisions - Part I: Beams and Slabs, *Structural Concrete*, Vol. 14, No. 3, pp. 195-203.

Thürlimann, B., 1976, Torsional strength of reinforced and prestressed concrete beams - CEB approach, ACI Symposium, Philadelphia, (ACI publication).

Rausch, E., 1929, *Berechnung des Eisenbetons gegen Verdrehung und Abscheren*, Julius Springer Verlag, Berlin (*Design of reinforced concrete for torsion and shear*).

CHAPTER 8

Anchorage

In prestressed construction, the ends of all tensioned tendons have to be anchored locally in the concrete. Usually, but not always, the anchorages are located at the ends of the member. Complex stress patterns develop in the concrete in the anchorage region, and large transverse tensile stresses are produced that can cause longitudinal cracking. This is a particular problem in the case of post-tensioned construction because of the large cable forces involved. Special reinforcing cages are therefore placed around the anchorages to control such cracking. In this chapter we consider design methods for the anchorage of the prestressing tendons in both post-tensioned and pretensioned construction.

8.1 Introduction

Prestressing tendons range in size from small wires with a tensile strength of just several kilonewtons up to large post-tensioned cables with capacities in excess of 10 000 kN. The tendon forces are transmitted directly to the concrete by bond in the case of pretensioned members, and through bearing-type end anchorages in the case of post-tensioned members. In either case, the compressive force on the concrete is initially concentrated in a relatively small region before fanning out into the cross-section of the member. This induces a complex three-dimensional stress pattern near the anchorage, which inevitably leads to transverse tensile stresses and vertical and horizontal cracks that may extend longitudinally along the beam.

A **transition length**, also referred to as a **lead length**, can be identified, over which the local stress perturbations progressively develop into the simple linear distribution of prestress that exists in the interior of the beam, as predicted by simple beam theory. The region of perturbation is also referred to as the **end zone** or **Saint-Venant region**, named after the French elasticity theorist Adhémar Jean Claude Barré de Saint-Venant.

SAINT-VENANT

Adhémar Jean Claude Barré de Saint-Venant (1797–1886) was a French mathematician who, at age 16, entered the École Polytechnique. Graduating in 1816, Saint-Venant went on to make significant contributions in the fields of structural and fluid mechanics and in mathematics. In fluid mechanics he is most remembered for his publication of the correct derivation of the Navier equations for viscous flow. In the field of structural mechanics, he made multiple important contributions but, arguably, his most significant was (in 1855) to the principle that bears his name, i.e. the St. Venant Principle. Here he had demonstrated mathematically that the influence of stresses caused by a localised disturbance (or load) in an elastic system dissipates rapidly with distance away from the disturbance; at a distance equal to the depth of the member, the influence of a concentrated or point load on a member has dissipated to the point where sectional stresses can be approximated, without significant error, to be that of the elastic system.

In 1868, at age 71, Saint-Venant was elected to the French Academy of Sciences for his contributions to the fields of engineering mechanics and mathematics and in 1869 he was given the title of Count by Pope Pius IX.

End zone conditions in pretensioned beams tend to be less critical than in post-tensioned construction. One reason for this is that the individual tendons are small, in comparison with post-tensioned cables, and are usually well distributed through the cross-section. In a pretensioned member the compressive force in the concrete is built up progressively by bond between the tendons and the surrounding concrete. The distance from the end of the member that is required for full prestress to develop is known as the **transmission length**.

Anchorages for post-tensioned cables

Various forms of anchorages have been developed for post-tensioned construction. In some anchorages, the cable or strand is held by grips that bear directly onto a steel plate, which in turn applies the prestressing force to the adjacent concrete. In another type, a stiff steel bearing element is embedded in the concrete and connected to a cone-shaped former-tube or 'trumpet' that provides a transition to the relatively small duct used to house the cable in the body of the member. Reference should be made to trade literature for descriptions and dimensional details of the anchorages supplied by the various prestressing companies (e.g. Freyssinet, VSL and others). These anchorages are normally housed within a recess at the end of the member, which is filled with mortar after the cables have been stressed and grouted. Occasionally the anchorage may be left exposed as an architectural feature.

With post-tensioned beams of I- or T-section, it is usually necessary to enlarge a small length of the beam at each end to a rectangular shape in order to accommodate the anchorages, and their associated hardware, and the reinforcement needed to resist transverse tensile forces that develop as a consequence of the anchorage system. Such a rectangular block is referred to as an **end block**. This term may also be used to refer to the end zone of a rectangular beam in which significant transverse reinforcing steel is placed to control cracking.

Importance of proper end zone design and construction

Many of the problems encountered in prestressed concrete construction occur in end zones as a result of faulty design or inadequate workmanship. The importance of proper design of the end zones cannot be over-emphasised.

End block distress, whether from uncontrolled cracking because the reinforcement is inadequate, or from bearing failure of the concrete behind the anchorage, is usually difficult and extremely expensive to remedy. If adequate transverse reinforcement is not provided to control the longitudinal cracks that may form both horizontally and vertically, these cracks may extend progressively along the beam.

Requirements for end zones and anchorage

The Australian Standard allows for two methods for the design of end zones in post-tensioned members:

- a simplified method of design that can be used for relatively simple cases where not more than two anchorages occur in plan or in elevation; and
- strut-and-tie modelling for both simple and more complex cases.

In this Chapter we will first consider the simplified design method for post-tensioned anchorages, and its theoretical basis. Next, anchorage and stress development for pretensioned members will be discussed briefly and in the final Sections of the Chapter we look at the strut-and-tie approach.

8.2 Simplified design approach for post-tensioned beams

The simplified approach to end zone design is based on the results of elastic stress analysis studies carried in the latter half of the last century. By way of introduction, we will first consider, in Section 8.2.1, the tensile stresses that develop in the end zones of post-tensioned beams and then review, in Sections 8.2.2 and 8.2.3, some simplified methods for evaluating tensile stresses and forces.

8.2.1 Tensile stresses in anchorage zones

Central anchorage in a rectangular beam section

The simple case of one centrally located anchorage in a beam with a rectangular section is shown in Figure 8.1. The stiff bearing plate used in the anchorage is small in comparison with the beam cross-section and hence applies a very concentrated compressive stress to the surface of the concrete. The compressive stress distributions on vertical sections at various distances away from the end of the beam are shown in Figure 8.1(b). They become progressively more uniformly distributed with increasing distance from the end surface. The length of the end zone is referred to as the **lead length**, *a*. Figure 8.1(c) shows schematically the compressive stress trajectories emanating from the bearing plate and fanning out over the distance *a*. Each trajectory may be thought of as a curved fibre which carries a constant fraction of the prestressing force from the bearing plate to the end of the lead length. Equilibrium considerations show that the curved compressed force lines require inward forces (i.e. transverse compression) close to the bearing plate, and outward forces (transverse tension) in the remainder of the lead length. Also shown in Figure 8.1(c) are the 'compatibility regions' of the end-zone. These regions are not required to contribute to the strength of the member but may be subjected to compatibility cracking and, if left uncontrolled, spalling of the concrete from the member in these regions.

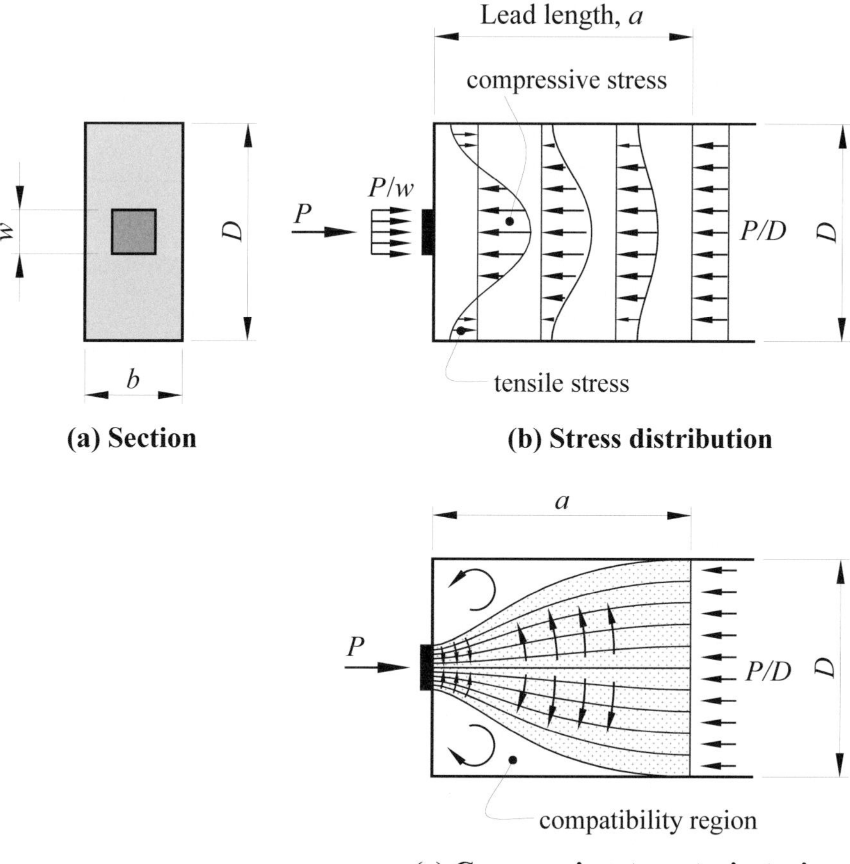

Figure 8.1 Compressive stresses in an end block

The end block of Figure 8.1 is redrawn using free bodies in Figure 8.2. The depths of the beam and the bearing plate are designated by D and w, respectively; the lead length is a and the prestressing force P acts horizontally.

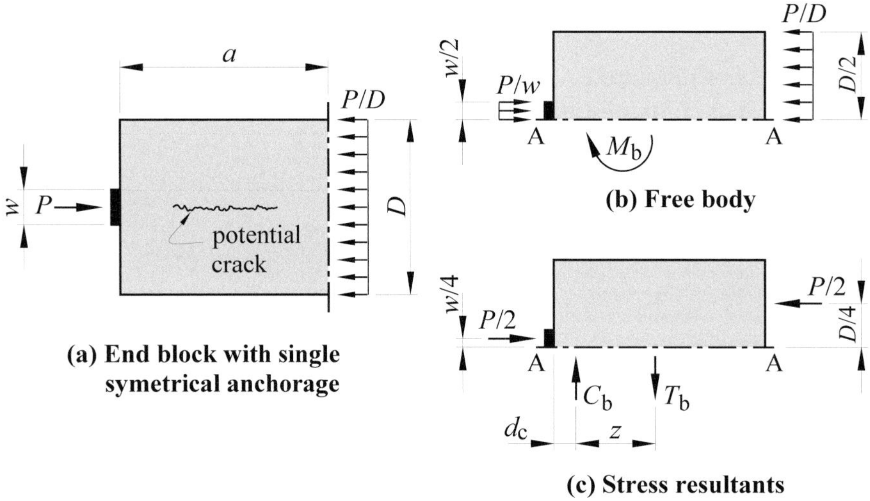

Figure 8.2 Transverse forces within end block

At the inner end of the anchorage zone a uniformly distributed force of magnitude P/D N/mm acts on the vertical plane, while at the bearing surface the concentrated distributed force is P/w N/mm.

Figure 8.2(b) shows the top half of the end block as a free body. For equilibrium, a moment M_b is required to act on the horizontal section A–A:

$$M_b = \frac{P}{2} \cdot \left(\frac{D}{4} - \frac{w}{4}\right) = \frac{PD}{8} \cdot \left(1 - \frac{w}{D}\right) \tag{8.1}$$

This moment induces transverse tensile and compressive stresses (σ_y) across A–A. The resultant of the tensile stresses is T_b, and the corresponding compressive force is C_b, as shown in Figure 8.2(c). As T_b is further from the loaded face than C_d, any horizontal cracking will be initiated at some distance away from the surface of the concrete, as indicated in Figure 8.2(a). A transverse force with the sense of T_b in Figure 8.2(c) is referred to as 'bursting'.

Two widely spaced anchorages

The situation is different in the end block shown in Figure 8.3, which has two symmetrically placed, but widely separated, anchorages. The eccentricity of each anchorage is e. If the top half of the end block is considered as a free body, as in Figure 8.3(b), equilibrium requires that a moment M_S acts on the cut section A–A:

$$M_S = \frac{P}{2}\left(e - \frac{D}{4}\right) \tag{8.2}$$

and the splitting tension force T_S, shown in Figure 8.3(c), is:

$$T_S = \frac{P}{2z}\left(e - \frac{D}{4}\right) \tag{8.3}$$

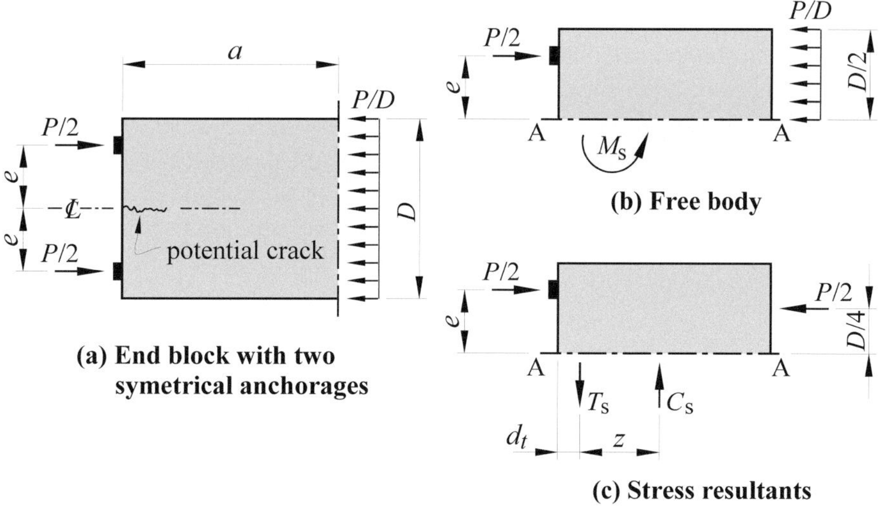

Figure 8.3 End block with two widely spaced anchorages

In this case the tensile stress-resultant T_S is close to the loaded face and there is therefore a tendency for a horizontal crack to occur at the face between the anchorages, as indicated in Figure 8.3(a). A transverse force of this type is referred to as a 'splitting' force. The terms bursting and splitting, although well established, have little physical significance. They both tend to cause cracking, but in different locations.

It will be noted that bursting moments can also occur behind the anchorages in Figure 8.3, especially if the anchorages are at some distance from the horizontal faces of the concrete.

It will be appreciated that when viewed in plan, the end zone will show a similar pattern of stresses and stress trajectories to those for the vertical plane in Figure 8.1. That is to say, the compressive stresses must also disperse in the direction transverse to the plane of the figure so that lateral compressive and tensile stresses are also present, which may cause vertical cracking in the end block behind the plate.

Beam analogy

The maximum static moments in an end block, and hence the locations of potential cracks, can be determined purely from the statics of selected free bodies. Such statical calculations can also be undertaken by regarding the end block as a beam, as in Figure 8.4, with supports wherever there are anchorages, and with a linearly varying load that corresponds to the elastic stress distribution in the interior of the beam, away from the end zone, due to the prestress. This beam concept can be useful in suggesting by inspection where the bursting and splitting moments occur, and may also be used to calculate their magnitude.

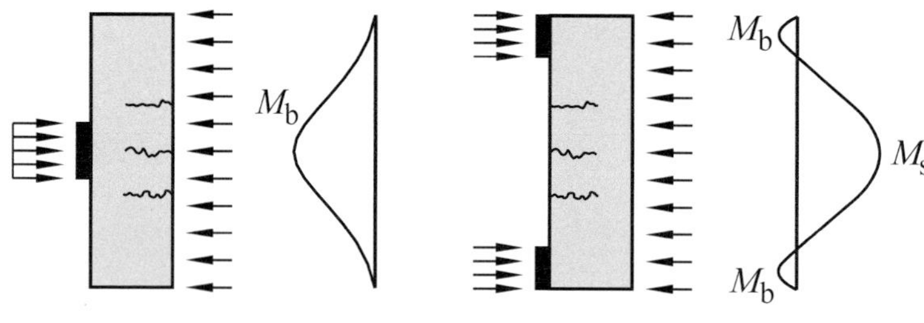

(a) one central anchorage **(b) two symmetrical anchorages**

Figure 8.4 Beam analogy for calculation of static moments

Generally, the bursting moments have the sense shown in Figure 8.2(b) and are greatest along or near the line of action of an anchorage, i.e. at a support of the analogous beam. Splitting moments have the sense shown in Figure 8.3(b) and occur between anchorages. In single-anchorage end blocks, splitting moments may occur if the anchorage is eccentric and located well away from the centroidal axis of the section. This is illustrated in Figure 8.5, where the eccentricity of the prestressing force is sufficient to cause horizontal tension in the top fibres. In the analogous beam, this amounts to a region subjected to negative distributed load. On a horizontal section near the top of the end zone, such as CD, splitting moments must exist as indicated in the free body ABDC in Figure 8.5(b). At lower horizontal sections such as EF bursting moments are required for equilibrium of the free body EFHG.

As already noted, bursting moments tend to induce cracking away from the end face, while splitting moments induce cracking at the end face.

8.2.2 Analysis of post-tensioned end blocks

We shall now consider how a simplified analysis of a post-tensioned end block can be made to obtain numerical design data. The moment acting on a horizontal section through the lead length of a post-tensioned beam, as in Figure 8.5, can be calculated from equilibrium requirements. It can be seen that:

1. the stress resultants T_b and C_b (T_s and C_s) are equal in magnitude, and
2. the force T_b (T_s) equals the static moment M_b (M_s) divided by the lever arm z.

For design it is necessary to determine, approximately, the magnitude and location of the total transverse tensile forces and the extent of the tensile stress region. We shall consider several typical cases.

Central anchorage in rectangular end block

The typical shape of the transverse stress distribution at the mid-depth of a centrally located anchorage on a rectangular end block and obtained from a linear elastic analysis is shown in Figure 8.6. This also includes the situation where several anchorages are located side by side at the mid-depth.

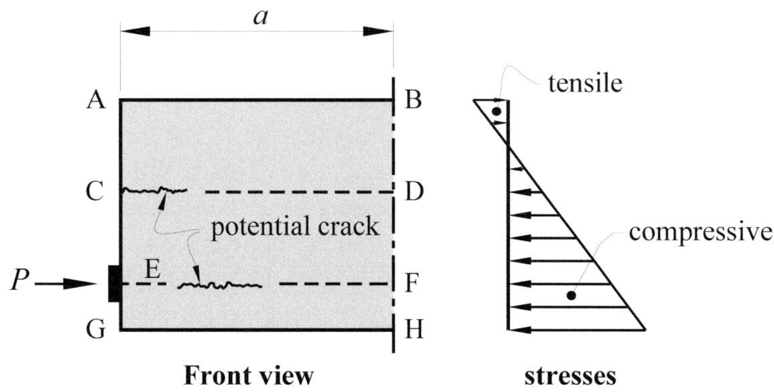

(a) End block with single eccentric anchorage

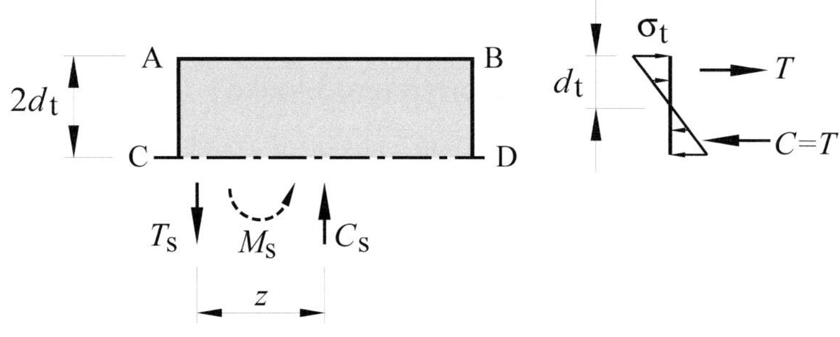

(b) Free body along line CD

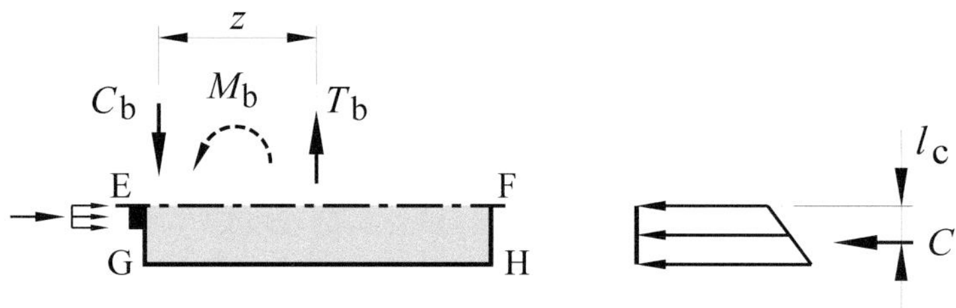

(c) Free body along line EF

Figure 8.5 Bursting and splitting moments for the case of an eccentric anchorage

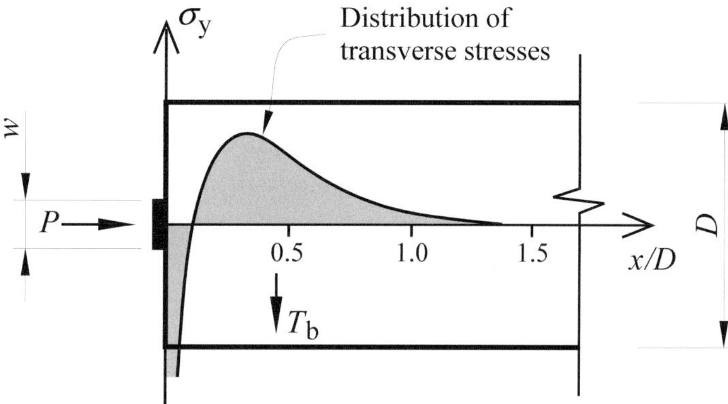

Figure 8.6 Distribution of transverse stresses for a central anchorage on a rectangular end block

The ratio w/D affects both the precise shape of the stress distribution and the magnitude and position of the maximum transverse tensile stress. Elastic analyses (Iyengar, et al, 1966) show that for all w/D ratios, the vertical stresses exist over a distance of approximately 1.3 D from the outer face, but are very small from a distance greater than D. The lever arm z between C and T remains almost constant at approximately $0.5D$ for values of w/D greater than 0.15. Except for very small and rarely used concentration ratios ($w/D <$ 0.15) the bursting tension force, T_b, is closely approximated by the equation of Mörsch (1924):

$$T_b = 0.25P\left(1 - \frac{w}{D}\right) \tag{8.4}$$

Using this expression and Equation 8.1 the lever arm z is found to be $0.5D$.

It should be noted that these equations, and the transverse stress distribution shown in Figure 8.6, apply only to the mid-depth section of the end block. Conditions at other sections are quite different and depending on the relative depth of the end block to the section depth w/D and the eccentricity e of the section being considered, the end block has pockets of transverse busting and splitting tension that need to be considered in design.

Rectangular end block with one eccentric anchorage

In the case of a single eccentric anchorage, the distribution of bursting and splitting moments can be determined using statics or the beam analogy. An approximate, symmetrical prism method was developed by Guyon (1953) for estimating the bursting tensile stresses behind an eccentric anchorage in a rectangular end block. This method, shown in Figure 8.7, assumes that the stresses in the real end block are approximately the same as those in an imaginary end block consisting of an equivalent prism with its centreline on the line of action of the prestressing force, *P*, and with the anchorage depth dimension D_e being twice the distance from the centreline of the anchorage to the nearest concrete face. The stress at the end of the lead-in length is taken as uniform over the depth of the equivalent prism and is equal to $P/(bD_e)$.

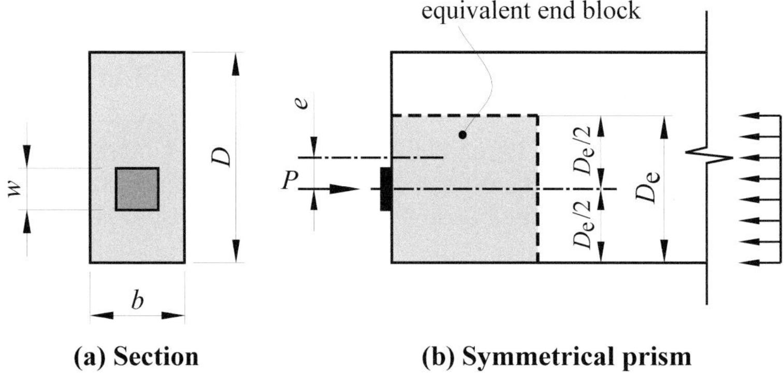

Figure 8.7 Symmetrical prism for analysis of eccentric anchorage

Rectangular end blocks with two or more anchorages

In the case of end blocks with two anchorages, the approach adopted will largely depend on the location of the anchorages relative to (1) the centroidal axis of the member and (2) the spacing between the anchorages. Where anchorages are sufficiently close, they may be treated as a single anchorage using Guyon's symmetrical prism approach. On the other hand, where the anchorages are well spaced, splitting forces can develop and the end block design may be treated using the model described in Figure 8.3.

Where an end block consists of more than two anchorages, a strut-and-tie approach, as described in Section 8.4, may be used to identify the transverse tensile (splitting and bursting) forces in each plane. The end block region is then detailed accordingly.

8.2.3 Simplified method for design of prismatic end blocks

The overall design requirement is that reinforcement must be provided in the anchorage zone to carry the tensile forces that develop due to the dispersion of the prestressing force, as already described. Since the dispersion occurs both laterally and vertically, reinforcement is required in vertical planes (parallel to the end face of the beam) in both the vertical and horizontal directions.

To obtain the magnitude of the design tensile forces, a simplified two-dimensional analysis is carried out in both the vertical x-y plane and the horizontal x-z plane. All possible critical loading conditions that can arise during the prestressing operation have to be considered, that is to say at any stage of the prestressing operation. For example, if there are several cables to be prestressed, then the critical transverse tensile force may occur when all of the cables have been tensioned, but is more likely to occur at an intermediate stage when one or more cables have been tensioned.

For each loading case, two potential transverse tensile forces should be considered:

1. the transverse tensile force located away from the external face but directly behind an anchorage, caused by a bursting moment, and
2. the transverse tensile force close to the exterior end face of the beam but remote from the line of an anchorage, caused by a splitting moment. Splitting and bursting moments are shown in Figure 8.5.

Transverse tensile design force due to bursting moments

For case (1), the total transverse tensile force due to bursting is determined using the equation of Mörsch (1924) in conjunction with Guyon's symmetrical prism approach. The bursting tension force is thus calculated from:

$$T_b = 0.25Pk\left(1 - \frac{w}{D_e}\right) \tag{8.5}$$

where P is the maximum force that occurs at the anchorage during prestressing (before any losses), w/D_e is the ratio of the dimension of the anchorage bearing plate in the relevant direction (vertical or horizontal) to that of the equivalent prism dimension in that direction and k is a factor. In AS 3600, this factor is taken as $k = 1.0$, which gives a non-conservative estimate of the bursting force. Guyon, in his approach, used $k = D/D_e$ and demonstrated that his model will always provide for a conservative estimate for the bursting force. In this book we recommend that Guyon's equation be adopted whenever the simplified approach is used in design.

Transverse tensile design force due to splitting moments

For case (2), where the splitting moment occurs in a section remote from an anchorage, two possibilities should be considered. The first is where there is a single eccentric anchorage, as shown in Figure 8.5, and the second is where there are two separated anchorages, as shown in Figure 8.3.

For a single eccentric anchorage, a crack will occur where the splitting moment is at its maximum and this occurs where the shear is zero (at depth $2d_t$ in Figure 8.5). The splitting moment is demonstrated in Figure 8.5 can be obtained from:

$$M_s = \frac{-\sigma_t b d_t}{2} \cdot \frac{4d_t}{3} = \frac{-2\sigma_t b d_t^2}{3} \tag{8.6}$$

$$T_s = M_s/z \tag{8.7}$$

where σ_t is the tensile stress in the concrete at the tensile fibre, b is the width of the cross-section and d_t is the depth from the tensile fibre to the axis of zero stress. The lever arm is taken as $z = 0.5D$, where D is the overall depth (or width) of the section. It is noted that by the convention adopted, the tensile stress is negative.

For a pair of anchorages spaced at s apart in a prismatic end block, the lever arm may be taken to be $z = 0.6s$. For this case the splitting force, T_s, is (Equation 8.3):

$$T_s = \frac{P}{1.2s}\left(e - \frac{D}{4}\right) \tag{8.8}$$

AS 3600 simplified design approach

The simplified design of post-tensioned anchorage zones covered in AS 3600 deals only with end blocks that do not contain more than two anchorages in plan and elevation. For more complex cases, a strut-and-tie approach is required.

To determine the area of transverse reinforcement required, AS 3600 requires that the calculated forces T_b and T_s be divided by a nominal stress of 150 MPa. This relatively low stress is indicative of the relatively tight crack control that is needed in regions behind prestressed anchorages, together with the simplifying nature of the assumptions made. It should be noted that all possible loading cases need to be considered, from when the first cable has been stressed progressively up to completion with all cables stressed.

In the case of bursting forces, where T_b is determined from Equation 8.5, the reinforcement is placed uniformly within a region between $0.2D_e$ and $1.0D_e$ from the face where the anchorage is located. In order to control cracking due to pockets of transverse tension that can give rise to '**spalling**' forces on the face of the end block, AS 3600 also requires that this reinforcement be continued at the same spacing through the region close to the end of the beam, i.e. in the region from $0.2D_e$ to as close as possible to the face.

For splitting forces T_s, the reinforcement is placed as close as practicable to the bearing surface, taking due account of cover and compaction limitations.

While the reinforcement is determined for the section of maximum bursting or splitting moments, the reinforcement determined needs to be continued throughout the full depth or breadth of the end block. This is to provide for the transverse tension stresses at other sections that cannot be calculated using the simplified procedures, such as surface cracking and spalling.

8.2.4 Surface cracking and spalling

Elastic analyses of stresses in end-block regions show that pockets of tensile stress can develop in regions remote from the main flow of forces that are additional to those that equilibrate the applied anchorage force. Such a region was shown in Figure 8.1(c), where it is referred to as a compatibility region. Figure 8.8 shows stress isobars in an end block for (a) a centrally located anchorage and (b) an eccentric anchorage, where the stress isobars represent contours of equal stress as a proportion of the average stress taken as $P/(bD)$. In Figure 8.8(a) a potential for spalling exists in the tensile regions in the top and bottom corners, while in Figure 8.8(b) there is a potential splitting region in the end face at some distance above the anchorage. In the case of splitting and bursting, these are equilibrium conditions with the forces determined from statics and the reinforcement determined accordingly. However, the spalling forces result from compatibility conditions from 'force whirls' that occur in dead regions of the end block.

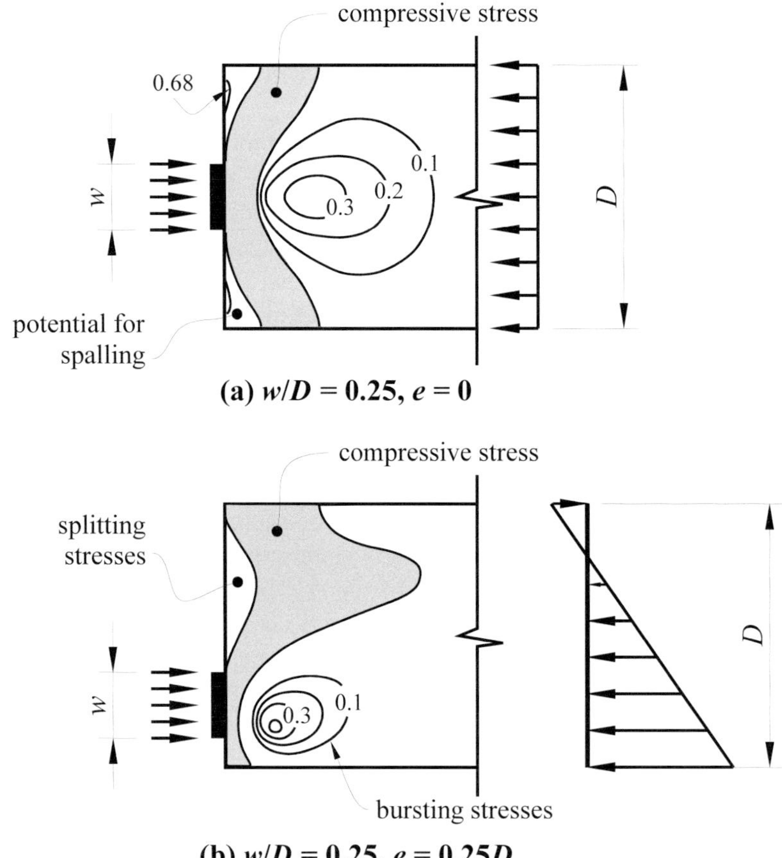

Figure 8.8 Transverse stress isobars for end zones of prestressed beam.

While such localised tensile regions do not affect strength, if left uncontrolled they can result in unsightly surface damage and lead to premature deterioration of the end region, notably with spalling at corners and edges. A light skin reinforcement, placed as close as is practicable to the concrete surface, can be used to contain such problems. The current edition of AS 3600 deals with the problem of surface cracking and spalling through detailing provisions, specifically by continuing the reinforcement needed to control the equilibrium forces of splitting and bursting to the face of the end block and with the first bars placed as close to the face as cover and compaction allow.

To control spalling cracks that may occur due to pockets of tension at the face of the end block, VSL (Rogowsky and Marti, 1991) suggest that reinforcement be provided as close to the face as practicable for a force equal to $0.02P_j$. This represents the upper limit of the spalling force in elastic analyses, for various cases, reported by Leonhardt (1964). Further to this provision, in the dead zones (e.g. the compatibility zone in Figure 8.1(c)) a minimum reinforcement ratio of 0.002 for building structures and 0.003 for bridge structures should be provided in each direction to control compatibility surface cracking in these areas.

8.2.5 Check on concrete bearing stresses

The possibility of concrete crushing behind the anchorage must also be considered in the design of the end block. Where confining reinforcement is not provided to the anchorage, the bearing stress on the concrete under the plate, σ_b, should not exceed f_b, where:

$$f_b = \min(\phi 0.9 f'_{cp} \sqrt{A_2/A_1},\ \phi 1.8 f'_{cp}) \tag{8.9}$$

where A_1 is the area of the bearing plate, A_2 is the largest area of concrete in the cross-section of the end zone that is geometrically similar to the area A_1 and $\phi = 0.6$. In calculating the bearing stress beneath the anchor plate, the force at jacking, P_j, is multiplied by the load factor 1.15 (see Section 3.4.1).

Anchorage systems are key elements in ensuring that the prestress can be transferred effectively to the member. They are proprietary elements of the system, together with other elements such sheathing and grouting systems. To enhance the bearing capacity immediately behind the anchor plate, such systems usually include helical confinement reinforcement. Manufacturers of prestressing components may test their anchorage system to AS/NZS 1314 and a higher value of allowable bearing stress determined than that obtained from Equation 8.9.

In practice, bearing failure usually results from poor concrete compaction behind the anchorage plate, resulting in honeycombed pockets into which the anchorage collapses. Thus careful supervision of the placing of concrete around anchorages is essential.

EXAMPLE 8.1

SINGLE ANCHORAGE IN A RECTANGULAR END BLOCK

A horizontal cable consisting of 16 strands of 12.7 mm diameter is anchored at the centre of a rectangular end block 900 mm deep and 450 mm wide, as shown in Figure 8.9. The maximum jacking force at transfer (before losses) is $P_j = 2000$ kN, the bearing plate is 265 mm square and the duct diameter is $d_d = 80$ mm. The concrete strength at transfer is $f'_{cp} = 35$ MPa. Design the end block reinforcement.

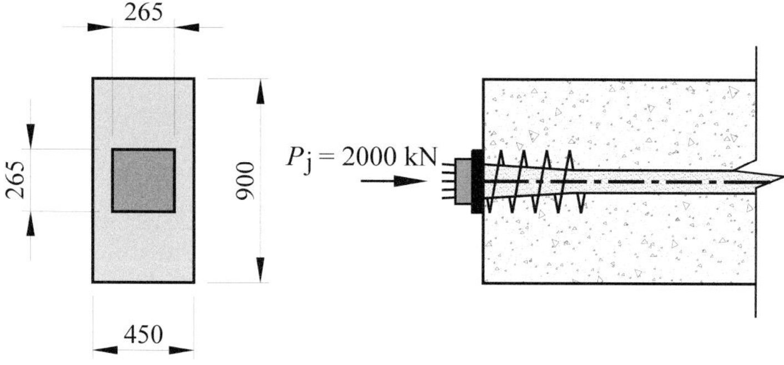

Figure 8.9 Single anchorage end block: Example 8.1

Solution

Vertical tensile force and area of reinforcement

In this standard case, bursting tension forces are developed along the centreline of the member and the total transverse tensile force may be obtained directly from either Equations 8.4 or 8.5.

In the vertical direction:

The ratio of the anchor plate depth to the depth of the end block is:

$$w/D = 265/900 = 0.29$$

and thus:

$$T_b = 0.25 \times 2000(1 - 0.29) = 355 \text{ kN}$$

Limiting the stress in the bursting reinforcement to 150 MPa, the required area of bursting steel is now calculated as:

$$A_b = \frac{355 \times 10^3}{150} = 2370 \text{ mm}^2$$

The length over which significant tension occurs is approximately $0.8D_e$; that is, $0.8 \times 900 = 720$ mm. Trying 4-legged N12 stirrups, the area of one set of stirrups in the web is:

$$A_w = 4 \times 110 = 440 \text{ mm}^2$$

and thus the maximum spacing of stirrups is:

$$s \leq 720 \times \frac{440}{2370} = 134 \text{ mm}$$

For detailing, the calculated bursting reinforcement is required to extend from the face of the end face to approximately $1.2D_e = 1080$ mm from the face.

In the horizontal direction:

The ratio of the anchor plate depth to the depth of the end block is:

$$w/D = 265/450 = 0.59$$

and thus:

$$T_b = 0.25 \times 2000(1 - 0.59) = 205 \text{ kN}$$

The required area of transverse reinforcement is:

$$A_b = \frac{205 \times 10^3}{150} = 1370 \text{ mm}^2$$

and is distributed over a length of $0.8 \times 450 = 360$ mm. We try 4–legs of N12 stirrups to cross in the cracking plane, giving $A_w = 440 \text{ mm}^2$ and the maximum spacing of stirrups is:

$$s \le 360 \times \frac{440}{1370} = 116 \text{ mm}$$

It is noted that two of the N12 legs are the horizontal legs of the stirrups calculated previously for the vertical bursting reinforcement, with an additional two legs to be provided via an additional stirrup. As before, the bursting reinforcement is to be extended over a minimum length of $1.2D_e = 540$ mm from the face.

Detailing of bursting reinforcement:

Comparing the solutions for the vertical and horizontal directions, for the vertical reinforcement we **adopt 11 sets of 4-legged N12 stirrups at 110 mm centres** with the first set placed as close to the end block face as cover permits, as shown Figure 8.10. The reinforcement ratio for the vertical reinforcement is:

$$p_w = \frac{440}{450 \times 110} = 0.0089$$

In the horizontal direction, two legs of reinforcement are provided via the horizontal legs of the reinforcement detailed for the vertical direction bursting steel. We thus provide for an additional **4 sets of 2-legged N12 stirrups at 110 mm centres** (see Figure 8.10) and, in doing so, any vertical plane cracking should be controlled well through the full section.

Compatibility reinforcement

The surface tensile spalling force is estimated as $0.02 \times 2000 = 40\ \text{kN}$. This requires a reinforcement area of:

$$A_s = 40{\times}10^3/150 = 270\ \text{mm}^2$$

placed close to the face. The first stirrup provides 440 mm^2 and is sufficient to control compatibility spalling at the end block face.

To control compatibility surface cracking, we shall adopt a minimum reinforcement ratio of 0.003 in each direction. In the vertical direction, the reinforcement supplied with $p_w = 0.089$ is more than adequate to control any horizontal surface cracking. In the horizontal direction in the elevation, we shall adopt 2–N12 bars at 160 mm centres (one bar on each face), giving a reinforcement ratio of:

$$p_w = \frac{220}{450 \times 160} = 0.0031$$

Check the bearing capacity of the concrete behind the anchor plate:

The area of the bearing plate is:

$$A_1 = 265 \times 265 = 70.23{\times}10^3\ \text{mm}^2$$

and the largest geometrically similar area of concrete that fits behind the anchorage plate is:

$$A_2 = 450 \times 450 = 202.5{\times}10^3\ \text{mm}^2$$

The maximum allowable bearing stress behind the anchor plate is determined by Equation 8.9 as:

$$f_b = \min(\phi 0.9 f'_{cp}\sqrt{A_2/A_1},\ \phi 1.8 f'_{cp})$$

$$= \min\left(0.6 \times 0.9 \times 35 \times \sqrt{\frac{202.5{\times}10^3}{70.23{\times}10^3}},\ 0.6 \times 1.8 \times 35\right)$$

$$= \min(32.1,\ 37.8) = 32.1\ \text{MPa}$$

The maximum stress on the concrete immediately behind the anchor plate at jacking is:

$$\sigma_b = \frac{1.15\ P_j}{A_1 - \pi d_d^2/4} = \frac{1.15 \times 2000\times10^3}{70.23\times10^3 - \pi 80^2/4} = 35.3\ \text{MPa}$$

which is greater than the limiting value of 32.1 MPa and thus confining reinforcement behind the bearing plate is needed.

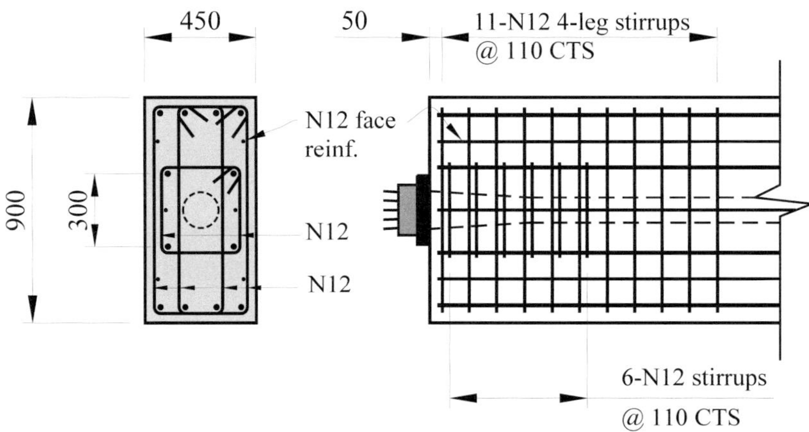

Figure 8.10 Reinforcement detail for end block: Example 8.1

EXAMPLE 8.2

TWO ANCHORAGES IN A RECTANGULAR END BLOCK

Two horizontal cables, each containing 14 strands of 12.7 mm diameter and each jacked to a maximum force at transfer of 1800 kN, are anchored 500 mm apart in a rectangular end block 1200 mm deep and 450 mm wide, as shown in Figure 8.11. The anchorages have bearing plates 280 mm square and the duct diameter is $d_d = 72$ mm. The concrete strength at transfer is f'_{cp} = 35 MPa.

Design the end block reinforcement for a stressing sequence where the lower tendon is stressed first to its full load, which is followed by stressing of the top tendon to its full load.

Other data: $A_g = 540 \times 10^3 \text{ mm}^2$; $I_g = 64.80 \times 10^9 \text{ mm}^4$

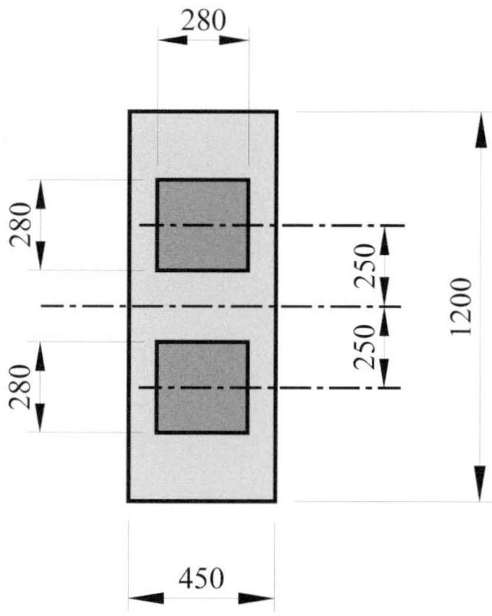

Figure 8.11 End block dimensions for Example 8.2

SOLUTION

Two loading conditions must be considered, (a) when only the lower cable is stressed and (b) when both cables are stressed. We consider each.

(a) lower cable only stressed:

For the case where the lower tendon is stressed alone, the jacking force is $P_j = 1800$ kN and it is applied at an eccentricity of $e = 250$ mm. As the external moment is zero, the stress at the top fibre (Equation 4.1):

$$\sigma_{ca} = \frac{P_j}{A_g} + \frac{P_j e y}{I_g}$$

$$= \frac{1800\times10^3}{540\times10^3} + \frac{1800\times10^3 \times 250 \times (-600)}{64.80\times10^9} = -0.83 \text{ MPa}$$

and at the bottom fibre:

$$\sigma_{cb} = \frac{1800\times10^3}{540\times10^3} + \frac{1800\times10^3 \times 250 \times 600}{64.80\times10^9} = 7.50 \text{ MPa}$$

It is noted that for the case of the lower tendon stressed and the upper tendon unstressed, the top fibre carries a small tensile stress but is less than the tensile strength of the concrete. The distribution of the longitudinal stresses at the end of the lead length is shown in Figure 8.12.

For the design of the end block for this load case, the bursting forces may be assessed either by (i) the simplified symmetrical stress block approach of Guyon (1953), or (ii) by calculating the bursting moment and detailing the reinforcement accordingly. We examine each approach in turn.

(i) Symmetrical stress block approach

For this eccentric loading the equivalent depth of the stressed prism is $D_e = 700$ mm, as shown in Figure 8.12. The bursting force is (Equation 8.5):

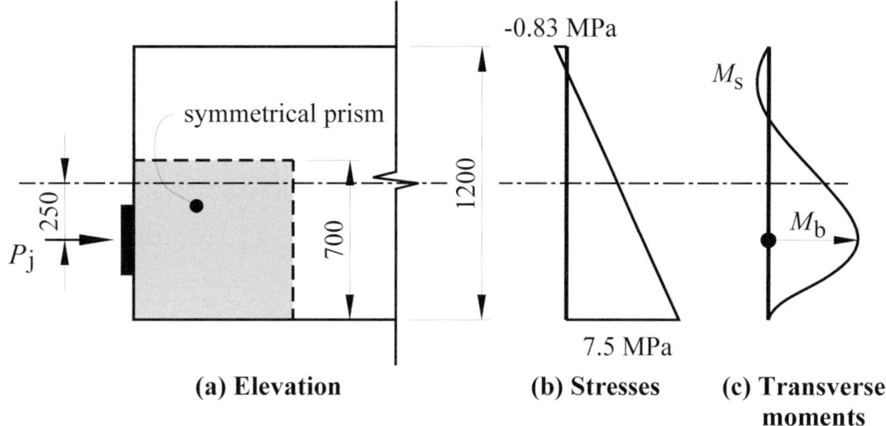

Figure 8.12 Longitudinal stresses and transverse moments: lower cable stressed

$$T_b = 0.25Pk\left(1 - \frac{w}{D_e}\right)$$

$$= 0.25 \times 1800 \times \frac{1200}{700}\left(1 - \frac{280}{700}\right) = 463 \text{ kN}$$

(ii) Transverse moment calculation method

Figure 8.13(a) shows the dimensions of the section for the case where the lower anchorage only is stressed. The resulting forces acting on the member at the end of the lead length (Figure 8.13(b)) are obtained by multiplying the stresses acting on the section by its width of 450 mm. With the forces on the section determined, the transverse shear force and bending moment diagrams are then determined (Figures 8.13(c) and (d)).

The transverse bending moments are a maximum along the axis where the shears are zero, which by calculation lies at a distances of 371.5 mm from the bottom fibre and 240 mm from the top fibre of the section (see Figure 8.13(c)). The latter provides for a splitting moment and is discussed subsequently.

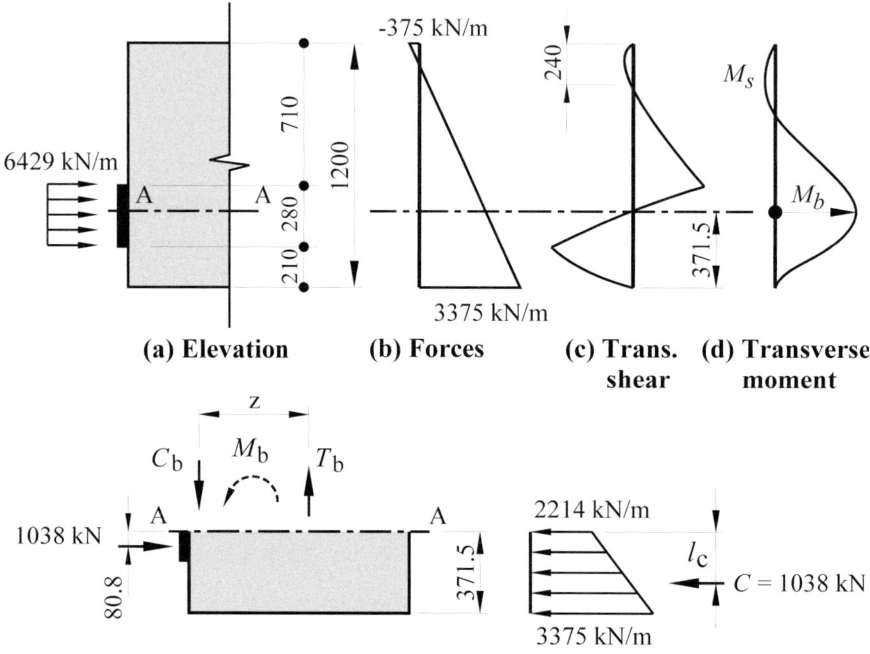

Figure 8.13 Free body along centreline of lower tendon

For the bursting moment behind the anchor plate, from the free body along the axis of zero shear (axis A–A in Figure 8.13(e)) the force C acting on the section is:

$$C = \left(\frac{2214 + 3375}{2}\right) \times 371.5 \times 10^{-3} = 1038 \text{ kN}$$

and its lever arm l_c is:

$$l_c = \frac{2214 \times \frac{371.5^2}{2} + \left(\frac{3375 - 2214}{2}\right) \times \frac{2}{3} \times 371.5^2}{1038 \times 10^3}$$

$$= 198.6 \text{ mm}$$

The bursting moment is thus calculated as:

$$\begin{aligned} M_b &= 1038\times10^3(198.6 - 80.8) \\ &= 122\times10^6 \text{ Nmm} \quad (122 \text{ kNm}) \end{aligned}$$

For a lever arm:

$$z = 0.5D_e = 0.5 \times 700 = 350 \text{ mm}$$

the bursting force is then:

$$T_b = \frac{M_b}{z} = \frac{122\times10^6}{350} \times 10^{-3} = 349 \text{ kN}$$

which is somewhat higher than that determined using the symmetrical prism approach.

The depth below the top fibre where the stress equals zero is $d_t =$ 120 mm (Figure 8.12(b)). The spalling moment at a depth of 240 mm from the top of the section (Equation 8.6) is:

$$\begin{aligned} M_s &= \frac{-2\sigma_t b d_t^2}{3} = \frac{2 \times 0.83 \times 450 \times 120^2}{3} \\ &= 3.6\times10^6 \text{ Nmm} \quad (3.6 \text{ kNm}) \end{aligned}$$

and the splitting tension force, for $z = 0.5\text{D} = 600$ mm, is:

$$T_s = M_s/z = 3.6\times10^6/600 = 6.0\times10^3 \text{ N} \quad (6 \text{ kN})$$

(b) both cables stressed:

The jacking force is $P_j = 1800$ kN in each tendon and the compressive stress at the end of the lead length is now constant at:

$$\sigma_c = \frac{2P_j}{A_g} = \frac{3600\times10^3}{540\times10^3} = 6.67 \text{ MPa}$$

and the uniform loading on the vertical section of the end block at the end of the lead length is $2P_j/D = 3000$ kN/m. This is shown in Figures 8.14(a) and (b).

The location of the point of zero shear is determined, as shown in Figure 8.14(d) and the maximum transverse moment is calculated as:

$$M_b = 1181\times10^3(197-92) = 124\times10^6 \text{ Nmm} \quad (124 \text{ kNm})$$

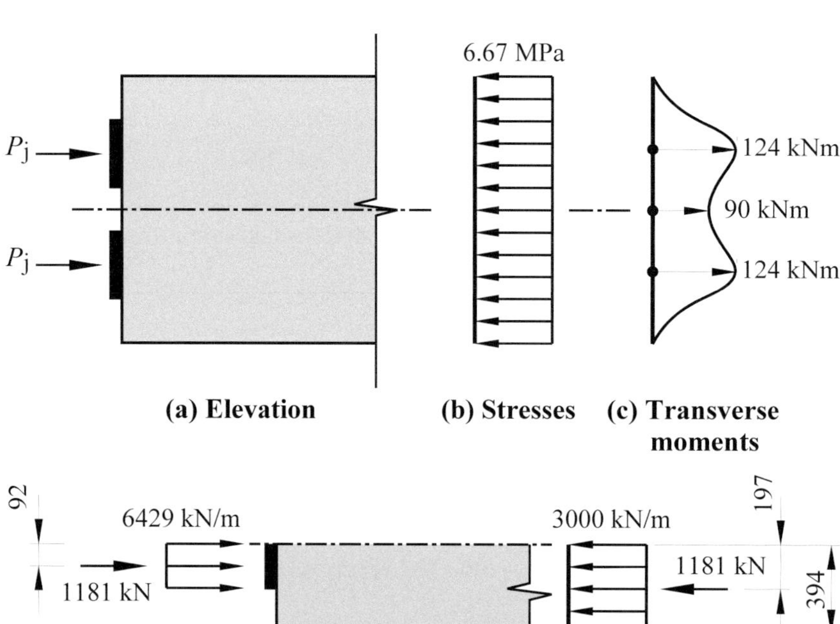

Figure 8.14 Stresses and moments with both cables stressed

At the centroidal axis, the bursting moment is:

$$M_b = Cl_c - P_j e$$

$$= 3000 \times 600 \times \frac{600}{2} - 1800\times10^3 \times 250$$

$$= 90\times10^6 \text{ Nmm} \quad (90 \text{ kNm})$$

The resulting bursting moment distribution is shown in Figure 8.14(c).

Vertical reinforcement requirements are examined for two sections: (1) the section behind each anchorage, where the maximum bursting moment of 124 kNm is considered to act on a symmetrical prism of depth 700 mm; (2) the centre section of the end block, where the bursting moment of 90 kNm arises from the combined action of both anchorages on the end block of depth 1200 mm (see Figure 8.14).

For case (1), the lever arm is:

$$z = 0.5D_e = 0.5 \times 700 = 350 \text{ mm}$$

and the bursting force is:

$$T_b = \frac{M_b}{z} = \frac{124\times10^6}{350} \times 10^{-3} = 354 \text{ kN}$$

For case (2), the lever arm is:

$$z = 0.5D = 0.5 \times 1200 = 600 \text{ mm}$$

and the bursting force is:

$$T_b = \frac{M_b}{z} = \frac{90\times10^6}{600} \times 10^{-3} = 150 \text{ kN}$$

Detailing the vertical reinforcement - section behind each anchorage

The calculations show that the simplified, symmetrical prism, approach provides a conservative estimate of the bursting force, exceeding the force determined using the refined calculation by 31 per cent. We shall take the lower more accurate value of $T_b = 354$ kN. The required area of bursting steel is thus:

$$A_b = \frac{354\times10^3}{150} = 2360 \text{ mm}^2$$

The length over which significant tension occurs is $0.8D_e = 0.8 \times 700 =$ 560 mm. Trying 4-legged N12 stirrups, the area of one set of stirrups in the web is:

$$A_w = 4 \times 110 = 440 \text{ mm}^2$$

and thus the maximum spacing of stirrups is:

$$s \le 560 \times \frac{440}{2360} = 104 \text{ mm}$$

For detailing, the calculated bursting reinforcement is required to extend from the end face to approximately $1.2D_e$ = 840 mm from the face. We shall adopt **9 sets of 4–leg N12 stirrups at 100 mm centres**.

Detailing the vertical reinforcement - centre section of the end block:

For the calculated bursting force of T_b = 150 kN, the required area of web reinforcement is:

$$A_b = \frac{150 \times 10^3}{150} = 1000 \text{ mm}^2$$

The length over which significant tension occurs is $0.8D = 0.8 \times 1200 =$ 960 mm. Trying 2-legged N12 stirrups, the area of one set of stirrups in the web is:

$$A_w = 2 \times 110 = 220 \text{ mm}^2$$

and the required spacing of stirrups is:

$$s \le 960 \times \frac{220}{1000} = 211 \text{ mm}$$

This reinforcement is required to extend from the end face to approximately $1.2D$ = 1440 mm from the face. The previously determined bursting reinforcement is more than sufficient to a distance of 840 mm from the end block face, beyond which we shall use an additional **3 sets of 2–leg N12 stirrups at 200 mm centres**. The reinforcement for the end block is shown in Figure 8.15.

Splitting moment reinforcement

For the splitting moment, the area of steel required is small ($A_s = 40\ \text{mm}^2$) and the 4–leg N12 stirrups provided is sufficient. This reinforcement is placed as close to the end block face as cover and detailing allow.

Horizontal bursting tension

As in Example 8.1, we must also consider the potential vertical cracking that occurs as a result of the dispersion of the compressive stress from the 280 mm wide bearing plate through the 450 mm width of the beam. The bursting tensile force is:

$$T_b = 0.25 \times 3600(1 - 280/450) = 340\ \text{kN}$$

requiring a reinforcement area of:

$$A_b = \frac{340 \times 10^3}{150} = 2270\ \text{mm}^2$$

and distributed over a length of $0.8D = 360$ mm. We shall try 2–legs of the N12 bars from the stirrups of the previously calculated vertical tension, plus four additional legs of N12 bars. Thus:

$$A_w = 6 \times 110 = 660\ \text{mm}^2$$

and the required spacing of the stirrup legs is:

$$s \leq 360 \times \frac{660}{2270} = 105\ \text{mm}$$

We **adopt 6 sets of 6–legs of N12 stirrups at 100 mm centres**, which extends a length beyond $1.2D = 540$ mm from the face of the end block. The reinforcement for the end block is shown in Figure 8.15.

Compatibility reinforcement

The surface tensile spalling force is estimated as $0.02 \times 3600 = 72\ \text{kN}$. This requires a reinforcement area of:

$$A_s = 72{\times}10^3/150 = 480\ \text{mm}^2$$

placed close to the face. The first two stirrups provides 880 mm^2 and is considered sufficient to control compatibility spalling at the end block face.

To control compatibility surface cracking, we take a minimum reinforcement ratio of 0.003 in each direction. In the vertical direction, the reinforcement supplied is more than adequate to control any horizontal surface cracking. In the horizontal direction we try 2–N16 bars (one each face) and the maximum spacing is determined as:

$$s \le \frac{400}{0.003 \times 450} = 296\ \text{mm}$$

The reinforcement placed in the corners of the of the bursting reinforcement, to anchor the stirrups, is sufficient for the control of compatibility cracking.

Check the bearing capacity of the concrete behind the anchor plate:

The area of the bearing plate is:

$$A_1 = 280 \times 280 = 78.40{\times}10^3\ \text{mm}^2$$

and the largest geometrically similar area of concrete that fits behind the anchorage plate is:

$$A_2 = 450 \times 450 = 202.5{\times}10^3\ \text{mm}^2$$

The maximum allowable bearing stress behind the anchor plate is determined by Equation 8.9 as:

$$f_b \le \min(\phi 0.9 f'_{cp}\sqrt{A_2/A_1},\ \phi 1.8 f'_{cp})$$

$$\le \min\left(0.6 \times 0.9 \times 35 \times \sqrt{\frac{202.5{\times}10^3}{78.40{\times}10^3}},\ 0.6 \times 1.8 \times 35\right)$$

$$\le \min(30.4,\ 37.8) = 30.4\ \text{MPa}$$

The maximum stress on the concrete immediately behind the anchor plate at jacking is:

$$\sigma_b = \frac{1.15P_j}{A_1 - \pi d_d^2/4} = \frac{1.15 \times 1800 \times 10^3}{78.40 \times 10^3 - \pi 72^2/4} = 27.8 \text{ MPa}$$

which is less than the limiting value of 30.4 MPa and, therefore, okay.

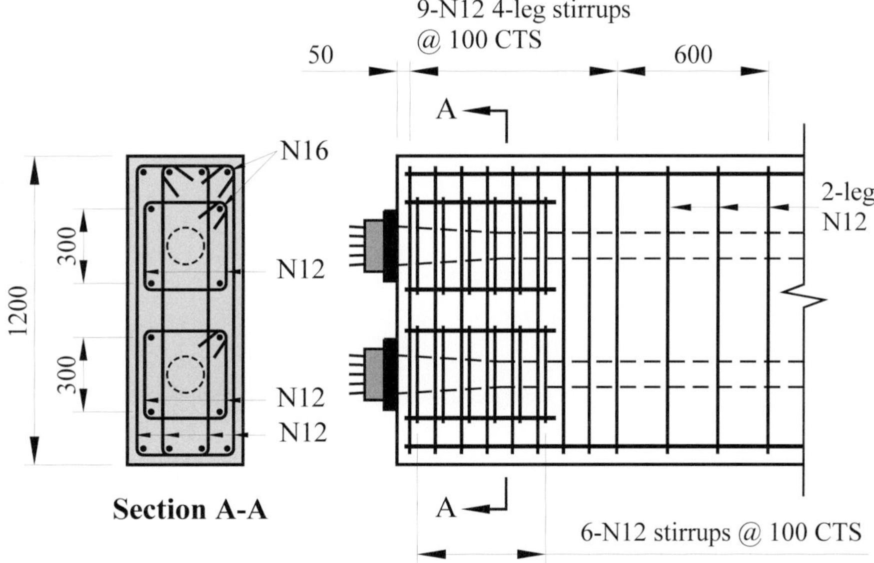

Figure 8.15 End block reinforcement for Example 8.2

8.2.6 Further design considerations

I-beams with rectangular end blocks

A rectangular end block on an I-section beam tends to act as a vertical deep beam supported by the I-section flanges and is liable to crack horizontally, as indicated in Figure 8.16.

Vertical reinforcement is required at the inner face of the end block to control such cracking, which becomes more pronounced as the size of the end block is decreased. However, provided that the length of the end block is at least 0.75D, and provided there is a gradually tapered transition from the rectangular to the I shape, the transverse tensile force may be estimated by the methods given previously for rectangular beams. Alternatively, the end block may be designed using the strut-and-tie method, as described in Section 8.4.

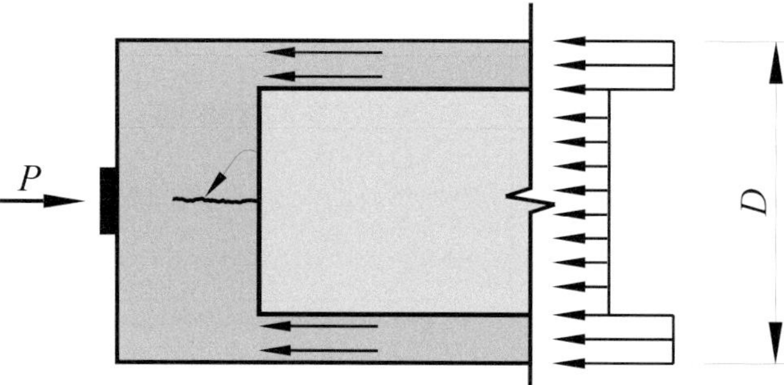

Figure 8.16 Rectangular end block in I-beam

Embedded anchorages

In the discussion so far the prestressing force has been applied through rigid steel plates bearing on the end face of the concrete. The effect of using embedded anchorages was investigated experimentally by Zielinski and Rowe (1962), who concluded that the type of anchorage made no significant difference to either the distribution or the magnitude of the transverse stresses.

Inclined anchorages

In most beams the tendons have a curved profile and are slightly inclined to the horizontal at the ends of the beam. The anchorages are therefore slightly out of vertical. The effect of the inclination was considered by Guyon (1953) who found no significant change in the bursting stresses for inclinations up to 1:10, although the splitting tensile stresses on the face of the end block were appreciably affected.

Interior anchorages

Post-tensioning cables are sometimes anchored at intermediate positions along the member. Intermediate anchorages are normally recessed into the beam, but may also be located in nibs projecting from the beam surface. The inclination of the cables at the intermediate anchorages, and the abrupt change in cross-section caused by the recess or nib, produce a more complex pattern of local stresses around the anchorages. 'Tearing' of the concrete behind an interior anchorage is a real possibility. Transverse cracks form close to and behind the anchorage, i.e. on the 'dead' or remote side of the anchorage. For a discussion of this and other forms of anchorage the reader is referred to Leonhardt (1964).

8.3 Anchorage of pretensioning tendons

In pretensioned construction the prestressing force is anchored by bond of the tendons to the adjacent concrete. The force in the tendon is zero at each end of the beam and increases progressively to a maximum value over a length that depends on the bond characteristics. For design purposes it is necessary to define two lengths:

1. the **transmission length** L_{pt} is the length required to build up the effective prestress $\sigma_{p.ef}$ in the tendon;
2. the **development length** L_p is the length required to develop the design stress σ_{pu} at ultimate.

Transmission length

The transmission length depends principally on the size and type of the tendon, its surface characteristics, its position within the member and to a lesser extent on the strength of the concrete at transfer.

As the tendon is stressed, according to Poisson its diameter is narrowed. Upon release, the cable expands at sections where its stress is reduced. This induces a radial compression on the strand from the surrounding concrete, in turn increasing the circumferential friction between the strand and the concrete (Figure 8.17). This is referred to as the **Hoyer effect**, after Hoyer (1939).

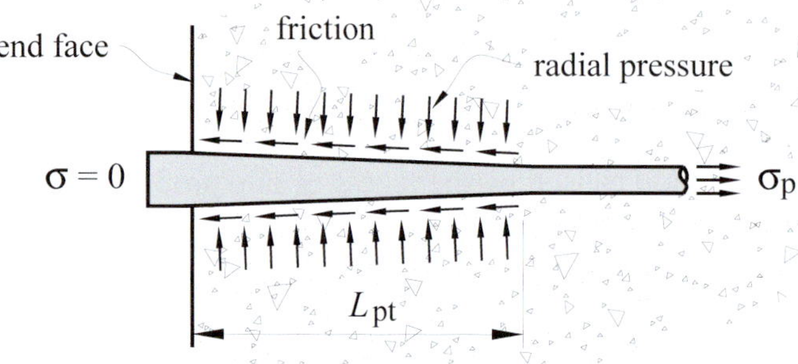

Figure 8.17 Transmission length for prestressing strand

Immediately after transfer the stress development in the tendon increases from zero at the beam end to $\sigma_{p.ef}$ at the end of the transmission length, following an exponential curve. Over time this shape changes as a result of concrete shrinkage, and there is evidence that a small length of the tendon at the end of the beam becomes completely unstressed. For tendons released gradually, the transmission length is as given in Table 8.1.

AS 3600 specifies that the tendon stress should be considered to be zero for a length of $0.1L_{pt}$ and to vary linearly from that point to $\sigma_{p.ef}$ at the end of the transmission length.

TABLE 8.1 Transmission lengths for pretensioned cables (AS 3600)

Type of tendon	Transmission length for gradual release, L_{pt},	
	$f_{cp} < 32$ MPa	$f_{cp} \geq 32$ MPa
Strand	60 strand diameters	60 strand diameters
Crimped wire	100 wire diameters	70 wire diameters
Indented wire	175 wire diameters	100 wire diameters

Care should be exercised if tendons are located towards the top of a member, where the bond capacity can be substantially reduced by the 'water gain' effect, which tends to create a film of water at the underside of the tendon, and by the settling of the concrete under the tendon as it sets. Plain prestressing wires towards the top of even a small section may have virtually no bond strength. Plain wires should never be used in pretensioned beams and their use is prohibited by AS 3600.

Development length

The development length for seven-wire strand at ultimate strength is specified in AS 3600 as not less than:

$$L_p = 0.145(\sigma_{pu} - 0.67\sigma_{p.ef})d_b \geq 60d_b \tag{8.10}$$

Consideration of transmission and development lengths is important in the design for shear of shallow members such as precast floor planks, where the critical sections for shear normally fall well inside both the transmission length and the development length.

Transverse tension in the end zone

Because of the gradual development of the prestressing force and the relatively small size and scattered distribution of pretensioning tendons, transverse tensions associated with bursting moments tend to be smaller in pretensioned members than post-tensioned members, and reinforcement is not normally required (Figure 8.18(a)). On the other hand, end cracking associated with splitting moments is often important in the webs of flanged pretensioned beams, especially if the pretensioning strands are concentrated in the top and/or bottom flanges.

Figure 8.18(b) shows the transmission of prestress in the flanges of an I-section girder with equal prestress top and bottom. In this instance, the prestress in the flanges sets up a condition for a potential splitting crack within the web, as shown. To determine the area of transverse reinforcement needed, the calculated splitting force T_s is divided by the nominal stress of 150 MPa. Because of the gradual introduction of the prestress over the transmission length, the spacing of the reinforcement should be calculated based on a length of $0.8L_{pt}$. This reinforcement is then placed uniformly from as close as possible to the face as cover and compaction allow and should extend to a distance of at least $1.2L_{pt}$ from the face.

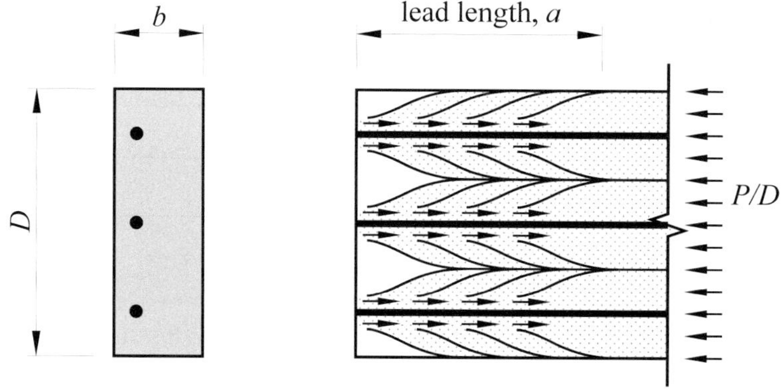

(a) Transmission of prestress for a line of pretensioned strands

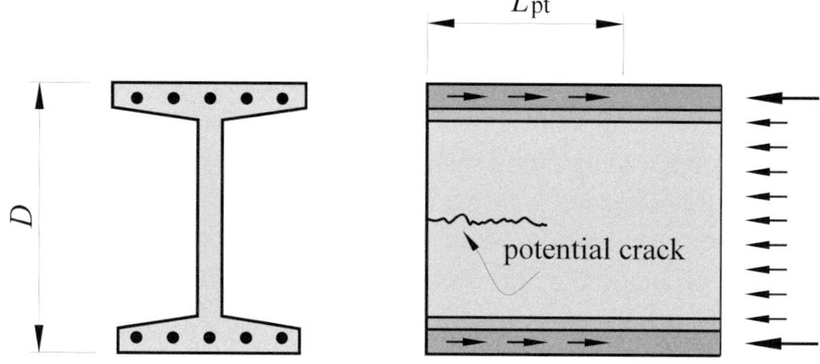

(b) Transmission of prestress in I-section girder

Figure 8.18 Dispersion of prestress in pretensioned members

8.4 Design of end blocks using strut-and-tie modelling

8.4.1 Introduction

In Section 3.7 of this book the principles for strut-and-tie modelling were introduced. In this section, we use these to show how to design prestressed concrete end-bocks using the strut-and-tie method to, generally, provide more efficient and flexible solutions than using the simplified procedures described in the previous section.

We again start with the simple case of one centrally located anchorage in a beam with a rectangular section, Figure 8.19(a), and consider the three-dimensional strut-and-tie model described in Figure 8.19(b). As before, the stiff bearing plate used in the anchorage is small in comparison with the beam cross-section and applies a concentrated compressive stress to the surface of the concrete. At some distance from the end block face, the stress disturbance has diminished and the stress on the section approximately equals P/A. The forces generated by this stress are distributed to four equally loaded struts that extend into the member to a first set of nodes. Each of these nodes is bounded by four members, all of which are in compression (i.e. struts). Thus, these nodes are in a state of tri-axial compression and their strength can be substantially greater than the uniaxial compressive strength of the concrete f_c', and

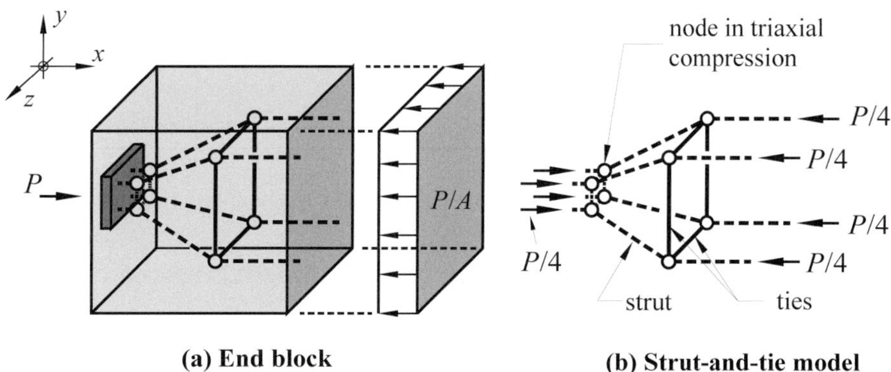

(a) End block (b) Strut-and-tie model

Figure 8.19 3D strut-and-tie model for a concentric anchorage

explains the higher allowable stress in the local zone immediately beneath the bearing plate given by Equation 8.9.

Beyond the compressive nodes, the struts open in both the x–y and x–z planes. This induces a tension in the section that is represented by the ties and needs to be reinforced for. We may simplify the modelling approach by treating each of the x–y and x–z planes separately, and reinforcing each to their statically enforced models accordingly.

In design, the D-region within the end block may be divided into two zones, the local zone and the general zone as shown in Figure 8.20. The local zone is directly adjacent to the anchorage device and consists of highly stressed confined concrete. The design of the local zone is normally left to the supplier of the anchorage and prestressing system and is generally designed to AS/NZS 1314.

In this Section we consider design of the general zone using the concepts of strut-and-tie modelling.

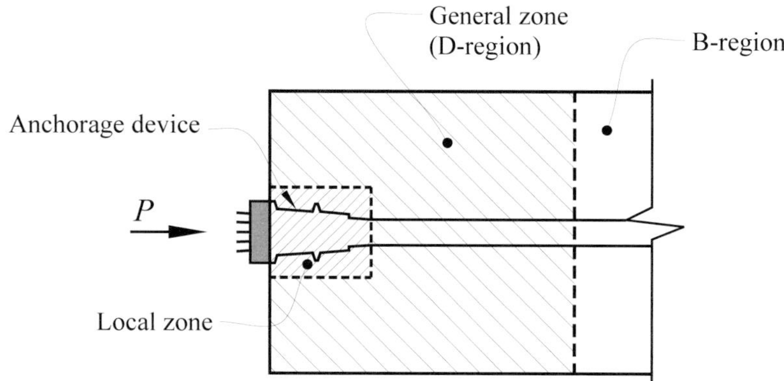

Figure 8.20 Definition of the local zone and the general zone

8.4.2 Central anchorage in a rectangular beam section

For a concentric concentrated load applied through a stiff bearing plate, Mörsch (1924) proposed using a model based on curved stress trajectories to define the compressive stress-field through the St Venant's region. Following the idea of Mörsch, the load path may be thought of as a funicular strut-and-tie model where the strut becomes a pair of arches connected at I (Figure 8.21(a)); the first through the range $0 \le y \le d_1$ and a second through the range $d_1 \le y \le d_2$. If these arches are taken as parabolic, then the distribution of bursting forces becomes uniform (Foster and Rogowsky, 1997a, 1997b). This is consistent with the uniform arrangement of the reinforcement that we detail to control these forces.

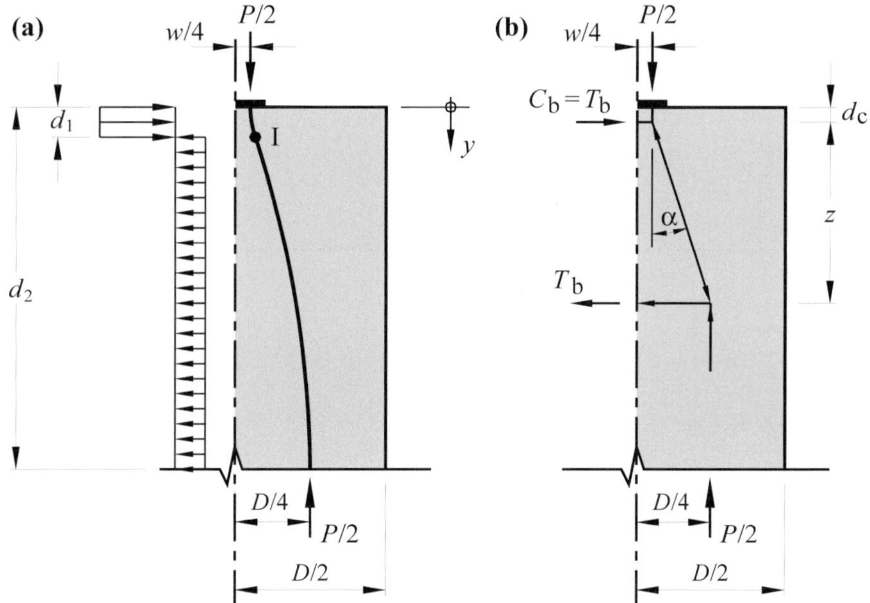

Figure 8.21 (a) Parabolic stress distribution; (b) statically equivalent strut-and-tie model

The funicular model shown in Figure 8.21(a) can be replaced with the statically equivalent strut-and-tie model shown in Figure 8.21(b). Taking z as the distance between the bursting compression force C_b to the centroid of the bursting tension force T_b, then the bursting force is:

$$T_b = \frac{D}{8z}\left(1 - \frac{w}{D}\right)P \tag{8.11}$$

If the internal level arm z is then taken as $z = D/2$, as is the case for the simplified model described in Section 8.2 above, the bursting force becomes that suggested by Mörsch and given by Equation 8.4.

The early design models of Guyon (1953) and Leonhardt (1964) are based on stress distributions obtained from linear-elastic analyses; however, the distribution of transverse tensile stress can look very different if non-linear behaviour is taken into account. In Figure 8.22 a comparison is made of the distributions obtained from the linear analysis and a non-linear analysis undertaken by Foster and Rogowsky (1997a). The analysis was carried out for a bearing plate width-to-depth ratio of $w/D = 0.2$ and for a uniform distribution of transverse reinforcing steel. The vertical axis of the graph represents the ratio of the transverse stress to the average prestress on the section (P/A), while the horizontal axis is the ratio of the distance from the anchorage face (x) to the depth of the section (D). In the non-linear analysis, the concrete had undergone significant cracking and the maximum stress in the steel was 220 MPa.

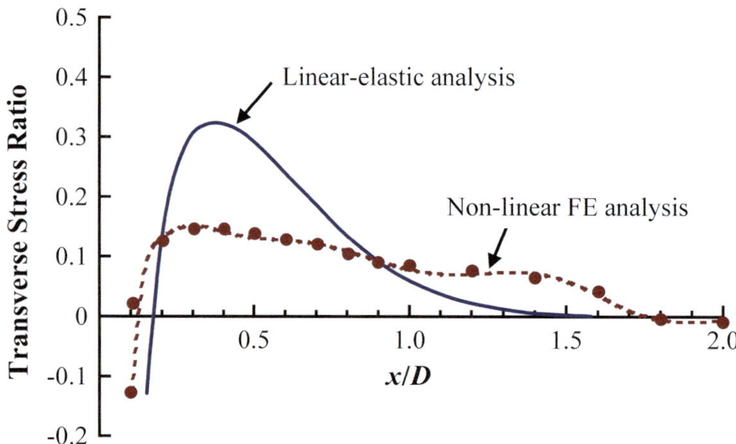

Figure 8.22 Transverse stress ratio relative to the distance from the bearing plate for w/D = 0.2.

It is seen that the behaviour is significantly different from the linear behaviour when cracking is considered. Figure 8.22 indicates that the disturbed region extends beyond a distance of D from the bearing surface which, according to the strut-and-tie model, can lead to lower quantities of bursting reinforcement (as the strut angle α is reduced). While the structure remains uncracked, the bursting stress distribution follows that of the linear analysis. When the peak stress reaches the tensile strength of the concrete, a substantial component of the tensile force is transferred from the concrete to the steel reinforcement. If the total bursting force in the region was transferred to the nearest reinforcing bars then a substantial increase in strain would be required. Rather, a proportion of the bursting force is transferred to the neighbouring concrete, resulting in further cracking. This process continues until the stress distribution is less than that required to cause further cracking in the concrete. It follows that the peaks shown in the non-linear analyses are spread over wider regions than for the linear analyses.

Whether a design is based on forces obtained from a linear stress analysis or a non-linear one, the behaviour of the member will adapt to the resulting arrangement of the transverse steel that is supplied. This steel must be detailed according to the equilibrium requirements of the load path assumed.

While in theory the angle α in the strut-and-tie model of Figure 8.21 may be freely chosen, control of cracking under service conditions and the ability of the member to redistribute the internal forces needs close consideration. An angle of $\alpha = 20$ degrees will generally prove acceptable in most situations, and corresponds to a constant bursting force of $T_b = 0.18\text{P}$ (Rogowsky and Marti, 1991).

Rather than selecting α, the lever arm in the strut-and-tie model may be chosen. As already mentioned, a lever arm of $0.5D$ will give a bursting force equivalent to that adopted in the simplified design procedures of AS 3600.

From their non-linear finite element (FE) analysis, Foster and Rogowsky (1997a) suggest a lever arm of:

$$z = 0.5D\left[\frac{1 - w/D}{1 - (w/D)^{2/3}}\right] \tag{8.12}$$

which gives a bursting force of:

$$T_b = 0.25P\left[1 - \left(\frac{w}{D}\right)^{2/3}\right] \tag{8.13}$$

The output from the design model in terms of the quantity or placement of the bursting reinforcement is not overly sensitive to the dimension d_c adopted (Figure 8.21) and it may be taken as (Foster and Rogowsky, 1997a):

$$d_c = w/4 \leq D/10 \qquad (8.14)$$

To determine the quantity of bursting steel, the design stress for the reinforcement must first be selected. For the strength limit state this stress may be taken as the yield stress multiplied by the strength reduction factor $\phi_s = 0.8$. Considering the load factor for bearing forces is 1.15, the effective limit on the stress in a 500 MPa grade bar for the strength limit state is about 350 MPa. Thus serviceability loads will almost always control the design of bursting reinforcement.

For serviceability, the VSL design guide for detailing for post-tensioning (Rogowsky and Marti, 1991) suggests that for moderately aggressive exposures, serviceability will be acceptable if the service load stresses are limited to 200 to 250 MPa. This value is significantly less restrictive than that of AS 3600, which requires a limit of 150 MPa for the strong degree of crack control needed in prestressed concrete end blocks. This value may be relaxed to 200 MPa where a moderate degree of crack control is acceptable.

The calculated reinforcement needs to be distributed through the stress-field. With reference to Figure 8.21 we see that the compressive stress-field extends a distance $2d_c$ from the face. The tensile stress-field starts at depth $2d_c$ and extends to a distance of $2z$ from the face (a length of z-d_c each side of the line of action of the bursting forces vector). The bursting reinforcement is distributed uniformly through the tensile stress-field and must extend to the edges of the member. For detailing purposes, AS 3600 requires that the bursting reinforcement calculated be extended to as close to the anchorage face as cover and compaction allow.

In addition to steel bars needed to maintain equilibrium, reinforcement is required in the dead (compatibility) zones that lie on the outside of the stress-fields and to prevent potential spalling of the concrete from the anchorage face. This has been discussed in Section 8.2.4 and similar provisions are required when designing using strut-and-tie models.

8.4.3 Eccentric anchorages in a rectangular beam section

Load eccentricity not greater than one-sixth of the section depth

We shall first consider the case where the eccentricity of the prestress is such that $e \le D/6$, that is where compression exists over the full depth of the section. In this case the strut-and-tie model is similar in form to that of the concentrically loaded anchorage and is shown in Figure 8.23. The statics of the model require that the lines of the forces $P/2$ pass through the centroids of the stress block that each represent. That is, the force represented by the hatched area ABEF in Figure 8.23 must equal $P/2$, as must the force given by area BCDE, and the centroids of these volumes define the lines of action of the forces.

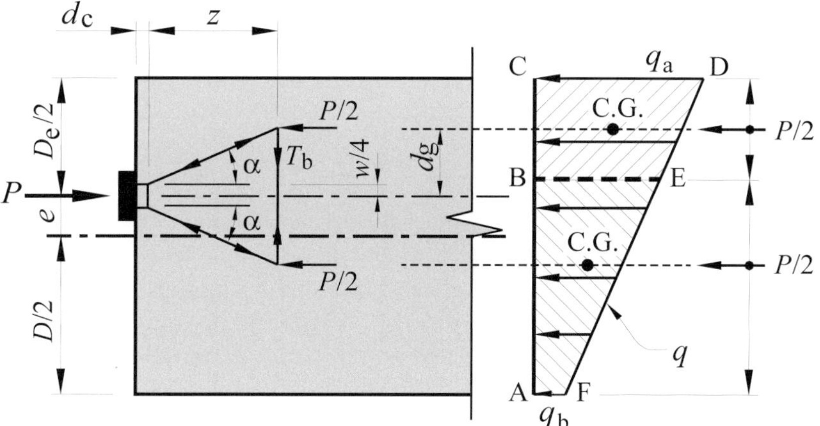

Figure 8.23 Strut-and-tie model for an eccentric anchorage ($e \le D/6$)

For the strut-and-tie model of Figure 8.23, the bursting force is:

$$T_b = \frac{P}{2}\tan\alpha = \frac{P}{2z}\cdot(d_g - w/4) \qquad (8.15)$$

To determine the bursting force T_b we may select a value for z, such as $z = 0.5D_e$ or, alternatively, we may substitute $D = D_e$ in Equation 8.12.

For the special case of a concentrically located anchorage, the bursting force is:

$$T_b = \frac{P}{8z} \cdot (D - w) \tag{8.16}$$

Where deemed sufficiently accurate, an approximate, and conservative, model may be adopted for the determination of the bursting forces. This model draws from the Guyon (1953) concept of a symmetrical prism, as previously described in Section 8.2.2. The model, demonstrated in Figure 8.24, takes an idealised stress distribution acting on the symmetrical prism section beyond the lead length of $P/(bD_e)$ and the member is then treated as for the case of a concentrically applied loading with D_e substituted for D in Equations 8.12 to 8.14.

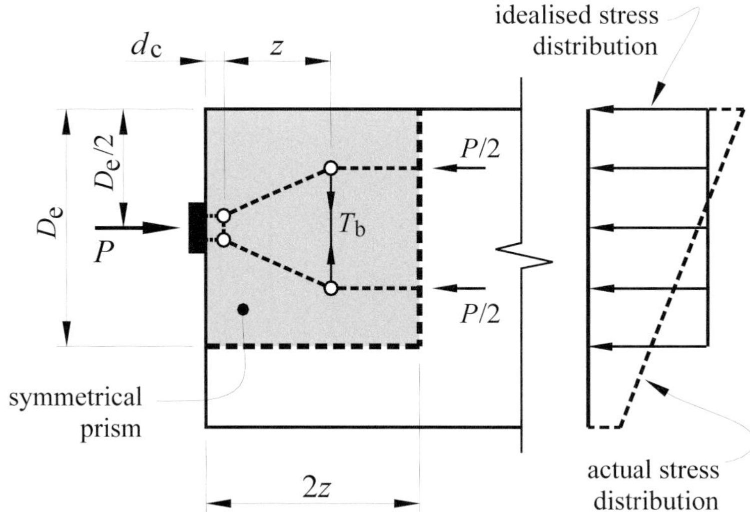

Figure 8.24 Approximate strut-and-tie model for an eccentric anchorage with $e \leq D/6$

Load eccentricity greater than one-sixth of the section depth

When the eccentricity is such that $e > D/6$, part of the section is in tension and this leads to the potential for splitting cracks (see Figure 8.5). For this case both the bursting tension in the line behind the anchorage and the splitting tension on the face away from the anchorage are described by the strut-and-tie model in Figure 8.25. In this model the stress block is divided into four regions: region I extending from the compressive fibre to a depth where the volume of the stress block gives a force equal to $P/2$; region II starting from the end of region I with a force of $P/2$; region III starting from the boundary of region II extending to the neutral axis with a compressive fore C equal in magnitude to the and tensile force T acting on the tensile side of the neutral axis; and region IV, the section on the tensile side of the neutral axis. These forces act at the centroids of the stress block regions that each represents and define the line of action of the forces on the boundary of the strut-and-tie model. The compressive forces each of $P/2$ generate a bursting force T_b, while the couple T and C generate the equilibrating splitting force of T_s.

To calculate the bursting force, the lever arm z needs to be selected. A value of $z = 0.5D_e$ may be adopted or, alternatively, we may substitute for $D = D_e$ in Equation 8.12. The bursting reinforcement is to be distributed from depth $2d_c$ and extending to a distance of $2z$ from the face.

The bursting and splitting forces are (Figure 8.25):

$$T_b = 0.5P\tan\alpha \tag{8.17}$$

$$T_s = T\cot\gamma \tag{8.18}$$

If the reinforcement needed to control splitting cracking is less than that needed for spalling (refer Section 8.2.4), the higher value for spalling should be supplied to the face. The reinforcement is to be distributed uniformly about the line of action of the force, with the first bars placed as near to the anchorage face as cover and compaction allow.

For a case where the eccentricity is large, the splitting force can be determined as shown in Figure 8.26.

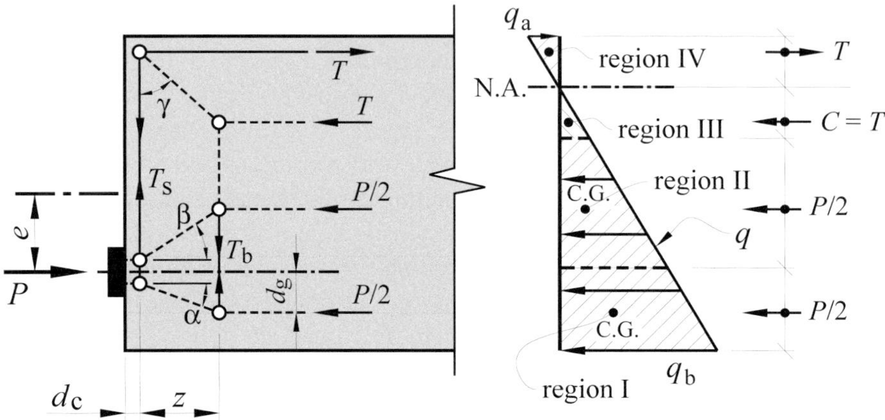

Figure 8.25 Strut-and-tie model for a eccentric anchorage, $e > D/6$

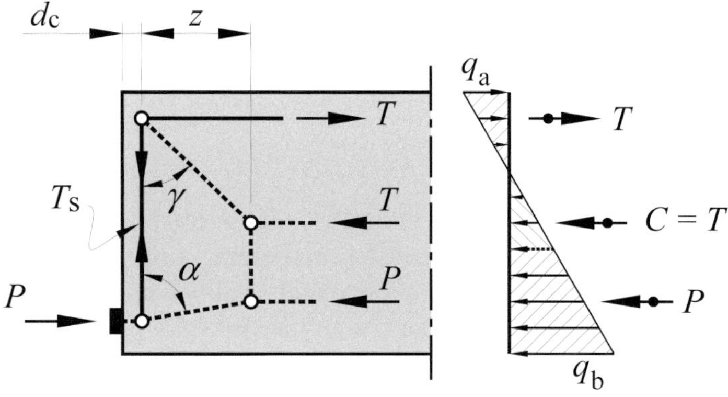

Figure 8.26 Strut-and-tie model for a single anchorage at a large eccentricity

8.4.4 Two widely spaced anchorages

We consider the case where we have two widely spaced anchors, for which we have seen previously (Figure 8.3) there is a potential for a splitting crack to occur from the face of the end section, midway between the anchorages. In this case, the end block may be treated as a deep beam with supports at the anchorage plates and loads at its boundary. An example is shown for the case of a symmetrical anchorage in Figure 8.27(a). Statics requires that the lines of the forces $P/2$ pass through the centroids of the stress block that each represent, which for this case is at depths of $D/4$ from each extreme fibre. For the strut-and-tie model of Figure 8.27(a), the splitting force is:

$$T_s = \frac{P}{2}\cot\alpha = \frac{P}{2z}\cdot\left(e - \frac{D}{4}\right) \tag{8.19}$$

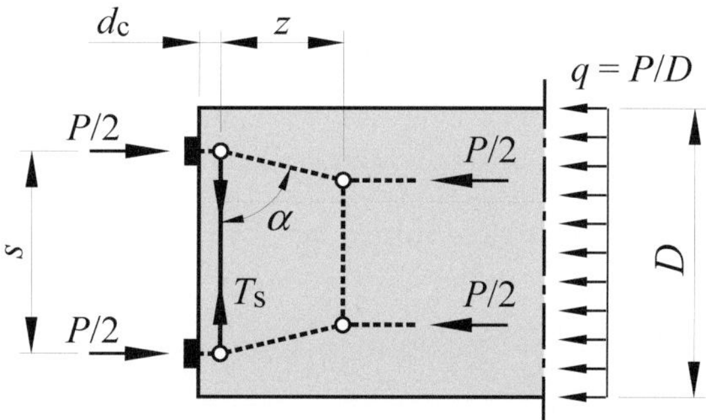

(a) End block with two symmetrical anchorage forces

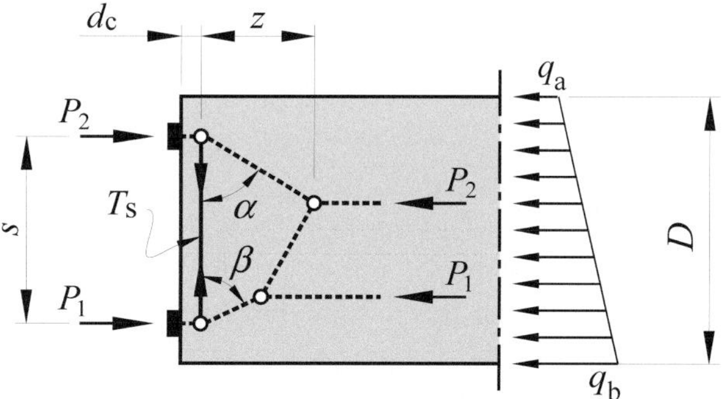

(b) End block with two non-symmetrical anchorage forces

Figure 8.27 Strut-and-tie models for two anchor system

To determine the splitting force T_s we first choose a value for the lever arm, z. If we choose $z = 0.6s$, as specified in AS 3600, the splitting force is:

$$T_s = \frac{P}{1.2s} \cdot \left(e - \frac{D}{4}\right) \tag{8.20}$$

and is the same as that obtained using the simplified method discussed in Section 8.2.3, with the resulting splitting force given by Equation 8.8.

If the anchorages are non-symmetric, or are not symmetrically loaded, the strut-and-tie model shown in Figure 8.27(b) may be applied. In this figure $P_1 \geq P_2$. The stress resultants at the end of the lead length are divided into two components P_1 and P_2, with each located at the centroid of their respective areas. With the lines of action of the boundary loads known, the splitting force can then be calculated:

$$T_s = P_2 \cot\alpha \tag{8.21}$$

The area of splitting steel is determined by dividing the calculated splitting force by the allowable stress in the reinforcement.

8.4.5 Flanged sections

For T- and I-section beams without rectangular end blocks, the bursting stresses can be found using a three-dimensional strut-and-tie model. An example is given in Figure 8.28 for the case of a cable anchored in the web of a T-section member and applying a force of P. This force is distributed first in the longitudinal direction to nodes a, b and c. The force C_3 is equal to one-half of the prestress force, i.e. $C_3 = P/2$. The other one-half of the prestressing force is distributed between C_1 and C_2 at nodes a and b, respectively. The force C_1 is the component that is distributed to the flange, whereas C_2 is the component distributed to the upper web. All forces C_1, C_2 and C_3 are located at the centroids of their respective stress blocks. Subsequently, the force C_1 at node a is distributed over the width of the flange to nodes d and e. This results in bursting tension T_{b2} that occurs in the flange, as shown. Once determined, sufficient reinforcing steel needs to be detailed to carry both bursting the forces T_{b1} and T_{b2}.

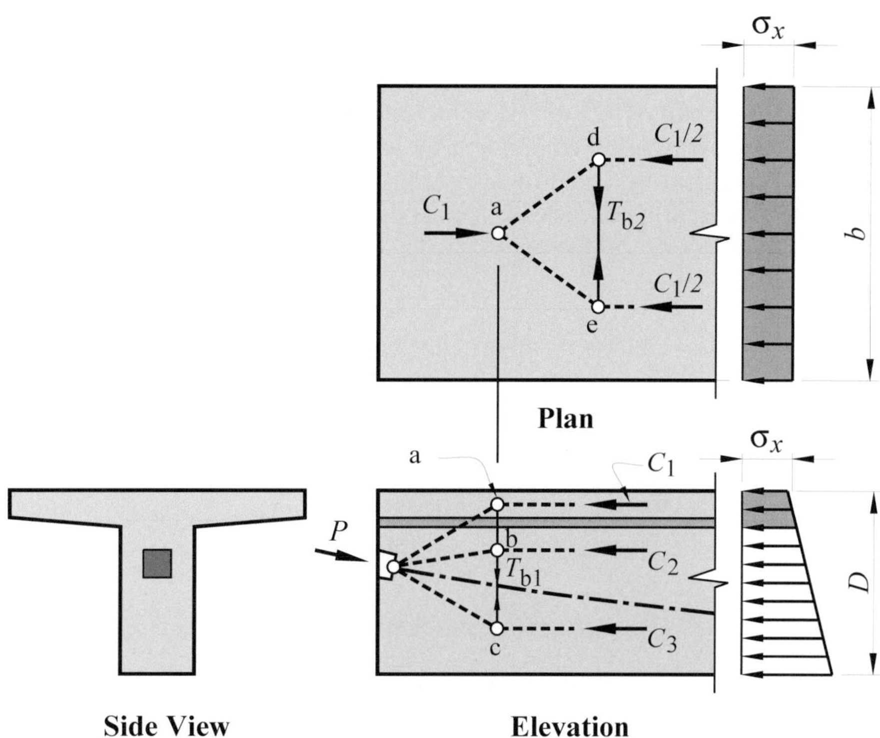

Figure 8.28 Strut-and-tie model for a T-section beam

8.4.6 The presence of support reactions

The presence of a support reaction can influence the internal load path and the location of the bursting forces. In Figure 8.29(a) the strut-and-tie model for a concentrically loaded rectangular end block is shown. In this case, the bursting force is determined and steel reinforcement detailed accordingly. Anti-spalling reinforcement will be provided to the end section near the anchorage face, as will reinforcement to control cracking in the compatibility regions.

Figure 8.29(b) shows the influence of the support reaction that results from the design loads. It is seen that the internal load path for equilibrium is changed from that of the end block that excludes this reaction. This needs to be considered in the final detailing of the member. For the bursting force ahead of the anchorage, the reinforcement designed and detailed using the model of Figure 8.29(a), together with stirrup reinforcement designed and detailed for the flexural shear, is sufficient to carry the forces that result from the altered load path. The force T_s in Figure 8.29(b) is an equilibrium requirement, not a compatibility requirement for which the spalling force is detailed. Face reinforcement designed to carry a load of $0.02P_j$ (see Section 8.2.4) is generally sufficient, provided that it is well detailed.

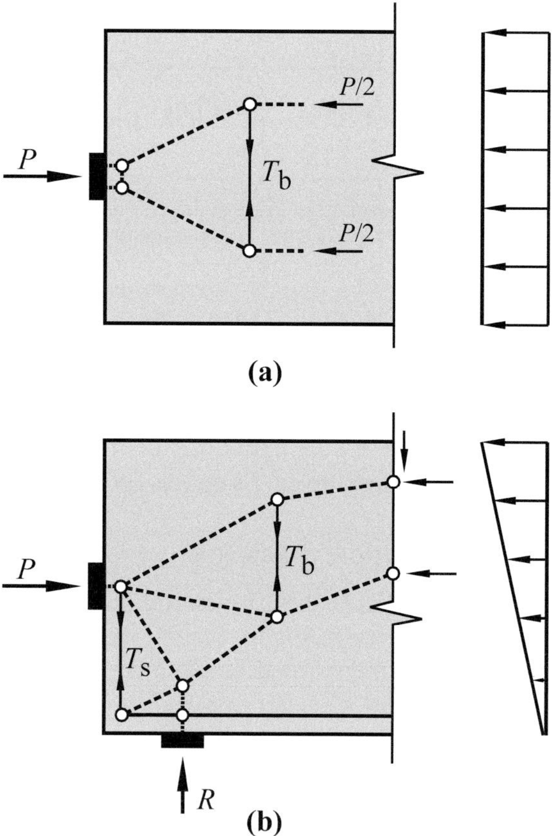

Figure 8.29 Influence of a support reaction

EXAMPLE 8.3

SINGLE ANCHORAGE IN A RECTANGULAR END BLOCK

In this example we shall design the end block of Example 8.2 using the strut-and-tie method.

The lower tendon is stressed first to a force of P_j = 1800 kN, followed by stressing of the top tendon also to a force of P_j = 1800 kN.

Other data: $e = \pm 250 \text{ mm}\,;\; A_g = 540\times10^3 \text{ mm}^2$

$$I_g = 64.80\times10^9 \text{ mm}^4$$

SOLUTION

Lower cable only stressed:

For the case where the lower tendon is stressed alone, the jacking force is P_j = 1800 kN and it is applied at an eccentricity of e = 250 mm. The strut-and-tie model for this case is described in Figure 8.25.

As the external moment is zero, the stress at the top fibre is (Equation 4.1):

$$\sigma_{ca} = \frac{P_j}{A_g} + \frac{P_j e y}{I_g}$$

$$= \frac{1800\times10^3}{540\times10^3} + \frac{1800\times10^3 \times 250 \times (-600)}{64.80\times10^9} = -0.833 \text{ MPa}$$

and at the bottom fibre:

$$\sigma_{cb} = \frac{1800\times10^3}{540\times10^3} + \frac{1800\times10^3 \times 250 \times 600}{64.80\times10^9} = 7.50 \text{ MPa}$$

As the section is of constant width, the distributed forces at the extreme fibres are obtained by multiplying the stresses by the width of the section, that is:

$$q_a = -0.833 \times 450 = -375 \text{ N/mm}$$

$$q_b = 7.50 \times 450 = 3375 \text{ N/mm}$$

The distributed loading at the boundary of the B- and D-regions (i.e. at the end of the lead length) is:

$$q = q_b - \frac{(q_b + q_a)}{D} = 3375 - \frac{(3375 + 375)\,y'}{1200}$$

$$= 3375 - 3.125y' \quad \text{N/mm}$$

where y' is the height measured from the bottom fibre (refer Figure 8.30).

The distance from the bottom fibre to top of the stress block described by region I in Figure 8.25 and representing a force of $P/2$ is calculated as:

$$\frac{1800 \times 10^3}{2} = \frac{1}{2} \times (3375 + 3375 - 3.125y_I')\,y_I'$$

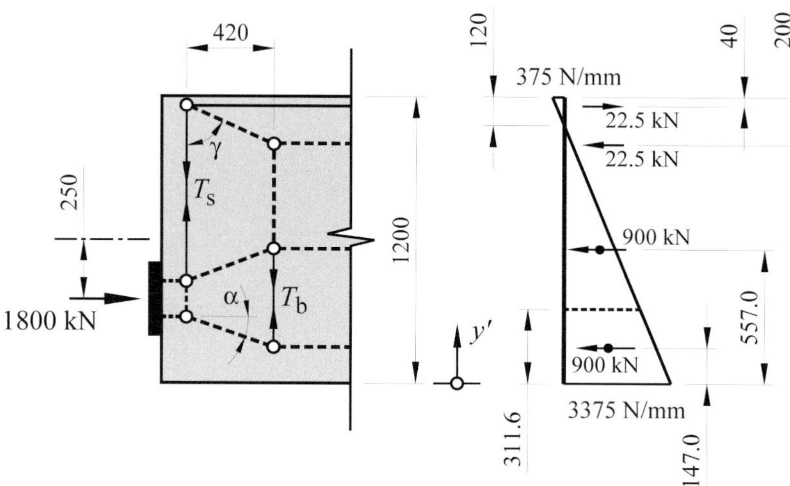

Figure 8.30 Strut-and-tie model for the lower cable stressed

$$y_I'^2 - 2.160\,y_I' + 576.0\times10^3$$

and solving the quadratic gives $y_I' = 311.6$ mm.

At the height of y_I', the distributed force is:

$$q(y_I') = 3375 - 3.125 \times 311.6 = 2401 \text{ N/mm}$$

and the dimension from the bottom fibre to the centroid of region I is:

$$\overline{y_I}' = \frac{2401 \times \dfrac{311.6^2}{2} + (3375\text{–}2401) \times \dfrac{311.6^2}{6}}{900\times10^3} = 147.0 \text{ mm}$$

Similarly, the dimension from the bottom fibre to the centroid of region II is determined as $y_{II}' = 557.0$ mm.

The depth from the top fibre to the centroid of the tensile stress block is (region IV) is $d_n/3 = 40$ mm, and the depth to its balancing compressive zone (region III) is 200 mm.

Next we choose the length of the lever arm z. In the simplified model discussed previously, $z = 0.5D_e = 350$ mm, while by Equation 8.12:

$$z = 0.5D_e \left[\frac{1 - w/D_e}{1 - (w/D_e)^{2/3}} \right]$$

$$= 0.5D_e \left[\frac{1 - 280/700}{1 - (280/700)^{2/3}} \right] = 0.66D_e = 460 \text{ mm}$$

The simplified model provides for somewhat conservative value, while that of Equation 8.12 is a reasonable upper limit. We choose $z = 0.6D_e = 420$ mm.

From the geometry:

$$\tan\alpha = \left(\left(350 - \frac{280}{4}\right) - 147.0\right)/420 = 0.317 \quad (\alpha = 17.6^\circ)$$

$$\cot\gamma = (200.0 - 40.0)/420 = 0.381 \quad (\gamma = 69.1^{\circ})$$

and thus:

$$T_{\mathrm{b}} = 0.5P\tan\alpha = 0.5 \times 1800 \times 0.317 = 285 \text{ kN}$$

$$T_{\mathrm{s}} = T\cot\gamma = 22.5 \times 0.381 = 8.6 \text{ kN}$$

Both cables stressed:

The jacking force is $P_{\mathrm{j}} = 1800$ kN in each tendon and the compressive stress at the end of the lead length is now constant at:

$$\sigma_{\mathrm{c}} = \frac{2P_{\mathrm{j}}}{A_{\mathrm{g}}} = \frac{3600\times10^{3}}{540\times10^{3}} = 6.67 \text{ MPa}$$

and the uniform loading on the vertical section of the end block at the end of the lead length is $2P_{\mathrm{j}}/D = 3000$ kN/m. This is shown in Figure 8.31.

To complete the strut-and-tie model, we take d_{c} as the lesser of $w/4$ and $D/10$ (Equation 8.14), giving $d_{\mathrm{c}} = 70$ mm.

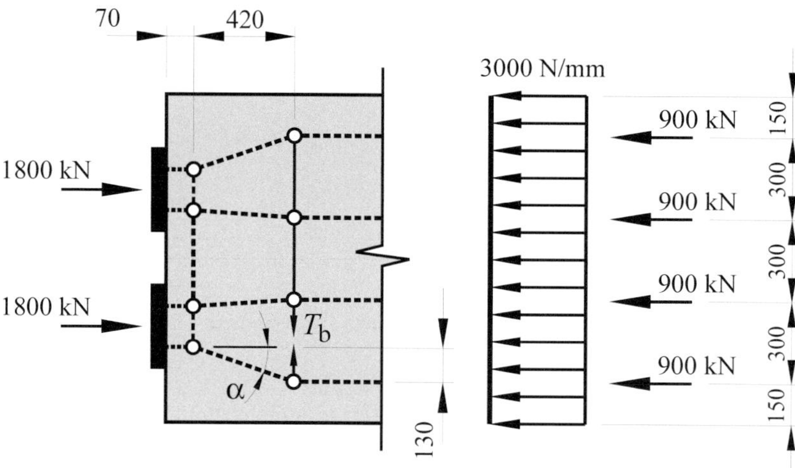

Figure 8.31 Strut-and-tie model with both cables stressed

As for the case of the single cable stress, we shall select $z = 420$ mm and from the geometry of the strut-and-tie model, we determine that:

$$\tan\alpha = 130/420 = 0.310 \quad (\alpha = 17.2^{\circ})$$

The bursting force is thus:

$$T_b = 0.5P\tan\alpha = 0.5 \times 1800 \times 0.310 = 279 \text{ kN}$$

Detailing the vertical reinforcement

Taking the higher of the bursting forces calculated from the different loading cases, i.e. $T_b = 285$ kN, and maintaining the maximum stress in the tie reinforcement as 150 MPa, the required area of bursting steel is:

$$A_b = \frac{285 \times 10^3}{150} = 1900 \text{ mm}^2$$

For detailing of the bursting steel, we shall try 4-legged N12 stirrups, giving an area for one set of stirrups of:

$$A_w = 4 \times 110 = 440 \text{ mm}^2$$

The stirrups are to be uniformly spaced over the bursting tension zone that extends from $2d_c = 140$ mm to $2z = 840$ mm, a length of 700 mm. Noting that the first stirrup should be placed at a half spacing from each of the bursting tensile zone boundaries (i.e. the number of spaces equals the number of stirrups), the total number stirrups needed is:

$$n \geq 1900/440 = 4.3$$

With five stirrups, the spacing is:

$$s \leq 700/5 = 140 \text{ mm}$$

For detailing, we shall extend the calculated bursting reinforcement from the end face to 840 mm from the face. We shall adopt **6 sets of 4–leg N12 stirrups at 110 mm centres**.

Splitting moment reinforcement:

As the splitting force is small, the 4–leg N12 stirrups provided near the face is sufficient.

Horizontal bursting tension:

In the horizontal direction (Figure 8.19), the compressive stress from the 280 mm wide bearing plate is to be distributed to the 450 mm width of the beam. Choosing a lever arm $z = 0.6b = 270$ mm, the bursting tensile force is (Equation 8.16):

$$T_b = \frac{P}{8z} \cdot (b - w) = \frac{3600}{8 \times 270}(450 - 280) = 283 \text{ kN}$$

requiring a reinforcement area of:

$$A_b = \frac{283 \times 10^3}{150} = 1890 \text{ mm}^2$$

We shall try 2–legs of the N12 bars from the stirrups of the previously calculated vertical tension, plus four additional legs of N12 bars. Thus:

$$A_w = 6 \times 110 = 660 \text{ mm}^2$$

Taking d_c as the lesser of $w/4$ and $D/10$, gives $d_c = 45$ mm. The stirrups are to be uniformly spaced over the bursting tension zone that extends from $2d_c = 90$ mm to $2z = 540$ mm, a length of 450 mm. The total number of stirrups needed is:

$$n \geq 1890/660 + 1 = 2.9$$

With three set of stirrups, the spacing is:

$$s \leq 450/3 = 150 \text{ mm}$$

We **adopt 6 sets of 6–legs of N12 stirrups at 110 mm centres**. The reinforcement for the end block is shown in Figure 8.32.

Compatibility reinforcement

The surface tensile spalling force is estimated as $0.02 \times 3600 = 72$ kN (see Section 8.2.4). This requires a reinforcement area of:

$$A_s = 72 \times 10^3 / 150 = 480 \text{ mm}^2$$

placed close to the face. The first two stirrups provides 880 mm^2 and is considered sufficient to control compatibility spalling at the end block face.

To control compatibility surface cracking, we take a minimum reinforcement ratio of 0.003 in each direction. In the vertical direction, the reinforcement supplied is more than adequate to control any horizontal surface cracking. In the horizontal direction we try 2–N16 bars (one each face) and the maximum spacing is determined as:

$$s \leq \frac{400}{0.003 \times 450} = 296 \text{ mm}$$

The reinforcement placed in the corners of the of the bursting reinforcement, to anchor the stirrups, is sufficient for the control of compatibility cracking.

Check the bearing capacity of the concrete behind the anchor plate:

The area of the bearing plate is:

$$A_1 = 280 \times 280 = 78.40 \times 10^3 \text{ mm}^2$$

and the largest geometrically similar area of concrete that fits behind the anchorage plate is:

$$A_2 = 450 \times 450 = 202.5 \times 10^3 \text{ mm}^2$$

The maximum allowable bearing stress behind the anchor plate is determined by Equation 8.9 as:

$$f_b = \min(\phi 0.9 f'_{cp}\sqrt{A_2/A_1},\ \phi 1.8 f'_{cp})$$

$$= \min\left(0.6 \times 0.9 \times 35 \times \sqrt{\frac{202.5\times10^3}{78.40\times10^3}},\ 0.6 \times 1.8 \times 35\right)$$

$$= \min(30.4,\ 37.8) = 30.4 \text{ MPa}$$

The maximum stress on the concrete immediately behind the anchor plate at jacking is:

$$\sigma_b = \frac{1.15P_j}{A_1 - \pi d_d^2/4} = \frac{1.15 \times 1800\times10^3}{78.40\times10^3 - \pi 72^2/4} = 27.8 \text{ MPa}$$

which is less than the limiting value of 30.4 MPa and, therefore, okay.

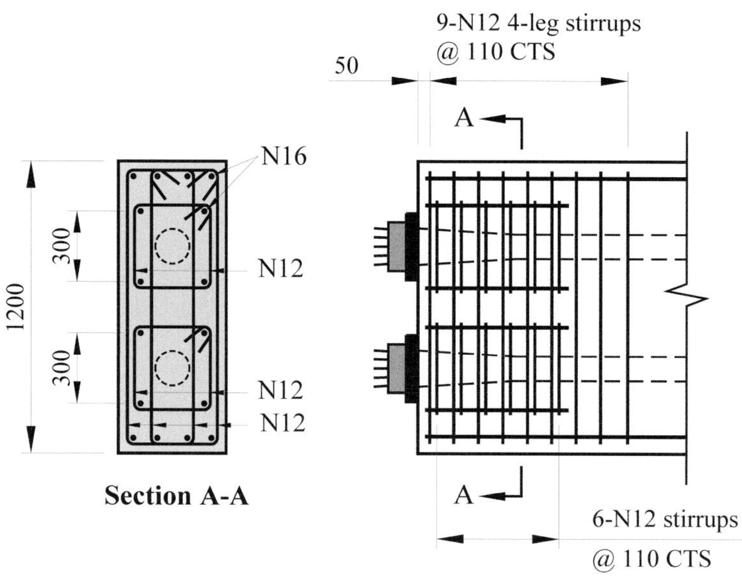

Figure 8.32 End block reinforcement for Example 8.3

EXAMPLE 8.4

SINGLE ANCHORAGE IN A T-SECTION END BLOCK

In this example we shall design the end block of the T-section member shown in Figure 8.33 using the strut-and-tie method.

The tendon is located on the centroidal axis, which is 540 mm from the bottom of the section, and stressed to a force of P_j = 2000 kN. The anchorage plate is 265 mm square and the diameter of the duct is d_d = 80 mm. The strength of the concrete at transfer is f'_{cp} = 35 MPa.

Other data: $A_g = 540{\times}10^3\ \text{mm}^2$;

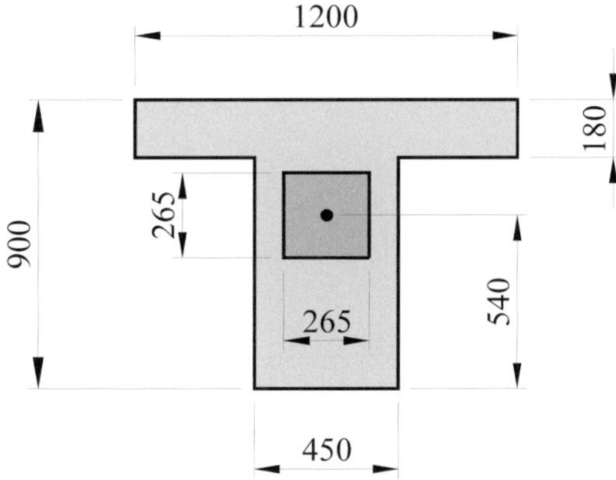

Figure 8.33 End block section for Example 8.4

Design in the *x-y* plane

Adopting the strut-and-tie model shown in Figure 8.34, we design the reinforcement for the bursting forces in the *x-y* plane. The depth of the equivalent prismatic stress block is:

$$D_e = 2 \times (900 - 540) = 720 \text{ mm}$$

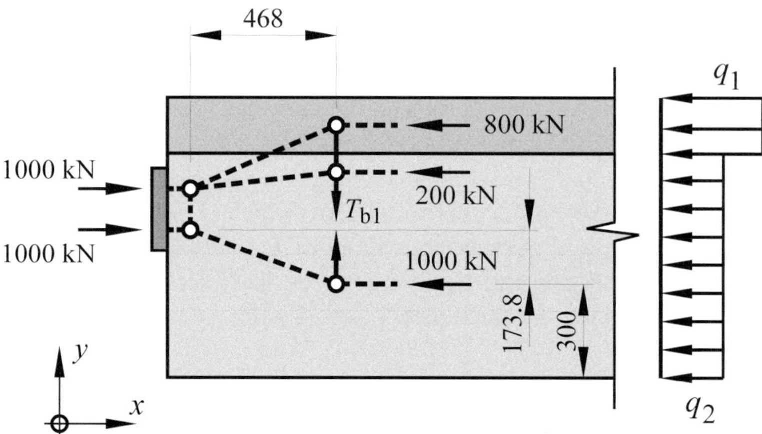

Figure 8.34 Strut-and-tie model for the *x-y* plane of Example 8.4

To optimise the reinforcement, we shall calculate the lever arm z from Equation 8.12; that is:

$$z = 0.5D_e \left[\frac{1 - w/D_e}{1 - (w/D_e)^{2/3}} \right]$$

$$= 0.5 \times 720 \times \left[\frac{1 - 265/720}{1 - (265/720)^{2/3}} \right] = 468 \text{ mm}$$

which is equal to $0.65D_e$. The depth from the bearing plate to the first nodes is taken as $d_c = w/4 = 66$ mm, which is less than $D_e/10$.

The average prestress on the section at the time of jacking is:

$$P_e/A_g = 2000{\times}10^3/540{\times}10^3 = 3.704 \text{ MPa}$$

and the boundary forces are:

$$q_1 = 3.704 \times 1200 = 4445 \text{ N/mm}$$

$$q_2 = 3.704 \times 450 = 1667 \text{ N/mm}$$

The component of the jacking force taken into the flange is:

$$C_F = 4445 \times 180 = 800 \text{ kN}$$

and the upper web component is 1000 – 800 = 200 kN.

The lower web component of 1000 kN is distributed over a web height of $1000 \times 10^3 / 1665 = 600 \text{ mm}$, with its centroid located 300 mm from the bottom of the section. The distance of this force to the quarter point on the bearing plate is $540 - 300 - 265/4 = 173.8 \text{ mm}$ and the bursting force is thus determined as:

$$T_{b1} = 1000 \times \frac{173.8}{468} = 371 \text{ kN}$$

Limiting the stress in the bursting reinforcement to 150 MPa, the area of bursting steel is:

$$A_b = 371 \times 10^3 / 150 = 2470 \text{ mm}^2$$

For detailing of the bursting steel, we shall try 4-legged N12 stirrups, giving an area for one set of stirrups of:

$$A_w = 4 \times 110 = 440 \text{ mm}^2$$

The stirrups are to be uniformly spaced over the bursting tension zone that extends from $2d_c = 122$ mm to $2z = 936$ mm, a length of 814 mm. Noting that the first stirrup should be placed at a half spacing from each of the bursting tensile zone boundaries, the total number stirrups needed is:

$$n \geq 2470/440 + 1 = 6.6$$

With seven stirrup sets through the bursting stress-field, the spacing is:

$$s \leq 936/7 = 134 \text{ mm}$$

For detailing, we extend the calculated bursting reinforcement to the end of the member and adopt **8 sets of 4–leg N12 stirrups at 130 mm centres**.

Design in the *x-z* plane - web

For the bursting steel in the *x-z* plane, we calculate the lever arm z from Equation 8.12 as:

$$z = 0.5b\left[\frac{1 - w/b}{1 - (w/(b))^{2/3}}\right]$$

$$= 0.5 \times 450 \times \left[\frac{1 - 265/450}{1 - (265/450)^{2/3}}\right] = 311 \text{ mm}$$

which is equal to $0.69b$.

The strut and tie model for the end *x-z* plane web is shown in Figure 8.35 and the bursting force determined as (Equation 8.16):

$$T_b = \frac{P}{8z} \cdot (b - w) = \frac{2000}{8 \times 311}(450 - 265) = 149 \text{ kN}$$

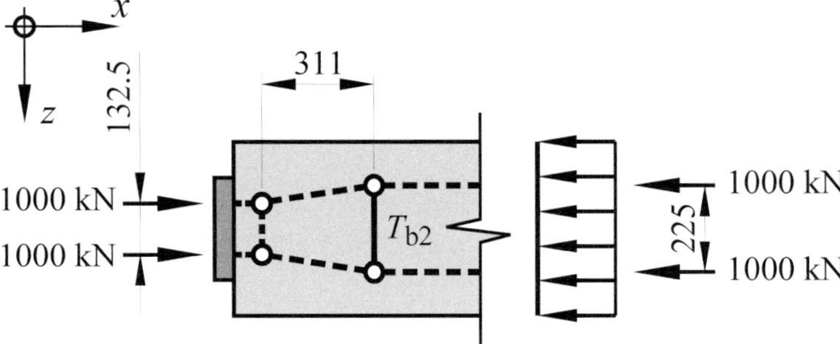

Figure 8.35 Strut-and-tie model for the *x*-z plane of Example 8.4

Limiting the stress in the bursting reinforcement to 150 MPa, the area of bursting steel is:

$$A_b = 149 \times 10^3 / 150 = 990 \text{ mm}^2$$

We shall try 4-legged N12 stirrups, giving an area for one set of stirrups of:

$$A_w = 4 \times 110 = 440 \text{ mm}^2$$

The stirrups are to be uniformly spaced over the bursting tension zone that extends from $2d_c = 90$ mm to $2z = 622$ mm, a length of 532 mm. The total number stirrups needed is thus:

$$n \geq 990/440 + 1 = 3.3$$

With four stirrup sets through the bursting stress-field, the spacing is:

$$s \leq 532/4 = 133 \text{ mm}$$

For detailing, we extend the calculated bursting reinforcement to the end of the member and adopt **5 sets of 4–leg N12 stirrups at 130 mm centres**.

Design in the *x*-z plane - distribution to flange

Next we distribute the web component of the bursting force that is attributed to the flange through the width of the flange. The forces at the web quarter points are 400 kN, with each located at a distance of 66 + 468 = 534 mm from the anchorage face. We shall take the lever arm as one-half the width of the flange $z = b/2 = 600$ mm and the bursting force T_{b3} (Figure 8.36) is:

$$T_b = 400 \times \frac{0.5(600 - 225)}{600} = 125 \text{ kN}$$

Again limiting the stress in the bursting reinforcement to 150 MPa, the area of bursting steel is:

$$A_b = 125 \times 10^3 / 150 = 830 \text{ mm}^2$$

The tie reinforcement is to be uniformly spaced over the bursting tension zone, which we shall take as approximately 500 mm each side of the centroid of the bursting force, located along the line of T_{b3} in Figure 8.36. Adopting a bar spacing of 130 mm we can fit 8 bars in this region. The area of one bar is 830/8 = 104 mm and thus we adopt 8-**N12 bars at 130 mm centres.** For detailing, we continue this reinforcement to the anchorage face, giving 12-N12 bars at 130 mm spacing.

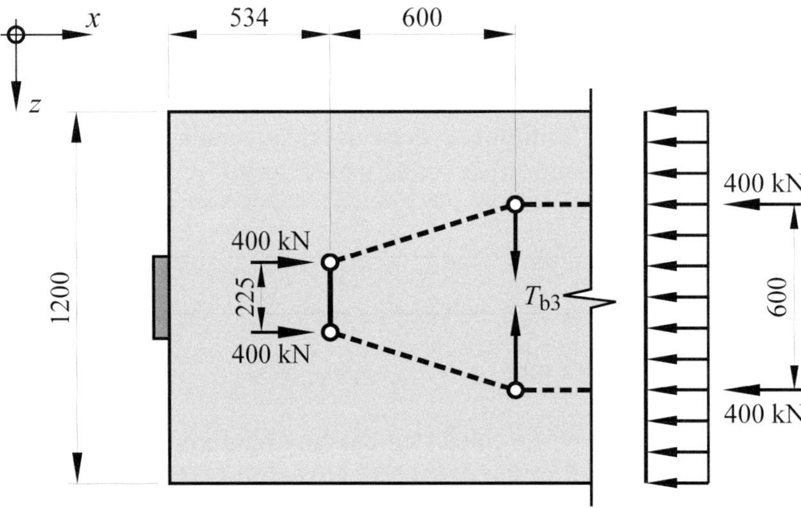

Figure 8.36 Strut-and-tie model for the *x*-z plane of Example 8.4

Compatibility reinforcement

The surface tensile spalling force is estimated as $0.02 \times 2000 = 40$ kN. This requires a reinforcement area of:

$$A_s = 40 \times 10^3 / 150 = 270 \text{ mm}^2$$

placed close to the face. The stirrup set provides 440 mm^2 and is sufficient to control compatibility spalling at the end block face.

To control compatibility surface cracking, we take a minimum reinforcement ratio of 0.003 in each direction. In the vertical direction, the reinforcement supplied is more than adequate to control any horizontal surface cracking. In the horizontal direction we try 2N16 bars, one bar close to each surface. The bar spacing is:

$$s \leq \frac{400}{0.003 \times 450} = 296 \text{ mm}$$

The reinforcement for the end block is shown in Figure 8.37.

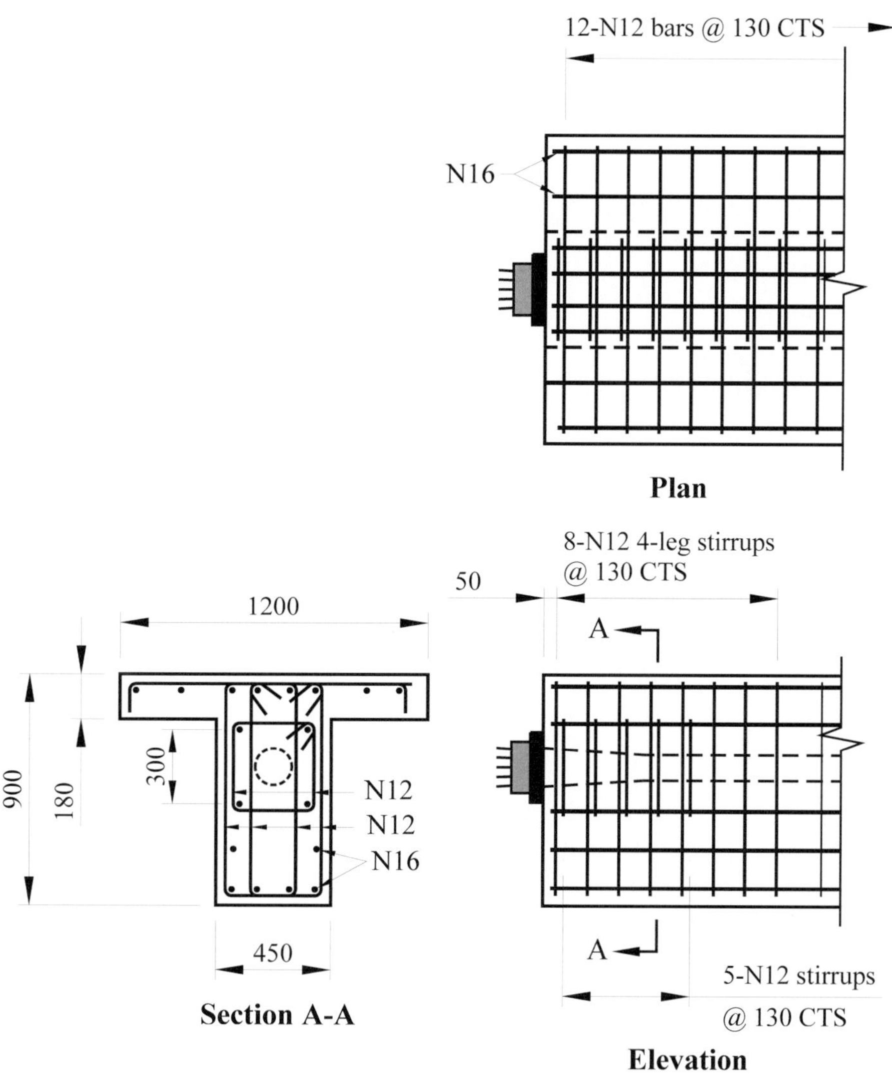

Figure 8.37 End block reinforcement for Example 8.4

Check the bearing capacity of the concrete behind the anchor plate:

The area of the bearing plate is:

$$A_1 = 265 \times 265 = 70.2\times10^3 \text{ mm}^2$$

and the largest geometrically similar area of concrete that fits behind the anchorage plate is:

$$A_2 = 450 \times 450 = 202.5\times10^3 \text{ mm}^2$$

The allowable bearing stress behind the anchor plate without testing is (Equation 8.9):

$$f_b = \min(\phi 0.9 f'_{cp}\sqrt{A_2/A_1},\ \phi 1.8 f'_{cp})$$

$$= \min\left(0.6 \times 0.9 \times 35 \times \sqrt{\frac{202.5\times10^3}{70.2\times10^3}},\ 0.6 \times 1.8 \times 35\right)$$

$$= \min(32.1,\ 37.8) = 32.1 \text{ MPa}$$

The stress on the concrete immediately behind the anchor plate at jacking is:

$$\sigma_b = \frac{1.15P_j}{A_1 - \pi d_d^2/4} = \frac{1.15 \times 2000\times10^3}{70.2\times10^3 - \pi 80^2/4} = 35.3 \text{ MPa}$$

which is greater than the limiting value of 32.1 MPa. The capacity of the anchor for bearing may be justified by the system manufacturer through testing in accordance with AS/NZS 1314. Alternatively, the concrete strength at transfer could be increased from 35 MPa to 38 MPa.

8.5 References

AS/NZS 1314, 2003, *Prestressing Anchorages*, Standards Australia/Standards New Zealand, 30 pp.

Foster S.J., Kilpatrick A.E. and Warner R.F. 2021, *Reinforced Concrete Basics: Analysis and design of reinforced concrete structures*, 3rd Ed., Pearson, 589 pp.

Foster S.J., and Rogowsky, D.M. 1997a, Bursting Forces in Concrete Members resulting from In-plane Concentrated Loads, *Magazine for Concrete Research*, Vol. 49, No. 180, pp. 231-240.

Foster S.J., and Rogowsky, D.M. 1997b, Splitting of Concrete Panels under Concentrated Loads, *Structural Engineering and Mechanics*, Vol. 5, No. 6, pp 803-815.

Guyon Y. 1953, *Prestressed Concrete*, Parsons, London.

Hoyer, E. 1939, Der Stahlsaitenbeton, Träger und Platten, Otto Elsner Verlagsgesellschaft, BerlinWien Leipzig.

Iyengar K.T.S.R., and Yogananda C.V. 1966, A Three Dimensional Stress Distribution Problem in the Anchorage Zone of a Post-tensioned Concrete Beam, *Magazine of Concrete Research*, Vol 18, No 55.

Leonhardt F. 1964, *Prestressed Concrete Design and Construction*, English Edn, Wilhelm Ernst, Berlin, 677 pp.

Mörsch E. 1924, Über die Berechnung der Gelenkquadcr, Beton Eisen, Vol. 23, No. 12, pp. 156-161.

Rogowsky D.M., and Marti P. 1991, *Detailing for Post-Tensioning*, No. 3, VSL Report Series, VSL International.

Zielinski J., and Rowe R.E. 1962, *The Stress Distribution Associated with Groups of Anchorages in Post-Tensioned Concrete Members*, CaCA London, Research Report No 13.

CHAPTER 9

Loss of prestress

There are two categories of prestress loss: immediate and deferred. Immediate losses occur during the prestressing operation. Deferred losses develop after the prestressing operation is complete, and are primarily due to inelastic behaviour of the concrete and the prestressing tendon. In this chapter we look at the underlying loss processes and then at methods for evaluating the various losses.

9.1 Types of losses

Loss of prestress begins to occur at the instant when the prestressing tendon is first tensioned and continues throughout the life of a prestressed concrete member. There are two main categories of losses. **Immediate losses** take place during the prestressing operation, while **deferred losses** occur after the prestressing operation is complete.

In post-tensioned members, the immediate losses are due to: (i) elastic compression of the concrete during the stressing of the tendons; (ii) friction along the member, between tendon and duct, during the stressing of the tendons; and (iii) slip in the grips during the final anchorage operation. In pretensioned construction there are no immediate losses due to duct friction or anchorage slip, and the main immediate loss of prestress is due to elastic compression of the concrete during the transfer of the tendon force to the concrete. The elastic deformations are greater than in post-tensioned construction because the concrete is younger when pretensioning is applied. The deferred losses due to creep and shrinkage also tend to be larger in pretensioned members, for the same reason.

Deferred losses in both pretensioned and post-tensioned members are mainly caused by time-dependent inelastic deformations in the concrete (creep and shrinkage) and the prestressing steel (stress relaxation).

Other, unusual types of losses can occur in special circumstances, such as deformation in construction joints in segmented members, temperature effects, long-term joint deformations, and deformations due to repeated loads. Such effects are uncommon and are not treated in this book.

9.2 Elastic loss

9.2.1 Pretensioned members

In an interior cross-section of a pretensioned member there is full bond between the tendon and the adjacent, hardened concrete when the prestressing force is transferred from the abutments to the concrete. During transfer, the concrete adjacent to the tendon undergoes a compressive strain of ε_{cp}, which is equal to the tensile strain decrement in the tendon, $\Delta\varepsilon_p$. The compressive stress in the concrete adjacent to the tendon, σ_{ci}, can be evaluated using Equation 4.1 from Chapter 4. The decrement in tendon tensile stress, $\Delta\sigma_p$, is then:

$$\Delta\sigma_p = \sigma_{ci} \times E_p / E_c \tag{9.1}$$

The concrete stress σ_{ci} can be obtained using the initial prestressing force P_i in Equation 4.1 and setting y to the cable eccentricity e. In order to achieve a predetermined prestressing force P_i just after transfer, the required force P_o in the cable prior to transfer is obtained by combining Equations 9.1 and 4.1:

$$P_o = P_i \times \left[1 + \left(\frac{1}{A_g} + \frac{e^2}{I_g}\right) \cdot \frac{E_p}{E_c} \cdot A_p\right] \tag{9.2}$$

If multiple tendons are present at different levels in the section, then the total area of the tendons can be used in Equation 9.2 with e taken as the mean eccentricity of the tendon area.

9.2.2 Post-tensioned members

In post-tensioned members, the cables are prestressed directly against the concrete, which compresses elastically under the prestressing force. If there is only one cable, there is no elastic loss because jacking is continued until the target force P_o is reached, and $P_i = P_o$. The cable extension is usually measured during the post-tensioning operation, either to evaluate the cable force or as a check on the cable force. In determining the required cable extension relative to the end of the concrete, the compressive strain in the concrete can be allowed for.

However, conditions are more complex when there is more than one prestressing cable. These are post-tensioned sequentially, and the tensioning of any one cable results in elastic losses in all the previously tensioned and anchored cables. The total loss in a cable thus depends on its place in the tensioning sequence: the first cable to be tensioned undergoes maximum loss, while the last cable has zero elastic loss. As the k-th cable at eccentricity e_k is tensioned to a force P_k, the elastic loss of force in a previously tensioned cable j, with area A_{pj} and at eccentricity e_j, is

$$\Delta P_{kj} = P_k \cdot \left(\frac{1}{A_g} + \frac{e_k e_j}{I_g}\right) \cdot \frac{E_p}{E_c} \cdot A_{pj} \tag{9.3}$$

Strictly speaking, Equation 9.3 is not applicable if the eccentricities vary along the beam; nevertheless, approximate values can be obtained by using average figures for e_k and e_j.

In practice, the cables tend to be bunched together in critical sections for bending, and an approximate estimate of the elastic loss can be obtained by using half of the value of $\Delta\sigma_p$ given by Equation 9.1. This is the average value that would be obtained from Equation 9.3 for a large number of tendons.

To allow for elastic losses, the cables can be over-stressed during prestressing, provided of course that the required individual jacking forces do not become excessive. AS 3600 requires that the jacking stress not exceed 85 per cent of the characteristic minimum breaking stress, f_{pb}, for stress-relieved post-tensioned tendons. This figure reduces to 80 per cent for pretensioned tendons, and 75 per cent for post-tensioned tendons not stress-relieved. If over-stressing is not used, the elastic loss has to be allowed for in the design calculations.

9.3 Duct friction loss

In post-tensioned members, friction between the cable and the surrounding duct occurs during the stressing operation, so that there is a progressive decrease in the prestressing force along the member, i.e. with increasing distance from the jacking point. The friction loss depends on the cumulative angle changes that occur in the cable direction as well as on local irregularities in the duct profile, which are referred to as duct 'wobble'. The duct friction loss is calculated using equations of exponential type. For example, AS 3600 expresses the cable stress σ_{pa} at distance L_{pa} from the jacking point as:

$$\sigma_{pa} = \sigma_{pj}\, e^{-\mu(\alpha_{tot} + \beta_p L_{pa})} \tag{9.4}$$

where σ_{pj} is the cable stress at the jacking point. The term β_p is the local angular deviation (wobble) in radians per metre, α_{tot} is the total change in angle in the cable, in radians, over the distance L_{pa} and μ is the coefficient of friction between cable and duct. The combined term $w = \mu\beta_p$ is sometimes referred to as the wobble coefficient.

It is important to note that α_{tot} is obtained by adding the absolute values of all angle changes in the cable over the distance L_{pa} being considered. The term α_{tot} therefore increases with increasing L_{pa}. For example, in Figure 9.1, the angle change over length L_{pa} is:

$$\alpha_{tot} = \theta_1 + 2\theta_2 + \theta_3 \tag{9.5}$$

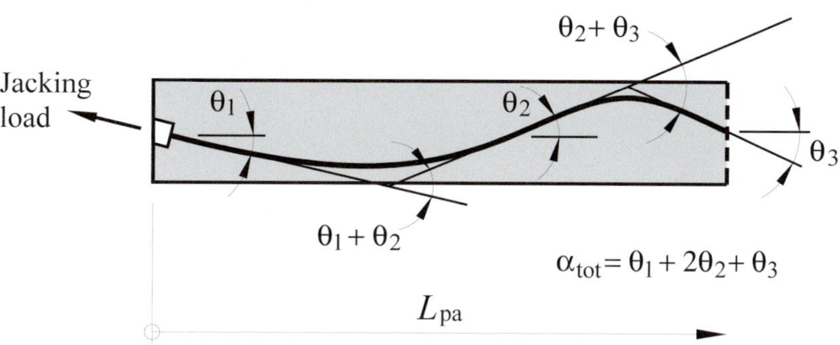

Figure 9.1 Summation of angle changes

If the prestressing cable follows a three-dimensional curve, then α_{tot} must include both the vertical and horizontal angular changes.

Provided the friction loss does not become very large, say in excess of about 20 per cent, the exponential term in Equation 9.4 can be approximated with good accuracy by the first term of its series expansion. Equation 9.4 then becomes:

$$\sigma_{pa} = \sigma_{pj}\,[1 - \mu(\alpha_{tot} + \beta_p L_{pa})] \tag{9.6}$$

In Equations 9.4 and 9.6, μ and β_p are coefficients that have to be evaluated empirically, and are available in the manufacturers' literature. Approximate values of μ and β_p are given in Tables 9.1 and 9.2, respectively.

TABLE 9.1 Friction curvature coefficients (AS 3600)

Details of duct	μ
Bright and zinc-coated metal sheathing	0.15 - 0.20
Bright and zinc-coated flat metal ducts	0.20
Greased-and-wrapped coating	0.15

TABLE 9.2 Values of wobble coefficient β_p (AS 3600)

Details of duct	β_p (raids/m)
Sheaths up to 50 mm internal diameter containing tendons of wires or strands	0.024 - 0.016
Sheaths of internal diameter between 50 mm and 90 mm containing tendons of wires or strands	0.016 - 0.012
Sheaths of internal diameter between 90 mm and 180 mm containing tendons of wires or strands	0.012 - 0.008
Sheaths up to 50 mm in diameter containing bars	0.016 - 0.008
Bars in greased-and-wrapped coating (any diameter)	0.008
Flat metal ducts containing tendons of wires or strands	0.024 - 0.016

9.4 Anchorage slip

During post-tensioning, the outer end of the jack exerts a tensile force on the cable and, at its inner end, an equal compressive force onto an end block that sits in or on the concrete. A special cable-gripping device comes into play at the end block when the prestressing force in the jack is released, with some slip inevitable in the gripping mechanism. The amount of slip depends on the device, and varies from almost zero in systems that use threaded bars and nuts, up to about 7 mm for wedge-type strand anchorages. The full loss of prestress due to **anchorage slip**, or **wedge slip**, occurs at the jacking end. However, owing to the friction in the duct the slip loss diminishes with distance away from the jacking end and becomes zero at a point which is determined by the friction properties of tendon and duct. This is shown in Figure 9.2(a). The loss due to anchorage slip is therefore often small at critical design sections near mid-span of a flexural member.

Nevertheless, the anchorage loss can be significant, especially in short cables, and needs to be allowed for in the design. Information on slip losses should be obtained from the manufacturers' literature of the anchorage system to be employed.

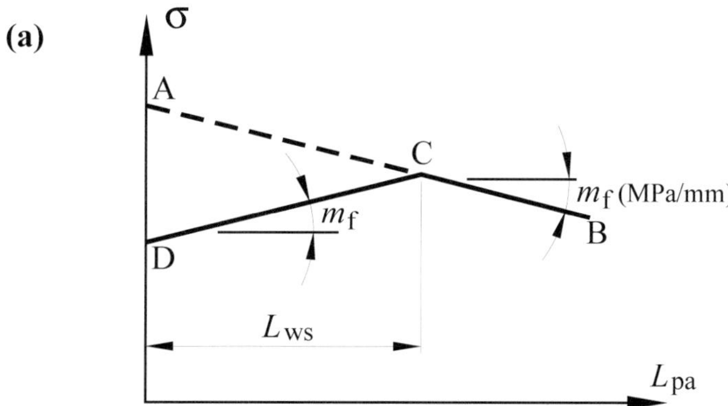

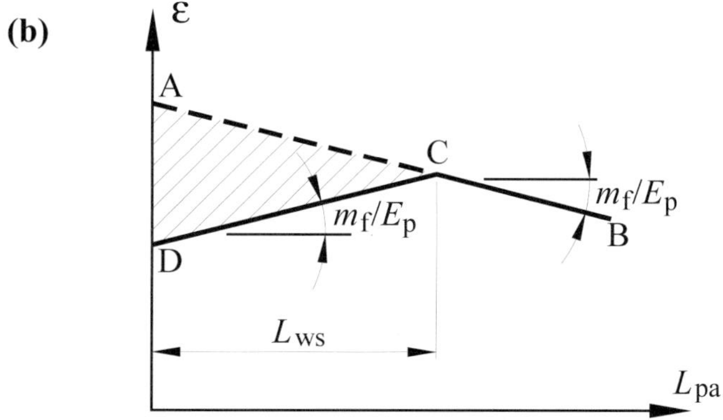

Figure 9.2 Anchorage slip loss

Calculation of anchorage slip (wedge slip) loss

In Figure 9.2(a), line AB represents the prestress σ_{pi} in the tendon. It is greatest at the jacking point A and reduces with distance from that point due to friction losses that can be calculated with Equation 9.4. The variation of the stress with distance is very close to being linear and has a slope m_f (MPa/mm).

When the prestressing force is transferred from the jack to the anchorage, the loss due to wedge draw-in is represented by line DC. It is usually assumed that AC and DC are both affected by friction and that these two lines have the same slope m_f, though of opposite sign. The length over which the cable force is affected by wedge slip is labelled L_{ws}.

Figure 9.2(a) is re-drawn in Figure 9.2(b), but in terms of steel strain rather than stress, so that the slope of lines AB and DC is now m_f/E_p. The area ACD, shown hatched, must be equal to the wedge slip at the anchorage, δ_{ws}. We can therefore express L_{ws} in terms of δ_{ws} as follows:

Strain loss AD: $$\Delta\varepsilon = 2\,m_f L_{ws}/E_p \tag{9.7}$$

Wedge-slip: $$\delta_{ws} = \Delta\varepsilon\, L_{ws}/2 = m_f L_{ws}^2/E_p \tag{9.8}$$

Wedge-slip length: $$L_{ws} = \sqrt{\frac{E_p \delta_{ws}}{m_f}} \tag{9.9}$$

The prestress loss at the anchorage is $2m_f L_{ws}$. For seven wire strand cables, the wedge slip (draw-in) is normally 6 to 7 mm. A calculation of losses due to wedge pull-in is included in Example 9.1.

9.5 Stress relaxation

When high stress is maintained in a tendon over a long period of time, relaxation occurs that can result in a loss of prestress of around 2 per cent or more. Information on the relaxation properties of different prestressing steels is provided in AS 3600–2018, together with a procedure for evaluating a *basic relaxation* and a *design relaxation*. Details have already been given in Section 2.2.4 of Chapter 2 for evaluating the basic relaxation and the design relaxation.

In calculating the prestress loss due to relaxation in the tendon, an adjustment to the basic relaxation is made to allow for the effect of other deferred losses (creep and shrinkage) on the sustained stress in the tendon. This interactive effect is allowed for in AS 3600 by adjusting the design relaxation R as follows:

$$R_{\text{adj}} = R\left(1 - \frac{\text{loss of stress due to creep and shrinkage}}{\sigma_{\text{pi}}}\right) \tag{9.10}$$

where σ_{pi} is the stress in the tendon immediately after transfer.

If elevated temperatures are used to accelerate the curing of a prestressed member, the Australian Standard requires that a part or all of the ultimate relaxation shall be deemed to have occurred by the end of the curing period. In this case, the loss of prestress due to relaxation is included as a component in the immediate loss of prestress calculations. This requirement will apply, for example, when low pressure steam curing is used for pretensioned members.

9.6 AS 3600 calculation of deferred losses

AS 3600 provides a simplified treatment of deferred losses due to creep and shrinkage, which has limited applicability. Specifically, the clauses for deferred losses focus on losses in the tensile tendon force. For design, it is necessary also to evaluate the losses in compression in the concrete. These are only equal to the tensile losses in the tendon when the beam section does not contain any normal reinforcement. When longitudinal reinforcement is present, the compressive losses in the concrete are always greater, and can be much greater than the tensile losses in the tendon.

According to AS 3600, the loss of tensile prestress in the tendon, due to concrete shrinkage in a section without longitudinal bar reinforcement, is obtained from the final design shrinkage strain, ε_{cs} (or $\varepsilon^*_{\text{sh}}$), as:

$$\Delta\sigma_{\text{p}} = E_{\text{p}}\varepsilon_{\text{cs}} \tag{9.11}$$

The loss of prestressing force in the tendon is then $A_{\text{p}}\Delta\sigma_{\text{p}}$. This calculation is similar to the order-of-magnitude approach described in Section B.2 of Appendix B.

AS 3600 specifically requires that the calculation for loss of stress in the tendon must be modified to take account of the restraining effect of reinforcement in the section, but it only deals specifically with the case in which the reinforcement is uniformly distributed over the cross-section. AS 3600 then allows $\Delta\sigma_p$ to be reduced by a factor related to the area of steel, A_s, and concrete, A_g. The reduced value is:

$$\Delta\sigma_{p.sh} = \frac{E_p \varepsilon_{cs}}{1 + 15(A_s/A_g)} \tag{9.12}$$

To deal with more realistic cases, the designer must use an appropriate method to analyse the long-term effects of creep and shrinkage.

In treating creep losses, the Standard requires an evaluation of creep strains, but in lieu of detailed calculations, and provided the sustained stress is less than $0.5f_c'$, the loss in tensile stress in the tendon due to concrete creep can be taken as:

$$\Delta\sigma_{p.c} = 0.8E_p\varepsilon_{cc} = 0.8E_p\varphi^*_{cc}\frac{\sigma_{ci}}{E_c} \tag{9.13}$$

Here, ε_{cc} is the free creep strain in the concrete at the level of the tendon, and 0.8 is a reduction factor to allow approximately for any partial restraint to free creep. The term σ_{ci} is the sustained concrete stress at tendon level, and E_c is the elastic modulus of the concrete at the time the sustained load is applied. Again, Equation 9.13 only gives the loss of prestress in the tendon. As in the case of shrinkage, creep will bring about a large transfer of compressive force from the concrete to any surrounding steel, so that the loss of prestress in the concrete will be considerably greater than in the tendon.

In Appendix B of this text, several analytic methods of varying complexity and accuracy are presented which can be used as a supplement to the AS 3600 provisions. They allow creep and shrinkage losses to be calculated in prestressed reinforced flexural members. They are briefly discussed in Section 9.7 of this chapter.

EXAMPLE 9.1

CALCULATION OF LOSSES IN A PRESTRESSED CROSS-SECTION WITHOUT LONGITUDINAL BAR REINFORCEMENT

Losses are to be determined according to AS 3600 for a simply supported post-tensioned beam that spans 20 metres and has the cross-section shown in Figure 9.3. The beam carries no dead load other than self weight. The 28-day characteristic concrete strength is $f_c' = 50$ MPa, while stressing takes place when the concrete is at age 14 days, with a strength of $f_{cp}' = 35$ MPa.

The prestressing tendon consists of two cables, each containing 14 super grade 7-wire low relaxation strands of 12.7 mm diameter. The strands are to be placed in a bright metal sheath of 81 mm diameter and stressed with an initial prestress of $\sigma_{pi} = 1280$ MPa. For each cable, the area is 1380 mm^2 and stressing is to be undertaken at one end, with the cables stressed in turn. The cables have parabolic profiles with eccentricities of zero at each end and 465 mm at mid-span. The beam is exposed in a near-coastal area and T = 20 ^{o}C. The losses will be calculated assuming that only the self-weight moment is acting.

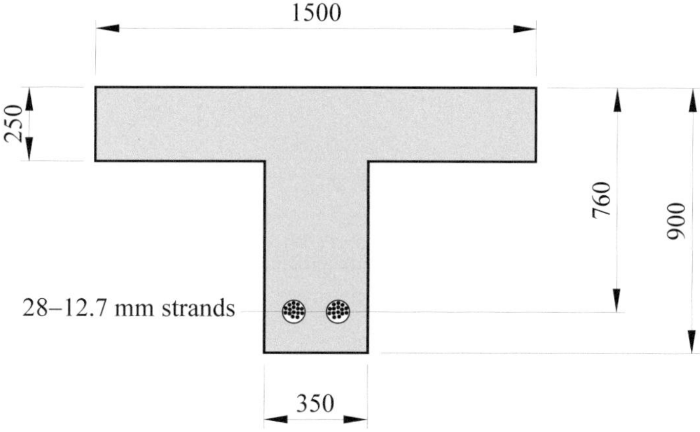

Figure 9.3 Beam section at mid-span – Example 9.1

Other data

At 14 days $f'_{cp} = 35$ MPa and $f_{cmi} = 38$ MPa

$$E_{ci} = 5060\sqrt{38} = 31.2\times10^{3}\ \text{MPa}; \quad \text{(Equation 2.9)}$$

$$E_{p} = 195\times10^{3}\ \text{MPa}; \ A_{g} = 602.5\times10^{3}\ \text{mm}^{2}; \ I_{g} = 38.6\times10^{9}\ \text{mm}^{4}$$

$$y_{t} = 295\ \text{mm}; \ y_{b} = 605\ \text{mm}; \ A_{p1} = A_{p2} = 1380\ \text{mm}^{2}$$

$$\sigma_{pi} = 1280\ \text{MPa}; \ A_{st} = 0$$

SOLUTION

Elastic loss

The concrete compressive stress increment, $\Delta\sigma_{i}$, at the tendon level at mid-span due to the stressing of a tendon to P_{i}, is obtained using Equation 4.1:

$$\Delta\sigma_{i} = P_{i}\left(\frac{1}{A_{g}} + \frac{e^{2}}{I_{g}}\right) = A_{p1}\sigma_{pi}\left(\frac{1}{A_{g}} + \frac{(d_{p}-y_{t})^{2}}{I_{g}}\right)$$

$$= 1380 \times 1280\left(\frac{1}{602.5\times10^{3}} + \frac{(760-295)^{2}}{38.6\times10^{9}}\right)$$

$$= 12.8\ \text{MPa}$$

As the first cable is fully stressed before the second tendon is stressed, its loss of prestress when the second cable is stressed is:

$$\Delta\sigma_{p} = \sigma_{cp} \times \frac{E_{p}}{E_{c}} = 12.8 \times \frac{195\times10^{3}}{31.2\times10^{3}} = 80\ \text{MPa}$$

The first cable will therefore be initially prestressed to 1280 + 80 = 1360 MPa to allow for the concrete elastic compression loss when the second cable is stressed. On stressing of the second cable, the total area of prestressing steel is $A_p = A_{p1} + A_{p2} = 2760$ mm^2 and the average elastic loss is 80/2 = 40 MPa. However, as cable 1 is over-stressed from its target stress, the nett elastic loss is zero.

Friction loss

Jacking is carried out at one live end. From Tables 9.1 and 9.2 we obtain:

$\beta_p = 0.014$ (sheaths up to 90 mm), and

$\mu = 0.2$ (bright metal sheathing)

At mid-span ($L_{pa} = 10.0$ metres):

$$\alpha_{tot} = \frac{4h}{L} = \frac{4 \times (760 - 295)}{20\ 000} = 0.093 \text{ rad}$$

$$\sigma_{pa} = \sigma_{pj} e^{-0.2(0.093 + 0.014 \times 10)} = 0.95\sigma_{pj} \tag{9.14}$$

This represents a loss of $0.05\sigma_{pi} = 0.05 \times 1280 = 64$ MPa.

At the dead end ($L_{pa} = 20$ metres):

$$\alpha_{tot} = 2 \times 0.093 = 0.186 \text{ rad}$$

$$\sigma_{pa} = \sigma_{pj} e^{-0.2(0.186 + 0.014 \times 20)} = 0.91\sigma_{pj} \tag{9.15}$$

Anchorage (wedge slip) loss

The amount of draw-in is obtained from manufacturers' information. For example, for VSL multi-strand systems a draw-in of 6 mm is given. We will use 6 mm for this example.

The loss in the first half-span, calculated above, is 64 MPa. Thus, the rate of loss is:

$$m_f = 64/10 = 6.4 \text{ MPa/m} \quad (0.0064 \text{ MPa/mm})$$

The length of cable affected by draw-in losses is then (Equation 9.9):

$$L_{ws} = \sqrt{\frac{E_p \delta_{ws}}{m_f}} = \sqrt{\frac{195\times10^3 \times 6}{0.0064}} = 13.5\times10^3 \text{ mm (13.5 m)}$$

The loss at the anchorage due to wedge slip is:

$$2m_f L_{ws} = 2 \times 6.4 \times 13.5 = 173 \text{ MPa}$$

This reduces along the tendon length because of friction and at the mid-span (L_{pa} = 10 metres) section the loss is approximately:

$$173 \times (13.5 - 10.0)/13.5 = 45 \text{ MPa}$$

which is about four per cent of the initial prestress.

Shrinkage Loss

From Equation 2.22, the hypothetical thickness is:

$$t_h = \frac{2A_g}{u_e} = \frac{2 \times 602.5\times10^3}{2(1500 + 900)} = 251 \text{ mm}$$

For a near coastal environment, the final design shrinkage strain is taken as:

$$\varepsilon^*_{sh} = 450\times10^{-6} = \varepsilon_{cs}$$

Adjusting for the presence of reinforcement, the prestress loss due to shrinkage becomes (Equation 9.12):

$$\Delta\sigma_p = \frac{E_p \varepsilon_{cs}}{1 + 15(A_s/A_g)} = \frac{195\times10^3 \times 450\times10^{-6}}{1 + 0} = 87.8 \text{ MPa},$$

The total shrinkage loss in the tendon is:

$$A_p \Delta\sigma_p = 2760 \times 87.8 = 242 \text{ kN}$$

Creep loss

The self-weight is $0.6025 \times 25 = 15.1$ kN/m, which produces a mid-span bending moment of 755 kNm, and a tensile concrete stress at the level of the prestressing steel of:

$$\sigma_{sw} = \frac{My}{I_g} = \frac{755\times10^6 \times (760 - 295)}{38.6 \times 10^9} = 9.1 \text{ MPa}$$

With $P_i = 2 \times 1380 \times 1280 = 3533\times10^3$ N, the compressive stress in the concrete due to the prestress at the level of the prestressing steel is:

$$\sigma_{pi} = \frac{P_i}{A_g} + \frac{P_i e_p y_p}{I_g}$$

$$= \frac{3533\times10^3}{602.5\times10^3} + \frac{3533\times10^3(760 - 295)^2}{38.6 \times 10^9} = 25.7 \text{ MPa}$$

The resultant concrete compressive stress at the cable level is:

$$\sigma_{ci} = 25.7 - 9.1 = 16.6 \text{ MPa}$$

For a hypothetical thickness of $t_h = 251$ mm and interpolating, the final creep coefficient is taken to be $\varphi^*_{cc} = 2.6$.

At the mid-span, the final creep strain at the cable level is estimated as (Equation 9.13):

$$\varepsilon_{cc} = 0.8\varphi^*_{cc}\left(\frac{\sigma_{ci}}{E_c}\right) = 0.8 \times 2.6 \times \frac{16.6}{31.2\times10^3} = 0.00111$$

The prestress loss in the tendon due to creep is therefore (Equation 9.11):

$$\Delta\sigma_p = E_p\varepsilon_{cc} = 195\times10^3 \times 0.00111 = 216 \text{ MPa}$$

Sum of losses due to creep and shrinkage

According to AS 3600, the deferred losses due to shrinkage and creep are 87.8 + 216 = 304 MPa, which is (304/1280) 100 = 24 per cent of the initial prestress.

Relaxation loss

For low relaxation strand, the basic relaxation is $R_b = 3.5$ per cent (refer Table 2.2). The coefficients k_7, k_8 and k_9 are then determined as follows:

Equation 2.2: $k_7 = 1.40$ (j = 10 000 days)

Table 2.3: $k_8 = 0.70$ ($\sigma_p/f_{pb} = 1280/1870 = 0.68$)

Equation 2.3: $k_9 = 1.0$ (T = 20 degrees)

Thus the design relaxation is $R = 1.40 \times 0.70 \times 1.0 \times 3.5 = 3.4$ per cent and the adjusted design relaxation loss is (Equation 9.10):

$$R_{adj} = 3.4 \times (1 - 304/1280) = 2.6 \text{ per cent}$$

Summary of per cent losses in tendon at mid-span section

Elastic loss: 40/1280 3%

Friction loss: 5%

Anchorage loss: 4%

Total immediate losses = 12 per cent

Creep and shrinkage loss *in tendon*: 24%

Relaxation loss *in tendon*: 3%

Total deferred losses = 27 per cent

The immediate losses can be compensated for by over-jacking. To account for the elastic loss in cable 1 on stressing of cable 2, the jacking stress is increased to 1360 MPa. To allow for friction and anchorage losses, the stress in cable 1 is increased by a further 9 per cent to 1482 MPa, 79 per cent of the breaking stress, and cable 2 is stressed to $1.09 \times 1280 = 1395$ MPa, 75 per cent of the breaking stress.

The required jacking forces are:

$$P_{j1} = A_{p1} \times \sigma_{pj1} = 1380 \times 1482 \times 10^{-3} = 2045 \text{ kN}$$

$$P_{j2} = A_{p2} \times \sigma_{pj2} = 1380 \times 1395 \times 10^{-3} = 1925 \text{ kN}$$

Discussion

This example shows how losses are calculated according to AS 3600 for a simple prestressed, unreinforced section. In this case the loss of compressive prestress in the concrete due to shrinkage and creep is equal to the tensile loss in the tendon. The situation is more complicated when the section contains reinforcement. This case is considered in Example 9.2.

9.7 Analytic methods for evaluating deferred losses

As we have seen, the AS 3600 treatment of deferred losses deals only with unreinforced prestressed sections and prestressed sections in which the reinforcing steel is uniformly distributed throughout the section. In more complex design situations, a comprehensive analytic treatment of creep and shrinkage losses is needed. For example, in the construction of a multi-span bridge by the launch method, the sustained concrete stresses change significantly over time in response to changes in the moments due to the self-weight and construction loads. Also, prestressing often has to be applied in stages, for example to coordinate with the application of dead loads.

To evaluate the loss of prestress that accompanies complex time histories, **the step-by-step method** of analysis can be used. Details of this method are given in Section B.4 of Appendix B. While its methodology is relatively simple in concept, this method provides good accuracy, but requires intensive computation.

In less complex situations, **the one-step, age-adjusted effective modulus method** is an appropriate, simpler, alternative. It is explained in Section B.3 of Appendix B. **Closed-form equations**, derived in Section B.5 of

Appendix B, provide an even simpler analytic alternative for evaluating losses due to creep and shrinkage. They require little computation and are ideal for use in preliminary design. The accuracy is usually sufficient for design calculations when the structure is not deflection sensitive.

Yet simpler order-of-magnitude equations are also presented in Section B.2 of Appendix B, which estimate the long-term loss of prestress in the concrete by first evaluating the compressive forces that are transferred over time to the steel reinforcement and tendon. The compressive force increments in the reinforcing steel, $\Delta X^*_{s.c}$, and in the tendon, $\Delta X^*_{p.c}$ (which is actually a decrease in tension), due to creep, are estimated as:

$$\Delta X^*_{s.c} = A_s E_s \varepsilon^*_{cc.s}\text{, and} \tag{9.16}$$

$$\Delta X^*_{p.c} = A_p E_p \varepsilon^*_{cc.p} \tag{9.17}$$

The terms $\varepsilon^*_{cc.s}$ and $\varepsilon^*_{cc.p}$ are the long-term free creep strains at the levels of the steel and tendon. For equilibrium, the tensile force increment (i.e. the loss of compressive prestress) in the concrete is equal to the sum of these compressive force increments. A similar approach, using the shrinkage strains $\varepsilon^*_{csh.s}$ and $\varepsilon^*_{csh.p}$, allows the prestress loss in the concrete due to shrinkage to be estimated.

It is appropriate here to restate some warnings in regard to deferred losses. In prestressed sections that contain steel reinforcement, the stresses in the cross-section change significantly over time with a transfer of compressive force from the concrete to the conventional steel reinforcement. This can result in a very large loss in the original prestress in the concrete, even though the loss of prestress in the tendon may remain small. It is necessary to evaluate such long-term losses in the concrete and to allow properly for them in design. Shrinkage and creep can also lead to a large loss of compressive stress in the extreme 'tensile' concrete fibres of flexural members, so that the actual cracking moment can be much smaller than the estimated value. Excessive cracking is too often the unwanted and unexpected result.

EXAMPLE 9.2

CALCULATION OF CREEP AND SHRINKAGE LOSSES FOR A BEAM CONTAINING LONGITUDINAL REINFORCING BARS

In the previous example we considered a section without longitudinal bar reinforcement. In this example a similar section is analysed but with a layer of 4N32 bars, as well as the prestressing tendon. The section is shown in Figure 9.4.

In this example, the deferred-loss calculations are undertaken using the simplified, closed-form equations derived in Section B.5 of Appendix B.

Other data:

As for Example 9.1 but with A_{st} = 3200 mm^2 and d_{st} = 840 mm.

.

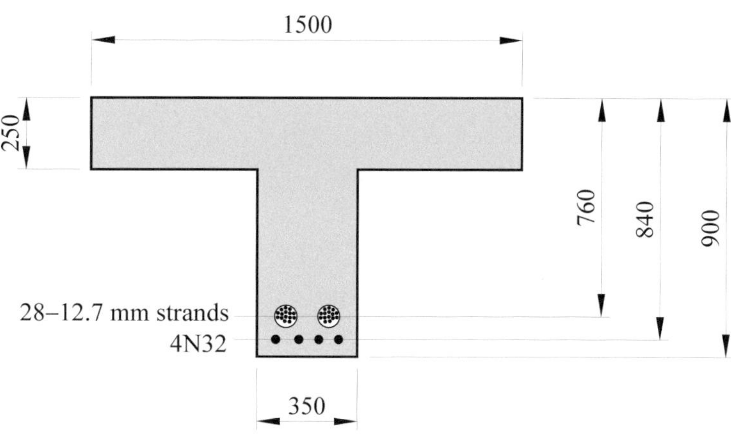

Figure 9.4 Beam section at mid-span – Example 9.2

SOLUTION

Immediate losses

The presence of normal reinforcement in the cross-section has only a slight effect on the elastic stiffness of the cross-section of the beam, and this is usually ignored in evaluating elastic behaviour. The immediate losses calculated for the section with reinforcement can therefore be taken to be as for the unreinforced section dealt with in Example 9.1. If necessary, the presence of the reinforcement can be allowed for using the information in Appendix A.

Calculation of shrinkage

We use Equation B.76 from Appendix B to evaluate the loss of prestress in the concrete due to shrinkage:

$$\Delta X^*_{\mathrm{c.sh}} = \gamma_3 A_{\mathrm{g}} k^* E^*_{\mathrm{c}} \varepsilon_{\mathrm{cs}}$$

where $k^* E^*_{\mathrm{c}}$ is the age-adjusted effective modulus for concrete, which will be taken to be 9.9×10^3 MPa.

The equivalent steel plus tendon area in the section is:

$$A_{\mathrm{eq}} = A_{\mathrm{p}} + A_{\mathrm{st}} = (2 \times 1380 + 3200) = 5960\ \mathrm{mm}^2$$

The equivalent depth is:

$$d_{\mathrm{eq}} = \frac{A_{\mathrm{p}} d_{\mathrm{p}} + A_{\mathrm{st}} d_{\mathrm{st}}}{A_{\mathrm{p}} + A_{\mathrm{st}}} = 803\ \mathrm{mm}\text{, and}$$

$$e_{\mathrm{eq}} = 803 - 295 = 508\ \mathrm{mm}$$

From Equations B.76 and B.77 we obtain:

$$\gamma_3 = \left[1 + \frac{602.5\times10^3 \times 9.9\times10^3}{5960 \times 200\times10^3} + \frac{508^2 \times 602.5\times10^3}{38.6\times10^9}\right]^{-1} = 0.0997$$

and hence:

$$\Delta X^*_{\mathrm{c.sh}} = 0.0997 \times 602.5\times10^3 \times 9.9\times10^3 \times 450\times10^{-6}$$
$$= 268\times10^3\ \mathrm{N} \quad (268\ \mathrm{kN})$$

Using the proportions of tendon and steel areas, we allocate the tensile force increments in tendon and steel as follows:

$$\Delta X_{\text{p}}^{*} = 268 \times 2760/5960 = 124\ \text{kN}$$

$$\Delta X_{\text{st}}^{*} = 268 - 124 = 144\ \text{kN}$$

Calculation of creep

For determining the loss of prestress in the concrete due to creep, we apply Equation B.58 of Appendix B, which is:

$$\Delta X_{\text{c.c}}^{*}(A_{\text{eq}}) = \gamma_3 A_{\text{g}} k^{*} E_{\text{c}}^{*} \varepsilon_{\text{cc.eq}}^{*}$$

The initial stress in the concrete at level d_{eq} is slightly larger than at the tendon level, and is calculated to be 17.8 MPa. The free creep strain at depth d_{eq} is the initial elastic strain there, multiplied by the creep coefficient (2.6):

$$\varepsilon_{\text{cc.eq}}^{*} = \frac{17.8}{31.2\times10^{3}} \times 2.6 = 1480\times10^{-6}$$

Values of γ_3, and $k^{*}E_{\text{c}}^{*}$ are as above, and the loss due to creep is thus:

$$\begin{aligned}\Delta X_{\text{c.c}}^{*}(A_{\text{eq}}) &= 0.0997 \times 602.5\times10^{3} \times 9.9\times10^{3} \times 1480\times10^{-6} \\ &= 880\times10^{3}\ \text{N} \quad (880\ \text{kN})\end{aligned}$$

The loss of prestress in the tendon and the increase in compressive force in the reinforcement are thus, respectively:

$$\Delta X_{\text{p}}^{*}(A_{\text{eq}}) = 880 \times 2760/5960 = 408\ \text{kN}\text{, and}$$

$$\Delta X_{\text{st}}^{*}(A_{\text{eq}}) = 880 - 408 = 472\ \text{kN}$$

Discussion

The total loss of compression in the concrete due to creep and shrinkage is 268 + 880 = 1148 kN. This is large, about 33 per cent of the initial prestressing force of 3532 kN. However, the loss of tensile prestress in the tendon is 124 + 408 = 532 kN, only a 15 per cent loss. The increase in compressive force in the reinforcing steel makes up the difference between these two losses.

9.8 References

AS 3600–2018, *Concrete Structures*, Standards Australia.

AS/NZS 4672.1 2007, *Steel prestressing materials - General Requirements*, Standards Australia, Sydney, Australia.

CHAPTER 10

Design procedures for statically determinate beams

A design procedure for statically determinate prestressed concrete beams is described in this chapter. Information is drawn from previous chapters on the behaviour and strength of prestressed concrete beams to provide the basis for the procedure, which is extended in Chapter 11 to deal with continuous beams.

10.1 Structural design

In the detailed design of a prestressed beam, values have to be chosen for a large number of design variables. These include the grade of concrete and the types of reinforcing steel and prestressing steel to be used, the dimensions of critical cross-sections along the beam, the amount of tensile and compressive reinforcement to be included in each critical section, the details of the transverse reinforcement, and the details of the prestressing cable including its area, its profile along the beam and the magnitude of the prestressing force. The values of these variables must be chosen so that all the design requirements are satisfied and the design is economically competitive.

If a quantitative cost function can be created to compare and evaluate alternative designs, then it is possible to treat the structural design process as an exercise in mathematical programming and optimization. The design requirements, including strength and serviceability, are used to formulate constraints that place limits on the possible values of the design variables. A feasible design space is thus created and a search within this space can then be undertaken to find the design point that minimises the cost function. The values of the design variables that define this point represent the optimal design.

Computer-based mathematical programming techniques have in fact been developed for the optimal design of various types of structures, including prestressed concrete beams. However, automated optimal structural design does not find wide practical use. This is partly because it is very difficult to formulate realistic and useful cost functions. To be useful, the cost function must account not only for the quantities of materials used and their unit costs, but also for important practical factors that are very difficult to quantify. Factors such as the knowledge and expertise of the contractor, the skills of the work force, the cost of plant hire, the cost of delays and maintenance, and depreciation costs for the completed structure can have an over-riding effect on costs. Such factors are best handled by intelligent engineering judgment based on extensive experience, rather than by mathematical modelling.

In practice, the process of structural design requires many iterative, trial-and-error calculations in order to identify an acceptable, economic design that satisfies all the design requirements. It is therefore important to use an efficient sequence of design steps in order to avoid unnecessary iterations. A detailed design procedure for statically determinate prestressed concrete members is presented in this Chapter. The sequencing of the steps has been chosen to make the method reasonably efficient. Before presenting this design method in detail, it will be useful to discuss some of the decisions that have to be made in the design of prestressed concrete members.

10.2 Choosing the type of construction

Before the detailed design of a structural member is undertaken, a number of broad design decisions have to be made, such as whether the construction is to be in concrete, in steel, or in some other structural material, and if in concrete, whether or not prestressing is to be used. Such decisions are essentially economic, and are best based on cost estimates for preliminary designs in the appropriate materials.

The decision on whether or not to use prestressing is of particular importance here. Broadly speaking, reinforced concrete will be more economical without prestressing when the spans are short and the imposed loads are small. As the spans becomes larger, self-weight has an increasingly important influence, and serviceability considerations start to dominate the design. At this stage it is economic to introduce prestressing to balance a portion of the design load and hence either control deflections and crack widths or reduce the overall size of the section. In very long flexural members, self-weight is the major design load and prestressing can be used to balance a large proportion of the self-weight.

The break-even point when it just becomes economic to prestress a concrete structure will vary significantly from job to job, from place to place and from time to time. Fluctuations in the relative costs of formwork, materials, labour and plant hire preclude any general guidelines from being given. The designer will usually have to make cost comparisons based on alternative preliminary designs.

For some designs, special performance requirements may determine the appropriate type of concrete construction. In some circumstances prestressing may be the only viable solution due to head height restrictions. A high level of prestress will be used if a crack-free structure is required, as for example in some liquid containing vessels or when biological screening is required. Higher levels of prestress may also be warranted if fatigue is a potential problem. A decision to use reinforced concrete without prestressing may be taken if expert labour and equipment are not available locally.

10.3 Choosing the cross-section

Good economy in prestressed beams can depend on repetition. The availability of standard precast beam sections should always be investigated before going to one-off solutions. Some standard sections are provided in the appendices of the Australian Standard Bridge Design Part 5: Concrete (AS 5100.5–2017). The standardisation of precast pretensioned beam sections in recent years has led to these sections becoming the dominant system for small to medium span bridges.

Shape of section

The most appropriate shape of cross-section for a beam depends on the details of loading and construction and on the cost structure for the job.

A rectangular section is not structurally efficient for members resisting bending action. The self-weight is high in comparison with other shapes, and the magnitude of the prestressing force required to accommodate a given bending moment is therefore greater than for, say, a thin-web flanged member. On the other hand, formwork is much simpler and cheaper for a rectangular beam, as is the placing of the tendons and reinforcement. Generally, rectangular members will be used for short spans where the ratio of self-weight to total load is small, and for one-off situations and special cases where little re-use of forms will occur.

Figure 10.1 shows three categories of non-rectangular sections that are commonly used for prestressed concrete beams. In the basic T section in Figure 10.1(a), the concentration of compressive concrete in the upper compressive flange is advantageous, while the ratio of dead load to live load is reduced by the use of a thin web. In such sections a check is needed to ensure that excessive compressive forces do not occur in the bottom of the section at transfer as a result of the prestress.

The I shape in Figure 10.1(b) can handle large initial compressive prestressing forces in the bottom fibres, and is also useful for long span beams and for continuous members that are subjected to both positive and negative bending. The box-section, a variation of the I shape, is particularly useful for slender beams when lateral stability becomes a consideration, and also for beams subjected to lateral loads or torsional moments. However, formwork costs are high for I, T and hollow box sections. Construction difficulties, such as placement of reinforcement, are also accentuated.

The inverted T section in Figure 10.1(c) has little concrete in the compressive flange and, by itself, is inefficient for resisting positive moment. However, it is suited to carrying large initial compressive prestress in the lower fibres at transfer and can be combined with a cast-in-situ upper concrete decking to form an effective composite structural system.

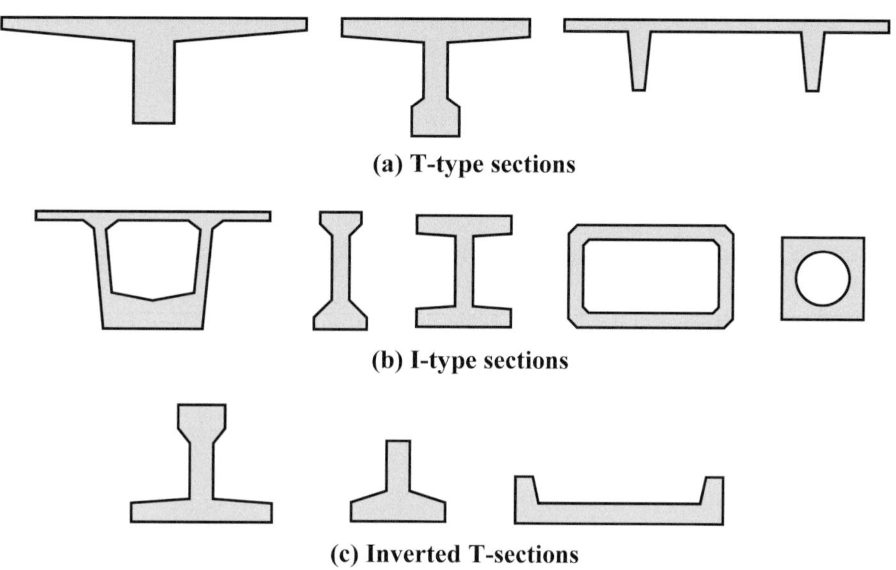

(a) T-type sections

(b) I-type sections

(c) Inverted T-sections

Figure 10.1 Non-rectangular sections

Depth of section

Collins and Mitchell (1991) suggest a range of values of span-to-depth ratios for standard deck elements, expressed in terms of the ratio *L*/*D*, where *D* is the overall depth of section. They also suggest typical values for simply supported prestressed concrete members. For a solid core, one-way floor slab they suggest a typical *L*/*D* value of 40; for a hollow-core one-way slab with light loading their values range from 40 to 50, and for heavier loading from 32 to 40. For a rectangular prestressed member they suggest a typical value of 20. For T beams placed side-by-side they recommend values of 20 to 30 for light design loads and 18 to 28 for heavy design loads.

For I beam bridge girders Collins and Mitchell recommend a typical *L*/*D* value of 18. The Lin and Zia (1974) suggestions for fully prestressed members are higher: 22 to 24, with 26 to 28 being near the upper limit. Leonhardt (1964) suggested an upper limit on the *L*/*D* ratio for bridge girders of 30, which was also the value used in the 1976 NAASRA Australian Bridge Specification. Leonhardt's recommendation for the economical design of simple beams was in the range 14 to 20. Libby (1971) suggested values varying between 16 and 22 for simple prestressed beams, but emphasised that values can be affected by load intensity, type of construction and allowable clearances.

Brock (2010) suggests that prestressed beams may initially be proportioned using L/D ratios of 20 for sections with a breadth of about $D/3$, to 30 for sections with breadth about $3D$.

In a discussion of the depths required for prestressed concrete beams compared with those for reinforced concrete beams, Leonhardt suggested that the depth of a prestressed girder can be reduced to about 65 per cent of that of a comparable reinforced concrete one. For I-section members, he suggested a lower figure of about 80 per cent of the depth of a comparable reinforced girder. At high dead-load to live-load ratios, his figure for T-section members is also 80 per cent but increases to 100 per cent when the dead-load to live-load ratio becomes smaller than about two.

Cross-section details

Span-to-depth ratios provide a guide for establishing a trial section when deflection considerations govern the design. If flexural strength governs, then there will be no significant difference in section depth between prestressed and reinforced concrete and a first estimate of section dimensions can be obtained from the required bending strength of critical cross-sections. The moment capacity of a ductile rectangular section without compressive reinforcement may be expressed as:

$$M_u = \alpha_2 f_c' \gamma k_u d^2 b \left[1 - \frac{\gamma k_u}{2}\right] \tag{10.1}$$

This expression applies also to a flanged section, provided that the stress block depth, $\gamma k_u d$, lies in the portion of the flange where the width, b, is constant. The depth d in Equation 10.1 is the depth to the resultant tensile force in the section, which is the total force in all tensile reinforcement and prestressing tendons that lie in the tensile zone at ultimate.

The design moment capacity ϕM_u must be at least as large as the design ultimate moment M^*, so that:

$$bd^2 \geq \frac{M^*}{\phi\, \alpha_2 f_c'\, \gamma k_u \left[1 - \frac{\gamma k_u}{2}\right]} \qquad (10.2)$$

A value should be chosen for the neutral axis depth parameter k_u that will ensure adequate ductility with yielding of the tensile reinforcement and tendon (see Section 5.7 in Chapter 5). A value for k_u significantly less than 0.36, say between 0.15 and 0.25, will usually be appropriate. The dimensions of the cross-section can then be obtained from Equation 10.2, having regard to the span-to-depth ratios discussed above. The value for M^* must of course include an estimate for the self-weight.

It will be noted that choosing a value for k_u fixes the total tensile yield force in the section at M^*. This force is carried jointly by the tensile reinforcing steel and prestressing tendons. The proportion of reinforcing steel to prestressing steel still remains to be chosen.

10.4 Choosing the prestressing details

With trial dimensions for the section decided, it is necessary to determine the details of the tendon. These can usually be determined from the required service load behaviour for the member, and may relate either to deflection control or crack control. The details of the prestress might alternatively be chosen on economic or practical considerations, although the serviceability requirements also need to be satisfied.

Deflection control

The designer may select a load at which the deflection will be, for practical purposes, zero. For example, a load between the self-weight and the full sustained dead load may be chosen for this purpose. The concept of load-balancing, introduced in Chapter 4, provides a simple way to determine the prestress details to satisfy this criterion.

We have previously seen that a uniformly distributed load w is balanced by a parabolic cable with a 'sag' of h and an effective prestressing force of P_e:

$$P_e = \frac{wL^2}{8h} \tag{10.3}$$

A value for the sag can be chosen from a consideration of the depth of the section and the cover that will be needed for the reinforcing steel and prestressing steel. With zero end eccentricities the sag h is equal to the mid-span eccentricity, $e = h$.

Non-zero end eccentricities can be used, but if the end eccentricity is negative (above the centroid), the benefits of the increased sag tend to be nullified by the opposing effect of the end eccentricity. Similarly if a positive (below the centroid) end eccentricity is used, the upward camber caused by the end couples tends to be nullified by the reduced sag. A large positive end eccentricity will induce tensile stresses in the upper fibres of the end region and a practical upper limit on the end eccentricity is that it should not cause top fibre cracking.

If a beam has cantilevered ends, negative eccentricities will be required over the supports and the in-span cable profile will then take on a shape similar to that in a continuous beam.

Crack control

The designer may wish to keep the structure crack-free under a selected service load that is a proportion of the full dead load. Assuming that the critical section of maximum moment has been pre-cracked during a previous application of the full design live plus dead load, we see that the cracks will start to open at decompression (i.e. when the bottom fibre stress is zero). To control cracking, the prestress can be chosen so that decompression occurs at the critical section when the selected service load acts. To apply the decompression criterion, the required effective prestressing force is:

$$P_e = \frac{M_G}{e + Z_b/A_g} \tag{10.4}$$

where e is the maximum available eccentricity in the section, A_g is the gross section area and Z_b is the section modulus for the bottom fibre. The moment M_G is that caused by the load at decompression.

If decompression is chosen to occur under the full sustained load, then the cracks will open as soon as any live load acts. If the service load selected for decompression is less than the sustained load, then cracks in regions of high dead-load moment will remain slightly open permanently, and will open further when live load acts.

Clearly, the choice of an appropriate decompression load will depend on the ratio of dead load to live load. As in the case of deflection control, several alternative check calculations can be made, using different prestress levels in order to achieve a good design outcome. These iterations will normally be done by computer. Referring to current design practice in Australia, Cross (2007) comments:

> "*The availability of computer software to carry out these (cracked section and deflection) calculations has meant that more often than not the amount of post-tensioning is selected to satisfy deflection criteria*".

Other considerations: cost and practicality

In some situations there may be reasons not directly related to service load behaviour that will lead the designer to substitute prestressing steel for part or all of the flexural reinforcement. For example, if it is difficult to satisfactorily fit the required quantity of conventional reinforcement into the cross-section, a small area of prestressing steel may be substituted for a larger area of reinforcing steel to provide the same total yield force. The ratio of yield strengths is such that the additional area of prestressing steel would be about a third of the area of the reinforcement replaced. The additional prestressing steel must of course be prestressed. In other cases the decision to use additional prestressing steel may be based purely on economic grounds, following costings of alternative designs.

If an upper limit on the prestressing force P_i is imposed by the need to control camber, then a convenient design procedure is to choose a value for the maximum long-term camber, i.e. Δ_h^*. The initial camber at transfer is then estimated, using the creep coefficient of the concrete:

$$\Delta_{hog} = \frac{\Delta_h^*}{1 + \varphi^*} \qquad (10.5)$$

The prestressing details can be chosen such that the initial camber, due to prestress and self-weight, is equal to Δ_{hog}. For example, in the case of a simply supported beam with a parabolic cable, we have:

$$\Delta_{hog} = -\frac{5}{384}\left((w_p - w_G)\frac{L^4}{EI}\right) \tag{10.6}$$

which allows w_p to be evaluated and, hence, the initial prestressing force:

$$P_i = \frac{w_p L^2}{8h} \tag{10.7}$$

10.5 Design steps

The design details for a prestressed concrete beam have to be determined by trial and error, and there is not any one 'correct' sequence of design steps. Nevertheless, a good design procedure should eliminate unnecessary iterations. The sequence of steps listed below satisfies this requirement and is intended as a starting point for persons new to prestressed concrete design. It may, of course, be adapted and improved on as personal experience is gained. It can also be modified as appropriate for the design of special types of members.

The detailed steps are as follows:

1. **Maximum design moments and shears:** Make a first estimate of the section depth using suggested depth-span ratios and hence an initial estimate of beam self-weight. Determine maximum bending moments and shear forces due to the self-weight, dead and live loads.
2. **Cross-section:** Calculate the design ultimate moments M^* at critical sections for bending; choose a value for k_u and hence obtain the cross-section details using Equation 10.2; check the self-weight against the estimate in Step (1). If necessary adjust, and repeat steps (1) and (2).
3. **Prestress level:** Choose a serviceability design criterion. For example:
   - the load that is to be balanced by the prestress, i.e. the load for which zero deflection is desired; or

- the load at which decompression is to occur; or
- full prestress to prevent cracking under the full service load.

4. **Cable details:** Estimate prestress losses and determine the cable force and cable profile to suit the requirements of Step 3; check cable details, duct sizes, cover, anchorage dimensions etc. to ensure that the prestressing tendons and anchorages fit within the section.
5. **Losses:** Calculate losses (Chapter 9) and adjust the details chosen in Step 4 if necessary.
6. **Moment capacity:** Check that ultimate strength in bending is adequate at all critical sections along the beam and provide additional tensile reinforcement if required.
7. **Transfer:** Check for the safety of critical cross-sections at transfer by calculating the section capacity against concrete crushing due to over-prestressing (Section 6.9.5 of Chapter 6).
8. **Deflections and crack control:** Check short-term and long-term deflections and crack widths under full service loading and also at transfer (Chapters 4 and 5). Adjust the design details as necessary.
9. **Shear:** Design for strength in shear at critical sections and provide stirrups where needed (Chapter 7).
10. **Anchorage:** Design the anchorage zones (Chapter 8).
11. **Other requirements:** Check for any other requirements such as torsion, fatigue etc., and adjust the design as necessary.

10.6 Discussion of key steps

The use of this procedure is illustrated in the design examples that are included in Section 10.8. Firstly, however, we elaborate on some of the key steps in the design method.

Choice of cross-section (Step 2)

The overall dimensions are chosen by means of Equation 10.2. However, to avoid subsequent adjustments to the cross-sectional dimensions, rough checks should be made at this stage on the width of the web and depth of the flange.

A simple first estimate of the minimum required web width is needed. This can be obtained from AS 3600–2018 formula for ultimate shear strength as limited by web crushing, and the design requirement $\phi V_u \geq V^*$. Assuming a value of θ_v = 31 degrees and α_v = 90 degrees, and with $\phi = 0.7$ (for web crushing) and $d_v = 0.9d$, the equation for minimum web thickness becomes:

$$b_v = \frac{7.2V^*}{f_c'd} \tag{10.8}$$

where d is the effective depth of the section.

The term b_w is the width of the web and is obtained from:

$$b_w = b_v + k_d \Sigma d_d \tag{10.9}$$

where Σd_d is the sum of the diameters (or maximum widths) of grouted tendon ducts that lie at any horizontal fibre in the web. The factor k_d has the value of 0.5 for grouted steel ducts, 0.8 for grouted plastic ducts and 1.2 for ungrouted ducts. It is usually desirable to set the web width appreciably greater then the minimum value given by Equation 10.9, in order to allow for web reinforcement in the section.

To check for the minimum thickness of the upper compressive flange, the total compressive force C at ultimate moment is first calculated, using a conservative estimate of $0.85d$ for the internal lever arm:

$$C = \frac{M_u}{0.85d} = \frac{M^*}{\phi 0.85d} \tag{10.10}$$

where for bending, $\phi = 0.85$. To keep the equivalent rectangular stress block depth within the flange, we need to place the following lower limit on flange thickness t:

$$t \geq \frac{C}{0.85f_c'b} = \frac{M^*}{0.72f_c'bd} \tag{10.11}$$

It is not strictly necessary for the equivalent stress block to be contained within the flange. Nevertheless any additional concrete needed to satisfy Equation 10.11 is usually small and, besides simplifying the ultimate moment calculations, this requirement gives improved section ductility.

Choice of load to be balanced (Step 3)

When load balancing is used to determine the prestress level, a choice has to be made of the load to be balanced; there are several things to consider. If the full sustained load (self-weight plus imposed dead load) is balanced, then the long-term deflections will effectively be kept to zero. However, if the live load is small in comparison with the full dead load, then this may well result in excessively high prestress, and it may be better then to balance only a proportion of the full dead load.

In the case of long-span members with high ratios of sustained load to live load, it will be appropriate to balance a proportion of the sustained load, as an uneconomic design may result from balancing the full sustained load.

If only a portion of the sustained load is to be balanced, it becomes necessary to calculate short-term and long-term downward deflection caused by the sustained load. An estimate of the initial short-term downward deflection Δ_e due to sustained load is:

$$\Delta_e = \Delta_o (M_s - P_e e)/M_s \qquad (10.12)$$

where Δ_o is the elastic deflection calculated for the uncracked beam when subjected to the full sustained load. The maximum bending moment in the beam due to the sustained load is M_s, and $P_e e$ is the negative moment due to prestress in the cross-section where M_s acts. The creep deflection is, approximately:

$$\Delta_c = \varphi_o^* \Delta_e \qquad (10.13)$$

A preliminary check for these deflections can be made when the balancing load is chosen. It should be remembered that Δ_e and Δ_c must be included when the total deflection is calculated in Step 8.

If the live load is a large proportion of the total load, it may be economic to balance not only the sustained load but also a portion of the live load; this will reduce the total deflection under full load. An upward camber under sustained load then results and initial and final values of the camber should be checked.

Alternative choice of decompression load (Step 3)

If the level of prestress is determined from the decompression load criterion, and the decompression load is taken as the sustained load, cracking will occur in the peak moment region where the full live load acts. Deflections under sustained load are not normally excessive in a member designed by the decompression criterion; nevertheless, a check should be made.

Cable detail and approximate load balancing (Step 4)

If the load-balancing method (Step 3) is used, the ideal cable profile would make the Pe diagram exactly match the moment diagram for the load to be balanced. In most practical situations, exact balancing of loads is unrealistic and unnecessary. For example, exact load-balancing of the cantilevered beam shown in Figure 10.2(a) would require the cable to be kinked at supports B and C and have zero slope and eccentricity at both cantilever ends. Furthermore, a variation from the parabolic shape would also be needed to allow for the variation in the P force along the beam, due to losses. None of these adjustments are of any practical importance.

Load-balancing is not an end in itself; the practical objective is to keep the deflections small throughout the beam. An approximate load balance is therefore adequate for practical design. In Figure 10.2 the load to be balanced is w_S and the moments due to w_S are denoted in Figure 10.2(c) as M_S. The peak negative values of M_S at supports B and C do not have to be balanced exactly by the prestressing moment. Indeed, it is advantageous to make $P_e e$ somewhat smaller than the peak M_S values. However, to achieve near-zero deflections, the inequality of moments over the supports should be compensated on either side of B and C with values of $P_e e$ somewhat larger than the values of M_S, as in Figure 10.2(c). A sketch diagram can be made of the unbalanced moments:

$$\Delta M = M_S - P_e e \tag{10.14}$$

as in Figure 10.2(d). The corresponding diagram of unbalanced curvatures along the beam has the same shape as ΔM, and Figure 10.2(d) can thus be used to make a free-hand sketch of the deflected shape of the beam, as in Figure 10.2(e). Such a sketch can serve as a guide in making adjustments to the cable profile.

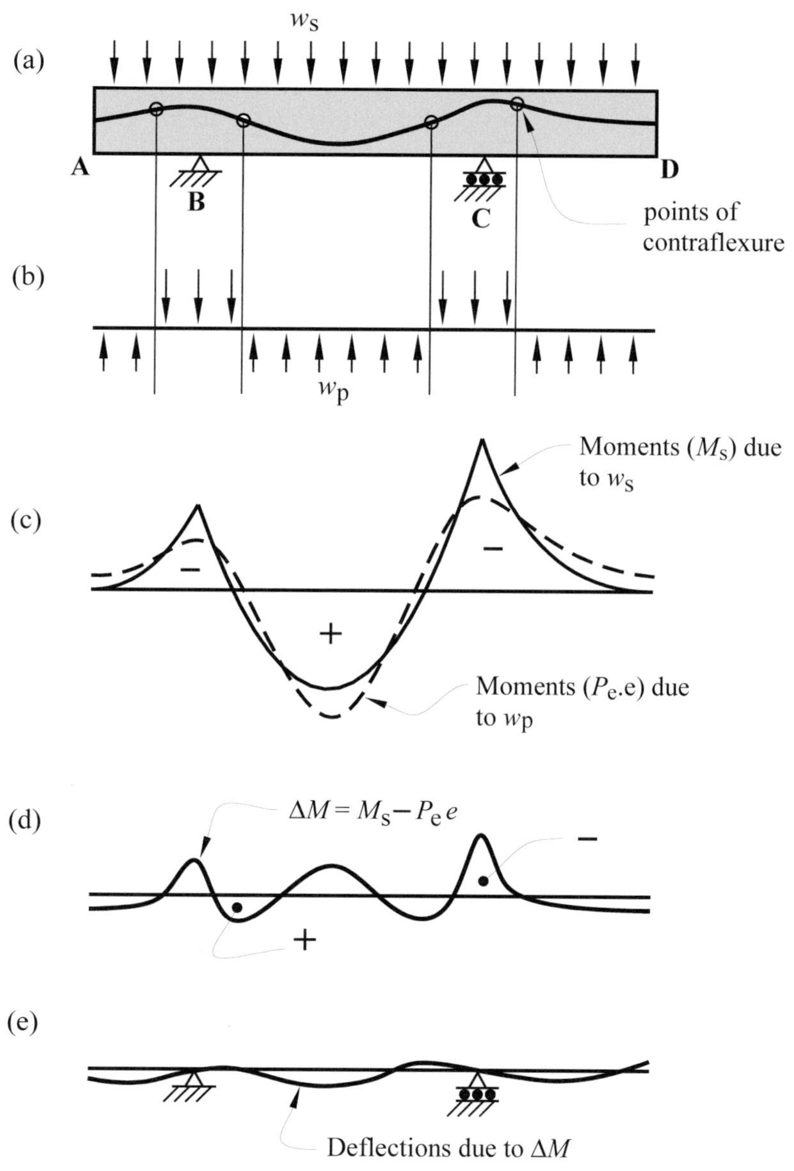

Figure 10.2 Approximate load balancing

Capacity at transfer (Step 7)

To check against failure during transfer by over-prestressing, the magnitude of the prestressing force is estimated that would cause failure by crushing of the concrete. Under the AS 3600 design requirements, the strength reduction factor to be used for this check is 0.6, which is consistent with that used for columns failing in compression. The prestressing force to be used is the maximum force applied during jacking, multiplied by a load factor $\gamma_p = 1.15$. For self-weight and any other dead load present during the stressing operation, the load factor is 0.9 for the normal case, whereby the moments due to these loads reduce the possibility of failure, but 1.15 if they increase it.

10.7 Design criteria for serviceability

As we have seen, prestressing improves the working load behaviour of a flexural member, and it will often be found that serviceability criteria are easily satisfied, especially when the load balancing method is being used. We will consider briefly the serviceability criteria of AS 3600.

Deflection limits

The deflection limits given in AS 3600 apply to both reinforced concrete and prestressed concrete members and are expressed as the ratio of deflection to effective span, Δ/L_{ef}, where L_{ef} is taken as the centre-to-centre distance between supports or the clear span plus overall member depth, whichever is smaller. An overall limiting value of 1/250 is placed on all members, but this is reduced to 1/500 in the case of transfer members. Further limits are placed on the increments in deflection that occur:

- after the addition or attachment of any partitions (1/1000); and
- after the addition or attachments of finishes (1/500).

For members such as bridge girders that are subjected to vehicular or pedestrian traffic, the limit on deflection under live load is 1/800. In the case of cantilevers these values are doubled, so that, for example, the total deflection limit becomes 1/125.

These limits provide a specific, but arbitrary, dividing line between acceptable and unacceptable structural behaviour in what is, in reality, a grey area and where personal judgment plays an important role. Such limits should be used by the designer with caution. In some cases they may prove to be inadequate. Deflection limits need to be evaluated for the particular structure and the special circumstances pertaining to the design. The values from AS 3600 are best regarded as useful lower bound guidelines for situations in which more specific information is unavailable and taking into consideration the variability of the material properties used in such calculations.

Crack width limits and crack control

According to AS 3600, cracking is to be controlled so that structural performance, durability and appearance are not compromised. However, it is difficult to set specific but reliable limits on acceptable cracking. There are various reasons for this. For example, the information available on the effect of crack widths on steel corrosion is still inadequate, while the question of visually acceptable crack widths is quite subjective. Nevertheless, limits on crack widths that are often chosen include: 0.2 mm when tight crack control is required, and 0.3 mm or 0.4 mm for moderate or mild control. For comparison purposes, Table 10.1 contains permissible crack widths that were proposed by ACI Committee 224. The figures are based on experience and are not justified on rational grounds.

According to AS 3600-2018, a check on crack control can be made either indirectly by ensuring that steel stresses in the cracked regions do not exceed safe limiting values, or directly by calculating crack widths. Methods for determining steel stress limits and crack widths as the basis for crack control checks have been provided, with examples in Chapter 5.

TABLE 10.1 Permissible crack widths (ACI Committee 224, 1972)

Exposure condition	Maximum allowable crack width (mm)
Dry air or protective membrane	0.40
Humid air, soil	0.30
De-icing chemicals	0.20
Sea-water and sea-water spray, wetting and drying	0.15
Water retaining structures	0.10

10.8 Design examples

Three examples are presented for the preliminary design of simply-supported prestressed concrete beams. All three examples concern the design of a beam that spans 23 metres and supports a permanent imposed uniformly distributed load of 25 kN/m plus self-weight, and a transient, live, load of 45 kN/m.

Three different criteria are used to determine the prestressing details. In the first example, the prestressing details are chosen to cause decompression under the permanent dead load. In the second example, load balancing of the dead load plus a component of the service load is adopted. In the third example a fully prestressed design is used. In all cases, the tendon has a parabolic profile with zero eccentricity at its ends.

Other data for the designs are:

Concrete: $f_c' = 40 \text{ MPa}$; $(\gamma = 0.87, \alpha_2 = 0.79)$

$E_c = 32.8\times10^3 \text{ MPa}$; $\psi_s = 0.7$; $\psi_l = 0.1$

$\varphi_{cc} = \varphi_o^* = 2.4$; $\varepsilon_{cs}^* = 600\times10^{-6}$

At transfer $f_{cp}' = 32 \text{ MPa}$

Prestressing steel: 12.7 mm strands with a breaking load of 184 kN ($f_{pb} = 1870 \text{ MPa}$); $E_p = 195\times10^3 \text{ MPa}$

Reinforcing steel: $f_{sy} = 500 \text{ MPa}$; $E_s = 200\times10^3 \text{ MPa}$

The results for the three designs are tabulated and compared at the end of this Section.

EXAMPLE 10.1

DESIGN FOR DECOMPRESSION UNDER PERMANENT LOAD

The member is to be designed so that the extreme tensile fibre at mid-span is at decompression under the permanent load, G.

SOLUTION

Estimate of self-weight (Step 1):

Since the loads are relatively high, a reasonably deep section will be needed. For consideration of self-weight, we assume an equivalent rectangular section of $L/D = 15$, $D = 23.0 \times 10^3 / 15 \approx 1500$ mm and with $b = D/3 = 500$ mm. The estimated loads are:

$$\text{estimated self-weight} = 1.5 \times 0.5 \times 25 = 18.7 \text{ kN/m}$$

$$\text{total dead load} = 18.7 + 25 = 43.7 \text{ kN/m}$$

Trial cross-section (Step 2):

$$\text{Factored load} \qquad w^* = 1.2 \times 43.7 + 1.5 \times 45 = 120 \text{ kN/m}$$

$$\text{At mid-span:} \qquad M^* = 120 \times 23.0^2 / 8 = 7935 \text{ kNm}$$

A flanged section will be used. Assuming that the top flange will accommodate the compressive force at M_u, the depth to the neutral axis will be small. We assume $k_u = 0.20$ and using Equation 10.2 we have:

$$bd^2 \geq \frac{7935 \times 10^6}{0.85 \times 0.79 \times 40 \times 0.87 \times 0.2\left(1 - \frac{0.87 \times 0.2}{2}\right)}$$

$$\geq 1.86 \times 10^9 \text{mm}^3$$

If $D = 1500$ mm and $d = 1350$ mm, the required flange width is 1021 mm. A top flange width of 1200 mm is tentatively chosen. The top flange shape as shown in Figure 10.3 is tentatively chosen and we check that the flange area can accommodate the concrete compressive stress resultant C at M_u. The lever arm and the compression force are:

$$z = d(1 - 0.5\gamma k_u) = 1350(1 - 0.5 \times 0.87 \times 0.2) = 1233 \text{ mm}$$

$$C = \frac{M^*}{\phi z} = \frac{7935 \times 10^6}{0.85 \times 1233} = 7571 \times 10^3 \text{ N}$$

and as $C = \alpha_2 f_c'(\gamma k_u d)b$, for the stress block to lie fully within a constant width flange, the minimum flange thickness is:

$$t_{min} = \gamma d_n = \frac{C}{\alpha_2 f_c' b} = \frac{7571 \times 10^3}{0.79 \times 40 \times 1200} = 200 \text{ mm}$$

This is slightly larger than the chosen full flange depth (150mm) but is less than the overall depth (250 mm) and should be adequate.

To choose the web width, we obtain from Equations 10.8 and 10.9, with d = 1350 mm:

At the support: $V^* = 120 \times 23.0/2 = 1380 \text{ kN}$

Minimum width: $b_v = \dfrac{7.2V^*}{f_c' d} = \dfrac{7.2 \times 1380 \times 10^3}{40 \times 1350} = 184 \text{ mm}$

Assuming that the tendon ducts are approximately 100 mm in diameter, the minimum web width b_w to avoid web crushing failure is:

$$b_{w.min} = 184 + 0.5 \times 100 = 234 \text{ mm}$$

We choose a somewhat larger value of 300 mm, which should be adequate. We adopt the trial section shown in Figure 10.3 with the section properties:

Cross-sectional area: $A_g = 720 \times 10^3 \text{ mm}^2$

Self-weight: $w_{sw} = 0.72 \times 25 = 18 \text{ kN/m}$

Depths to centroid: $y_a = 662$ mm (top); $y_b = 838$ mm (bottom)

Second moment of area: $I_g = 187.7\times10^9$ mm^4

Section moduli: $Z_a = 283.5\times10^6$ mm^3;

$Z_b = 224.0\times10^6$ mm^3

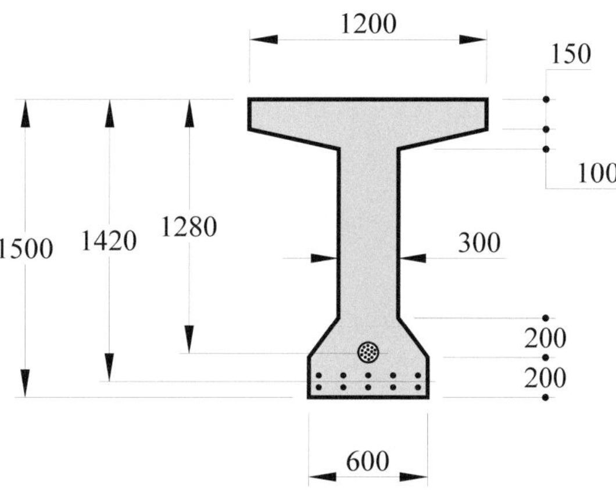

Figure 10.3 Trial cross-section for Example 10.1

Serviceability criterion (Step 3):

We design for decompression at mid-span under full dead load, so that the beam will be free of flexural cracks under this load. The total dead load is:

$$G = 18.0 + 25.0 = 43.0 \text{ kN/m}$$

The required decompression moment at mid-span is therefore:

$$M_o = M_G = 43.0 \times 23.0^2/8 = 2843 \text{ kNm}$$

Prestressing force and eccentricity (Step 4)

A single tendon will be used as a trial so that the maximum available eccentricity at mid-span will be approximately 618 mm (i.e. $d_p = 1280$ mm). The required effective prestressing force is then (Equation 10.4):

$$P_e = \frac{M_o}{e + \dfrac{Z_b}{A_g}} = \frac{2843 \times 10^6}{618 + \dfrac{224 \times 10^6}{720 \times 10^3}} = 3060 \times 10^3 \text{ N}$$

Prestress losses and cable selection (Step 5):

Assuming that deferred losses total 20 per cent ($\eta = 0.8$), we determine the initial prestressing force required at mid-span to be:

$$P_i = 3060/0.8 = 3825 \text{ kN}$$

For a cable with a parabolic profile having an eccentricity that varies from zero at each support to 618 mm at mid-span, the cable slope at the support is:

$$\theta = \frac{4h}{L} = \frac{4 \times 618}{23.0 \times 10^3} = 0.107 \text{ radians}$$

We estimate friction losses from Section 9.3 with $\mu = 0.2$ and $\beta = 0.016$. The stress in the tendon at mid-span is then:

$$\sigma_p = \sigma_{pj} e^{-\mu(\alpha_{tot} + \beta L_{pa})}$$

$$= \sigma_{pj} e^{-0.2(0.107 + 0.016 \times 11.5)} = 0.943 \sigma_{pj}$$

The required initial prestress at the support (the jacking force) is therefore:

$$P_{ij} = 3825/0.943 = 4056 \text{ kN}$$

With the stress in the cables at the jack at $0.80 f_{pb}$, the required breaking load of the cable is 4056/0.80 = 5070 kN. The breaking load for one 12.7 mm diameter strand is 184 kN, and the number of strands required is 5070/184 = 28.

We choose a tendon with 28–12.7 mm strands, having area $A_p = 2760 \text{ mm}^2$. For this tendon, the duct diameter is 106 mm and the anchorage bearing plate is 315 mm square. Thus the web will need to be widened at the ends of the beam.

The effective stress at mid-span is:

$$\sigma_{pe} = P_e / A_p = 3060 \times 10^3 / 2760 = 1109 \text{ MPa}$$

Reinforcement required for M^* at mid-span (Step 6):

The design load for strength is:

$$w^* = 1.2 \times 43 + 1.5 \times 45 = 119.1 \text{ kN/m}$$

and the corresponding design moment and the required ultimate moments are, respectively:

$$M^* = 119.1 \times 23.0^2/8 = 7876 \text{ kNm}$$

$$M_u \geq M^*/\phi = 7876/0.85 = 9266 \text{ kNm}$$

Although for the section chosen the lower part of the stress block will lie within the tapered region of the flange, this will not change the lever arm significantly and we shall take $z \approx 1233$ mm . The tensile force required at M_u is then determined as:

$$T_{py} + T_{sy} = 9266 \times 10^3/1233 = 7515 \text{ kN}$$

To evaluate the stress in the tendon at the ultimate condition we will use the approximate AS 3600 expression as in Equation 6.31. However, we first need to estimate the quantities of both the prestressing and reinforcing steels. Assuming that the tendon at ultimate is at yield, and from AS 3600 for strand $f_{py} = 0.82 f_{pb} = 1530$ MPa. The tension carried by the tendon is:

$$T_{py} \approx A_p f_{py} = 2760 \times 1530 \times 10^{-3} = 4223 \text{ kN}$$

The remainder of the force is taken by reinforcing steel and is thus:

$$T_{sy} \approx 7515 - 4223 = 3292 \text{ kN}$$

and the required area is estimated as:

$$A_{st} \approx T_{sy}/f_{sy} = 3292 \times 10^3/500 = 6584 \text{ mm}^2$$

With $A_p = 2760 \text{ mm}^2$ and our estimate of 6584 mm^2 for A_{st}, from Equation 6.31 we obtain (ignoring any compressive reinforcement):

$$k_1 = 0.4$$

$$k_2 = \frac{A_{pt} f_{pb} + (A_{st} - A_{sc}) f_{sy}}{b_{ef} d_p f_c'}$$

$$= \frac{2760 \times 1870 + (6584 - 0) \times 500}{1200 \times 1280 \times 40} = 0.138$$

$$\sigma_{pu} = f_{pb}\left[1 - \frac{k_1 k_2}{\gamma}\right] = 1870\left[1 - \frac{0.4 \times 0.138}{0.87}\right] = 1751 \text{ MPa}$$

The tension carried by the tendon is then:

$$T_p = A_p \sigma_{pu} = 2760 \times 1751 \times 10^{-3} = 4833 \text{ kN}$$

The force taken by the reinforcing steel is:

$$T_s = 7515 - 4833 = 2682 \text{ kN}$$

and the required area is:

$$A_{st} = T_s / f_{sy} = 2682 \times 10^3 / 500 = 5364 \text{ mm}^2$$

We **trial 10-N28 bars**, giving $A_{st} = 6200$ mm^2, at average depth of 1420 mm.

An analysis of the trial section to AS 3600 with $\sigma_{pu} = 1751$ MPa gives the following results:

Decompression moment:	$M_o = 3010$ kNm
Ultimate moment:	$M_u = 10{,}276$ kNm
Neutral axis depth at M_u:	$d_n = 241$ mm ($k_u = 0.19$)

Capacity of mid-span section at transfer (Step 7):

The self-weight moment at mid-span is $M_G = 18 \times 23.0^2 / 8 = 1190$ kNm. The initial prestressing force at mid-span is 3825 kN. We first check the top fibre stress to see whether the section is uncracked:

$$\sigma_a = \frac{P_i}{A_g} - \frac{P_i\,e}{Z_a} + \frac{M_G}{Z_a}$$

$$= \frac{3825\times10^3}{720\times10^3} - \frac{3825\times10^3\times(1280-662)}{283.5\times10^6} + \frac{1190\times10^6}{283.5\times10^6}$$

$$= +1.17 \text{ MPa (compressive)}$$

Since the top fibre stress is compressive, the section does not crack at transfer and the bottom fibre stress can be calculated using gross section properties. The bottom fibre stress is:

$$\sigma_b = \frac{P_i}{A_g} + \frac{P_i\,e}{Z_b} - \frac{M_G}{Z_b}$$

$$= \frac{3825\times10^3}{720\times10^3} + \frac{3825\times10^3\times(1280-662)}{224.0\times10^6} - \frac{1190\times10^6}{224.0\times10^6}$$

$$= +10.6 \text{ MPa (compressive)}$$

Since the bottom fibre stress is less than $0.6f'_{cp} = 19.2\ \text{MPa}$, the section satisfies the AS 3600 deemed-to-comply requirements for capacity at transfer.

Crack control at full service load (Step 8):

The maximum service load is 18 + 25 + 45 = 88 kN/m, and the corresponding bending moment at mid-span is 5819 kNm. At this bending moment the stress in the bottom reinforcement is found from a cracked section analysis to be 166 MPa, which is within the AS 3600 limit of 200 MPa at which crack widths may be deemed to be acceptable.

Check deflections (Step 8):

A check of deflections similar to that undertaken in Example 5.5 in Chapter 5 is required to ensure that the service limit state for deflections are met. As a rough first check on our trial section, we shall assume that $I_{ef} = 0.4I_g$.

The upwards load from the prestress is:

$$w_p = 8\times3060\times0.618/23.0^2 = 28.6 \text{ kN/m (upwards)}$$

which balances only a part of the dead load. The deflection due to the unbalanced dead load is then:

$$\Delta_G = \frac{5}{384} \cdot \frac{(w_G - w_p) L^4}{E_c I_{ef}}$$

$$= \frac{5 \times (43 - 28.6) \times (23.0\times10^3)^4}{384 \times 32.8\times10^3 \times (0.4 \times 187.7\times10^9)} = +21 \text{ mm (downwards)}$$

The short-term deflection due to the live load only is estimated as:

$$\Delta_Q = \frac{5}{384} \cdot \frac{\psi_s Q L^4}{E_c I_{ef}}$$

$$= \frac{5 \times 0.7 \times 45 \times (23.0\times10^3)^4}{384 \times 32.8\times10^3 \times (0.4 \times 187.7\times10^9)} = +47 \text{ mm (downwards)}$$

The sustained component of the live load deflection, $\Delta_{Q.sus}$, is obtained in a similar manner for the load $\psi_1 Q = 0.1 \times 45 = 4.5$ kN/m and is $\Delta_{Q.sus} =$ 6.7 mm. The total initial deflection due to the sustained loads are:

$$\Delta_{sus} = \Delta_G + \Delta_{Q.sus} = 21 + 6.7 = +28 \text{ mm (downwards)}$$

As the section contains a significant quantity of reinforcing steel in the tensile zone, a braking coefficient of $\alpha = 2.0$ is assumed and the creep deflection estimated as:

$$\Delta_c = \varphi_o^* \Delta_{sus} / \alpha = 2.4 \times (+28)/2.0 = +34 \text{ mm (downwards)}$$

For shrinkage, again given the significant quantity of reinforcing steel, a conservative value of $k_r = 0.5$ is taken (Equation 5.37) and, assuming a constant shrinkage warping along the length of the member, the shrinkage curvature is estimated to be:

$$\kappa_M = k_r \varepsilon_{sh}^* / D = 0.5 \times 600\times10^{-6} / 1500 = 200\times10^{-9} \text{ mm}^{-1}$$

and the resulting deflection:

$$\Delta = \frac{\kappa_M L^2}{8} = \frac{200 \times 10^{-9} \times (23.0 \times 10^3)^2}{8} = +13 \text{ mm (downwards)}$$

The total deflection for our trial section is then:

$$\Delta_{tot} \approx 21 + 47 + 34 + 13 = 115 \text{ mm} \quad \text{(span/200)}$$

These rough approximations for the trial cross-section sizing indicate that deflections may be an issue and the section size may require adjustment. However, as refined calculations in the final design could also show the section to be satisfactory, we stay with this trial section.

Shear reinforcement (Step 9):

Calculations similar to those carried out for Example 7.4 show that shear reinforcement consisting of N12 stirrups at 300 mm spacing throughout the beam satisfies the AS 3600 requirements for shear.

EXAMPLE 10.2

DESIGN FOR ZERO DEFLECTION FOR THE PERMANENT LOAD

In this example we shall design the girder of Example 10.1 for zero deflection when the load is equal to the permanent load $w_G = 43$ kN/m.

SOLUTION

We shall adopt a similar shaped section to that of the previous example with a a parabolic profile used to balance the load.

Prestressing force and tendon requirements (Step 4):

The prestressing force will be greater than that used in the previous example and we will assume that two cables will be required, aligned vertically. Considering concrete cover and constructional issues, we take the maximum available mean eccentricity to be approximately 618 mm. The required prestressing force at mid-span is then (Equation 10.3):

$$P_e = \frac{w_G L^2}{8h} = \frac{43.0 \times 23.0^2}{8 \times 0.618} = 4601 \text{ kN}$$

$$P_i = 4601/0.8 = 5751 \text{ kN}$$

As in the previous example, the factor for friction losses is 0.943 and so the jacking force is:

$$P_{ij} = 5751/0.943 = 6099 \text{ kN}$$

Limiting the stress in the strands to 80 per cent of their breaking load, the needed breaking capacity of the tendon is:

$$P_{pu} = 6099/0.80 = 7624 \text{ kN}$$

and the number of 12.7 mm diameter strands is 7624/184 = 41.

We try two cables each with 21–12.7 mm diameter strands. The area of each tendon is $A_{pi} = 2070$ mm^2. The effective stress at mid-span is:

$$\sigma_{pe} = 4601 \times 10^3/(2 \times 2070) = 1111 \text{ MPa}$$

The trial section is shown in Figure 10.4.

Reinforcement required for M_u at mid-span (Step 6):

As this design has more prestressing steel, it will need less reinforcing steel than in Example 10.1. Initially assuming an area of 1000 mm^2 of tensile reinforcing steel, we evaluate k_2 and hence use Equation 6.31 to obtain the stress in the strands at ultimate. We obtain $\sigma_{pu} = 1755$ MPa. Assuming $k_u = 0.2$, the moment carried by the prestress is then:

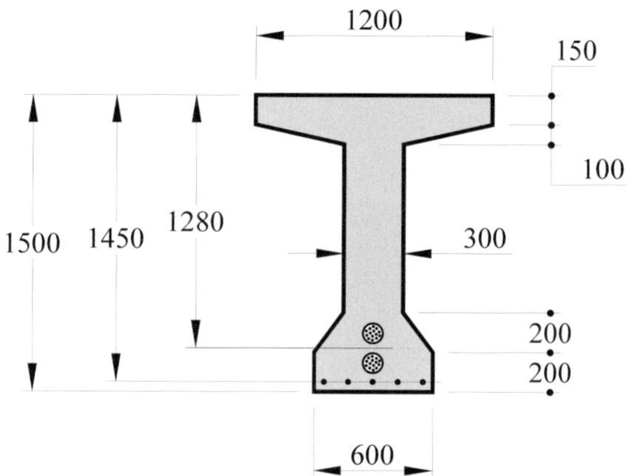

Figure 10.4 Trial cross-section for Example 10.2

$$M_{pu} = A_p \sigma_{pu} z_p$$

$$= 2 \times 2070 \times 1755 \times 1280\left(1 - \frac{0.87 \times 0.2}{2}\right)$$

$$= 8491 \times 10^6 \text{ Nmm} \quad (8491 \text{ kNm})$$

The portion of the ultimate moment to be taken by the tensile reinforcing steel is:

$$M_{su} = M_u - M_{pu} = 9266 - 8491 = 775 \text{ kNm}$$

and area of tensile reinforcement needed is:

$$A_{st} = \frac{M_{su}}{f_{sy} z} = \frac{775 \times 10^6}{500 \times 1450\left(1 - \frac{0.87 \times 0.2}{2}\right)} = 1170 \text{ mm}^2$$

As a **trial**, we take **4-N20** bars, giving $A_{st} = 1240 \text{ mm}^2$.

An analysis of the trial section to AS 3600 with $\sigma_{pu} = 1755$ MPa gives:

Decompression moment: $M_o = 4810$ kNm

Ultimate moment: $M_u = 10\ 137$ kNm

Neutral axis depth at M_u: $d_n = 239$ mm $(k_u = 0.19)$

Since this section has more prestress than that of Example 10.1, it is clear that there is no problem with crack control, and that minimum shear reinforcement will suffice. The additional prestress makes conditions for capacity at transfer more critical, and this is now checked.

Capacity at transfer (Step 7):

The initial prestressing force at mid-span is 5751 kN. We first check the extreme fibre stresses due to this force and the self-weight moment of 1190 kNm. The top fibre stress is:

$$\sigma_a = \frac{P_i}{A_g} - \frac{P_i\ e}{Z_a} + \frac{M_G}{Z_a}$$

$$= \frac{5751\times10^3}{720\times10^3} - \frac{5751\times10^3\times(1280-662)}{283.5\times10^6} + \frac{1190\times10^6}{283.5\times10^6}$$

$$= -0.35 \text{ MPa (tensile)}$$

As this is smaller that the tensile strength $f'_{ct.f} = 0.6\sqrt{32} = 3.4$ MPa, the section is assumed to be uncracked and calculation using gross section properties is valid. The bottom fibre stress is then:

$$\sigma_b = \frac{P_i}{A_g} + \frac{P_i\ e}{Z_b} - \frac{M_G}{Z_b}$$

$$= \frac{5751\times10^3}{720\times10^3} + \frac{5751\times10^3\times(1280-662)}{224.0\times10^6} - \frac{1190\times10^6}{224.0\times10^6}$$

$$= +18.5 \text{ MPa (compressive)}$$

The AS 3600 limit is $0.6f'_{cp} = 0.6\times32 = 19.2$ MPa, so the section just satisfies this requirement for capacity at transfer.

Check deflections (Step 8):

As for the previous example, we undertake a rough first check on our trial section for deflection. As the section has more prestress than before, we assume that $I_{ef} = 0.5I_g$. It is important to note here that a number of judgement based decisions are being made in this analysis and that for the final design, more detailed calculations will be needed to check that the service limit states are indeed satisfied.

As the dead load is balanced by the prestress, the short-term deflection is a result of the live load only and is estimated as:

$$\Delta_s = \frac{5}{384} \cdot \frac{\psi_s Q L^4}{E_c I_{ef}}$$

$$= \frac{5 \times 0.7 \times 45 \times (23.0{\times}10^3)^4}{384 \times 32.8{\times}10^3 \times (0.5 \times 187.7{\times}10^9)} = +37 \text{ mm (downwards)}$$

The sustained component of the live load deflection, Δ_{sus}, is obtained in similar manner for the load $\psi_l Q = 0.1 \times 45 = 4.5$ kN/m and is $\Delta_{sus} = 5.3$ mm. As the section contains some reinforcing steel in the tensile zone, a braking coefficient of $\alpha = 1.5$ is assumed and the creep deflection estimated as:

$$\Delta_c = \varphi_o^* \Delta_{sus}/\alpha = 2.4 \times (+5.3)/1.5 = +8 \text{ mm (downwards)}$$

For shrinkage, a value of $k_r = 0.5$ is assumed (Equation 5.37) and, for a constant shrinkage warping along the length of the member, the shrinkage curvature is estimated to be:

$$\kappa_M = k_r \varepsilon_{sh}/D = 0.5 \times 600{\times}10^{-6}/1500 = 200{\times}10^{-9} \text{ mm}^{-1}$$

and the resulting deflection:

$$\Delta = \frac{\kappa_M L^2}{8} = \frac{200 \times 10^{-9} \times (23.0{\times}10^3)^2}{8} = +13 \text{ mm (downwards)}$$

The total deflection for our trial section is then:

$$\Delta_{tot} \approx 37 + 8 + 13 = 58 \text{ mm} \quad \text{(span/400)}$$

which appears to be satisfactory, subject to checking with a refined calculation.

EXAMPLE 10.3

Design for a Fully Prestressed Section

In this third example, the girder is to be designed so that cracking does not occur under the loading $G + Q$ (i.e. fully prestressed). At transfer the beam carries its self-weight only.

Solution

As for the previous examples, a parabolic profile is adopted for the tendon, with the overall section dimensions previously used.

Prestressing force and eccentricity (Step 4):

We choose the prestressing force and eccentricity so that, at the mid-span section, the top fibre tensile stress at transfer does not exceed $0.6\sqrt{32}$ = 3.39 MPa, and the bottom fibre tensile stress under full load does not exceed $0.6\sqrt{40}$ = 3.79 MPa.

For the uncracked section, the stresses in the extreme fibres are determined from:

$$\sigma_a = \frac{\eta P_i}{A_g} - \frac{\eta P_i e}{Z_a} + \frac{M}{Z_a} \tag{10.15}$$

$$\sigma_b = \frac{\eta P_i}{A_g} + \frac{\eta P_i e}{Z_b} - \frac{M}{Z_b} \tag{10.16}$$

At transfer the initial prestressing force P_i applies and $\eta = 1$. As calculated previously, the self-weight of the girder is 18 kN/m and at the mid-span section the self-weight bending moment is $M_G = 1190$ kNm, which causes a top fibre compressive stress of $M_G/Z_a = +4.2$ MPa. The resultant top fibre stress is:

$$\sigma_a = \frac{P_i}{720 \times 10^3} - \frac{P_i \times 618}{283.5 \times 10^6} + 4.2 \geq -3.39 \text{ MPa} \tag{10.17}$$

and solving gives $P_i \leq 9595$ kN

Under the full service load and after losses, the effective prestress is P_e and the service load bending moment is 5819 kNm, which results in a bottom fibre tensile stress of $M_t/Z_b = -26.0$ MPa. We again assume 20 per cent deferred losses, so $\eta = 0.8$ and:

$$\sigma_b = \frac{0.8P_i}{720\times10^3} + \frac{0.8P_i \times 618}{224.0 \times 10^6} - 26.0 \geq -3.79 \text{ MPa} \qquad (10.18)$$

Solving gives $P_i \geq 6693$ kN and thus $6693 \text{ kN} \leq P_i \leq 9595 \text{ kN}$. Selecting the lower value of $P_i = 6693$ kN, then $P_e = 0.8 \times 6693 = 5354$ kN. With these values, the extreme fibre stresses are (Equations 10.15 and 10.16):

At transfer: $\sigma_a = -1.1$ MPa (tensile)

$\sigma_b = +22.2$ MPa (compressive)

After all losses and under a load of $G + Q$:

$\sigma_a = +16.3$ MPa (compressive)

$\sigma_b = -3.79$ MPa (tensile)

As in the previous examples, accounting for friction in the ducts, the jacking force is:

$$P_{ij} = 6693/0.943 = 7098 \text{ kN}$$

Limiting the stress in the strands to 80 per cent of their breaking load, the needed breaking capacity of the tendon is:

$$P_{pu} = 7098/0.80 = 8873 \text{ kN}$$

and the number of 12.7 mm diameter strands is 8873/184 = 48.

We shall try two cables each with 24–12.7 mm diameter strands. The area of each tendon is $A_{pi} = 2366$ mm^2. The effective stress at mid-span is:

$$\sigma_{pe} = 5354\times10^3/(2 \times 2366) = 1130 \text{ MPa}$$

Reinforcement required for M_u (Step 6):

This design has more prestressing steel and therefore needs less reinforcing steel than in Example 10.1. Assuming no tensile reinforcing steel, the stress in the strands at ultimate is estimated to be (Equation 6.31) $\sigma_{pu} = 1746$ MPa. Setting $k_u = 0.2$, the moment carried by the prestress is then:

$$\begin{aligned} M_{pu} &= A_p \sigma_{pu} z_p \\ &= 2 \times 2366 \times 1746 \times 1280(1 - 0.87 \times 0.2/2) \\ &= 9655 \times 10^6 \text{ Nmm} \quad (9655 \text{ kNm}) \end{aligned}$$

As the moment capacity with the tendons is greater than M^*/ϕ = 7876/0.85 = 9266 kNm, no tensile reinforcement is needed for strength.

Capacity at transfer (Step 7):

As calculated above, the bottom fibre stress at transfer is 22.2 MPa, which exceeds the AS 3600 deemed-to-comply limit of 19.2 MPa. Thus, the capacity at transfer needs to be improved. This could be achieved by increasing the concrete strength, by increasing the width of the bottom flange or, but less attractively, by substantially increasing the amount of bottom reinforcement. We shall increase the concrete strength to $f_c' = 50$ MPa with a minimum stress at transfer of $f_{cp}' = 37$ MPa.

Check deflections (Step 8):

As the section remains uncracked through its full service load range, $I_{ef} = I_g$. At transfer, the upwards loading from the tendon is:

$$w_p = \frac{-8P_i h}{L^2} = \frac{-8 \times 6693 \times 10^3 (1280 - 662)}{23000^2} = 62.5 \text{ kN/m}$$

and the resulting initial camber is:

$$\Delta_{hog} = -\frac{5}{384} \frac{(62.5 - 18.0)\,23000^4}{32800 \times 187.7 \times 10^9} = -26.3 \text{ mm (upwards)}$$

The prestressing losses occur over time, and the prestress reduces to P_e, so that the upwards loading from the tendon reduces to $w_p = 0.8 \times 62.5 = 50.0$ kN/m.

Adding the superimposed dead load and the sustained part of the live load, $\Delta G + \psi_l Q = 25.0 + 0.1 \times 45.0 = 29.5$ kN/m, the resulting nett sustained load is $w_{sus} = -50.0 + 29.5 = -20.5$ kN/m and the initial deflection is:

$$\Delta_o = \frac{5}{384} \frac{(-20.5) \times 23000^4}{32.8 \times 10^3 \times 187.7 \times 10^9} = -12.1 \text{ mm (upwards)}$$

As the deflection is upwards, the braking parameter is taken as 1.0 and creep deflection is estimated as:

$$\Delta_c^* = \varphi_o^* \Delta_s = 2.4 \times (-12.1) = -29.0 \text{ mm (upwards)}$$

The initial deflection plus creep deflection is then $-12.1 - 29.0 = 41.1$ mm (span/560). The shrinkage deflection will be downwards, as will the deflection due to the short term component of the live load over that of the long-term component $(\psi_s - \psi_l)\,Q$. Thus, the deflections for our trial section appear to be satisfactory, subject to check by refined calculation.

10.8.1 Comparison of results from Examples 10.1 to 10.3

In Table 10.2, the results of the three examples are summarised and compared. The prestressing requirements of course depend on the design criterion used, and it should be noted that superimposed dead load is rather high.

It is clear from Table 10.2 that the decompression criterion leads to the most economical design, but long-term deflections would need to be carefully checked. If they were found to be excessive, the prestress could be increased.

Load balancing of the full permanent load results is a design that is not much different from the fully prestressed design. Camber problems may occur for both of these designs at transfer, especially if there is a delay in applying the long-term sustained load ($G+\psi_l Q$). For the fully prestressed design, long-term hogging of the girder due to creep needs careful consideration.

TABLE 10.2 Comparison of designs, Examples 10.1, 10.2 and 10.3

	Example 10.1	Example 10.2	Example 10.3
Criterion for prestress level	Decompression under the permanent load *G*	Permanent, dead load, has a zero contribution to the deflection.	No cracking at any service load
Area of prestressing steel (mm^2)	2760	4140	4730
Area of reinforcing steel (mm^2)	6200	1550	0
Comments	Most economical design. Deflections may be a problem.	Close to fully prestressed design. Consideration of hogging deflection after transfer	Increased concrete strength, or increased size of bottom flange, is needed for capacity at transfer. Consider long-term hogging deflection.

10.9 References

ACI Committee 224 1972, Control of cracking in concrete structures, *Concrete International*, American Concrete Institute, pp. 35–75.

AS 3600–2018, *Concrete Structures*, Standards Australia, Sydney, Australia.

AS 5100.5–2017, *Australian Standard Bridge Design Part 5: Concrete*, Standards Australia, Sydney, Australia.

Brock, G. (2010), *RAPT user manual*, Prestressed Concrete Design Consultants Pty. Ltd., Australia.

Collins, M. P., and Mitchell, D. 1991, *Prestressed Concrete Structures*, Prentice-Hall, Inc., Englewood Cliffs, New Jersey, 766 pp.

Cross E. 2007, Post-tensioning in Building Structures, *Concrete in Australia*, Vol 33 No.2 Dec.

Leonhardt F. 1964, *Prestressed Concrete Design and Construction*, English Edn, Wilhelm Ernst, Berlin.

Libby J. R. 1971, *Modern Prestressed Concrete*, Van Nostrand Reinhold, New York.

Lin T.Y. and Zia P. 1974, Strength and Deformation of Prestressed Concrete Elements, Chapter 6 in *Reinforced Concrete Engineering*, Bresler B. (Ed), John Wiley & Sons, New York.

NAASRA 1976, *Highway Bridge Design Specifications*, National Association of Australian State Road Authorities, Section 6, Prestressed Concrete Design.

CHAPTER 11

Continuous beams

The behaviour, analysis and design of continuous prestressed concrete members is dealt with in this chapter. The design advantages of using continuous construction are first discussed, and the relevance of hyperstatic reactions and secondary moments and shears is explained. The design procedure introduced in Chapter 10 is extended to cover post-tensioned continuous members.

11.1 Advantages of continuous construction

Continuity in prestressed construction can bring significant economic and aesthetic benefits. In a design situation where a prestressed beam is required to have several adjacent spans, the two obvious alternatives to consider are:

- adjacent but structurally separated simply supported beams, and
- a single, multi-span continuous beam.

A comparison of these alternatives shows that the bending moments in the continuous beam are significantly smaller, and a smaller depth of beam can thus be used to satisfy the strength and serviceability requirements. There are further cost savings in the continuous solution through the reduced number of beam supports, fewer anchorages, less maintenance and streamlining of the prestressing and construction operations.

In a continuous girder an efficient cable profile will follow, approximately, the shape of the bending moment diagram, with the maximum positive eccentricities located at the mid-spans and the maximum negative eccentricities over the supports. This was shown in Figure 1.9 of Chapter 1. Good design economy can be achieved by using the prestressing cable to balance a proportion of the service load, and providing additional reinforcing steel in the critical regions of maximum bending moment to ensure adequate strength in flexure.

On the debit side, friction losses during the prestressing operation can become large in continuous members because of the more complex geometry of the cable. Friction losses must therefore be carefully checked. Axial shortening can also become an issue when long continuous members are prestressed, for example in large slabs on ground. The effects of shortening must then be evaluated and may require special joint details.

The behaviour of continuous prestressed members is more complex than for statically determinate members, not only because of the statical redundancy but also because of the hyperstatic reactions and secondary moments that may develop as a result of prestressing. Deflection calculations also tend to be more complex in the case of continuous members. A clear understanding of the full effects of prestressing on continuous construction, and of the behaviour of continuous members under load, is a necessary prerequisite for designers. However, the designer who has this understanding will find that the design process can be relatively straightforward for continuous beams, especially if the load balancing technique is employed. The tedium of iterative calculations can be largely eliminated through the use of suitable software for the detailed design work. The end result can be an economic and elegant solution to the structural design problem.

11.2 Effects of prestress in continuous beams

11.2.1 Hyperstatic reactions and secondary moments and shears

Continuous, statically indeterminate prestressed concrete beams differ from determinate prestressed beams, and also from non-prestressed indeterminate beams, in one important aspect: when the prestress is applied it induces flexural deformations along the beam and, if these deformations are incompatible with the support constraints, they induce reactions at the supports. These are called **hyperstatic reactions**, and they in turn induce **secondary bending moments** and **secondary shear forces** throughout the beam.

This is demonstrated using a simple numerical example. The two-span continuous post-tensioned beam shown in Figure 11.1 has a straight prestressing tendon at a constant eccentricity of 100 mm, which is tensioned to 1200 kN. The spans are both 12 metres and the cross-section is 600 mm deep and 300 mm wide. The modulus of elasticity of the concrete is 30×10^3 MPa.

Intuitively we can see how the eccentric prestress tends to lift the beam off the central support B. To maintain compatibility (the condition of zero deflection at support B) a downward reaction R_B must develop at B during the prestressing operation. For equilibrium, the two equal upward reactions at A and C must equal the downward reaction at B; that is, $R_B = R_A + R_C$.

To evaluate the reactions caused by the prestress, we ignore the self-weight and any applied loads and imagine the support at B to be removed, as in Figure 11.1(b). The moment due to prestress in all sections between A and C is $Pe = 1200 \times 0.1 = 120$ kNm, and is constant for all sections. The upwards deflection due to this moment is 53.3 mm. To evaluate the hyperstatic reaction R_B we calculate the magnitude of a point load applied at B on the 24 metre span AC that gives a downwards deflection of 53.3 mm, as in Figure 11.1(c). The value obtained is $R_B = 30$ kN and thus $R_A = R_C = 15$ kN. These hyperstatic reactions produce secondary moments that are zero at A and C and rise linearly to a maximum of 180 kNm at B, as in Figure 11.1(f). The secondary moments are all positive in this simple case.

The hyperstatic reactions and the accompanying secondary moments are present from the time of prestressing and affect the subsequent behaviour of the beam under service load conditions and at overload. The total moment due to prestress in any concrete section of the beam, M_P, is made up of a **primary moment,** M_1, and a **secondary moment**, M_2:

$$M_P = M_1 + M_2 \tag{11.1}$$

The primary moment due to the prestress is determined from the prestressing force P and the eccentricity e in the section: $M_1 = Pe$. The secondary moment M_2 is induced by the hyperstatic reactions and therefore must always vary linearly between supports. The moments M_1 and M_2 are shown in Figures 11.1(e) and 11.1(f), respectively, and M_p in Figure 11.1(g).

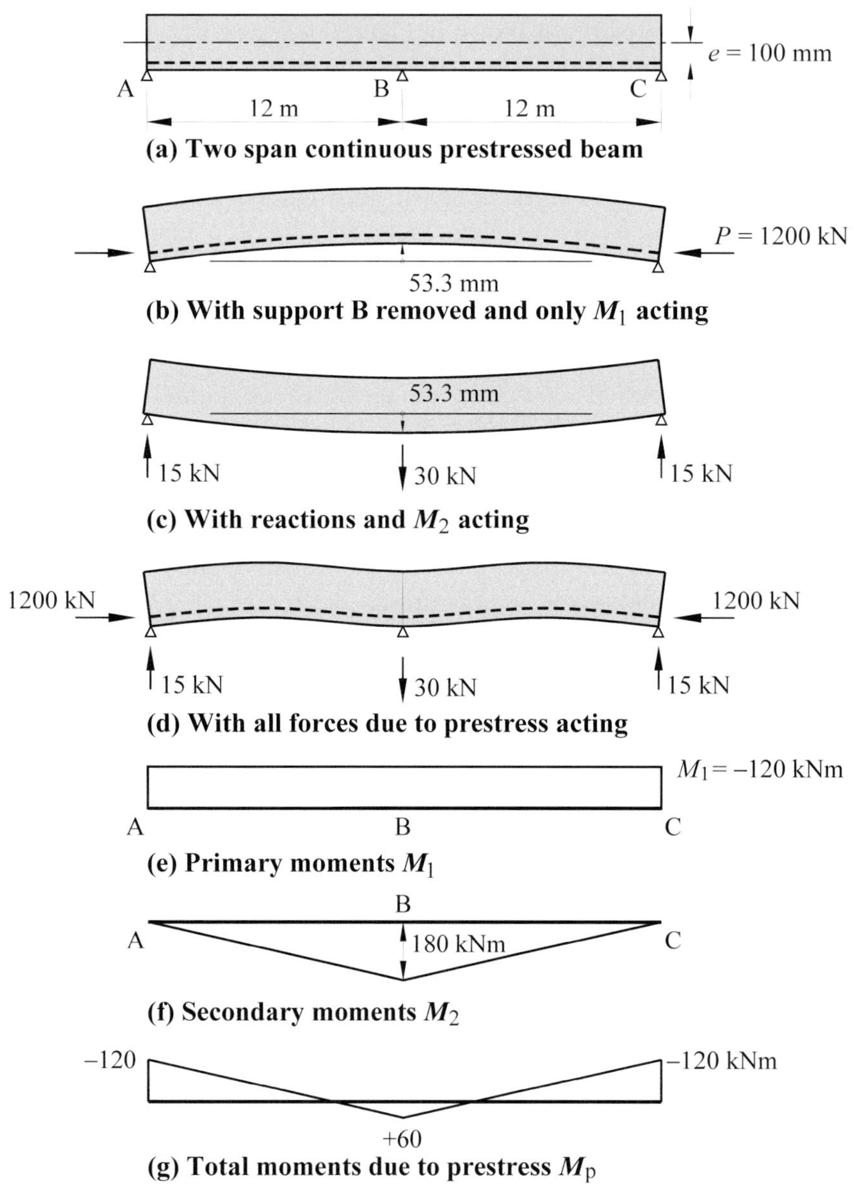

Figure 11.1 Continuous beam: hyperstatic reactions and secondary bending moments

The hyperstatic reactions induce shear forces as well as bending moments in the indeterminate beam. In general, at a concrete cross-section where the tendon has a slope of θ, the **primary shear force** caused by the prestress is $V_1 = P\theta$. The shear force induced by the hyperstatic reactions is called the **secondary shear force** due to prestress, V_2. In the case of the beam in Figure 11.1 the primary shear due to prestress is zero, because the tendon is straight and $\theta = 0$. The secondary shear, $V_2 = 15$ kN, is constant along the beam but changes sign at B. Because they are caused solely by the hyperstatic reactions acting at the supports, the secondary shears must be constant between support points.

Hyperstatic reactions do not occur inevitably in all indeterminate prestressed beams. It is possible, though unusual, for the flexural deformations induced by the prestress to be compatible with the support restraints. To take a trivial example, if the tendon in the beam in Figure 11.1 were located along the centroidal axis, the curvature due to prestress in all sections would be zero and the hyperstatic reactions, R_A, R_B and R_C, would also be zero. If a cable profile in an indeterminate beam does not induce hyperstatic reactions it is said to be a **concordant cable**. Methods of analysis to evaluate hyperstatic reactions and secondary moments and shears in continuous beams are discussed in Section 11.3.

11.2.2 Moment in the cross-section and moment in the concrete

In the analysis of prestressed concrete members we sometimes need to consider the cross-section of the full beam (with the tendon included), while at other times we consider the cross-section of the concrete beam (with the tendon removed). For example, in applying the equivalent load concept we consider the concrete beam minus the tendon, and represent the influence of the tendon on the concrete by means of equivalent loads. In dealing with continuous beams, where primary and secondary moments can act simultaneously, it is important to distinguish between the moments applied to the concrete section and the moments applied to the entire section.

For a statically determinate beam subjected only to prestress:

- there is no moment acting on the *complete* cross-section (concrete plus tendon); and
- the moment acting on the concrete in the cross-section is equal to the product of the prestressing force P and the eccentricity e, i.e. Pe.

Considering the continuous beam in Figure 11.1, we see that Figure 11.1(d) is a free body diagram for the concrete beam with the tendon removed. The end forces $P = 1200$ kN are exerted on the concrete by the tendon. In Figures 11.1(e), (f) and (g), the moments M_1, M_2 and the total moment M_P along the two-span beam are shown.

For a statically indeterminate beam subjected only to prestress:

- the primary moment at a cross-section is $M_1 = Pe$;
- the moment acting on the complete cross-section (concrete plus tendon) is the secondary moment M_2.
- the total moment acting on the concrete cross-section, M_P, is the sum of the primary and secondary moments; that is $M_P = M_1 + M_2$.

Any reinforcement present is taken here to be part of the concrete section.

11.3 Calculating the effects of prestress by the equivalent load method

The hyperstatic reactions and the secondary moments in the simple two-span beam in Figure 11.1 were calculated using first principles. This approach is not practical for many realistic cases, and computer analysis is normally used. Whatever method of analysis is employed, the equivalent load concept is useful in studying the effects of prestress.

In the equivalent load method, the forces exerted by the prestressing tendon on the concrete beam are treated as external loads and the beam is analysed to determine the moments caused by these loads. If the beam is continuous, the moment at any cross-section is the total moment due to prestress, M_P. Having evaluated M_P using the equivalent load method, and knowing the primary moment $M_1 = Pe$, we obtain the secondary moment as $M_2 = M_\text{P} - M_1$.

The total moment acting in the full cross-section (concrete plus tendon) is M_2. The hyperstatic reactions are determined from the M_2 moments by considering a free body for each span.

To illustrate this procedure we apply it to the beam in Figure 11.1. The equivalent loads here consist only of forces $P = 1200$ kN, applied at each end, A and C, at an eccentricity of 100 mm. This is statically equivalent to an axial force of 1200 kN and a couple of 120 kNm applied at each end. Analysing beam ABC for the effect of the end couples we obtain the reactions $R_B = 30$ kN and $R_A = R_C = 15$ kN, as indicated in Figure 11.1(c). The primary moments, secondary moments and total moments are shown in Figures 11.1(e), (f) and (g).

As a second example we consider the two-span beam in Figure 11.2(a). The cable profiles are parabolic, with a sag of 150 mm at each mid-span location and an abrupt change of slope (kink) at support B with an upwards eccentricity at B of 100 mm. The cable force is assumed to be constant at 300 kN throughout.

The tendon exerts on the concrete beam an upwards uniformly distributed load, which is evaluated using Equation 4.10:

$$w_p = 8Ph/L^2 = 8 \times 300 \times 0.15/6^2 = 10.0 \text{ kN/m}$$

There are also vertical forces (equivalent point loads) acting at A, B and C but these forces flow directly into the supports and, provided the supports are unyielding, do not affect the bending moments. Analysing the two-span concrete beam for the upward load of 10 kN/m we find that the bending moment at the interior support is +45 kNm, with moments in the spans as shown in Figure 11.2(c). These are the total moments due to prestress, M_P.

The secondary moment at B is:

$$M_2 = M_P - M_1 = 45 - (300 \times 0.1) = 15 \text{ kNm}$$

To find the hyperstatic reaction R_A at A, we consider equilibrium of span AB and equate $M_2 = 15$ kNm to $6 \times R_A$. We thus obtain $R_A = R_B = 2.5$ kN (acting upwards), so that the downward hyperstatic reaction at B is 5 kN.

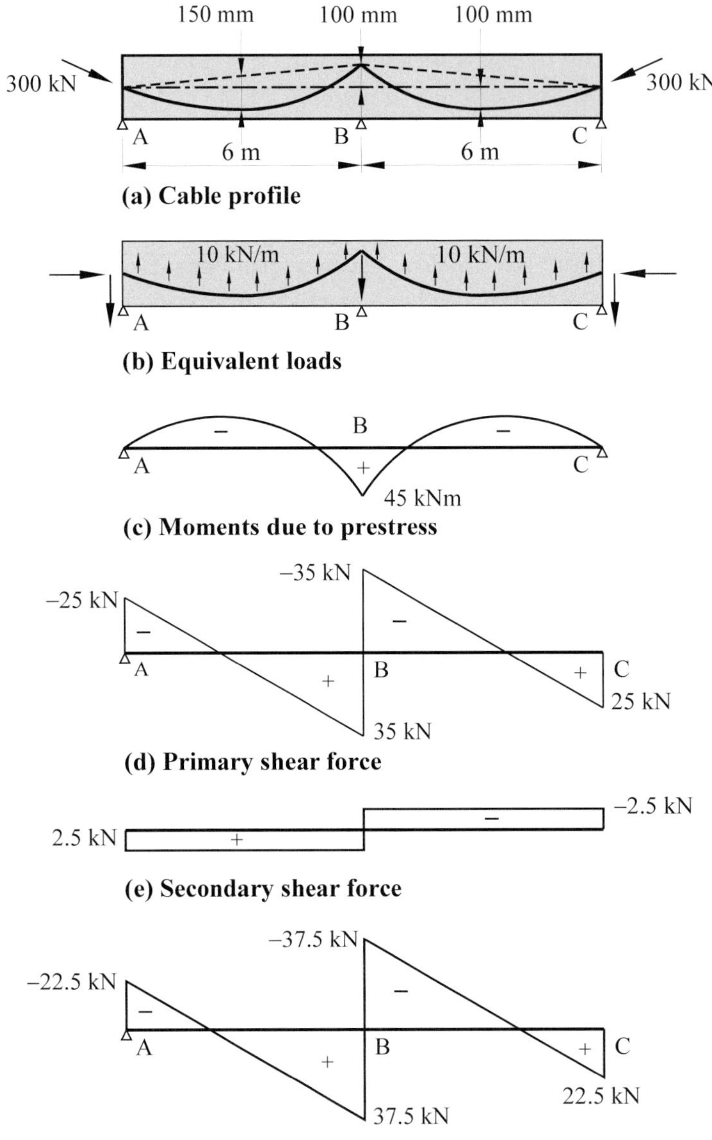

(a) Cable profile

(b) Equivalent loads

(c) Moments due to prestress

(d) Primary shear force

(e) Secondary shear force

(f) Total shear force due to prestress

Figure 11.2 Continuous beam, moments and shears

The primary, secondary and total shears due to prestress can be obtained similarly. The total shear force diagram for V_P in Figure 11.2(f) is obtained from the free body diagram in Figure 11.2(b) including the reactions already determined above. The difference between the total shear force and the secondary shear force is the primary shear force, which at any section is simply the vertical component of the prestressing force, i.e. $P_v = P\,\theta$.

EXAMPLE 11.1

EFFECTS OF PRESTRESS IN A TWO-SPAN BEAM

For the two-span beam shown in Figure 11.3(a), use first principles to determine the moments, support reactions and shear forces induced by the prestress. The force in the tendon is assumed to be constant at 300 kN throughout its length.

SOLUTION

In Figure 11.3(b) the forces (equivalent loads) exerted by the tendon on the concrete act at the anchorages and at the points D, B and E where the tendon changes direction. These forces represent the total effect of the prestress on the concrete; the moment that they cause in a cross-section is the total moment due to prestress, M_p. The vertical forces at the supports have no effect on the moments, so that the total moments due to prestress are those caused by the vertical upward loads of 30 kN at the centre of each span.

Analysis of the beam for the effects of these loads shows that a positive moment of $M_p = 33.75$ kNm acts in the concrete section at B, in addition to the axial compression of 300 kN. The resultant of the longitudinal stresses in the concrete at B is therefore a compressive force of C = 300 kN which acts at a distance above the centroid equal to 33.75 kNm/300 kN = 0.1125 m (112.5 mm), as shown in Figure 11.3(c).

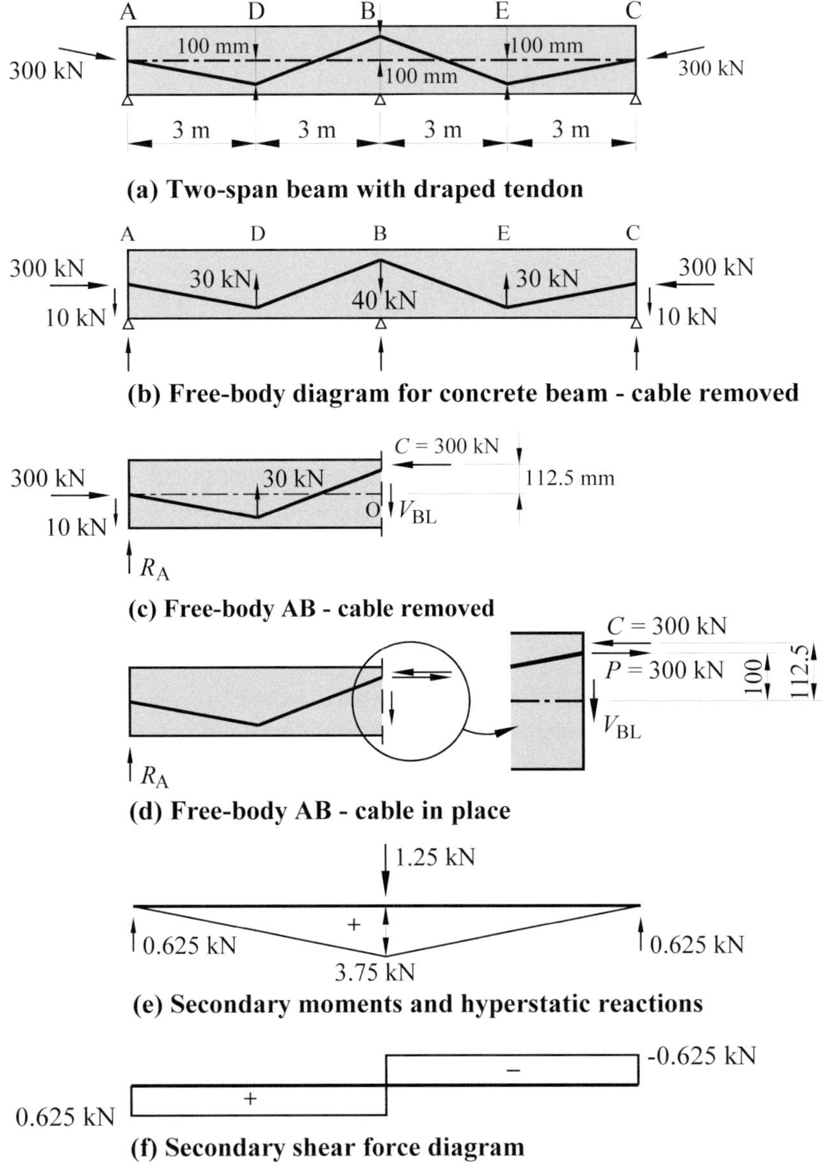

Figure 11.3 Two-span beam - Example 11.1

The reaction at A caused by prestress may be calculated by considering the equilibrium requirements for Figure 11.3(c), which is a free body diagram of span AB. This diagram is drawn for the concrete, with the cable removed, and therefore shows the equivalent loads exerted by the cable at A and D. The section at B is taken just to the left of the support, so that neither the support reaction nor the 40 kN force at the cable 'kink' act on the free body. The force V_{BL} is the shear force acting on the concrete section.

Taking moments about O, shown in (Figure 11.3c):

$$6R_A - 10 \times 6 + 30 \times 3 - 300 \times 0.1125 = 0 \text{, and } R_A = 0.625 \text{ kN}$$

It is however simpler, once the moments due to prestress have been determined, to consider the beam as a whole (i.e. with the cable included) in order to calculate the hyperstatic reactions. The corresponding free body diagram for span AB is shown in Figure 11.3(d). Again taking moments about O:

$$6R_A - 300 \times 0.1125 + 300 \times 0.1 = 0 \quad .... \qquad R_A = 0.625 \text{ kN}$$

Considering vertical equilibrium of either Figures 11.3(c) or 11.3(d), the shear force V_{BL} just to the left of B is found to be 0.625 kN. The hyperstatic reactions are 0.625 kN upwards at each exterior support and 1.25 kN downwards at the centre support. The bending moments resulting from these reactions, i.e. the secondary bending moments, are shown in Figure 11.3(e), and the secondary shear forces in Figure 11.3(f).

EXAMPLE 11.2

EFFECTS OF PRESTRESS IN A THREE-SPAN BEAM

The three-span prestressed concrete beam, shown in Figure 11.4(a), has a uniform rectangular section 800 mm deep and 300 mm wide. A preliminary analysis is required for an idealised cable that has a constant force of 1200 kN. The cable has a parabolic profile in spans BC and CD and a kinked draped profile in AB, as shown. Calculate the moments and reactions due to prestress.

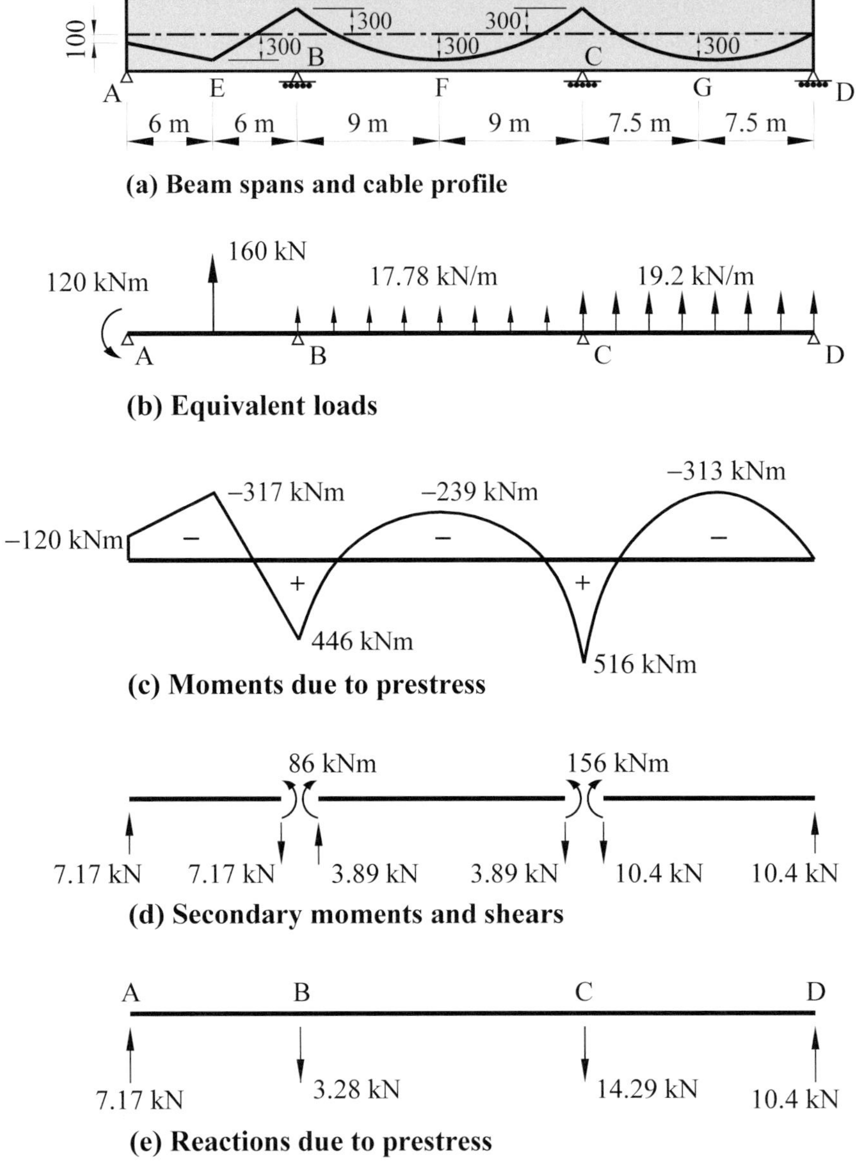

Figure 11.4 Three-span continuous beam for Example 11.2

SOLUTION

Moments due to prestress

The loads exerted by the cable on the concrete are first calculated. At E the angular change in radians is $(0.2 + 0.6)/6 = 0.133$, and so the cable exerts an upwards force of $0.133 \times 1200 = 160\ \text{kN}$.

The uniformly distributed upward loads produced by the prestress in spans BC and CD are:

Span BC: $$w_p = 8 \times 1200 \times 0.6/18^2 = 17.78\ \text{kN/m}$$

Span CD: $$w_p = 8 \times 1200 \times 0.45/15^2 = 19.2\ \text{kN/m}$$

In addition to these loads the 100 mm eccentricity at A gives rise to an end couple of 120 kNm.

Figure 11.4(b) shows these equivalent loads acting on the beam. The diagram does not include the vertical components of the prestressing forces acting at the supports, since these forces do not cause moments in the beam. An analysis of the beam under these loads is used to determine the bending moments in Figure 11.4(c). These are the total moments due to prestress.

Reactions due to prestress

The secondary moments, M_2, are calculated by subtracting the primary (Pe) moments from the total moments due to prestress:

At B: $$M_2 = 446 - (1200 \times 0.3) = 86\ \text{kNm}$$

At C: $$M_2 = 516 - (1200 \times 0.3) = 156\ \text{kNm}$$

The reactions are calculated from these moments. For example, if we consider a free body of span AB in Figure 11.4(d), summation of moments about A gives the shear force immediately to the left of B as:

$$V_{BA} = 86/12 = 7.17\ \text{kN} \quad \text{(downward)}$$

Consideration of the free body of span BC gives the shear force immediately to the right of B as:

$$V_{BC} = (156 - 86)/18 = 3.89 \text{ kN} \quad \text{(upward)}$$

The reaction at B is thus:

$$R_B = 7.17 - 3.89 = 3.28 \text{ kN (downward)}$$

The reactions due to prestress are shown in Figure 11.4(e).

11.4 Cable profiles for continuous post-tensioned beams

11.4.1 Concordant and non-concordant cables

The hyperstatic reactions and secondary moments caused by prestress in a continuous beam depend on the profile of the prestressing cable. If no hyperstatic reactions and hence no secondary moments are induced by the prestress, then the cable profile is said to be **concordant**. Cable profiles that induce hyperstatic reactions, and hence secondary moments, are **non-concordant**.

From a design point of view, and considering economy and beam behaviour at service load and overload, there is no good reason why a concordant cable should be preferred to a non-concordant cable.

11.4.2 Idealisations and simplifications

Various idealisations and simplifications are made with regard to the cable profile and prestressing details in order to simplify the calculations, especially at the preliminary design stage. For example, in Figures 11.2, 11.3 and 11.4 the cables are shown with kinks over the interior supports. Such kinks can be assumed for preliminary calculations but in practice a sharp curvature is used over a short region on either side of the support. This was explained in Chapter 1 and is shown in Figure 1.9.

In previous discussions, we have assumed that the cable force is constant in magnitude over the length of the continuous beam. In reality, the cable force varies considerably along the beam because of frictional forces that occur during the prestressing operation. The cable force decreases progressively from a maximum value at the jacking end to a minimum at the far end, or in the case of jacking at both ends, to a minimum near the middle of the beam.

Gross section properties of the concrete section are used in routine analysis and also to calculate stresses and strains in a section, ignoring the reinforcing steel and, in post-tensioned construction, areas of ducts and tendons. This has been discussed in Section 4.2. For most design calculations this simplification is acceptable. Appendix A explains how more accurate calculations can be made, if needed, to allow for reinforcement, ducts and tendons in the section.

The approximations mentioned here simplify the analysis and the results obtained are normally satisfactory, certainly for preliminary design calculations and often also for the final detailed design checks.

11.4.3 Practical cable shapes

Until now we have discussed idealised cables with parabolic shapes, straight segments and kinks. Such idealised profiles suit external loads that are either uniformly distributed or concentrated. In practice, cable profiles in continuous beams have a varying curvature within each span and the equivalent loads may be more complex than simple uniformly distributed loads and point loads.

For a curved cable profile, the equivalent load at any cross-section is the product of the cable force, P, and the curvature, κ. To calculate the equivalent loads on the beam we thus need to evaluate both P and κ at points along the member. If the eccentricity of the cable is defined algebraically as a function of distance x along the span, we can determine the curvature as the second derivative of the cable eccentricity, $e(x)$.

Cable profiles are frequently made up of piece-wise parabolic segments. Even if the cable is non-parabolic, it is usually sufficient to approximate the cable shape by separating it into separate segments with a parabolic approximation in each segment. A different value of uniformly distributed load is then obtained as the equivalent load within each segment.

The beam shown in Figure 11.5(a) is the same as that in Figure 11.2(a) but the theoretical cable profile has been replaced by a more realistic shape. It consists of three parabolic segments in each span: AB, BC and CD. The cable has zero slope at B (at mid-span) and again at D (at the interior support), while segments BC and CD have a common slope at the intersection point C. The cable eccentricity at C is calculated to be 66.6 mm, and the equivalent loads in the three parabolic segments are:

In region AB: $w_{AB} = 6.67$ kN/m

In region BC: $w_{BC} = 16$ kN/m

In region CD: $w_{CD} = -80.0$ kN/m

Using these equivalent loads, the total moment due to prestress at D is calculated to be $M_P = 42.6$ kNm. In previous calculations using the idealised profile of Figure 11.2(a) the moment at the interior support D was found to be 45 kNm, a difference of only about 6 per cent.

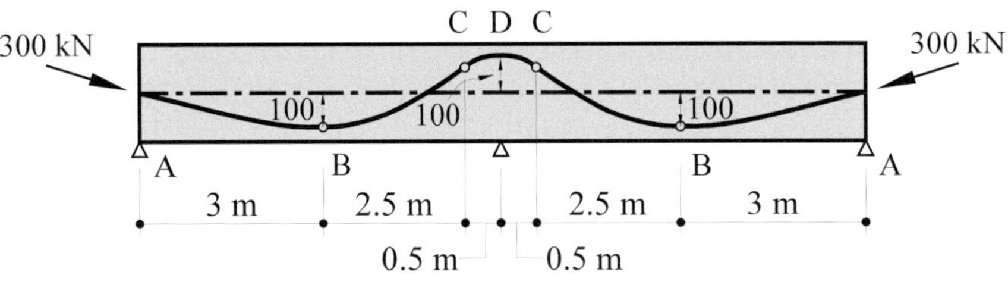

(a) Cable profile

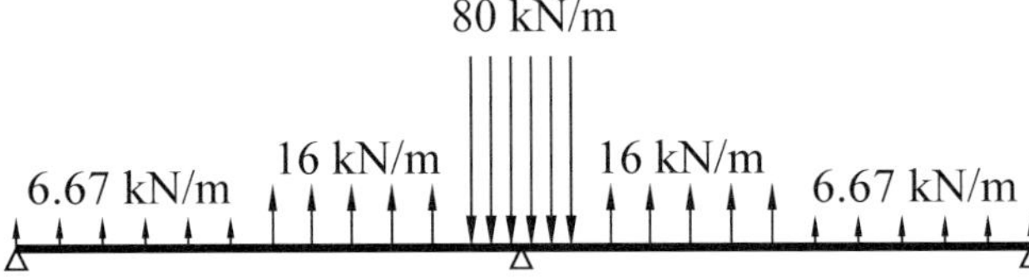

(b) Equivalent loads

Figure 11.5 Practical cable profile

11.4.4 Friction losses

As a result of friction the cable force P decreases with distance from the jacking end, so that the equivalent load also decreases with distance from the jacking end. For design, two procedures may be used to take account of this effect, an approximate calculation method or a segmental analysis.

Approximate calculation method

If the friction losses are reasonably small, the analysis may be based on a constant cable force within each span, taken as the value at mid-span. A more conservative approach is to use the maximum value of the prestressing force to calculate the conditions at transfer and the minimum value for conditions at full design load.

Calculation by segmental analysis

Greater accuracy can be obtained by dividing the beam into segments, with the assumed constant force in each segment determined at the middle of the segment. While the accuracy can be improved by taking a larger number of segments, it would be rarely necessary to use more than four segments per span to account for friction losses.

11.4.5 General method for calculating prestress effects

A more general approach can be used to determine the effects of prestress for real cable profiles and friction losses when high accuracy is required in the detailed design calculations. Values for both the cable force P and the cable eccentricity e can be determined at cross-sections along the member, and a diagram of the primary moment due to prestress, $M_1 = Pe$, can be constructed with good accuracy. The equivalent load diagram is then obtained by twice differentiating the Pe diagram. These equivalent loads may then be used (together with any couples at the free ends due to eccentricity of the cable there) in a segmental analysis of the continuous beam to determine the moments due to prestress.

Beams that have curved or inclined centroidal axis lines can also be analysed using this procedure. In calculating the moments due to prestress, the eccentricity is measured from the centroidal axis at each section, as indicated previously in Figure 4.7.

11.5 Service load behaviour of continuous beams

11.5.1 Pre-cracking, short-term behaviour

Continuous prestressed concrete members are normally cracked when the full design service live load acts in conjunction with the dead load. Nevertheless, the cracks will often be closed when only the sustained loads act. If the external load is not large enough to cause the cracks to open, the short-term response will be very close to being linear and elastic. The flexural stresses and elastic strains can then be determined by normal linear structural analysis of the indeterminate structure followed by an analysis of the individual cross-sections as described in Chapter 4. The effects of hyperstatic reactions on overall structural behaviour must of course be taken into account and the hyperstatic moments included when the stresses in individual section are determined.

11.5.2 Post-cracking, short-term behaviour

With increasing load, cracks first open in the peak moment regions near mid-span and over the interior supports, and then progressively further away from the peak-moment sections. As the cracks open in any region, there is a progressive decrease in the bending stiffness from the initial uncracked value I_g. Furthermore, the bending stiffness of the cracked section, I_{cr}, progressively decreases as the moment increases. With increasing load there is a progressive decrease in the bending stiffness in each peak moment region along the beam. Some redistribution of the internal moments therefore occurs and the behaviour is non-linear, even though the load is still in the service range.

An iterative procedure is needed if a precise analysis is to be undertaken of the behaviour of a continuous member in the post cracking range. An initial linear structural analysis based on the gross concrete section stiffness I_g can be used to identify those peak moment regions where the cracks will be open at a particular load level. The values obtained for the moments in the cracked critical regions from this first analysis can be used to estimate trial section stiffnesses for the cracked regions. The modified bending stiffnesses are then used in a second linear analysis to re-evaluate the moments in the critical sections. Several iterations may be needed to achieve convergence. The moments, stresses, strains and deformations can then be evaluated. Such an analysis is tedious but becomes necessary if the deflections are to be calculated from the bending stiffnesses of cracked critical sections. The calculation of deflections is discussed further in Section 11.6.

11.5.3 Long-term behaviour: creep and shrinkage

Creep and shrinkage play a decisive role in the long-term behaviour of continuous members. The level of the sustained load is also an important consideration, in particular whether or not cracks remain open under the sustained load.

The cracks in critical sections will usually be closed under sustained loading so that the creep curvature in each critical section can be estimated using the information given in Section 4.6 of Chapter 4. If some regions are cracked under sustained loading then the creep curvature can be evaluated using the information in Section 5.8.3 of Chapter 5.

A simplified calculation of creep curvature in an uncracked section can be made by assuming free creep under the conditions at the start of the process. The free creep curvature, κ_{co}^*, ignores the effects of the reinforcing steel and tendons in the section but it will usually be acceptable for preliminary calculations. An improved estimate can be obtained by adding the correction factor, $\Delta\kappa_c^*$, as explained in Appendix B and Section 4.6 of Chapter 4.

A layer of complexity is added by the structural indeterminacy, because the creep process will usually result in a progressive redistribution of the moments in individual sections. If the creep deformations are incompatible with the support conditions, then incremental changes occur in the hyperstatic reactions and the secondary moments. The situation is similar to that when the deformations due to prestress are incompatible with the support conditions and produce hyperstatic reactions. An elastic analysis of the continuous member at the end of the creep process can be used to estimate the increments in hyperstatic reactions and secondary elastic moments that are needed to maintain compatibility. Such a refined analysis will rarely be warranted in normal design situations when the uncertainties and inaccuracies of the creep analysis and the original creep data are considered.

Shrinkage also produces progressive changes in curvature in the critical sections. As with creep, the increments in shrinkage curvature in an individual critical section can be estimated using the information provided in Section 4.6 of Chapter 4 or Section 5.8.3 of Chapter 5, as appropriate. The increments in shrinkage curvature will also affect deflections and may, as for creep, bring about incremental changes in the hyperstatic reactions and secondary moments. If required, a refined analysis can be undertaken to investigate the incremental changes in the hyperstatic reactions and secondary elastic moments due to shrinkage.

The changes in stress levels following sustained loading are of relevance in design if checks are to be made on stress increments for crack control. The main effect of the creep and shrinkage, however, will be to increase the long-term deflections.

11.6 Deflection calculations for continuous beams

11.6.1 Introduction

Deflection calculations for continuous prestressed beams are more complicated than for statically determinate beams. As we have already seen, the flexural stiffness of a cracked section decreases with increasing moment, and the moment has to be obtained from a preliminary structural analysis. To undertake such an analysis, the stiffnesses of the critical sections must first be known. An iterative calculation process is therefore unavoidable, even when simplified deflection calculation methods are employed.

11.6.2 Short-term deflections of continuous beams

We have seen in Chapters 4 and 5 that AS 3600 allows two methods for calculating short-term deflections: one by refined calculation and one by simplified equations for the bending stiffness. We look at these in turn, in relation to continuous beams.

Deflections by refined calculation

The refined calculation method requires a first principles approach that takes into account cracking and tension stiffening and other influences such as load history and construction details. A non-linear step-by-step analysis of beam behaviour is required. This approach may be unavoidable in some complex design situations, for example in the case of push-out bridge construction, where many different loading conditions arise. Design evaluations are needed not only as a check on deflections but also to evaluate conditions in the critical cross-sections that are subjected to positive and negative moments at different stages of construction.

In a refined deflection calculation for a specific load case, iterative elastic analyses are carried out using trial values of the bending stiffness for each segment, following the sequence already described in Section 11.5.2. In summary, the first analysis uses the stiffness of the uncracked concrete sections and shows which regions are cracked and provides first trial values of the moments in critical sections that are used to estimate values for the effective stiffnesses of the cracked segments. The second analysis is undertaken with an update of the cracked section stiffnesses.

Simplified calculation of deflections

We saw in Section 4.7 of Chapter 4 that the deflection curve for a beam, $y(x)$, is obtained by double integration of the curvature $\kappa(x)$ and, as was explained in Section 5.8.2 of Chapter 5, this is a purely geometric relationship that applies irrespective of whether the beam sections behave in a linear or non-linear manner.

In the simplified method, Equation 4.17 is approximated as:

$$\Delta = \frac{1}{E_c I_{ef.av}} \iint M(x)\, dx\, dx \tag{11.2}$$

where E_c is the elastic modulus of the concrete and $I_{ef.av}$ is a weighted average effective second moment of area for the member.

The average bending stiffness for a span ($E_c I_{ef.av}$) is calculated using the effective stiffness determined at a few critical sections along the member and applying a weighted average calculation. In AS 3600, the average effective second moment of area for a member, $I_{ef.av}$, is determined from the values of I_{ef} calculated using the short-term serviceability load $G + \psi_s Q$ at nominated cross-sections as follows:

(a) For a simply supported span, the value at mid-span.

(b) For a continuous beam:

(i) for an interior span, half the mid-span value plus one-quarter of each support value; or

(ii) for an end span, half the mid-span value plus half the value at the continuous support.

(c) For a cantilever, the value at the support.

For calculating the effective stiffness, $E_c I_{ef}$, at a given cross-section, the method set out in Section 5.8.2 of Chapter 5 is used. The method can be used with any elastic model, such as those in Appendix D or those available in computing software packages, by replacing EI by $E_c I_{ef.av}$.

For a member supporting a uniformly distributed load on span L, the mid-span deflection can be determined with reasonable accuracy using the approximation:

$$\Delta = \frac{L^2}{96\, E_c I_{ef.av}} [M_L + 10M_M + M_R] \tag{11.3}$$

where M_L, M_M and M_R are the moments at the left, middle and right ends of the span, respectively. If the moment at the section is negative, its sign in Equation 11.3 is also negative.

11.6.3 Long-term deflections in continuous beams

The effects of creep and shrinkage have been discussed in Section 11.5.3. We have seen how the additional long-term curvatures in individual sections due to creep and shrinkage can be estimated using the information given in Chapters 4, 5 and 10 for statically determinate beams. We have also seen how creep and shrinkage may cause incremental changes in the hyperstatic reactions and secondary moments.

In many practical design situations it will be sufficient first to determine the free creep curvature $\kappa^*_{co} = \varphi^*_o \kappa_o$ at critical sections and, if necessary, to apply a curvature correction $\Delta\kappa^*_c$, determined as appropriate from Equations B.53, B.61 or B.70, as explained in Appendix B. For cracked sections, Equation 5.32 can be used. For shrinkage deflection calculations, curvatures may be approximated using Equation B.78 in Appendix B for uncracked sections and Equation 5.37 for cracked sections.

The long-term deflections can then be estimated by integrating these curvatures, ignoring any structural incompatibilities and incremental changes in hyperstatic reactions. Likewise, shrinkage curvatures can be determined and integrated to estimate the deflection due to shrinkage.

11.6.4 Deflections - design considerations

The control of deflections in prestressed concrete continuous beams presents somewhat of a paradox to the designer. On the one hand, continuity and prestressing both lead to reduced overall deflections, in particular if load balancing is employed to create a condition of zero deflection for the main design service load combination. Excessive deflection should not therefore be a design problem. On the other hand, complex deflection calculations are required by AS 3600 for a continuous prestressed member. Furthermore, these complex calculations have limited accuracy.

Some simplification of the calculation procedure is warranted. For example, simplifying assumptions can be introduced to avoid repeated structural analyses of the indeterminate structure. By assuming the locations of the points of contraflexure in each span, the regions between these zero moment points can be analysed statically. The complexity of such deflection computations is then the same as for a statically determinate beam.

It is emphasised that the iterations referred to in this discussion arise purely from the analytic process. They are separate and additional to the iterations that are part of the normal design process. In these circumstances it becomes worthwhile for designers to look at the available options. AS 3600 allows a range of alternative design methods that use various types of structural analysis. The possibility of carrying out design checks for both strength and serviceability using a non-linear full-range computer analysis becomes an attractive option. The one full-range analysis provides all the information needed for all the strength checks and deflection checks. Strength design using non-linear analysis was mentioned briefly in Chapter 3 and further information can be found in the literature (*fib* Bulletin 45, 2008).

11.7 Overload behaviour and flexural strength

11.7.1 Continuous beams at overload

The behaviour of a continuous beam in the overload range depends on the deformation characteristics of the peak moment regions and, in particular, on their ductility. In the case of ductile under-reinforced beams under increasing overload, yielding of the tendon and the steel first occurs in the maximum moment regions and this results in an increase in the rate of deflection with increasing load.

Progressive yielding then occurs in other high moment regions with rapidly increasing deflections for modest increases in load. For an under-reinforced section, the yield moment is usually not much lower than the moment capacity so that large deformations have to occur (but with only small increments in moment) to bring the section from yield to failure. For a typical continuous beam at high overload, large deformations occur in the yielded peak moment regions and these result in large additional mid-span deflections, but only marginal increases in load in the final stage preceding failure. There is also a progressive redistribution of the moments in a continuous beam with increasing load so that, as the load capacity is reached, the moment capacity of each critical section is approached.

In the case of prestressed beams with reinforcing steel and prestressing steel placed at various levels in the critical cross-sections, the flexural deformations at high overload may not show the ideal elastic-plastic behaviour displayed by an under-reinforced prestressed beam in which all the tensile steel is placed at one level. Nevertheless, the pattern of overload behaviour will include large plastic deformations in peak moment regions and large deflections at mid-span.

Non-flexural and non-ductile modes of failure are also possible. For example, shear failure may occur by diagonal tension splitting failure if the stirrup reinforcement is inadequate, or by crushing of the web concrete if there is too much web reinforcement. If the peak moment sections contain too much tensile steel, then non-ductile failure by concrete crushing will occur before large deflections can develop. The designer has to ensure that such failures will not intervene before the flexural capacity is reached.

11.7.2 Strength design considerations

As already mentioned, AS 3600 allows several alternative methods for flexural strength checks. For continuous beams, the strength check can be based on non-linear analysis. The load capacity of the system calculated by non-linear analysis, P_{ult}, must be at least as large as the factored design ultimate load. If there are alternative load configurations to be considered, then the load capacity of the system must in all cases exceed the factored design ultimate load.

Alternatively, an ultimate strength check may be used. In this approach, elastic analysis is used to evaluate the moments at design ultimate load in critical sections, such as at the mid-spans and at the interior supports. A check is then made to ensure that the moment capacity of each critical section is sufficient to carry the design ultimate moment, calculated from the factored design loads.

The question arises as to what stiffnesses should be used in an elastic analysis of the continuous beam under the design ultimate load combinations. In fact, the choice of stiffness values is not a major issue, provided the structure is ductile. Plasticity theory shows that any internal moment distribution that satisfies the requirement of equilibrium of internal and external forces will lead to a safe design for a ductile structure and, thus, analyses based on the gross uncracked section properties can be used for the strength limit states. The indeterminate structural analyses required for ultimate strength design are therefore much simpler than those needed for deflection calculations.

An ultimate strength design check, based on an elastic structural analysis, ignores the beneficial effects of the moment redistribution that occurs at high overload. AS 3600 therefore allows a reduction to be made to the elastically calculated maximum moment at a critical section (usually in the negative moment region over a support), provided that the other moments are adjusted to ensure that equilibrium requirements are satisfied. The amount of redistribution allowed depends on the ductility of the critical moment sections. The neutral axis parameter k_u is used as an approximate measure of section ductility and provided k_u does not exceed 0.2 in any critical region, a reduction in the peak moment of up to 30 per cent is allowed. If the value of k_u reaches 0.4 in any section, then no redistribution at all is allowed. For intermediate values the maximum percentage redistribution allowed is given by AS 3600 as:

$$\Delta M = 75(0.4 - k_u) \text{ per cent} \tag{11.4}$$

11.7.3 Treatment of secondary effects at ultimate load

AS 3600 states that in design calculations for strength, the secondary bending moments and shears due to prestress shall be included with a load factor of 1.0 when the design moments and shears for load combinations are calculated. In the following discussion it will be seen that this is a quite conservative, but understandable, approach to strength design for continuous prestressed concrete members.

Hyperstatic reactions, secondary moments and secondary shears are strictly only meaningful in the context of elastic behaviour, prior to the application of loads that bring the system into the non-linear behaviour range. In Section 11.3 of this Chapter, we evaluated these quantities using linear elastic analysis and the concept of superposition. At overload, when the behaviour has become non-linear, it is meaningless to try to separate out the various components of the moment at a critical section, i.e. to separate the secondary moment due to prestress from the moments due to dead and live load. Since ductile structures redistribute moments at overload, including the secondary moments induced at the time of prestressing, the secondary moments can and should be ignored in the strength design calculations for ductile continuous beams and slabs (Warner and Faulkes, 1983).

The situation is more complicated for structures where the critical sections have limited ductility. In the extreme situation of a fully brittle structure in which linear behaviour persists up to failure, then the secondary moments and the hyperstatic reactions remain physically meaningful up to failure. In this unrealistic situation, it would be correct to determine them by superposition and to take them into account in the strength design calculations.

However, there is a range of behaviour between a fully brittle structure and the ductile behaviour we assume in design. In real situations, even for an over-reinforced beam, there is significant non-linear behaviour at high overload with considerable moment redistribution. The AS 3600 requirement, that secondary moments be considered in ultimate strength calculations, apparently aims to covers this in-between range of behaviour by treating it as brittle. This approach is quite conservative. For ductile structures, redistribution is allowed such that these secondary effects can be re-balanced to adjacent regions, and may be effectively ignored at the strength limit state.

11.8 Design procedure for continuous beams

The design steps proposed in this section take into account the behaviour previously described. The sequence of steps is a modification of the one already described for statically determinate members in Chapter 10, and has been chosen to reduce the iterations to a reasonable level.

11.8.1 Design steps

(1) Trial section details, self-weight, moment envelopes

Make a first estimate of overall beam depth using available span-depth information and estimate the beam self-weight. Using an elastic analysis with the gross cross-section stiffness of the member, calculate the design moments and design shears for the unfactored design loads. Considering the various possible load combinations, prepare moment envelopes and shear envelopes.

(2) Trial cross-sections

Using the moments from Step 1, identify critical cross-sections along the beam for both negative and positive bending. For each of the critical sections in turn, calculate the required design ultimate moment M^*, choose a value for k_u that will provide adequate ductility and determine appropriate cross-section details by applying Equation 10.2. A small value for k_u is chosen so that sections are ductile and the ϕ factor for bending will be 0.85. If the beam is to have a constant cross-section, then this will be chosen to suit the most critical section. With the section shape and dimensions chosen, check the self-weight against the original estimate and, if necessary, adjust and repeat Steps 1 and 2.

(3) Serviceability criterion

Choose a serviceability design criterion for determining the prestressing details. The main possibilities include:

(a) **deflection control:** choose an appropriate load to be balanced by the prestress and for which there will be zero deflection (see Section 11.8.2 below); or

(b) **crack control:** choose an appropriate load at which decompression will occur at one or more critical sections along the beam; or

(c) **crack prevention:** ensure that cracks do not occur under full service load.

Whichever option is chosen, this step is simply a starting point for choosing the trial prestress details and will be fine-tuned in later steps.

(4) Prestressing force and cable profiles

Treating each span separately, determine trial values for the prestressing force P (after losses) to satisfy the criterion chosen in Step 3. In calculating P for each span, the maximum available cable eccentricities at mid-span and over the supports are used. These are evaluated from the trial section depth and the required cover to the cable and reinforcing steel. Choose the maximum value for P for the critical span and adjust the value of P in the other spans to allow for losses. The profile in the other spans is then adjusted, usually by decreasing the maximum eccentricities to suit the larger value of P. As the losses depend on the eccentricities, this may become a trial and error calculation.

(5) Friction losses

Estimate the friction losses along the member, and hence determine appropriate (different) values of P for each span. Then adjust the cable eccentricities in each span (at mid-span and over the supports) to suit the value of the cable force P and hence adjust the cable profile over the full length of the beam. Re-calculate friction losses and adjust the results from Steps 4 and 5 as necessary. Calculate other losses (Chapter 9), and adjust details in previous steps as necessary.

In friction calculations for a parabolic profile with different eccentricities at each end of a span, the eccentricity and angle may be expressed as follows:

For continuous beams:

$$y = 4h\left(\frac{x}{L}\right)^2 + (e_2 - e_1 - 4h)\frac{x}{L} + e_1 \qquad (11.5)$$

$$\theta = \frac{dy}{dx} = \frac{8hx}{L^2} + \frac{e_2 - e_1 - 4h}{L} \qquad (11.6)$$

where e_1, e_2 and h are shown in Figure 11.6(a).

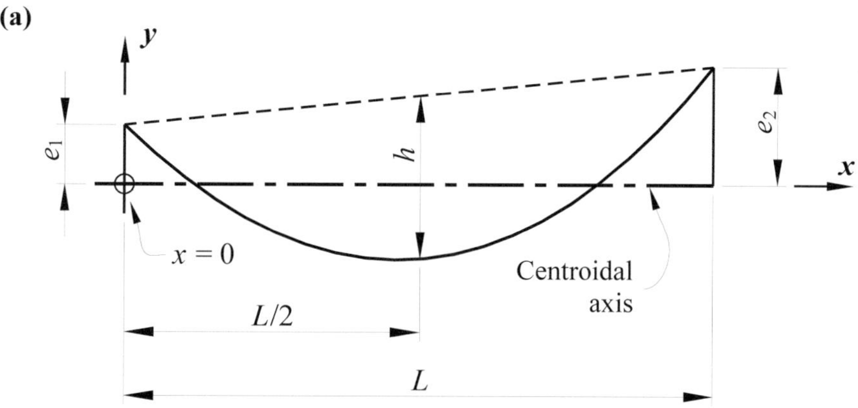

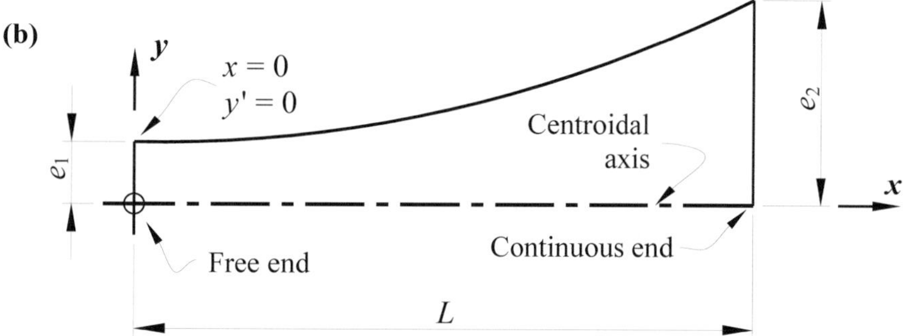

Figure 11.6 Parabolic cable profiles for (a) continuous and (b) cantilever beams

For cantilever beams:

$$y = (e_2 - e_1)\left(\frac{x}{L}\right)^2 + e_1 \tag{11.7}$$

$$\theta = \frac{dy}{dx} = \frac{2(e_2 - e_1)x}{L^2} \tag{11.8}$$

where e_1 and e_2 are shown in Figure 11.6(b).

(6) Hyperstatic reactions

Calculate the hyperstatic reactions, secondary moments and shears due to the prestressing cable (Section 11.2).

(7) Moment capacity

Using the appropriate factored design loads and the analyses from Step 1, evaluate the required moment capacity at critical design sections, and hence determine the appropriate amount of reinforcement needed. Check that ductility is adequate for each section by evaluating k_u. If it is advantageous, moment redistribution of up to 30 per cent may be used to reduce the peak negative moments over the interior supports, provided that the ductility requirements (involving the relevant k_u values) are satisfied. If achieving adequate flexural strength is a problem, try increasing the size of the section and repeat previous steps as necessary. Check for safety at transfer (Section 6.9 of Chapter 6).

(8) Deflections and crack control

Check deflections and crack widths for the relevant service load combinations (Chapter 5). If necessary adjust the details of the prestress and/or the cross-section and repeat previous steps as necessary.

(9) Shear strength

Check for shear strength and design the web reinforcement (see Chapter 7).

(10) End block design

Design the end block reinforcement (see Chapter 8).

(11) Special requirements:

Check for any special design requirements such as torsion, fatigue and vibration.

11.8.2 Discussion of step 3 in Section 11.8.1

The choice between deflection control and crack control in Step 3 of Section 11.8.1 will depend on the nature of the construction and its proposed use. The choice of the loads at which the serviceability check is made will depend on the required performance of the structure, and importantly also on the relative magnitudes of self-weight, imposed dead load and live load.

In the case of deflection control, the load to be balanced will usually be considerably less than the full service load, otherwise excessive camber in the beam spans will result when the live load is absent. In many practical situations the self-weight will be a reasonably large proportion of the total load and balancing the self-weight is then a reasonable first choice. A further point to be considered regarding deflection control is that the spans can be pre-cambered during construction to reduce the impact of the total deflection.

In many situations the most appropriate load to balance is governed by cost considerations rather than directly by serviceability requirements. The first choice of the load to be balanced may well be made on the basis of past experience of similar design situations and then modified, as appropriate, to seek the most economical solution.

When crack control is a serviceability criterion, it should be remembered that cracks start to open in a region as soon as the decompression moment is reached, as cracking will have already occurred during a previous application of the full live load.

In practice the preliminary design will usually be carried out using the load balancing approach because of its simplicity. Once a trial design is established, the prestress level can be fine tuned to ensure that crack control and deflection requirements are met.

The option of no cracking under service load is an unusual design situation for building construction but may be desirable in some bridges and special containment structures due to their longer design service life criteria.

EXAMPLE 11.3

DESIGN OF A 3-SPAN CONTINUOUS BEAM

A preliminary design is required for a post-tensioned continuous beam with three spans as shown in Figure 11.7. The design is to be for a constant I-shaped cross-section. The beam will carry a uniform dead load of 20 kN/m, applied after stressing, in addition to its own weight. A variable live load has a maximum value of 15 kN/m. The concrete to be used has a 28 day strength of $f_c' = 40$ MPa and strength at transfer $f_{cp}' = 30$ MPa.

Other data: $\gamma = 0.87$; $\alpha_2 = 0.79$

SOLUTION

Bending moments (Step 1)

The bending moments for a distributed load w with various loading combinations for each span are calculated assuming linear elastic conditions. They are summarised in Figure 11.7.

The beam self-weight is initially estimated as 10 kN/m. The critical section for the bending moment is at the interior support, where from Figure 11.7(b) the design ultimate bending moment is:

$$M^* = (10 + 20) \times 35.15 \times 1.2 + 15 \times 39.41 \times 1.5 = 2152 \text{ kNm}$$

Trial section (Step 2)

We select $k_u = 0.2$ to ensure good ductility and from Equation 10.2:

$$bd^2 \geq \frac{2152 \times 10^6}{0.85 \times 0.79 \times 40 \times 0.87 \times 0.2 \times \left[1 - \frac{0.87 \times 0.2}{2}\right]} = 504 \times 10^6$$

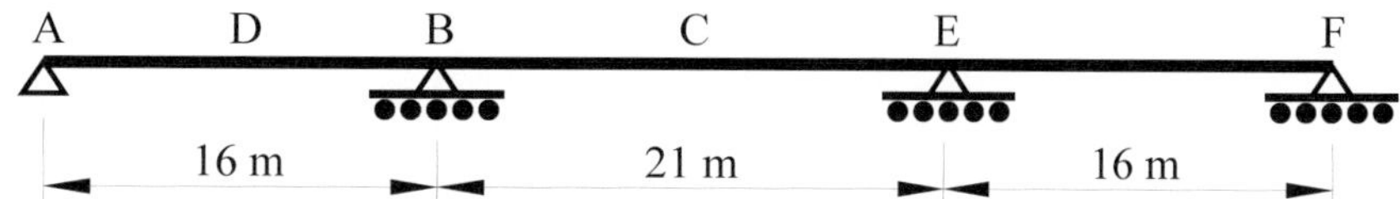

(a) Three-span continuous beam

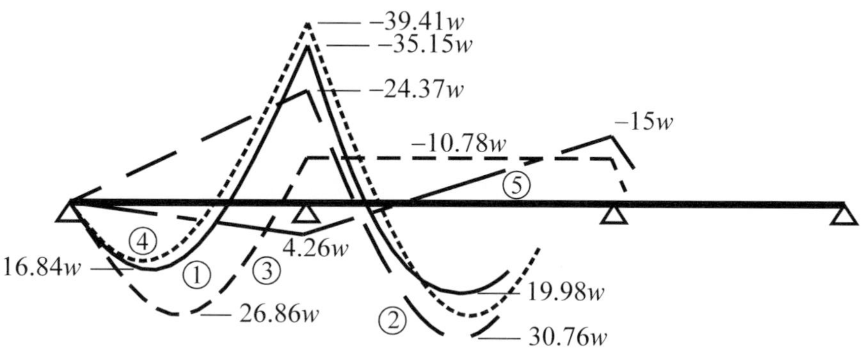

(b) Live load bending moments

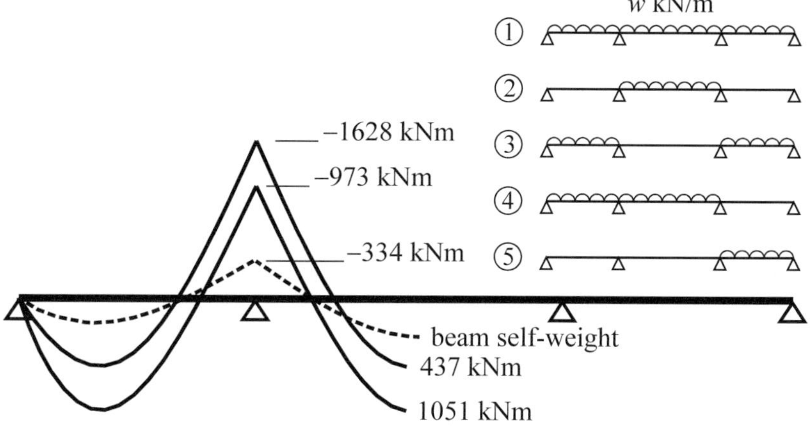

(c) Bending moment envelopes

Figure 11.7 Design of continuous beam

We try a section with:

flange breadth $b = 600$ mm

effective depth $d = 975$ mm

overall depth $D = 1100$ mm

The flange thickness t required to accommodate the concrete compressive force at M_u is estimated from Equation 10.11:

$$t = \frac{M^*}{0.72 f_c' bd} = \frac{2152 \times 10^6}{0.72 \times 40 \times 600 \times 975} = 128 \text{ mm}$$

We try a flange thickness tapering from 150 mm at the edge to 250 mm at the web.

The maximum shear force is approximately $V^* = 605$ kN. The lower limit on nett web width is obtained from Equation 10.8:

$$b_v = \frac{7.2 V^*}{f_c' d} = \frac{7.2 \times 605 \times 10^3}{40 \times 975} = 112 \text{ mm}$$

If a cable with a duct diameter of approximately 90 mm is used, the minimum web width required (Equation 10.9) is $112 + 0.5 \times 90 = 157$ mm, which, as explained earlier, is a low estimate. We selected $b_w = 200$ mm and try the section shown in Figure 11.8.

The self-weight of the section is $w_{sw} = 9.5$ kN/m and the revised design bending moment is $M^* = 2131$ kNm.

Prestressing cable details (Steps 3 and 4)

The total sustained load is 29.5 kN/m. As a first trial, we design for the effective prestress to balance 18 kN/m. The minimum distance from the top or bottom surface of the beam to the centre of the tendon is estimated as 130 mm, giving a maximum available eccentricity of 420 mm.

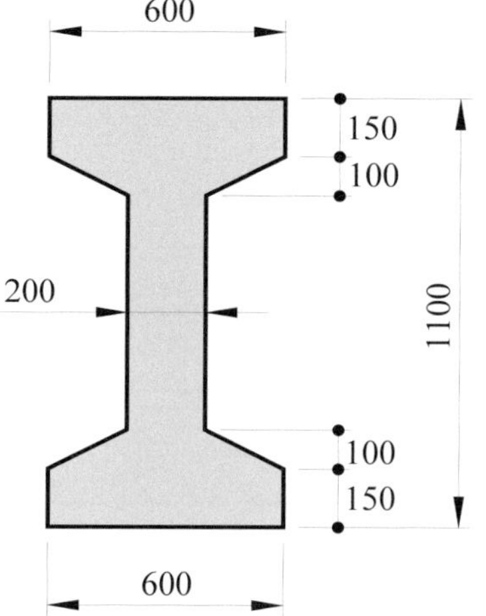

Section properties:

$A_g = 380 \times 10^3 \text{ mm}^2$

$I_g = 54.9 \times 10^9 \text{mm}^4$

$Z = 99.8 \times 10^6 \text{mm}^3$

Figure 11.8 Beam cross-section

Span BE: The maximum available cable drape is $2 \times 420 = 840$ mm. The effective prestressing force to balance a uniform load of 18 kN/m is calculated from Equation 10.3:

$$P_e = \frac{wL^2}{8h} = \frac{18 \times 21^2}{8 \times 0.84} = 1181 \text{ kN}$$

Span AB: Friction losses in the continuous beam will be significant. For load balancing calculations, we assume that the cable force in each span is constant at its mid-span value. In order to establish a preliminary cable profile to be used in the calculation of friction losses, we first assume that the friction loss between D and C is 15 per cent and, hence, determine the required cable drape in span AB:

$$P_e = 1.15 \times 1181 = 1358 \text{ kN}$$

The drape required to balance 18 kN/m is then:

$$h = \frac{wL^2}{8P_e} = \frac{18 \times 16^2}{8 \times 1358} = 0.424 \text{ metres}$$

and the required eccentricity at mid-span is:

$$e_D = 424 - 420/2 = 214 \text{ mm, say } 215 \text{ mm}$$

Friction losses (Step 5)

The idealised cable profile shown in Figure 11.9 is used for the calculation of friction losses.

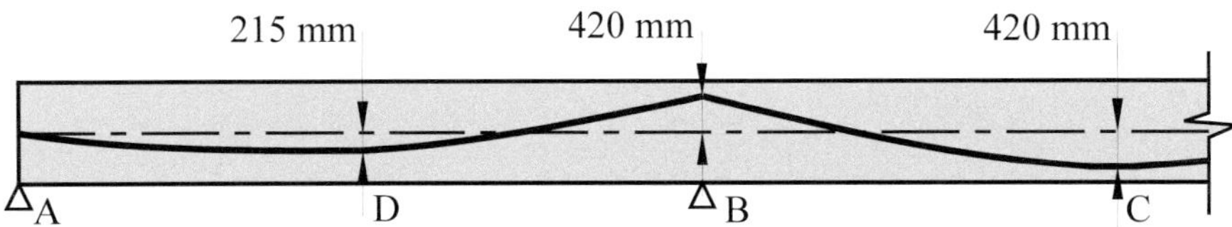

Figure 11.9 Idealised cable profile

The slopes at the ends and mid-span points of the parabolic profiles are determined from Equation 11.6 as:

At A:	– 0.080 rad
At D:	+ 0.026 rad
At B (LHS):	+ 0.132 rad
At B (RHS):	– 0.160 rad
At C:	zero

Friction losses are calculated from Equation 9.4 with the coefficients taken as $\mu = 0.20$, $\beta_p = 0.016$:

$$\frac{\sigma_{pa}}{\sigma_{pj}} = e^{-\mu(\alpha_{tot} + \beta_p L_{pa})}$$

It is assumed that the cable is to be jacked from each end and the results of friction calculations are shown in Table 11.1. These calculations are approximate since they are based on an idealised cable profile, but they are adequate for preliminary design.

TABLE 11.1 Friction loss calculations

POINT	Total angular deviation α_{tot}	Distance L_{pa}	σ_{pa}/σ_{pj}
A	0	0	1.00
D	0.106	8.0	0.954
B (LHS)	0.212	16.0	0.910
B (RHS)	0.504	16.0	0.859
B (average)			0.885
C	0.664	26.5	0.804

The prestress values at critical points are:

At A: $P_e = 1181/0.804 = 1469 \text{ kN};$ $P_i = 1836 \text{ kN}$

At D: $P_e = 1469 \times 0.954 = 1401 \text{ kN};$ $P_i = 1751 \text{ kN}$

At B: $P_e = 1469 \times 0.885 = 1300 \text{ kN};$ $P_i = 1625 \text{ kN}$

At C: $P_e = 1181 \text{ kN};$ $P_i = 1476 \text{ kN}$

If we assume that time-dependent losses total 20 per cent, the required initial prestressing force at A is $1469/0.8 = 1836 \text{ kN}$. If we limit the initial prestress at the jack to 80 per cent of breaking stress, we require a cable with a breaking load of $1836/0.8 = 2295 \text{ kN}$. The breaking load of one 12.7 mm strand is 184 kN (see Table 2.1), so we require a total of $2295/184 = 12.5$ strands.

We shall try a cable with 12–12.7 mm strands. The area of the tendon is:

$$A_p = 12 \times 98.6 = 1183 \text{ mm}^2$$

Although the cable stress at the jack prior to transfer will be 83 per cent of breaking stress, this will be immediately reduced at transfer by wedge pull-in.

Losses due to wedge pull-in (Step 5)

Prior to transfer, the initial prestressing force in span AB diminishes with distance from the anchorage, as calculated from Equation 9.4, in a closely linear manner at a rate of 10.6 kN/m, or 8.92 MPa/m. Assuming a wedge pull-in δ_{ws} of 6 mm, the length of cable affected by pull-in losses is, from Equation 9.9:

$$L_{ws} = \sqrt{\frac{E_p \delta_{ws}}{\alpha_f}} = \sqrt{\frac{200 \times 10^3 \times 6}{8.92 \times 10^{-3}}} = 11.6 \times 10^3 \text{ mm}$$

The loss at the anchorage is $2 \times 8.92 \times 11.6 = 207$ MPa, or 245 kN. After transfer the initial prestressing force in span AB varies as shown in Figure 11.10. The maximum force in the cable is 1713 kN, which is 77 per cent of the cable breaking load of $12 \times 184 = 2208$ kN.

For load balancing purposes we estimate the average force in span AB as the value at mid-span, $P_i = 1675$ kN, with $P_e = 1340$ kN. The latter value is close to our estimate of 1358 kN assumed in determining the trial cable profile.

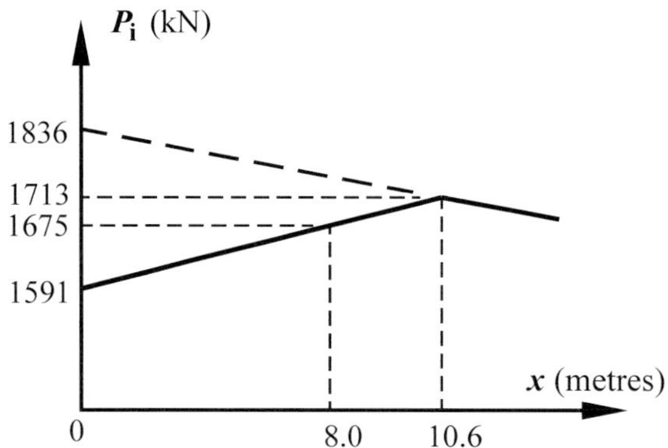

Figure 11.10 Variation of initial prestressing force in span AB

Secondary moments (Step 6)

The effective prestress is designed to produce a uniform upward equivalent load of 18 kN/m in all spans. From Figure 11.7, the moments due to prestress at B are:

Total moment: $M_p = 18 \times 35.15 = 633$ kNm

Primary moment: $M_1 = 1300 \times 0.420 = 546$ kNm

Secondary moment: $M_2 = 633 - 546 = 87$ kNm

The hyperstatic reactions have magnitude $87/16 = 5.44$ kN upward at the exterior supports, and 5.44 kN downward at the interior supports. The secondary moment diagram is shown in Figure 11.11.

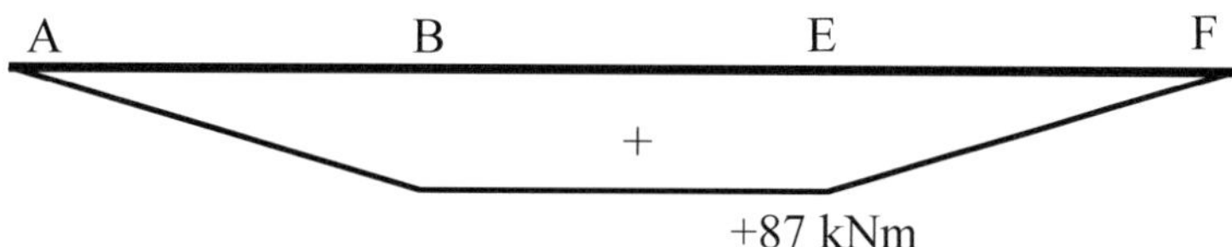

Figure 11.11 Secondary bending moment diagram

Reinforcement required for flexural strength (Step 7)

At the section at support B, the applied self-weight, dead and live loads produce a factored negative moment of 2131 kNm. The secondary moment at B is +87 kNm, which is included for strength calculations with a load factor of unity. The design moment at B is thus:

$$M^* = -2131 + 87 = 2044 \text{ kNm}$$

With $\phi = 0.85$, the required moment capacity at B is:

$$M_u \geq 2044/0.85 = 2405 \text{ kNm}$$

The lever arm distance at M_u is:

$$z = d(1-\gamma k_u/2) = 975(1 - 0.87 \times 0.2/2) = 890 \text{ mm}$$

The required total steel tensile force at M_u is:

$$T_p + T_{sy} = 2405{\times}10^6/890 = 2702{\times}10^3 \text{ N (2702 kN)}$$

Taking $f_{py} = 0.82 f_{pb} = 0.82 \times 1870 = 1533$ MPa, the force carried by the tendon is:

$$T_{py} = 1183 \times 1533 = 1814{\times}10^3 \text{ N} \quad (1814 \text{ kN})$$

and thus the force required to be carried by conventional reinforcement is:

$$T_{sy} = 2702 - 1814 = 888 \text{ kN}$$

$$A_{st} = 888{\times}10^3 / 500 = 1776 \text{ mm}^2$$

We try **4–N24** bars (A_{st} = 1800 mm^2).

The trial section shown in Figure 11.8 is checked for moment capacity with 4–N24 bars placed at a depth of 60 mm from the top of the section and a tendon consisting of 12–12.7 mm diameter strands at a depth of 130 mm from the top of the section (A_p = 1183 mm^2, f_{py} = 1533 MPa, σ_{pe} = 1100 MPa).

An analysis of the section gives M_u = 2501 kNm, which is slightly greater than the demand of 2405 kNm, and thus okay.

We now check the moment capacity of the positive moment sections in spans AB and BC, with each section containing the prestressing cable only (Table 11.2). Span AB is checked at the section of maximum moment, 6.0 metres from the support, where the cable eccentricity is 241 mm.

Check on conditions at cracking (Step 8)

We check conditions at full service load after all losses have occurred. The effective prestress produces a uniform upward equivalent load of 18 kN/m in all spans. The total dead load is 29.5 kN/m, so that the unbalanced portion of the sustained load is $29.5 - 18 = 11.5$ kN/m. The total unbalanced load is therefore a constant load of 11.5 kN/m plus a variable live load of 15 kN/m.

We check first at support B, where the unbalanced negative moment is:

$$M_{ub} = 11.5 \times 35.15 + 15 \times 39.41 = 995 \text{ kNm}$$

TABLE 11.2 Positive moments in spans

Component	Moment (kNm)	
	Span AB	Span BC
Factored dead, live & self-weight moment	1200	1399
Secondary moment M_2	33	87
Design moment M^*	1233	1486
Ultimate moment M_u	1541	1934
Effective capacity ϕM_u	1310	1644

The extreme fibre stresses due to M_{ub} are:

$$\sigma = M_{ub}/Z = \pm 995\times10^6/99.8\times10^6 = \pm10.0 \text{ MPa}$$

The effective prestress in the tendon at B is 1300 kN and, thus, the compressive stress at the balanced load is:

$$\sigma = P_e/A_g = -1300\times10^3/380\times10^3 = -3.4 \text{ MPa}$$

The extreme fibre tensile stress at the top of the section is thus:

$$\sigma_{top} = 10.0 - 3.4 = +6.6 \text{ MPa}$$

The tensile stress exceeds the flexural tension limit of $0.6\sqrt{40} = 3.8$ MPa, so a cracked section analysis is needed to check that the crack control criterion is satisfied. The criterion adopted is that the increment of steel stress between decompression and full service load does not exceed 200 MPa.

The full service load moment at B is:

$$M = -(35.15 \times 29.5 + 39.41 \times 15) = -1628 \text{ kNm}$$

The secondary moment is M_2 = +87 kNm and thus the nett service load moment is:

$$M_s = -1628 + 87 = -1541 \text{ kNm}$$

An elastic cracked section analysis with P_e = 1300 kN, shows that for a moment of –1541 kNm the tensile stress in the top reinforcing bars is 175 MPa. This is within the limit of 200 MPa, indicating that crack control is OK.

Table 11.3 summarises the results of analyses for the positive moments in spans AB and BC. The table shows that the positive moment regions of both spans remain uncracked at full service load.

Check conditions at transfer (Step 7)

At transfer the cable exerts an upward equivalent load of 18/0.8 = 22.5 kN/m. The only other load acting is the self-weight of 9.5 kN/m, so that the unbalanced load is an upward load of 22.5 – 9.5 = 13 kN/m in all spans.

At support section B, the unbalanced load produces a positive moment of $35.15 \times 13 = 457$ kNm. The initial prestressing force at B is 1625 kN, so that the extreme fibre stresses at transfer are:

TABLE 11.3 Conditions in positive moment sections

Forces, Moments and Stresses	Span AB	Span BC
Effective prestress P_e (kN)	1401	1181
Max unbalanced positive moment M_{ub} (kNm)	597	691
Extreme fibre tensile stress due to M_{ub} (MPa)	+6.0	+6.9
Compressive stress P_e/A_g (MPa)	–3.7	–3.1
Nett extreme fibre tensile stress (MPa)	+2.3	+3.8

Bottom: $\sigma_b = \dfrac{1625 \times 10^3}{380 \times 10^3} - \dfrac{457 \times 10^6}{99.8 \times 10^6}$

$= 4.3 - 4.6 = -0.3$ MPa (tension)

Top: $\sigma_a = 4.3 + 4.6 = 8.9$ MPa (compression)

The section remains uncracked at transfer, and as this is the most critical section it is clear that no flexural cracking occurs anywhere in the continuous beam at transfer. Also, the low value of the extreme fibre compressive stress indicates that there is no problem with strength at transfer.

Discussion

This example illustrates an approach to the preliminary design of continuous prestressed beams based on load balancing. The level of prestress was chosen to balance about 60 per cent of the total sustained load. This was little more than a guess of what would produce an economic design, and was based on previous experience. Alternative approaches would include balancing a larger or smaller proportion of the sustained load and using more or less reinforcement, or choosing the prestress so that decompression occurs under the full sustained load.

Irrespective of the approach adopted, the preliminary design should deliver a feasible but not necessarily optimal design. With suitable software available, it is a simple matter to fine-tune the design by trying different values for the key variables and comparing the alternative designs in terms of economy, cracking levels, and deflections at transfer and at full load.

11.9 References

Warner, R.F. and Faulkes, K.A., 1983, Overload behaviour and design of continuous prestressed concrete beams, *International Symposium on Nonlinearity and Continuity in Prestressed Concrete*, University of Waterloo, Ontario.

fib Bulletin 45, 2008, *Practitioners' Guide to Finite Element Modelling of Reinforced Concrete Structures*, (Eds, Maekawa, K., Vecchio, F., and Foster, S.), Fédération Internationale du Béton, 337 pp.

CHAPTER 12

Slab systems

This chapter describes typical prestressed slab floor systems that are used in Australia. The methods of analysis and design used for prestressed concrete slabs are explained, with emphasis placed on the load balancing method.

12.1 Introduction

Prestressed slab systems are widely used as floors in structures such as office blocks, apartment buildings, car parking stations, warehouses, shopping centres and schools. The systems most commonly used include the flat slab floor, in which the slabs span in two directions and are supported by columns only, usually on a roughly square grid; and band-beam construction, in which wide shallow beams carry the loads to the columns, usually in one direction. These systems are shown in Figure 12.1.

The advantages of flat slab construction include simple and economical formwork, minimum floor-to-floor height and an unobstructed soffit that allows planning flexibility while facilitating the installation of air-conditioning ducts and other services. Prestressing allows the designer to control slab deflections and thereby makes possible the construction of larger span floors and more slender slabs. It also inhibits cracking at service loads, improves punching shear strength, simplifies steel layout and fixing and may permit reduced formwork stripping times. Prestressed construction may be competitive for flat slab spans as small as seven metres, and becomes increasingly attractive for larger spans.

Band-beam construction is widely used for structures such as car parking stations, which normally require much longer spans in one direction than in the other. The system most commonly chosen uses wide shallow prestressed continuous beams for the long spans, with either one-way prestressed or reinforced concrete slabs in the short direction.

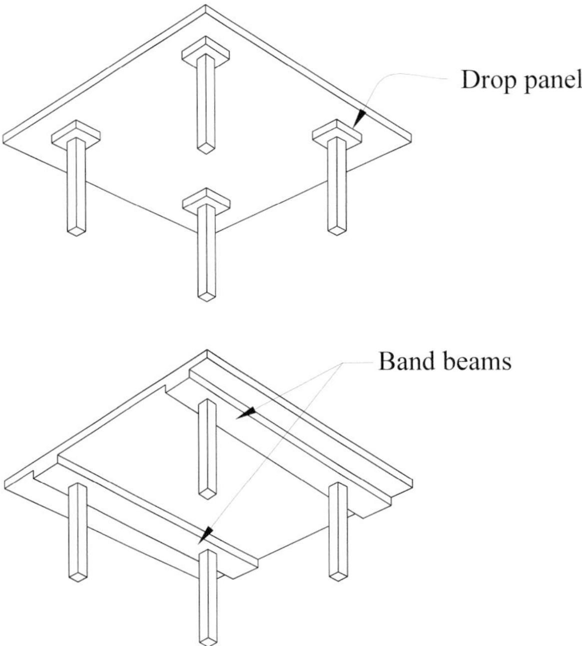

Figure 12.1 Flat slab and band beam floors

In the United States and Europe, prestressed slabs are often constructed with unbonded tendons, each consisting of one heavily greased 12.7 mm or 15.2 mm strand encased in plastic. This system allows shallow slab depths, reduces friction losses and eliminates the grouting operation. However, there are significant disadvantages associated with the use of unbonded construction. These include reduced flexural moment capacity, the need to provide additional reinforcement for crack control, future demolition problems, and the safety questions that arise with a system in which the failure of a tendon at any point – due to corrosion, fire, anchorage failure or any other cause – leads to loss of the entire tendon. Australian Standards do not permit the use of unbonded construction for suspended slabs.

Bonded prestressed slab construction in Australia is made economical by the availability of specially developed flat anchorages and ducts. The ducts are approximately 75 mm wide and 19 mm deep and hold either four or five 12.7 mm or 15.2 mm strands side by side. The anchorages have a maximum width of 230 mm and depth of 90 mm. This system allows grouted cables to be accommodated within a shallow slab and is used for almost all prestressed slab construction in Australia.

12.2 Effects of prestress

The treatment of the effect of prestress on continuous beams in the previous chapter applies equally to continuous slab systems. However, there are additional factors to be considered that arise from (i) the two-dimensional nature of slab systems, and (ii) the effect of changes in slab depth, e.g. those occurring at drop panels or at band beams.

It is necessary to consider separately the two types of forces that prestressing tendons exert on slab floors:

- the forces applied at the anchorages that cause axial compression within the slab, but may also give rise to bending moments if the anchorages are eccentric to the centroid of the slab, or where changes in thickness occur as at drop panels or band beams; and
- the transverse forces exerted wherever curvature exists along the cable that result in bending moments and shear forces in the slab.

12.2.1 Stress dispersion from anchorages

In analysing any prestressed slab system, the designer needs to be aware that there are edge areas of slab between anchorages that are unprestressed. This is shown in Figure 12.2. The dispersion angle α, on either side of the tendon centre line is taken in AS 3600 to be 30 degrees; in Eurocode 2 (2004) an angle of arctan(2/3) = 33.7 degrees is adopted. It is particularly important to recognise, when calculating the punching shear strength at a corner column such as that shown in Figure 12.2, that the area of slab surrounding the column is not prestressed in either direction.

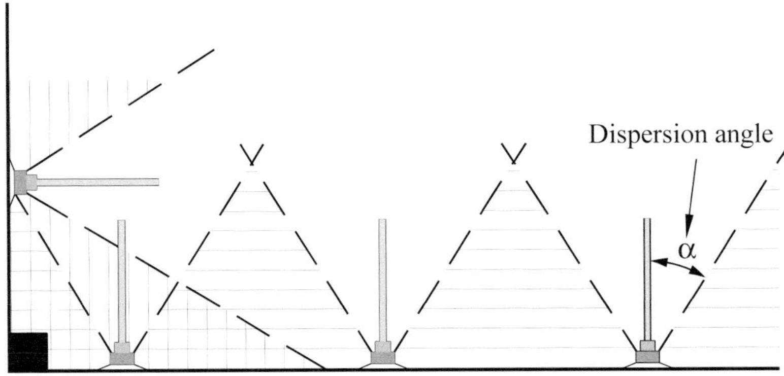

Figure 12.2 Stress dispersion from anchorages

It should also be noted that the stresses arising from the anchorage forces are distributed uniformly across interior sections of the slab, regardless of the spacing of the anchorages, provided that the anchorages are located symmetrically with respect to the floor's longitudinal centreline.

12.2.2 Effects of prestress on slab systems

We now explain the effects of prestressing on each of the following systems in turn:

- one-way slabs, with depth variation;
- two-way slabs supported on walls or beams;
- flat plates;
- flat slabs; and
- band beam floors

One-way slabs

The effects of prestress on one-way slabs, whether simply supported or continuous, may be analysed by considering a strip of slab of unit width to be a beam and using the procedures outlined in previous chapters. Obviously the prestress has no effect in the transverse direction that will need to be provided with temperature and shrinkage reinforcement.

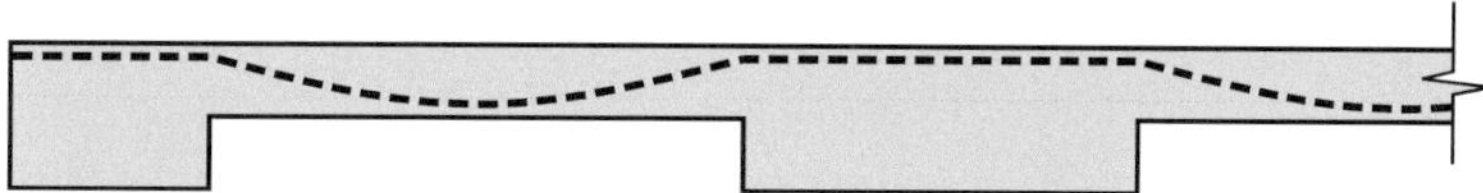

Figure 12.3 Slab spanning across band beams

To determine the moments and stresses due to prestress we use the equivalent loads, both those caused by cable curvature and those due to the forces applied at the anchorages. Any variation in the slab depth affects these moments. In Figure 12.3 the continuous slab is supported on wide shallow beams in the transverse direction. The slab thickness, and hence its stiffness, is much greater in these regions.

The stresses arising from the prestressing forces applied at the anchorages (often referred to as '*P*/*A*' stresses) are affected by the change in eccentricity that occurs at the beams. As shown in Figure 12.4, this has the effect of introducing a couple of magnitude Pe_{d} on each side of the beam, where e_{d} is the change in centroidal axis depth that occurs at the change in section. The effects of these couples should be analysed and the resulting bending moments included when designing the slab.

A typical bending moment diagram, obtained from analysing the floor system for the effect of the *P*/*A* moments induced by the changes in slab cross-section at the beams, is shown in Figure 12.5. These moments are usually small in the slab compared with moments due to transverse loads, but they are not negligible. The moments in the sections across the supporting beams are positive and normally help to counteract moments due to the imposed loads.

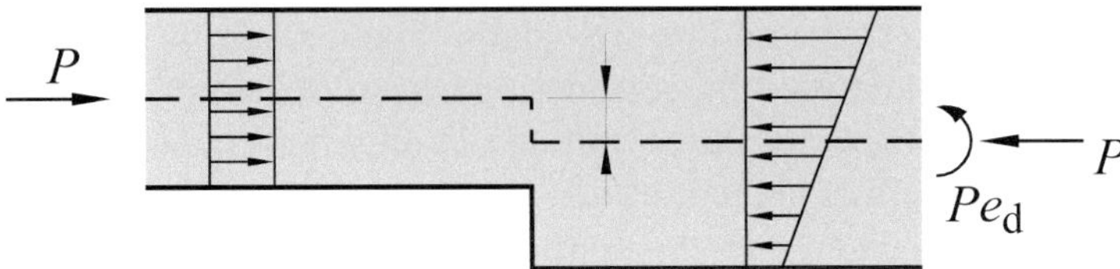

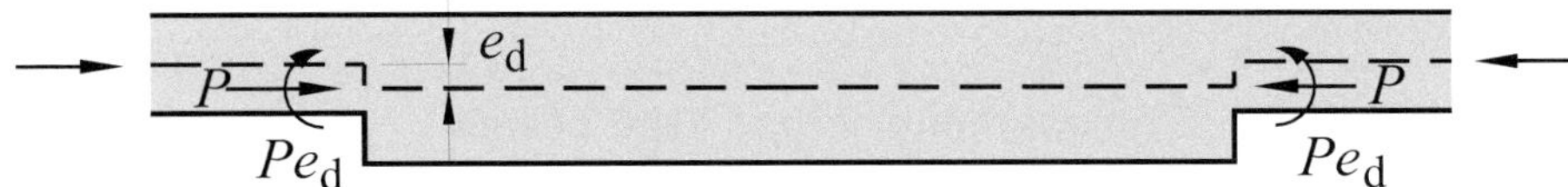

Figure 12.4 Effect of step-change in eccentricity

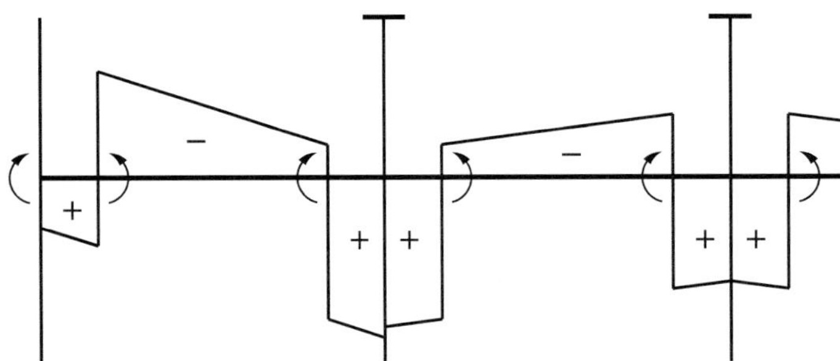

Figure 12.5 Bending moments due to slab depth changes

Two-way slabs supported on edges

The single-panel rectangular slab shown in Figure 12.6 is simply supported by walls on all four sides. It is prestressed by parabolic tendons in both the x- and y-directions, with the x- and y-directions selected such that $L_x \leq L_y$. All tendons are anchored at their ends at the mid-depth of the slab.

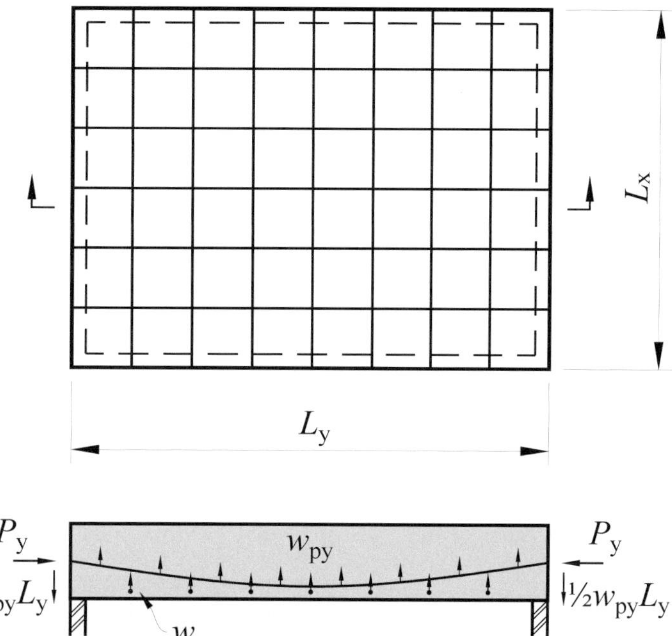

Figure 12.6 Single-panel prestressed slab

The upward loads per unit area exerted on the concrete by the tendons are:

$$w_{px} = 8P_x h_x / L_x^2 \tag{12.1}$$

$$w_{py} = 8P_y h_y / L_y^2 \tag{12.2}$$

where P_x and P_y are the prestressing forces per unit width of slab in the x- and y-directions, respectively, and h_x and h_y are the corresponding cable sags.

The total upward load per unit area on the slab is:

$$w_p = w_{px} + w_{py} \tag{12.3}$$

The moments induced in the slab by prestress are therefore those caused by an upward uniformly distributed load of magnitude w_p. In addition, the slab is subjected to axial compressive forces of magnitude P_x and P_y per unit width in the x- and y-directions, respectively.

We consider next a typical interior panel, as shown in Figure 12.7, of a two-way slab floor that is continuous in both directions and supported by beams or walls at the column lines. The slab is prestressed by parabolic tendons in each direction.

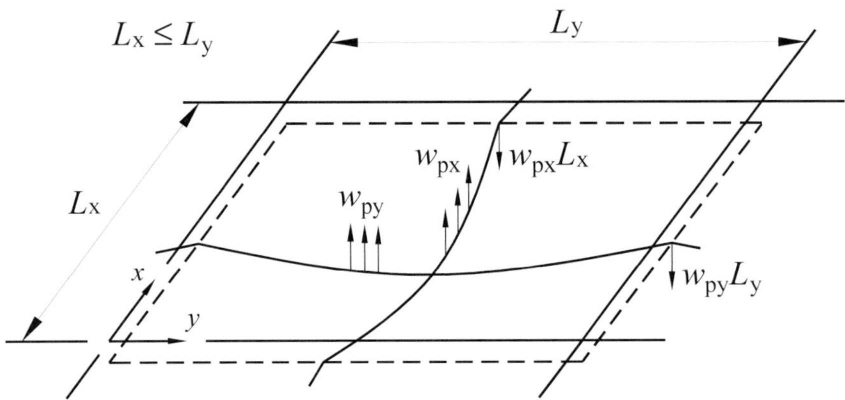

Figure 12.7 Equivalent loads for tendons in two-way slab

The total upward load per unit area due to the curvature of the tendons is given by Equation 12.3. If the objective of design is to balance a uniformly distributed load w kPa, the prestressing forces P_x and P_y kN/m would be chosen to satisfy Equations 12.1, 12.2 and 12.3. Any combination of P_x and P_y that satisfy these three equations would meet the balanced load requirement. The minimum total quantity of prestressing steel would result from placing all the tendons in the short direction, none in the long direction. Although this arrangement of tendons would satisfy the load balancing criterion, other design considerations, such as the need to control shrinkage cracking in the long direction, would make it unsuitable.

The supporting beams or walls of an interior panel must supply line reactions that are equal to the downward forces exerted by the tendons where their curvature is reversed over the supports. These forces per unit length are:

$$W_x = w_{py} L_y \qquad (12.4)$$

$$W_y = w_{px} L_x \qquad (12.5)$$

Flat plates

A flat plate is a two-way slab floor without any drop panels or beams spanning between the columns. The slab panel edge reactions may be considered to be carried to the columns by 'virtual beams' or 'strong bands' contained within the depth of the slab. In the idealised case shown on the left of Figure 12.8, a single large prestressing cable along the y-direction column line provides an upward 'equivalent load' equal to the downward loads imparted by the 'kinks' in the x-direction slab tendons. The magnitudes of the equivalent loads required from the column line cables are given by Equations 12.4 and 12.5.

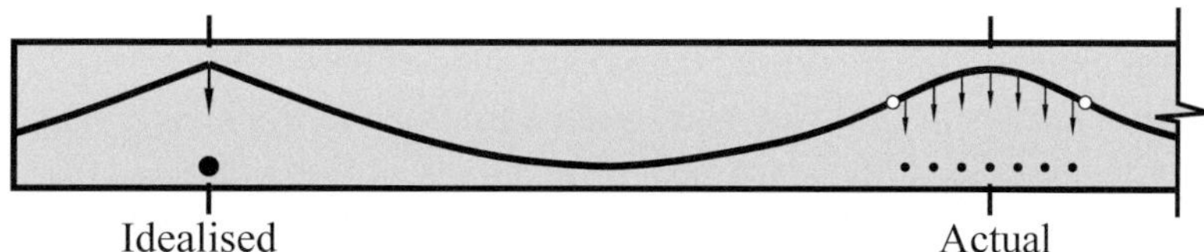

Figure 12.8 Slab cable profiles

Real post-tensioned tendons, of course, are not kinked but have reverse curvature over a small finite length, as shown on the right of Figure 12.8. For 'perfect' load balancing, the downward loads imparted by the slab tendons should be counteracted by column line cables located within the width over which the slab tendons have reverse curvature. In practice, cable duct size and spacing considerations determine the minimum width required to accommodate the column line cables. There is little point in striving to achieve strict load balancing in the placement of prestressing tendons, as considerable variation in spacing can occur without having a noticeable effect on slab behaviour.

Provided that the tendon anchorages are located at the mid-depth of the slab, so that the anchorage forces induce only uniform compressive stresses throughout the floor, the moments due to prestress are caused solely by the equivalent loads resulting from cable curvature. Provided that the cables are parabolic in profile, and that the downward forces associated with reverse cable curvature over the column lines are counteracted by appropriately placed and sized column line cables, the nett equivalent loads on the slab are a series of uniformly distributed upward line loads at the tendon locations. For practical purposes these line loads can be replaced by an appropriate load per unit area for the calculation of bending moments due to prestress. The analysis can be carried out using any of the standard procedures for obtaining bending moments in uniformly loaded flat plates.

Flat slabs

With regard to the moments arising from equivalent loads caused by cable curvature, the previous discussion relating to flat plates applies equally to flat slabs, except that the increased stiffness of the slab in the drop panel regions should be taken into account in the calculations.

When considering the effect of the forces applied at the anchorages, consideration should be given to the moments arising from the eccentricity induced at each drop panel by the increased thickness. This effect is the same as that discussed previously for one-way slabs with variations in thickness. The discussion applies equally to flat slabs with drop panels, which of course experience moments in both directions caused by the drop panel-induced eccentricity.

The procedures for calculating the P/A moments associated with drop panels are covered in detail in Example 12.1.

Band beam floors

Band beam floors are commonly used where floor spans in one direction are much greater than in the other direction. Wide shallow beams are used (Figure 12.3), usually with a total depth of two to three times the slab depth, and a breadth in the range 15 to 25 per cent of the transverse span. The beam may have tapered sides to assist with concrete placement and to provide a more gradual transition from the band beam depth to the slab depth.

If the slab is prestressed (transversely), the forces imparted at the anchorages induce moments at the step from slab to beam, as discussed above. The slab tendons, which are placed after the band beam tendons, usually have constant eccentricity over the width of the band-beams and drape parabolically through the effective span of the slab. This is to maximise the tendon curvature and hence the upward equivalent load. Placement of cables in band beam floors is simpler than in flat slabs, where the cables from the two directions have to be interwoven.

Although the prestressing tendons in the band beam direction are contained within the beams, the prestressing forces imparted at the anchorages disperse throughout the width of the floor so that slabs between the band beams are also prestressed in the transverse direction. The stresses at interior sections should be calculated using the full bay-width cross-section. Calculation of the effects of the equivalent loads induced by cable curvature involves an assumption about the effective flange width to be used for the T-section. This may be taken as the smaller of:

- the effective flange width for beams; and
- the column strip width as if it were a two-way slab.

12.2.3 Effect of restraint from vertical elements

In a large prestressed floor system, the connections between the floor slab and any stiff vertical elements (columns or walls) need careful consideration. The slab must be able to shorten if it is to be effectively prestressed, and also to avoid unacceptable shrinkage cracking. The location and orientation of shear walls, and of any stiff columns, may be critical. It is often necessary to design permanent movement joints to allow shortening to occur. For a detailed discussion of these effects, see Aalami and Barth (1988).

12.3 Effects of prestress plus service load

In the previous section, the effect of prestress alone acting on slab systems was discussed. We now consider the effect of prestress acting together with applied loads in the service load range. Within this range prestressed slab structures are predominantly uncracked and display essentially linear elastic behaviour. Superposition procedures are therefore valid. It is convenient to analyse conditions at a non-balanced service load by considering the load increment between the service load and the balanced load.

12.3.1 The balanced load condition

Load balancing has been discussed in Chapter 4 (Section 4.4) for statically determinate beams and in Chapter 11 for continuous beams. It is almost universally used as the starting point for the design of prestressed slabs, since it offers a simple and powerful means for controlling deflections that normally govern the design of slabs, and it greatly simplifies the analysis in the service load range.

Perfect load balancing would be achieved if the upward forces due to cable curvature exactly counteracted a selected service load; at that load the bending moments, shear forces and deflections would then all be zero. Although perfect balancing is never possible, in most cases it can be approximated to a degree that is adequate for practical purposes.

The two-way slab panel shown in Figure 12.7 would be balanced for a uniformly distributed load w_b equal in magnitude to the total equivalent load w given by Equation 12.3, assuming that the spacing of the prestressing tendons provides loads that reasonably approximate a uniformly distributed load.

For the case of a flat slab, additional tendons are required at the column lines, as discussed in Section 12.2. To highlight the principles of the load balancing approach for flat slabs, we consider the case of the internal panel of the flat slab floor shown in Figure 12.9 with equal spans in each direction. Extension to other cases is straightforward once the principles are understood.

We assume that the slab is to be prestressed to balance a uniformly distributed load w_b kPa. To simplify the discussion we consider that the cables in the mid-panel area are spaced in each direction at one metre intervals, or that the prestressing force P in each direction is expressed in kN per metre width of slab. The cable sag, h, is assumed to be the same in each direction. Each cable exerts an upwards force on the concrete of $8Ph/L^2$ kPa throughout the span, so that the total upwards force from the cables in both directions is $16Ph/L^2$ kPa. Therefore, in order to balance a load of w_b kPa, the required prestressing force per metre width in each direction in the mid-panel area is:

$$P = w_b L^2/(16h) \text{ kN/m} \tag{12.6}$$

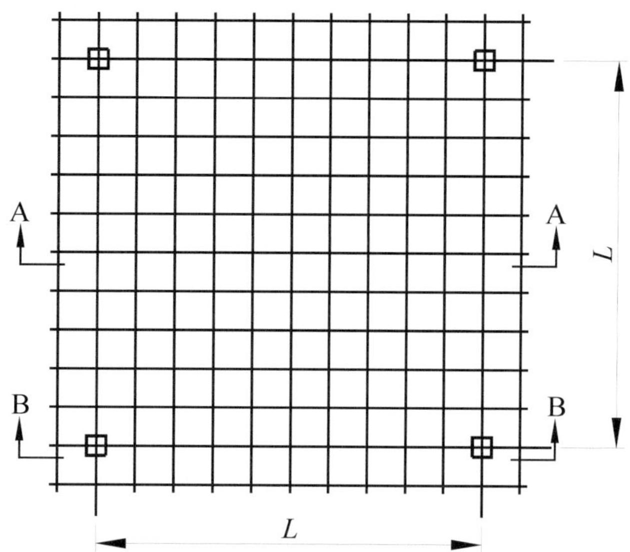

(a) Plan of internal panel

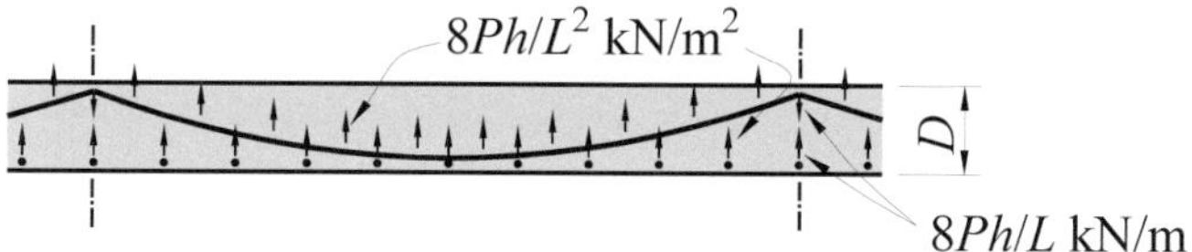

(b) Section A-A typical cable: idealised profile

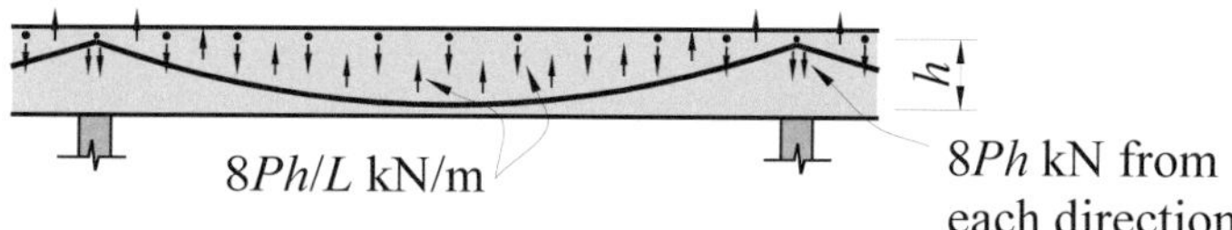

(c) Section B-B column line cable: idealised profile

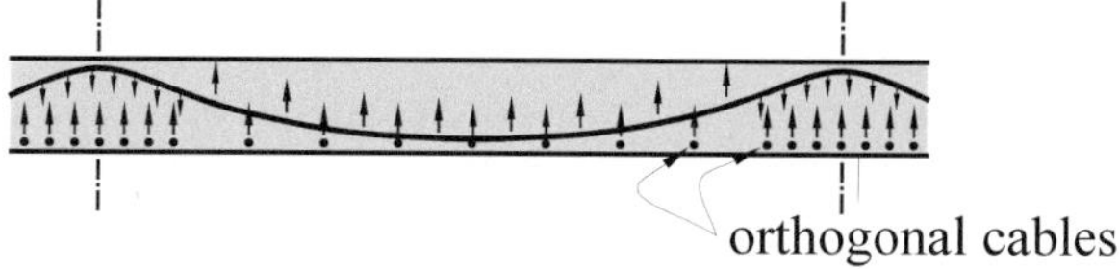

(d) Section A-A real cable profile

Figure 12.9 Flat slab design

At the kinks on the internal column lines (with the idealised tendon profile shown in Figure 12.9(b)) the tendons exert on the concrete a downward force equal to $8Ph/L$ kN/m.

These forces in turn need to be balanced by transverse cables on the column lines, which effectively transfer the load to the columns. The required pre-stressing force in the column line cables is PL kN. The cables also have a sharp change in direction over the columns, where they exert a downward force of $8Ph$ kN. The total downwards force from the two directions at each column is $16Ph = w_b L^2$ kN.

Loads and reactions

As noted above, if a load w_b kPa is to be balanced, the cables in each direction in the mid-panel region may be chosen to supply an upward equivalent load of $0.5w_b$ kPa. At the column lines, these cables exert downward forces of $0.5w_b L$ kN/m that must be counteracted by the column line cables in the orthogonal direction. The total upward load that must be supplied over the area of the panel by the cables in each direction is therefore $0.5w_b L^2 + 0.5w_b L.L = w_b L^2$ kN. This is, of course, a statement of the equilibrium requirement that in any two-way system the total load must be fully accounted for in each direction.

The total prestressing force, P_T, required over the panel width in each direction is $P_T = (w_b L)L^2/(8h) = w_b L^3/(8h)$ kN per panel, each way.

For rectangular panels with spans of L_x and L_y, and cable sags of h_x and h_y, the total prestressing forces required are:

$$x\text{-direction:} \quad P_{Tx} = w_b L_y L_x^2/8h_x \tag{12.7}$$

$$y\text{-direction:} \quad P_{Ty} = w_b L_x L_y^2/8h_y \tag{12.8}$$

In this discussion we have assumed, for convenience, an idealised cable profile as shown in Figures 12.9(b) and 12.9(c). A more realistic cable profile is shown in Figure 12.9(d). For theoretically perfect load balancing, the column line cables should be concentrated over the width where the slab cables have reverse curvature. However, as mentioned earlier, details of duct width and spacing tend to determine the minimum practical width required to place the column-line cables, and considerable variation in cable distribution may occur without having a significant effect on service load behaviour.

From the load balancing point of view, the panel tendons may be distributed between the x- and y-directions arbitrarily, provided that the column line tendons satisfy Equations 12.4 and 12.5. Load balancing may be achieved by either of the tendon arrangements shown in Figure 12.10. However, consideration of slab behaviour at loads other than the balanced load, and especially at high overload, suggests that the tendons in each direction should be concentrated in column-line bands, as shown in Figure 12.10(b).

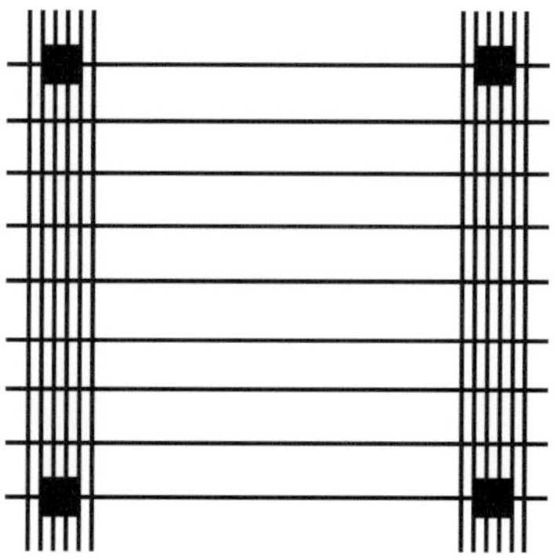

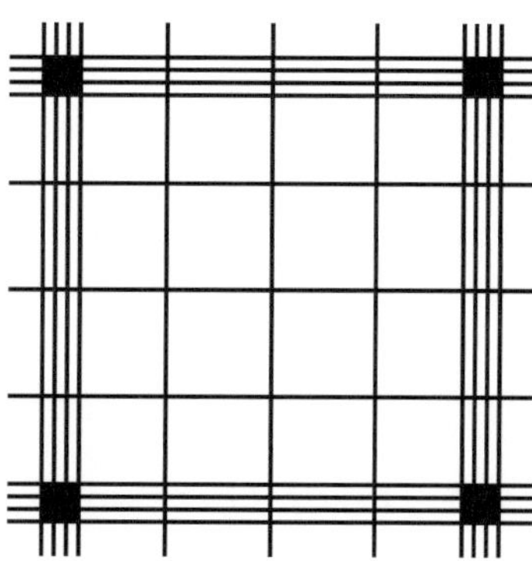

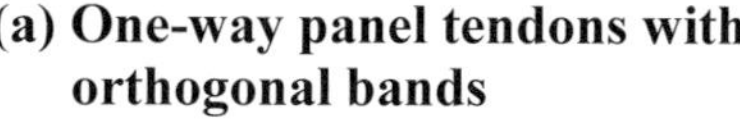

(a) One-way panel tendons with orthogonal bands

(b) Two-way panel banded in x- and y-directions

Figure 12.10 Tendon distribution

12.3.2 Unbalanced service loads

Prestressed slabs are usually designed so that they remain uncracked under service loads. They are assumed to respond to short-term loads in a linear elastic manner, so that superposition methods can be used for analysis.

At the balanced load, a prestressed slab has no deflection and is subjected only to the stresses resulting from the prestressing forces imparted at the anchorages. The deflections and the stresses at any other load within the service range may, therefore, be obtained by calculating the effects of the unbalanced portion of the load and adding these effects to the conditions at the balanced load.

Conditions are normally checked at the two service load extremes of transfer and full live load. At transfer, the unbalanced load (usually upward) is the difference between the equivalent load induced by the initial prestress and the self-weight plus any other loads applied prior to stressing. The possibility of unacceptably high camber, and of cracking in peak moment regions, should be checked. After all of the prestress losses have occurred, and the full live load is applied, the unbalanced (downward) load is the difference between the total dead plus live load and the equivalent load due to the effective prestress. Deflections and cracking need to be checked.

12.4 Cracking

Prestressed slabs will not normally experience significant flexural cracking at service loads, especially when designed by the load balancing method. Nevertheless, two types of cracking occur and both need to be considered by the designer: one is due to flexure, and the other is due to shrinkage and temperature effects. The latter can cause severe cracking if the slab is restrained against movement.

12.4.1 Flexural cracking

In AS 3600–2018, the control of flexural cracking in slabs is treated similarly to that in beams. Cracking is deemed to be under control if the maximum tensile stress in the extreme concrete fibre does not exceed $0.25\sqrt{f_c'}$. If this limit is exceeded, then reinforcement has to be placed near the tensile face with a centre-to-centre spacing of not more than 300 mm or $2.0D$, and a check needs to be made that the stress in the reinforcement does not exceed the values given in Table 12.1 for the crack widths w'_{max}. As an alternative to checking the steel stress increment, a crack width calculation can be made that is similar to the calculation for beams. Crack width calculations will rarely be needed in the design of prestressed slabs and the procedure will not be discussed here. However, it is dealt with in detail for both reinforced concrete beams and slabs in the companion text by Foster et al. (2021).

TABLE 12.1 Maximum stress increment for control of flexural cracks

Bar diameter (mm)	Maximum stress in bars and/or tendons (in MPa)					
	w'_{max} = 0.2 mm		w'_{max} = 0.3 mm		w'_{max} = 0.4 mm	
	$D_s \leq 300$ mm	$D_s > 300$ mm	$D_s \leq 300$ mm	$D_s > 300$ mm	$D_s \leq 300$ mm	$D_s > 300$ mm
10	245	280	300	340	350	390
12	210	240	270	300	310	340
16	180	200	240	260	280	300
20	160	170	220	220	255	260
24	145		195		225	
≥ 28 and bonded tendons	120		185		210	

12.4.2 Stress calculations for flexural crack control

To check on cracking in a slab at transfer and at full service load, the peak moment is determined and the extreme fibre tensile stress in the assumed uncracked concrete is calculated to see whether it is less than either $0.25\sqrt{f'_c}$ or $0.6\sqrt{f'_c}$, as already explained.

For two-way slabs supported on beams or walls, moments due to the unbalanced loads may be calculated using bending moment coefficients derived from elastic analyses.

For flat slabs, the moments due to unbalanced loads may be calculated using the idealised frame method (see Foster et al., 2021). The peak column strip moment is compared with the cracking moment (using the tensile stress limits listed above) to estimate whether significant cracking is likely to occur. It will be realised that such methods give only average moments in the column strip, and these are appreciably smaller than the peak moments that occur at the column. Nevertheless a calculation procedure based on average column strip values is adequate to indicate whether cracking occurs to an extent sufficient to appreciably affect the behaviour of the slab. Regardless of the result of this calculation, it is good practice to provide some conventional reinforcement each way in the top of the slab close to the column to control potential cracking caused by the peak moments that are not captured by the average column strip moment calculations discussed here.

In the uncommon case where a cracked section analysis is required for estimating steel stresses, the following procedure is suggested:

- Use an idealised frame analysis to calculate the maximum negative moment across the width of the panel. Note that this calculation must be based on the total load;
- Include any secondary moments due to prestress, and moments due to increased depth at drop panels;
- Determine the column strip moment by multiplying the panel moment by the appropriate coefficient (usually 0.75 for interior columns);
- Using this moment, and the areas of prestressing and reinforcing steel contained within the column strip, carry out a cracked section analysis (see Chapter 5) to determine the steel stresses.

Clearly this procedure gives only approximate average stresses. However it will usually be found that the calculated steel stresses are well below the limiting values as given in Table 12.1.

12.4.3 Crack control for shrinkage and temperature

Control of cracking in prestressed slabs due to shrinkage and temperature effects is necessary to ensure adequate durability and fire resistance, as well as appearance. Cracking can occur as the result of external loading, as well as shrinkage and temperature change, and is also affected by any restraints that affect free movement of the slab.

The provisions of AS 3600 for prestressed slabs are similar to those for reinforced slabs. Minimum areas of reinforcement are specified for both the primary main direction of bending and the secondary (orthogonal) direction to control shrinkage and temperature cracking, which depend on the degree of restraint against in-plane movement, and the exposure.

For slabs within enclosed buildings, values of the minimum areas of reinforcement are given in Table 12.2 for both the primary and secondary directions and for fully restrained slabs and slabs that are free to move. If the slab is partially restrained in the secondary direction, the minimum required steel area required has to be assessed, taking into account the minimum values for the unrestrained and restrained cases.

In regard to the values in Table 12.2, it should be noted that prestressed flat slabs normally have average prestress levels (prestressing force divided by the section area) of between 1.4 and 2.0 MPa that will usually preclude cracking provided this stress level is actually achieved. However, adjacent structural members such as stiff columns and walls can provide restraint to the slab and reduce the level of prestress actually entering the slab. If the prestress is absorbed by adjacent elements, instead of providing compression in the slab, severe shrinkage cracking may result.

TABLE 12.2 AS 3600 minimum reinforcement for cracking in slabs

Minimum reinforcement in the primary direction

In the direction of the span of a one-way slab, or in each direction in a two-way slab, the area of reinforcement should be not less than:

(i) the amount required for minimum flexural strength p_{min}, and

(ii) 75 per cent of that required in the secondary direction, as appropriate.

Minimum reinforcement in the secondary direction

For slabs unrestrained against in-plane shortening the minimum reinforcement in the secondary direction is $(1.75 - 2.5\sigma_{cp})bD_s \times 10^{-3}$.

For restrained slabs fully enclosed within a building, the area of reinforcement should be not less than:

$(1.75 - 2.5\sigma_{cp})bD_s \times 10^{-3}$ for a minor degree of crack control;

$(3.5 - 2.5\sigma_{cp})bD_s \times 10^{-3}$ for a moderate degree of crack control; or

$(6.0 - 2.5\sigma_{cp})bD_s \times 10^{-3}$ for a strong degree of crack control

Note: The second term in the bracket allows for the beneficial effect of the prestress, and σ_{cp} is the average compressive stress (P_e/A_c) in the concrete due to prestress.

For the prestress to be effective, the slab must be free to shorten. Movement joints may need to be provided at shear walls to allow the slab to move. Consideration needs to be given to the size and stiffness of columns, the arrangement and orientation of shear walls, and the provision of movement joints in the slab. A detailed discussion of these matters is outside the scope of this book, but an excellent treatment is given in Aalami and Barth (1988), and advice can be sought from specialist prestressing companies.

12.5 Deflections

12.5.1 Short-term deflections

Deflection is commonly the governing limit state for concrete slabs and the facility for controlling deflections, which prestressing provides, is a major reason for the use of prestress in slabs. Slab deflections may be approximated by considering an appropriate strip of slab as an equivalent beam and using the provisions for calculating beam deflections, as discussed in Chapters 10 and 11.

Most prestressed slabs are designed to remain essentially uncracked at service loads. At the balanced load, the slab is level and for the calculation of short-term deflections it is necessary to consider only the effect of the unbalanced portion of the service loads. If the design is based on balancing the full sustained load, creep deflections will also be effectively eliminated. Deflections due to concrete shrinkage are also significantly smaller than for reinforced concrete slabs because of the relative absence of cracking and because steel quantities are much smaller. Nevertheless, deflections should be checked both at transfer and at full service load. If the balanced load criterion is used to determine the effective prestress level, an unbalanced upward load will be acting in the period after transfer before the prestress losses have occurred and the possibility of excessive upward deflections, both elastic and creep, should be considered.

The Australian Standard requires that deflections on the column strips of the idealised (equivalent) frame be checked, which is convenient where a frame analysis is being undertaken using computer software. This also allows for easy inclusion of pattern loading, which can be important in prestressed construction where a substantial proportion of the dead load is balanced and the unbalanced loads are predominately transient (live) loads.

A number of methods have been proposed for estimating the mid-panel deflections of flat slab floors. One of the more useful methods is that of Nilson and Walters (1975), which is referred to as the wide beam method. In this method, illustrated in Figure 12.11, the calculated column strip deflection in the x-direction, Δ_{cx}, is added to the middle strip deflection in the y-direction, Δ_{my}, (or vice versa) to obtain an estimate of the mid-panel deflection Δ_{mp}. Thus:

$$\Delta_{mp} = \Delta_{cx} + \Delta_{my} \tag{12.9}$$

or

$$\Delta_{mp} = \Delta_{cy} + \Delta_{mx} \tag{12.10}$$

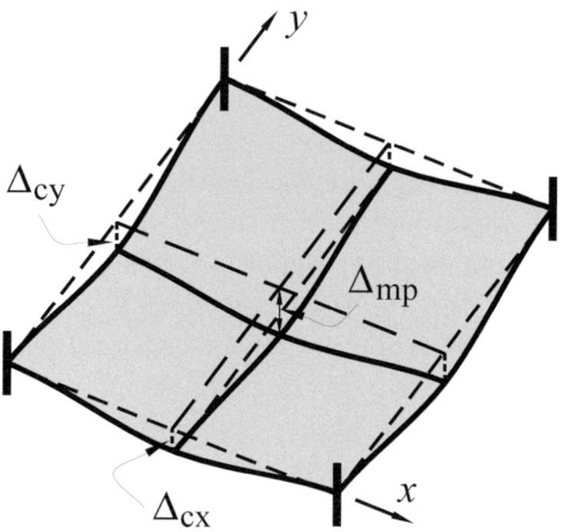

Figure 12.11 Wide beam method for calculation of deflections

Four steps are involved in the calculation:

(a) An idealised model is considered in which the slab panel spanning in the x-direction acts as a wide beam resting on line supports at the y-direction column lines (see Figure 12.11). The mid-span deflection is calculated as the resultant of the downward deflection caused by the panel-wide load acting on a simple span and the upward deflection caused by the negative moments at the column lines.

This calculation gives in effect a mid-span x-direction deflection Δ_{av}, which is the average deflection across the width of the wide beam.

(b) Recognising that the bending moments and curvatures are greater in the column strips than in the middle strips, the averaged deflection Δ_{av} is now adjusted to get estimates of the column strip and middle strip deflections. This is done by multiplying Δ_{av} by the ratio of the curvature of the strip to that of the wide beam. For example, the mid-span deflection of the column strip is calculated as:

$$\Delta_{cs} = \Delta_{av} \times \frac{M_{cs}/(E_c I_{cs})}{M_{wb}/(E_c I_{wb})} = \Delta_{av} \times \frac{M_{cs}}{M_{wb}} \cdot \frac{I_{wb}}{I_{cs}} \qquad (12.11)$$

In Equation 12.11, I_{wb} is the second moment of area of the wide beam, averaged between the value at the support and that at mid-span; I_{cs} is the average second moment of area of the column strip and M_{cs} and M_{wb} are the bending moments in the column strip and the panel-width wide beam, respectively, considering both positive and negative bending moments. For flat plate floors the ratio M_{cs}/M_{wb} is usually taken as 0.7, representing an approximate average of the proportions of total panel moment allocated to the column strip for the positive and negative moment regions. Similarly the ratio M_{ms}/M_{wb} would be taken as 0.3.

(c) Repeat Steps 1 and 2 for the other direction to get estimates of the column strip and middle strip deflections in the y-direction.

(d) Obtain the mid-panel deflection as either the sum of the x-direction column strip deflection and the y-direction middle strip deflection (Equation 12.9) or the sum of the y-direction column strip deflection and the x-direction middle strip deflection (Equation 12.10). Because of the approximations involved in the method, these two alternatives normally yield somewhat different results and the greater is chosen as the deflection estimate.

The method has the advantage for the designer that the values of moments, section properties etc. that are needed for the estimation of deflections are the same as those already calculated for the slab in the determination of bending moments.

The application of this method is illustrated in Example 12.1.

12.5.2 Long-term deflections

In prestressed slabs, most of the sustained load is normally balanced by the equivalent load due to prestress. Creep deflections are therefore small and may be estimated by multiplying the elastic deflection due to the sustained portion of the unbalanced load by the creep coefficient φ^*.

Shrinkage curvatures on the column and middle strips may be determined from one-way beam analyses, for each direction, using the methodologies outlined in Section 4.6 for uncracked members or Section 5.8.3 for cracked members.

12.6 Ultimate strength analysis

12.6.1 Flexural strength

Prestressed slabs must possess adequate flexural strength such that:

$$\phi M_u \geq M^* \tag{12.12}$$

where M_u is the calculated flexural strength and M^* is the design moment. When using either the idealised frame or simplified methods, the design moment on a panel is divided into column and half-middle strips, with each strip checked separately (as demonstrated in Example 12.1).

The full self-weight, dead and live loads must be used in calculating the design ultimate moments in Equation 12.12, since superposition techniques such as load balancing have no validity beyond the elastic stage of behaviour. The proportion of steel in slabs is normally low so that ductile failure is assured, and secondary moments may be neglected in strength calculations. If the strength at critical sections is found to be deficient, the required additional moment capacity may be provided by adding conventional reinforcing bars.

The design may also be checked by using the yield-line method to calculate the strength of the slab, in which case Equation 12.12 is applied over the yield line being considered.

12.6.2 Shear strength

Two modes of shear failure in slabs need to be considered. The first is referred to as 'wide beam' failure and involves diagonal cracks extending across the entire width of the slab. This mode may be checked by considering a unit strip of the slab as a beam and using the procedures discussed in Chapter 7. Wide beam shear is rarely critical for slabs.

The second failure mode is applicable to flat plates and flat slabs and is referred to as 'punching shear' failure. In this mode, diagonal cracks extend from the junction of the column and the slab soffit around all four sides of the joint. The inverted truncated pyramid so formed separates from the surrounding slab and 'punches' through the floor. For reinforced concrete flat slabs the inclined surface of the pyramid makes an angle of approximately 45° with the top surface of the slab. For prestressed slabs the effect of the axial prestressing tends to reduce the angle, giving a somewhat flatter pyramid.

For design purposes, the punching shear strength of prestressed slabs is calculated in a similar manner to that used for reinforced concrete slabs. In the absence of significant moment transfer, the shear resistance is assumed to be governed by an equation of the form:

$$V_u = \tau u d \tag{12.13}$$

where d is the effective depth to the tension steel; u is the perimeter of a nominal critical shear zone which is essentially parallel to the column perimeter but a distance $d/2$ outside it; and τ is a nominal ultimate shear stress which for prestressed slabs is given in AS 3600 as:

$$\tau = f_{cv} + 0.3\sigma_{cp} \tag{12.14}$$

In this equation f_{cv} is the nominal shear strength attributed to reinforced concrete slabs and σ_{cp} is the average intensity of effective prestress in the concrete. The second term in Equation 12.14 is an empirical allowance for the beneficial effect of prestressing on shear strength. The value of σ_{cp} needs to be evaluated separately for corner, edge and internal columns. Except for the addition of this term, the design of prestressed slabs for shear, with or without moment transfer effects, is treated in the same manner as the design of reinforced concrete slabs, as outlined in the companion text of Foster et al. (2021).

Tests on prestressed flat plates (e.g. Gerber and Burns, 1971) have indicated that tendons located in the immediate vicinity of columns contribute more to shear capacity than do tendons remote from columns. This is indirectly recognised in AS 3600, which requires that at least 25 per cent of the steel required to resist the panel negative moment shall be provided within a strip whose width is equal to the column width plus twice the depth of the slab or drop panel.

In considering the shear strength at edge or corner columns, it is particularly important that the designer consider the location of the slab anchorages in relation to the column positions and the areas of the slab near the edges that do not receive any effective prestress. For example, the critical shear perimeter around the corner column in Figure 12.2 receives no prestress from either direction and must be designed using reinforced concrete, not prestressed concrete, provisions for shear strength.

12.7 Design steps for prestressed slabs

Design is necessarily an iterative process. The following numbered steps are suggested as a reasonably efficient sequence to follow for design based on load balancing. Some changes in the sequence may be appropriate, depending on the design problem.

The guidelines included here under Step 1 for the initial choice of design parameters, such as the span-to-depth ratio and the load to be balanced, are suggestions to start the design process. When checks are later made for strength, serviceability and other design criteria, modifications will usually have to be made. Successive runs, carried out with the aid of computer programs, will lead to an efficient, and economical design.

1. Selection of slab thickness

For a design based on load balancing, deflections are controlled by the prestress and the slab thickness can be selected on the basis of span-to-depth ratios suggested by previous experience. Extra thickness around the columns may be required to resist punching shear, but this can be provided locally in the form of drop panels or column capitals. Vibration problems may emerge if the slab is too thin. The FIP Recommendations (FIP 1980) suggest that for solid flat plates continuous over two or more spans, the ratio of span to total slab depth should not exceed 42 for floors and 48 for roofs. Similar ratios could be applied as a guide for two-way edge supported slabs. The span-to-depth ratios are applied to the long span for flat slabs and to the short span for two-way edge supported slabs.

Span-to-depth guidelines from recent Australian publications are summarised in Table 12.3 (C&CAA, 2003; Cross, 2007). These L/D ratios assume live loads of 3 to 5 kPa and an applied dead load additional to self-weight of 1 kPa. They apply to continuous spans only; for end spans, the ratios should be reduced by 10 to 15 per cent. The figures for flat slabs assume drop panels of normal proportions, with depths of 50 to 100 per cent greater than the slab depth and with widths equal to one-third of the span.

TABLE 12.3 Guidelines for span to depth ratios

Structural System	L/D ratio
Flat plates	35 - 45
Flat slabs	40 - 50
Beam depth in band beam slab floors	21 - 27

Cross (2007) also suggests that economical spans for flat plates range from seven to nine metres, for flat slabs up to 13 metres and for band beam systems from eight to 15 metres.

2. Selection of load to be balanced

If the floor is required to be flat under long-term loading then the full self-weight plus all or part of any permanent applied dead load should be balanced. A more economical solution may be obtained by balancing for a lower load and using conventional reinforcement in peak moment regions. Precamber is also an option that will allow less than the full sustained load to be balanced. A guide to the amount of load to balance for an economical structure is given in Table 12.4 (adapted from Cross, 2007).

TABLE 12.4 Suggested load to balance.

Occupancy	Superimposed Dead Load (SDL), kPa	Live Load (Q) kPa	Load to Balance kPa
Car parks	Nil	2.5	0.7 SW to 0.85 SW
Shopping centres	0.0 - 2.0	5.0	0.85 SW to 1.0 SW
Residential (check transfer carefully)	2.0 - 4.0	1.5	SW + 0.3 SDL
Office buildings	0.5 - 1.0	3.0	0.8 SW to 0.95 SW
Storage	Nil	2.4 kPa/m height	1.0 SW + 0.2 Q

Note: SW = self-weight.

3. Determination of available cable sag

The covers to tendon centre lines at the bottom of a slab are normally governed by fire protection requirements. Values given in AS 3600 for a 200 mm thick slab vary from 35 mm for a fire resistance period of 90 minutes, to 60 mm for a period of four hours. In negative moment regions, the concrete cover above the tendons is normally governed by corrosion protection rules and values in AS 3600 vary from 20 mm to 45 mm or more, depending on the exposure classification. Once the slab thickness and the cover requirements are known, the maximum available tendon sag (h) can be determined.

4. Prestress requirements

The required prestressing force in each direction is calculated from load-balancing formulae. At this stage a check should be made on the average prestress in each direction. If the prestress level is too low it may be inadequate to control shrinkage cracking; if it is too high it may lead to problems arising from excessive axial shortening due to elastic and creep deformations. Estimates of losses should be included in the calculations.

Prestress levels should normally be in the range 1.4 to 2.4 MPa after losses, with 1.4 to 2.0 MPa being the most common. Levels below 1.4 MPa are considered too low to control shrinkage, while those above 2.4 MPa may be prone to excessive shortening and should be carefully checked.

5. Tendon arrangement

As discussed in Section 12.3, load balancing requirements can be satisfied by a wide variety of tendon arrangements. Both balanced load and unbalanced load stages, including conditions at high overload, are well met by a distribution that places approximately 75 per cent of the tendons in each direction in the column strip, with the remaining 25 per cent in the middle strip. In addition, as mentioned in Section 12.4, AS 3600 requires that 25 per cent of the steel area required for negative moment in a panel must be located in a cross-section centred on the column and of width equal to the column width plus twice the overall depth of the slab or drop panel, if any.

6. Punching shear strength

The punching shear capacity of the slab is checked as indicated in Section 12.6. If the shear capacity is not adequate, drop panels or column capitals can be provided or shear reinforcement in the form of shear studs may be added.

7. Frame analysis for moments

A frame analysis needs to be undertaken to determine the moments due to prestress (which include those due to equivalent loads and those induced by drop panel eccentricity), the moments caused by service loads and the moments to be used for checking flexural moment capacity. The idealised frame analysis method, as detailed in AS 3600, can be used for this purpose.

8. Flexural strength

The moment capacity of critical cross-sections is checked using factored elastic moments as indicated in Section 12.6. For flat slabs this check may be based either on the panel moment applied to the panel-width cross-section, or by considering the column strips and middle strips separately.

9. Serviceability conditions at unbalanced loads

The serviceability conditions of deflection and cracking should be checked both at transfer, when the prestressing forces have their initial value, and after all prestress losses have occurred and the full live load acts. Problems with either deflection or cracking may lead to a decision to adjust the prestress level and/or slab thickness.

10. Anchorage zone reinforcement

Conditions in the anchorage zones are investigated using the procedures discussed in Chapter 8. It is necessary to consider the possibility of cracking occurring both horizontally and vertically through the slab. Figure 12.12(a) shows an edge elevation of a flat plate of depth D with mid-depth anchorages at a spacing S apart. The anchorage is assumed to have an effective bearing plate with width of h_s and depth of h_d. The transition from the concentrated longitudinal compressive stresses at the anchorage plate to uniformly distributed stresses of magnitude $\sigma_o = P/SD$ within the slab gives rise to transverse bursting tensile stresses and potential horizontal cracking behind the anchorage as shown in Figure 12.12(b). Two-dimensional analytical investigations (see Section 8.2) indicate that the peak transverse tensile stress, averaged over the width S, is approximately:

$$\sigma_y = 0.5\sigma_o(1 - h_d/D) \qquad (12.15)$$

Three-dimensional finite element analyses to investigate how the peak transverse tensile stress varies throughout the breadth of a beam cross-section (Yettram and Robbins, 1969), indicated that the true peak stress values are some 60 per cent higher than the averaged values obtained from two-dimensional analyses. Since the S/D ratios typical for slabs are considerably greater than the breadth to depth ratios investigated in the finite element studies, it is recommended that, for slabs, the peak vertical tensile stresses be estimated as:

$$\sigma_{y.max} = \sigma_o(1 - h_d/D) \qquad (12.16)$$

The total vertical tensile force associated with each anchorage is given (from Equation 8.4) as:

$$T = 0.25P(1 - h_d/D) \qquad (12.17)$$

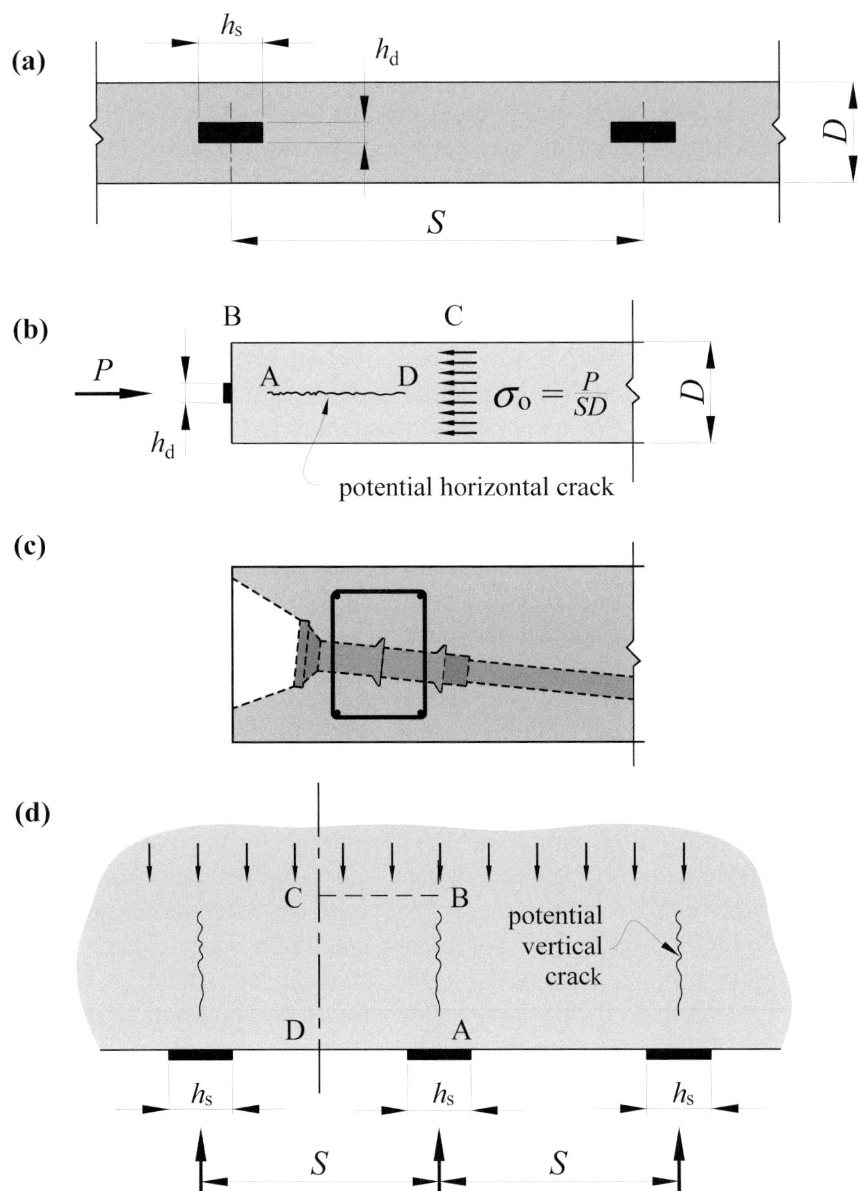

Figure 12.12 Anchorage zone cracking

The necessary reinforcement can be supplied in the form of closed stirrups on each side of the anchorage, as indicated in Figure 12.12(c). In recent years, however, it has become more common practice to instead use rectangular helixes around the anchorage, which are routinely supplied by the prestressing companies. It should be noted that tensile forces are also generated between anchorages and it is good practice to detail longitudinal U-bar reinforcement along the edge of the slab to control these forces and to reinforce the non-prestressed areas between anchorages.

Just as the vertical dispersion of the anchorage stresses gives rise to vertical tensile stresses, so the horizontal dispersion of longitudinal compressive stresses sets up horizontal tensile stresses leading to potential vertical cracking, as shown in Figure 12.12(d). The maximum tensile stress may be estimated from:

$$\sigma_{y.max} = \sigma_o(1 - h_s/S) \tag{12.18}$$

In a slab prestressed in both directions these tensile stresses are reduced and may be nullified by the compressive P/A stresses imparted by prestressing anchorage forces in the orthogonal direction. Particular attention should however be paid to the horizontal tensile stresses near the corners of the slab, where the orthogonal prestress may not be fully established.

When reinforcement is required to control vertical cracking, transverse reinforcing bars may be used in the top and bottom faces of the slab. The distance over which the horizontal tensile stresses exists is from approximately $0.2S$ to $1.0S$ measured from the effective bearing surface of the anchorage.

EXAMPLE 12.1

DESIGN OF A PRESTRESSED CONCRETE FLAT SLAB

A prestressed concrete flat slab floor is be designed. It consists of four 9 metre square bays in each direction, with 2.5 metre cantilevers on all sides. The storey height is 4.0 metres and all columns are 400 mm square in cross-section. A quarter of the slab is shown in Figure 12.13.

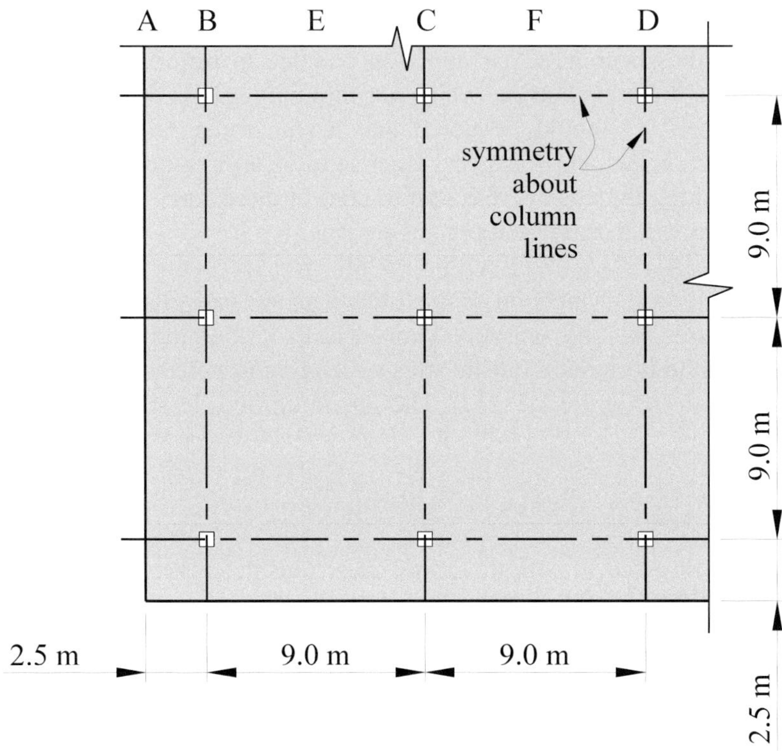

Figure 12.13 Floor plan for Example 12.1

The floor forms part of a non-residential complex and the cover to tendons is 20 mm for durability, and an axis distance of 45 mm for fire. The slab is to be designed for a live load of 4 kPa and a partition load of 0.5 kPa.

Other data: $f_c' = 40$ MPa; $(\gamma = 0.87, \alpha_2 = 0.79)$

$E_c = 32.8\times10^3$ MPa; $\psi_s = 0.7$; $\psi_l = 0.4$

$f_{pb} = 1870$ MPa; $E_p = 195\times10^3$ MPa

$\varphi_o^* = 2.0$; $\varepsilon_{cs}^* = 600\times10^{-6}$

SOLUTION

1. Selection of trial slab depth

To start the design we estimate a slab depth of 210 mm, which corresponds to a span/total depth ratio $L/D = 9000/210 = 43$ and is consistent with experience (see Table 12.3). For a minimum fire resistance period of 2 hours, the Standard requires a minimum slab depth of 200 mm, which is satisfied.

The self-weight and dead loads are thus:

$$\text{Slab self-weight} = 0.21 \times 25.0 = 5.25 \text{ kPa}$$

$$\text{Total dead load} = 5.25 + 0.50 = 5.75 \text{ kPa}$$

2. Selection of load to be balanced

For load balancing, we shall balance (long-term) 4.5 kPa, which corresponds to 86 per cent of the self-weight (see Table 12.4).

3. Tendons and available sag

We choose tendons each comprising four 12.7 mm strands housed in rectangular ducts 70 mm wide by 19 mm high. The maximum tendon sag available depends on cover requirements and positioning so as to avoid a clash in the location of tendons crossing in x- and y-directions. In addition the thickness of concrete outside the 70 mm wide slab ducts must be sufficient to ensure the integrity of the concrete cover. In the placing of the tendons, we will lay those in the x-direction first and in the y-direction second.

The tendons in the y-direction rest on top of those in the x-direction. The sags are therefore the same in both directions and this means that the prestress requirements will also be the same in both directions. In the following, we only consider the x-direction tendons.

Top cover: To ensure the integrity of the concrete above the tendons, a minimum of 20 mm cover above the ducts is needed. With bar chairs produced in 5 mm increments, we select a 150 mm chair. This gives a minimum distance from the top of the slab to the centre of the x-direction strands of $210 - 150 - 12.7/2 = 54$ mm (ignoring the wall thickness of the duct, see Figure 12.14(a)) and a cover above the y-direction ducts of 210 – 150 –2(19) = 22 mm, which is satisfactory.

Bottom cover: For a two-hour fire resistance period, AS 3600 requires a minimum axis distance of 45 mm (and 35 mm for any conventional reinforcement). With a 35 mm bar chair, the height from the slab soffit to the centre of the x-direction strands is 35 + 19 – 12.7/2 = 48 mm (Figure 12.14(b)). The cover for corrosion protection and minimum concrete requirement below the ducts are also satisfied.

The maximum available sag in the x-direction in the interior spans is therefore:

$$h = 210 - 54 - 48 = 108 \text{ mm}$$

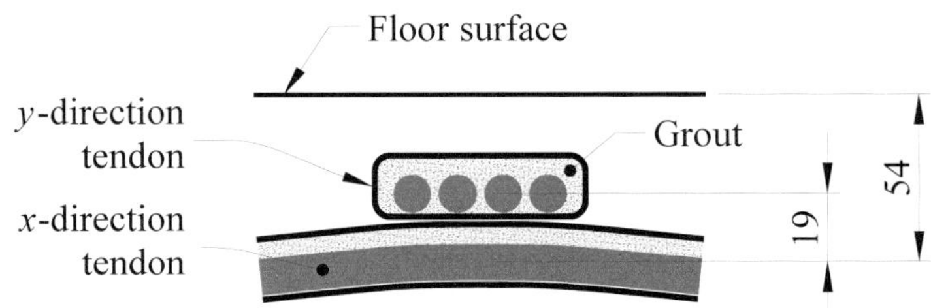

(a) Minimum distance from floor surface for x-direction strands

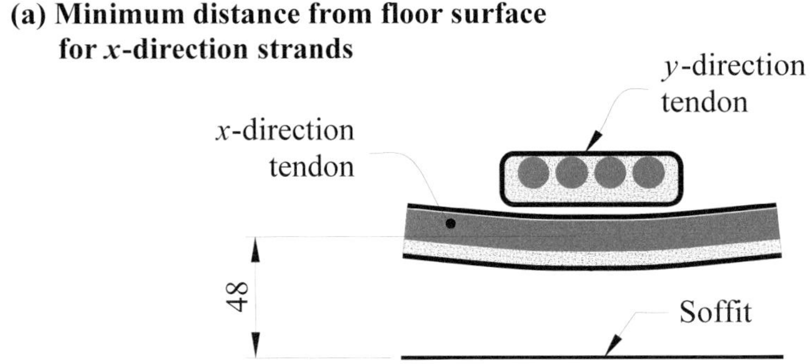

(b) Minimum distance from soffit for x-direction strands

Figure 12.14 Minimum distance from nearest surface to the centre of strands placed in the x-direction

4. Prestress requirements (including loss estimates)

To balance 4.5 kPa at the mid-span of CD (point F in Figure 12.13), the required effective prestressing force $P_{e.F}$ per metre width of slab, using the full sag of 108 mm, is (Equation 12.1):

$$P_{e.F} = \frac{w_b L^2}{8h} = \frac{4.5 \times 9^2}{8 \times 0.108} = 422 \text{ kN/m}$$

Estimating the losses of prestress due to friction between points F and A to be 12 per cent, the jacking force is estimated as:

$$P_j = 422/0.88 = 480 \text{ kN/m}$$

Allowing 15 per cent for wedge draw-in, the effective prestressing force at A is:

$$P_{e.A} = 0.85 \times 480 = 408 \text{ kN/m}$$

The cable sag required in the cantilever span AB to balance 4.5 kPa, is:

$$h_{AB} = \frac{w_b L_c^2}{2P_{AB}} = \frac{4.5 \times 2.5^2}{2P_{AB}} = \frac{14.06}{P_{AB}} \text{ m}$$

Setting P_{AB} equal to $P_{e.A}$, the required sag in the cantilever AB is:

$$h_{AB} = 14.06/408 = 0.034 \text{ m}$$

We shall select a cable sag in span A-B of 34 mm.

The trial cable profile is shown in Figure 12.15.

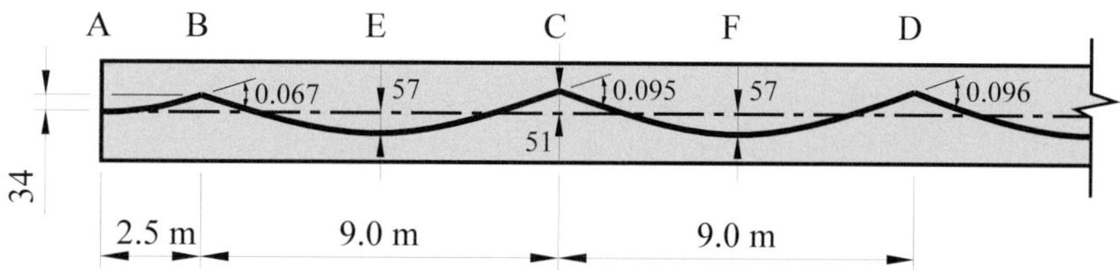

Figure 12.15 Trial cable profile

Friction losses

The cable profiles are given by Equations 11.5 and 11.7 for the continuous and cantilever spans, respectively. The slopes are calculated with Equations 11.6 and 11.8. Friction losses are calculated from Equation 9.4. The values of the friction coefficients are taken as $\mu = 0.20$ and $\beta = 0.016$ and thus:

$$\sigma_{pa}/\sigma_{pj} = e^{-0.2(\alpha_{tot} + 0.016 L_{pa})}$$

Cables are to be jacked from both ends; the friction loss calculations are summarised in Table 12.5.

TABLE 12.5 Friction loss calculations

POINT	A	B		E	C		F	D
		LHS	RHS		LHS	RHS		LHS
Slope θ (rad)	0	0.027	–0.040	0.004	0.047	–0.048	0	0.048
Angle sum α_{tot}	0	0.027	0.094	0.130	0.181	0.276	0.324	0.372
α_{tot} at support		0.067			0.229			0.420
L_{pa} (m)	0	2.5		7.0	11.5		16.0	20.5
σ_{pa}/σ_{pj}	1.0	0.979		0.953	0.921		0.890	0.861

Prestressing force required for each panel

The balanced load for the full panel width of 9 metres of segment AB is:

$$w_p = 4.5 \times 9 = 40.5 \text{ kN/m}$$

For span CD, the sag is 51 +57 = 108 mm. The required effective prestress at the mid-span (F) is:

$$P_{e.F} = \frac{40.5 \times 9^2}{8 \times 0.108} = 3797 \text{ kN}$$

Considering the friction loss (Table 12.5), the forces at the jacking points (A) at each end are:

$$P_{e.A} = 3797/0.890 = 4266 \text{ kN}$$

For span BC, the sag is:

$$h_{BC} = 57 + 0.5(51 + 34) = 99.5 \text{ mm}$$

The required effective prestress at the mid-span (E) is:

$$P_{e.E} = \frac{40.5 \times 9^2}{8 \times 0.0995} = 4121 \text{ kN}$$

and considering friction loss, the force required at the jacking section is:

$$P_{e.A} = 4121/0.953 = 4324 \text{ kN}$$

Thus, span BC governs. The prestressing cables must provide an effective prestressing force at A of 4324 kN. With consideration of the friction losses, the effective prestressing forces across the panel width are given in Table 12.6.

TABLE 12.6 Effective prestress at critical sections

Section	A	B	E	C	F	D
σ_{pa}/σ_{pj}	1.0	0.979	0.953	0.921	0.890	0.861
P_e (kN) per panel	4324	4233	4121	3982	3848	3723

Wedge draw-in losses at transfer

If we estimate deferred losses at 18 per cent, the initial prestressing force at A is:

$$P_{i.A} = 4324/0.82 = 5273 \text{ kN}$$

Limiting the stress in the strands to $0.85 f_{pb}$ at jacking, the minimum number of strands needed is $5273/(0.85 \times 184) = 34$.

Selecting nine tendons per panel width, each consisting of four 12.7 mm strands, the total area of the tendons is:

$$A_p = 9 \times 4 \times 98.6 = 3550 \text{ mm}^2$$

From Table 12.5 the average friction loss calculated between A and C (over a length of 11.5 metres) is:

$$= (1 - 0.921)/11.5 = 0.0069 \text{ per metre } (6.9{\times}10^{-6} \text{ per mı}$$

and the slope of the friction loss line is:

$$m_f = f \times \frac{P_{i.A}}{A_p} = 6.9{\times}10^{-6} \times \frac{5273{\times}10^3}{3550} = 10.2{\times}10^{-3} \text{ MPa/mm}$$

From Equation 9.9, assuming 6 mm wedge slip, the length affected by wedge slip losses is:

$$L_{ws} = \sqrt{\frac{E_p \delta_{ws}}{m_f}} = \sqrt{\frac{195{\times}10^3 \times 6}{10.2 \times 10^{-3}}} = 10.7 \text{ m}$$

which is reasonably close to the 11.5 metres used to determine the average friction loss.

The draw-in loss at A is (Section 9.4):

$$2 m_f L_{ws} = 2 \times 10.4{\times}10^{-3} \times 10.7{\times}10^3 = 223 \text{ MPa}$$

which corresponds to a force of $223 \times 3550 = 792{\times}10^3$ N (792 kN).

The prestressing force at A after transfer is therefore:

$$P_{t.A} = 5273 - 792 = 4481 \text{ kN}$$

This force increases by 792/2/10.7 = 37.0 kN/m until it meets the original prestress line, at x = 10.7 m (Figure 12.16). The prestressing forces across the 9.0 m panel width are summarised in Table 12.7.

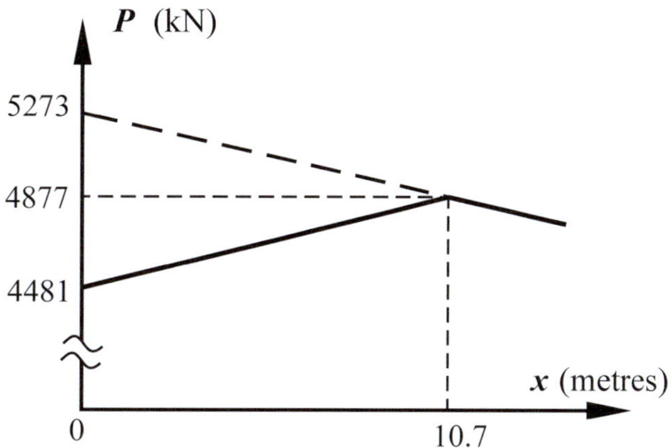

Figure 12.16 Variation of initial prestressing force

TABLE 12.7 Prestress at critical sections

Section	A	B	E	C	F	D
x (m)	0	2.5	7.0	11.5	16.0	20.5
P_j (kN) per panel	5273	5181	5014	4848	4681	4515
P_i (kN) per panel	4481	4574	4740	4848	4681	4515
P_e (kN) per panel	3674	3751	3887	3975	3838	3702

After transfer, the maximum prestress occurs at x = 10.7 metres and is:

$$P_{i.x=10.7} = 4481 + 37.0 \times 10.7 = 4877 \text{ kN}$$

The maximum average prestress on the section is thus:

$$\sigma_{cp} = \frac{0.82 \times 4877 \times 10^3}{9000 \times 210} = 2.12 \text{ MPa}$$

The minimum average prestress on the section is (at A):

$$\sigma_{cp} = \frac{3674 \times 10^3}{9000 \times 210} = 1.94 \text{ MPa}$$

The average prestress lies in the range of 1.9 MPa to 2.1 MPa and is okay (refer Section 12.7 Step 4).

5. Cable arrangement

The maximum force in the strands at jacking is 5273/36 = 146 kN ($0.79f_{pb}$) and at transfer is 4877/36 = 135 kN ($0.73f_{pb}$). These are acceptable. The distribution of the prestressing cables in a typical slab panel is shown in Figure 12.17. Six of the nine tendons per panel are located in the column strip, three in the middle strip.

6. Punching shear strength

Shear capacity at internal column, without moment transfer

The critical shear perimeter, u, around the square columns is defined by a square surrounding the column and distant $d_{om}/2$ from it.

With $d_{om} = 210 - (54 - 19/2) = 166$ mm, the critical shear perimeter is:

$$u = 4 \times (400 + 166) = 2264 \text{ mm}$$

and the concrete shear strength is (AS 3600) $f_{cv} = 0.34\sqrt{f_c'}$.

The factored load is $w^* = 1.2 \times 5.75 + 1.5 \times 4 = 12.9$ kPa and the shear force on the critical perimeter at an interior column is therefore:

$$V^* = 12.9(9.0^2 - 0.566^2) = 1041 \text{ kN}$$

At the internal column at D, the effective prestress is:

$$\sigma_{cp} = \frac{3702 \times 10^3}{9000 \times 210} = 1.96 \text{ MPa}$$

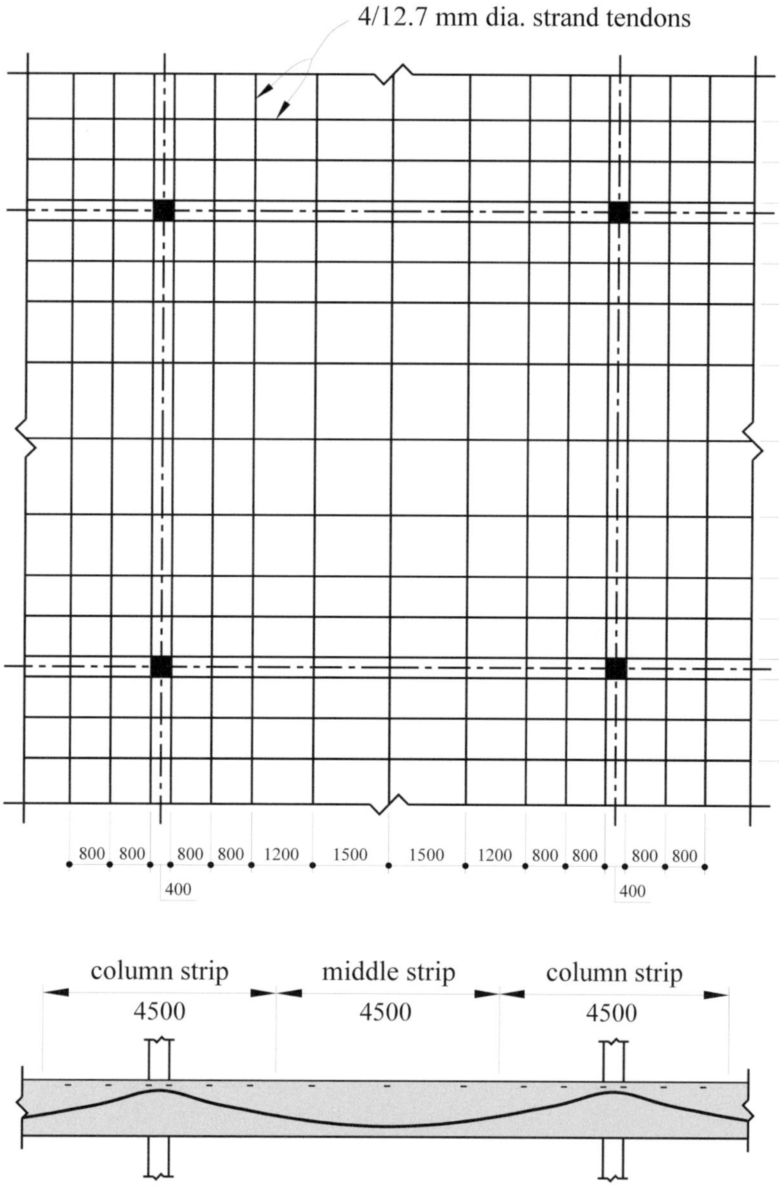

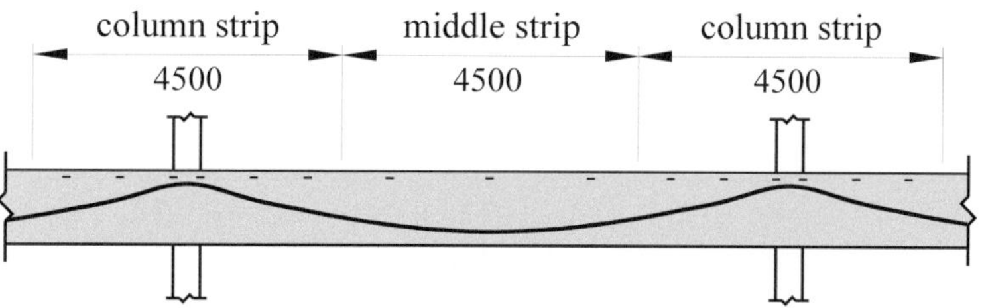

Figure 12.17 Distribution of tendons

From Equations 12.13 and 12.14, the effective shear resistance is:

$$\begin{aligned}\phi V_{uo} &= \phi[ud_{om}(f_{cv} + 0.3\sigma_{cp})] \\ &= 0.75[2264 \times 166(0.34\sqrt{40} + 0.3 \times 1.96)] \\ &= 772 \times 10^3 \text{ N} \quad (772 \text{ kN}) \ < V^* = 1041 \text{ kN}\end{aligned}$$

Since $\phi V_u < V^*$, the 210 mm thick slab does not provide adequate strength against punching shear. We therefore increase the slab thickness around the column by providing a 3000 mm wide drop panel with a trial depth of 110 mm below the slab soffit.

With consideration of the drop panel:

$$d_{om} = 166 + 110 = 276 \text{ mm}$$

$$u = 4 \times (400 + 276) = 2704 \text{ mm}$$

and the design shear at the critical perimeter is:

$$V^* = 12.9(9.0^2 - 0.676^2) = 1039 \text{ kN}$$

The average prestress level in the region of the drop panel is:

$$\sigma_{cp} = 1.96 \times \frac{9000 \times 210}{(9000 \times 210) + (110 \times 3000)} = 1.67 \text{ MPa}$$

and the factored shear capacity is:

$$\begin{aligned}\phi V_{uo} &= 0.75 \times 2704 \times 276(0.34\sqrt{40} + 0.3 \times 1.67) \\ &= 1484 \times 10^3 \text{ N} \ (1484 \text{ kN}) \ \geq 1039 \text{ kN} \ (\therefore \text{OK})\end{aligned}$$

We trial a drop panel of dimensions $3000 \text{ mm} \times 3000 \text{ mm} \times 320 \text{ mm}$ deep and check for its strength with consideration of unbalanced moments.

Capacity at interior column with unbalanced moment transfer

According to AS 3600, the slab in the vicinity of an interior column must be able to transfer an unbalanced bending moment not less than:

$$M_v^* = 0.06[(1.2q_G + 0.75q_Q)L_t L_o^2 - 1.2q_G L_t (L_o')^2]$$

where q_G and q_Q are the permanent and imposed slab service loads in kPa, respectively, L_t is the width of the design strip and L_o' is the smaller value of L_o for the adjoining spans and determined as shown in Figure 12.18.

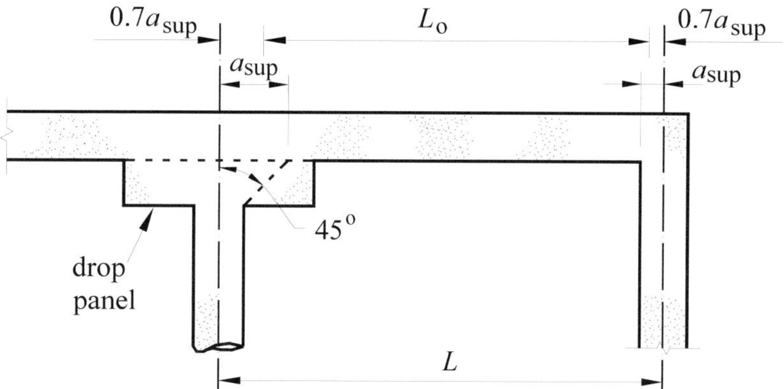

Figure 12.18 Definition of span lengths

For our slab, $L_t = 9.0$ m and the span length is:

$$L_o' = L_o = 9.0 - 2[0.7(0.200 + 0.110)] = 8.57 \text{ m}$$

The out-of-balance moment is therefore:

$$\begin{aligned} M_v^* &= 0.06[(1.2 \times 5.75 + 0.75 \times 4) \times 9.0 \times 8.57^2 \\ &\quad - 1.2 \times 5.75 \times 9.0 \times 8.57^2] \\ &= 119 \text{ kNm} \end{aligned}$$

From AS 3600:

$$\phi V_u = \frac{\phi V_{uo}}{1 + (uM_v^*)/(8V^* a d_{om})}$$

where the width of the torsion strip in the direction of bending is $a = 400 + 274 = 674$ mm. Thus:

$$\phi V_u = \frac{1484}{1 + \frac{2704 \times 119 \times 10^3}{8 \times 1041 \times 676 \times 276}} = 1229 \text{ kN} \geq V^* = 1041 \text{ kN } (\therefore \text{OK})$$

Shear capacity with unbalanced moment at edge column

It is shown later (see Figure 12.27) that the unbalanced moment M_v^* at edge column B is 400 – 303 = 97 kNm. The design shear force is:

$$V^* = 12.9[9.0 \times (4.5 + 2.5) - 0.676^2] = 807 \text{ kN}$$

In this example, since the slab is cantilevered past the columns, the critical perimeter at the edge column is identical to that for an interior column calculated above. From above, ϕV_{uo} = 1484 kN.

For consideration of the unbalanced moment, it is required that (AS 3600):

$$\phi V_u = \frac{\phi V_{uo}}{1 + (uM_v^*)/(8V^* a d_{om})} \geq V^*$$

$$= \frac{1484}{1 + \frac{2704 \times 97 \times 10^3}{8 \times 807 \times 676 \times 276}} = 1219 \text{ kN} \geq 807 \text{ kN} \quad (\therefore \text{OK})$$

By observation, punching shear is not critical at the corner column location and we **adopt drop panels of dimensions 3000 mm square in plan by 320 mm deep** throughout.

7. Equivalent (idealised) frame analysis

A frame analysis is carried out to determine the moments due to the unbalanced service loads, the secondary moments induced by prestress, and the moments to be used for checking flexural moment capacity.

The equivalent frame method of analysis is used. For more detail on this method the reader is referred to the companion text *Reinforced Concrete Basics* (Foster et al., 2021). The frame to be analysed is shown in Figure 12.19.

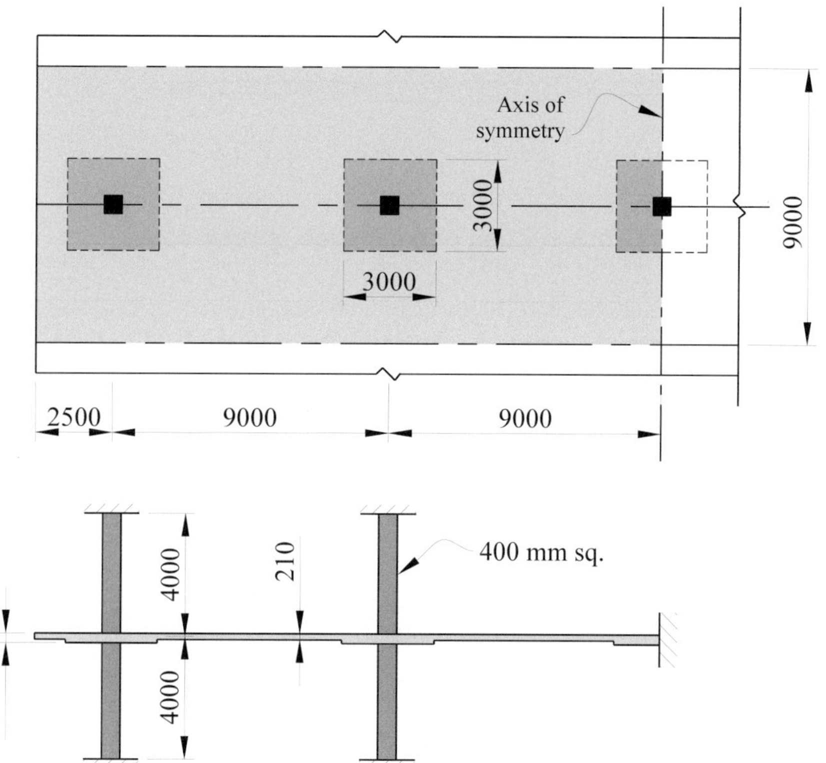

Figure 12.19 Equivalent frame for analysis

The properties of the equivalent beam used in the idealised frame analysis are calculated from a panel-width strip, as shown in Figure 12.19. The beam-slab cross-section changes at the drop panel, and within the slab-column joint. The moment of inertia of the 9000 mm wide by 210 mm slab section is:

$$I_b = 9000 \times 210^3 / 12 = 6.9 \times 10^9 \text{ mm}^4$$

Within the region of the drop panel (Figure 12.20), the second moment of area is determined as $I_b = 14.4 \times 10^9 \text{ mm}^4$. The centroidal axis is 129 mm from the top.

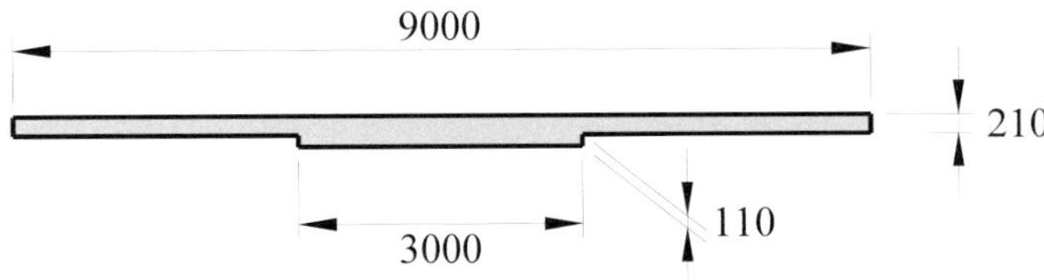

Figure 12.20 Cross-section of beam-slab at drop panel

The equivalent slab-beam to be used in analysis, and the second moments of area over each region, is shown schematically in Figure 12.21. In this figure, the equivalent depth of the drop panel strip to give $I_b = 14.4 \times 10^9 \text{ mm}^4$ is calculated to be 268 mm.

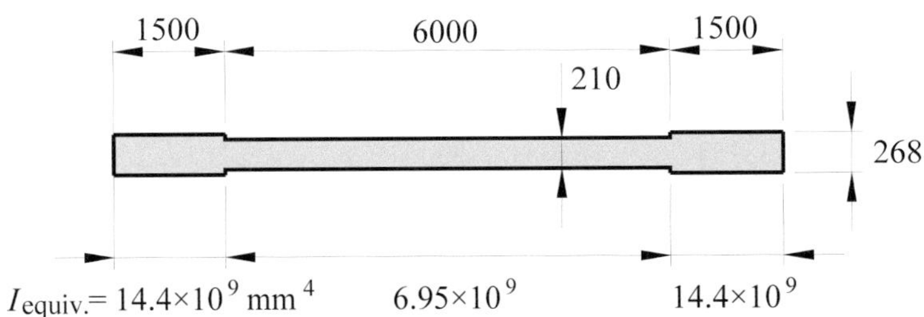

Figure 12.21 Variation in second moment of area of slab-beam

For the columns, the equivalent second moment of area, including the influence of the torsion strip, is determined as $I_{equiv} = 600 \times 10^6 \text{ mm}^4$.

Moments due to prestress:

The equivalent load due to the curvature of the tendons is:

$$w_p = 9 \text{ m} \times 4.5 \text{ kPa} = 40.5 \text{ kN/m}$$

The calculated moments due to these loads are shown in Figure 12.22.

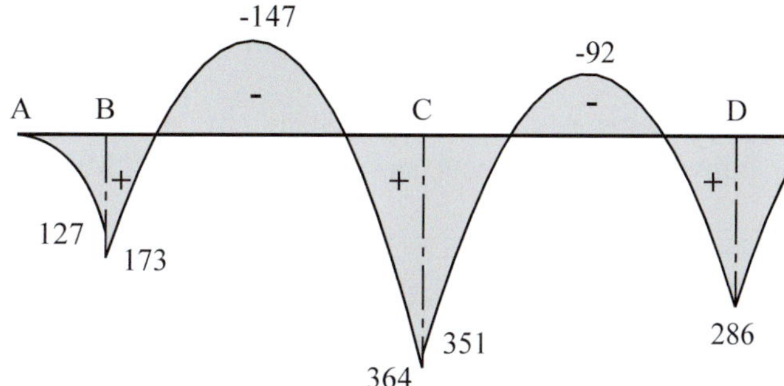

Figure 12.22 Moments due to curvature of prestressing tendons

Moments due to eccentricity at drop panels

All tendon anchorages will be located at the mid-depth of the slab, i.e. 105 mm from the top surface. Where the slab section includes a drop panel, the depth to the centroid becomes 129 mm. The horizontal prestressing forces applied at the anchorages therefore have an eccentricity e_d of 24 mm at the drop panel sections, so that a couple of $0.024P_e$ acts at these sections.

The couples calculated each side of the drop panels are calculated in Table 12.8 and shown in Figure 12.23.

TABLE 12.8 Calculation of couples each side of drop panels

Joint	P_e (kN)	e_d (mm)	$P_e e_d$ (kNm)
B	3751	24	90
C	3975	24	95
D	3702	24	89

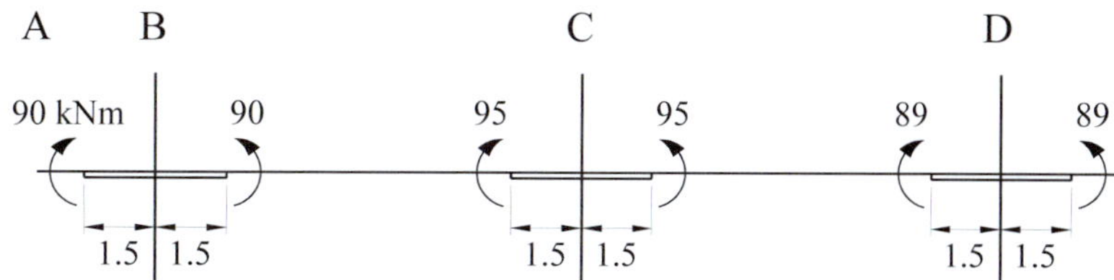

Figure 12.23 Couples due to eccentricity at drop panels

The moments due to these couples are shown in Figure 12.24 and the total moments due to prestress, obtained by adding these moments to the tendon curvature moments of Figure 12.22, are shown in Figure 12.25.

The secondary moments M_2 may be calculated by subtracting the primary moments M_1 from the total moments, as shown in Table 12.9 and Figure 12.26.

TABLE 12.9 Calculation of secondary moments

Joint	P (kN)	e # (mm)	Pe (kNm)	M_t (kNm)	M_2 (kNm)
B	3751	58	218	260	42
C Left	3975	75	298	438	140
C Right	3975	75	298	426	128
D	3702	75	278	363	85

Note: # eccentricities calculated to centroidal axis of section with drop-panel.

8. Flexural strength check

The design ultimate loading is 12.9 kPa, which is equal to 116.1 kN/m over the 9.0 metre width of the panel. The resulting bending moments are given in Figure 12.27(a). Note that the moments are capped at the critical section for bending, taken at a distance of $0.7a_{sup} = 0.217$ m from the support centreline (see Figure 12.18).

The peak negative moments at joint C may be reduced slightly with consideration of some redistribution of the bending moments. We shall reduce the peak moments each side of C to 820 kNm and adjust the adjacent positive bending moments (at E and F) accordingly. This represents an 11 per cent redistribution of moments in span BC and an eight per cent redistribution in span CD. This redistribution is allowable by AS 3600 provided that the neutral axis depth parameter does not exceed 0.25 (i.e. $k_u \leq 0.25$). This will be checked later.

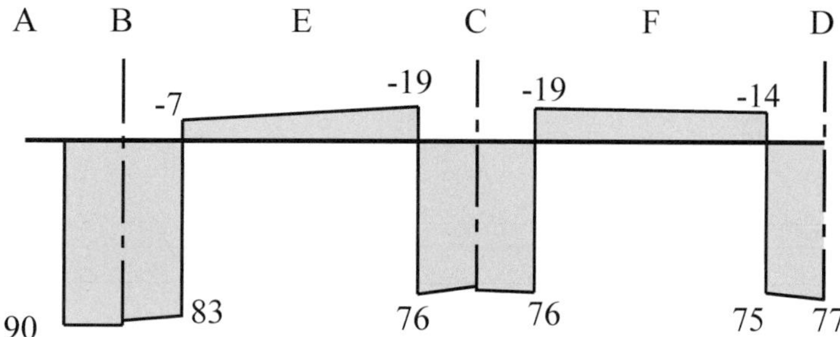

Figure 12.24 Moments due to drop panel eccentricity

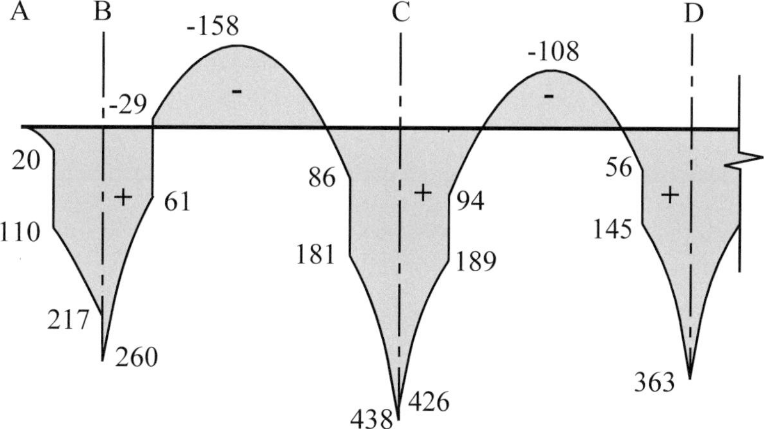

Figure 12.25 Total moments due to prestress

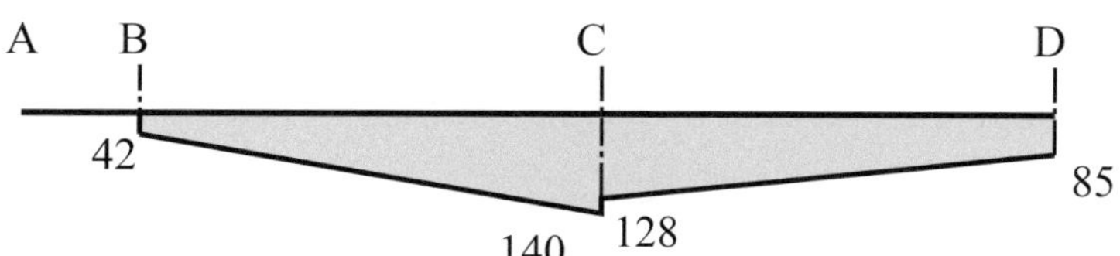

Figure 12.26 Secondary moments

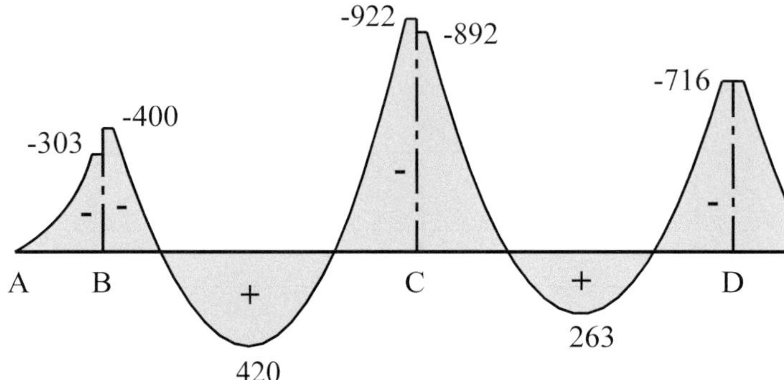

(a) Design moments - no moment redistribution

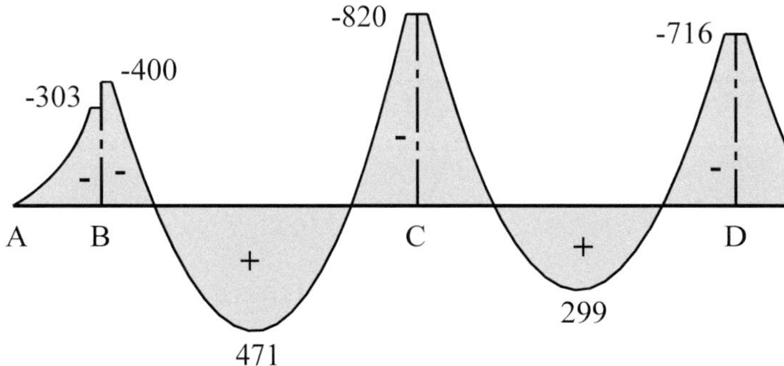

(b) Design moments - with moment redistribution

Figure 12.27 Design bending moments for 9.0 metre wide panel

Negative moments - column strip

The section for the design of the column strip is shown in Figure 12.28. Considering the negative moment at joint C, 75 per cent of the moment is distributed to the column strip and 12.5 per cent to each adjacent half-middle strips (that is, a total of 25 per cent to the middle strips). The column strip moment is:

$$M^* = 0.75 \times 820 = 615 \text{ kNm}$$

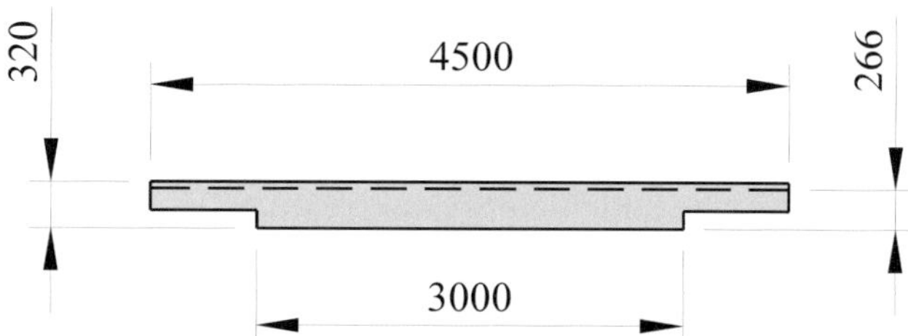

Figure 12.28 Column strip cross-section at drop panel

In the column strip there are six tendons with a total area of:

$$A_p = 6 \times 4 \times 98.6 = 2370 \text{ mm}^2$$

at effective depth to the tendons of $d_p = 320 - 54 = 266 \text{ mm}$.

By Equation 6.31 the stress in strands at ultimate is determined as:

$$k_1 = 0.4; \quad k_2 = \frac{2370 \times 1870 + 0}{3000 \times 266 \times 40} = 0.139$$

$$\sigma_{cp} = 1870\left(1 - \frac{0.4 \times 0.139}{0.87}\right) = 1750 \text{ MPa}$$

Assuming the lever arm between the internal forces is $0.9d_p$, the ultimate moment is determined as:

$$M_u = 2370 \times 1750 \times 0.9 \times 266 \times 10^{-6} = 993 \text{ kNm}$$

$$\phi M_u = 0.85 \times 993 = 844 \text{ kNm}$$

As $\phi M_u = 844 \text{ kNm} \geq M^* = 615 \text{ kNm}$, the section is satisfactory and no bar reinforcement is needed for strength.

For the M_u determined, the depth of the neutral axis is:

$$d_n = k_u d = \frac{A_p \sigma_{cp}}{\alpha_2 f_c' \gamma b} = \frac{2370 \times 1750}{0.79 \times 40 \times 0.87 \times 3000} = 50.3 \text{ mm}$$

and $k_u = 50.3/266 = 0.19$, which is less than both the value of 0.25 needed for 11 per cent redistribution of moments and the value of 0.36 needed to ensure adequate ductility, and therefore okay.

Lastly, with $\gamma d_n = 0.87 \times 50.3 = 43.8\ \text{mm}$ our assumption that $0.9d_p$ is seen to be sufficiently accurate and slightly conservative.

Negative moments - half-middle strips

For the two half-middle strips, the design moment is:

$$M^* = 0.25 \times 820 = 205\ \text{kNm}$$

The two half middle strips have the equivalent of three tendons (total), giving $A_p = 1180\ \text{mm}^2$. The effective depth to the tendons is:

$$d_p = 210 - 48 = 162\ \text{mm}$$

By Equation 6.31 the stress in strands at ultimate is determined as:

$$k_1 = 0.4;\quad k_2 = \frac{1180 \times 1870 + 0}{4500 \times 162 \times 40} = 0.0757$$

$$\sigma_{cp} = 1870\left(1 - \frac{0.4 \times 0.0757}{0.87}\right) = 1805\ \text{MPa}$$

Again assuming the lever arm between the internal forces is $0.9d_p$, the ultimate moment is determined as:

$$M_u = 1180 \times 1805 \times 0.9 \times 162 \times 10^{-6} = 311\ \text{kNm}$$

$$\phi M_u = 0.85 \times 311 = 264\ \text{kNm}$$

As $\phi M_u = 264\ \text{kNm} \geq M^* = 205\ \text{kNm}$, the section is satisfactory and no bar reinforcement is needed for strength.

Positive moments

It is obvious from inspection that the flexural moment capacity of the slab is adequate for the positive moments.

Top steel required near column

As stated in Section 12.7 Step 7, AS 3600 requires that 25 per cent of the steel required to resist the panel negative moment be located within a distance of 200 + 320 = 520 mm either side of the column centre line.

From Figure 12.27, the maximum panel negative moment (at C) is 820 kNm, so the required capacity near the column is $0.25 \times 820/\phi \geq 241$ kNm, with $\phi = 0.85$. The cable distribution, shown in Figure 12.17, has two tendons close to the column. These tendons provide a moment capacity of $2/6 \times 993$ = 331 kNm, so this requirement is satisfied.

9. Serviceability checks

The design service loads are:

$$\text{short-term:} \qquad w_s = 5.25 + 0.50 + 0.7(4.0) = 8.6 \text{ kPa}$$

$$\text{long-term:} \qquad w_l = 5.25 + 0.50 + 0.4(4.0) = 7.4 \text{ kPa}$$

The unbalanced loads for the short-term service load are:

$$(8.6 - 4.5) \times 9 = 36.9 \text{ kN/m}$$

Note that at the balanced load the slab is subjected not only to the axial compression P/A but also to moments induced by drop panel eccentricity. These moments must be considered here in addition to the moments caused by the unbalanced loads.

The moments due to the unbalanced loads are shown in Figure 12.29.

Crack control

We first check whether cracking is likely to occur at the section of maximum negative moment, which occurs at C. At the critical section, the magnitude of the moment is –293 kNm. The moment due to the drop panel eccentricity Figure 12.24 is +76 kNm. The resultant moment at this section under full service load is therefore –293 + 76 = –217 kNm.

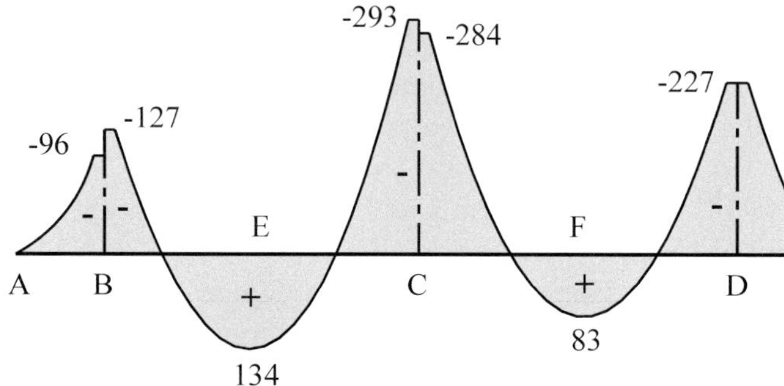

Figure 12.29 Unbalanced moments for short-term service load

We consider conditions in the column strip:

Moment: $M = 0.75 \times (-217) = -163$ kNm

Prestress: $P_e = 3975 \times 0.5 = 1988$ kN

The section properties for the T-section through the drop panel in the column strip are:

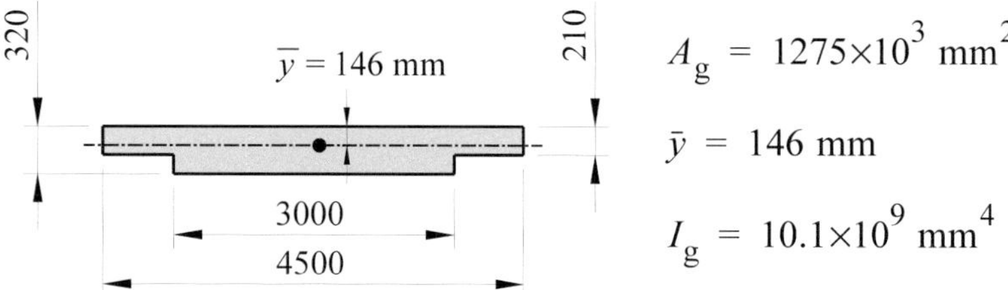

$$A_g = 1275 \times 10^3 \text{ mm}^2$$

$$\bar{y} = 146 \text{ mm}$$

$$I_g = 10.1 \times 10^9 \text{ mm}^4$$

The tensile stress in the top fibre is:

$$\sigma_a = \frac{-1988 \times 10^3}{1275 \times 10^3} + \frac{163 \times 10^6 \times 146}{10.1 \times 10^9} = 0.80 \text{ MPa (tension)}$$

As the tensile stress in the concrete is less than $0.25\sqrt{f_c'} = 1.6 \text{ MPa}$, we conclude that the slab is uncracked due to negative moment.

The maximum positive moment at E (Figure 12.29) is 134 kNm; the negative moment due to drop panel eccentricity at E is –13 kNm (Figure 12.24). The resultant moment is therefore 134 – 13 = 121 kNm.

For positive moments, AS 3600 requires that 60 per cent of the moment be distributed to the column strip; thus, the column strip moment is $0.6 \times 121 =$ 73 kNm and acts on a rectangular cross-section that is 4500 mm wide and 210 mm deep.

For this section:

$$A_g = 945{\times}10^3 \text{ mm}^2; \quad Z = 4500 \times 210^2/6 = 33.1{\times}10^6 \text{ mm}^3$$

$$P_e = 0.5 \times 3887 = 1944 \text{ kN}$$

and the stress in the bottom fibre is:

$$\sigma_b = \frac{-1944{\times}10^3}{945{\times}10^3} + \frac{73{\times}10^6}{33.1{\times}10^6} = 0.15 \text{ MPa (tension)}$$

which is less than the limit of $0.25\sqrt{f'_c}$ and thus the section remains uncracked.

Deflections

Deflection will be calculated at the column lines and at mid-panel using the idealised (equivalent) frame described above. Load patterning is considered with the unbalanced short-term service load applied to the critical end span column strip and with unbalanced dead loads on the column strips only applied to the interior spans and cantilever. The secondary bending moments induced by the eccentricity of prestress at the drop panels is also considered and, as the sections are uncracked under service conditions, gross stiffnesses are used.

Short-term deflections

The unbalanced short-term service load on the 9.0 metre wide beam is:

$$w_s = 9(5.25 + 0.50 + 0.7 \times 4 - 4.5) = 36.5 \text{ kN/m}$$

and the unbalanced dead load on the wide beam is:

$$w_g = 9(5.25 + 0.50 - 4.5) = 11.3 \text{ kN/m}$$

The service design loads are shown in Figure 12.30 with $w_{AB} = w_{CD} =$ 11.3 kN/m and $w_{BC} = 36.5$ kN/m.

The analysis gives the deflections as:

Mid-span of BC: $\Delta_E = 5.1$ mm (downwards)

Cantilever end: $\Delta_A = -5.0$ mm (upwards).

If secondary bending moments are ignored, $\Delta_E = 5.5$ mm and $\Delta_A = -4.2$ mm.

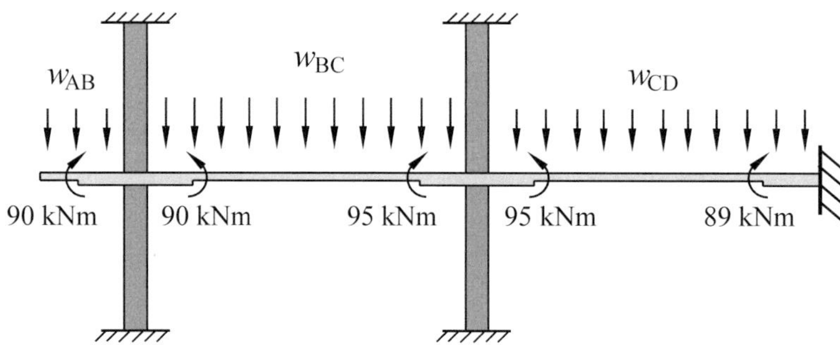

Figure 12.30 Loads for calculation of Δ on column-lines

For determination of the long-term elastic deflection, the unbalanced service load on the column strip is:

$$w_s = 9(5.25 + 0.50 + 0.4 \times 4 - 4.5) = 25.7 \text{ kN/m}$$

The service design loads for the long-term deflection calculations are shown in Figure 12.30 with $w_{AB} = w_{BC} = w_{CD} = 25.7$ kN/m and an analysis gives:

Mid-span of BC: $\Delta_E = 2.1$ mm (downwards)

Cantilever: $\Delta_A = -2.2$ mm (upwards).

Creep deflections

Ignoring the presence of the tendon, we use Equation 4.14 to calculate the creep deflection due to sustained load as:

$$\text{Mid-span of BC:}\quad \Delta^*_{E.c} = \varphi^*_o\,\Delta_E$$

$$= 2.0 \times 2.1 = 4.2 \text{ mm (downwards)}$$

$$\text{Cantilever:}\quad \Delta^*_{A.c} = 2.0 \times (-2.2) = -4.4 \text{ mm (upwards)}$$

As the deflections are small, a correction for the presence of the tendon is not needed.

Shrinkage deflections

For the determination of shrinkage deflections, we shall use the method outlined in Section 4.7 of Chapter 4. For the calculation of the section properties, the drop panels are ignored. The section properties are:

$$n_p = E_p/E^*_c = 195{\times}10^3/32.8{\times}10^3 = 5.95$$

$$A_g = 9000 \times 210 = 1890{\times}10^3 \text{ mm}^2$$

$$I_g = 9000 \times 210^3/12 = 6.9{\times}10^9 \text{ mm}^4$$

$$p_p = A_p/A_g = 3550/1890{\times}10^3 = 0.00188$$

At section E:

$$\gamma_3 = \frac{1}{1 + \dfrac{1}{5.95 \times 0.00188} + \dfrac{57^2 \times 1890{\times}10^3}{6.9{\times}10^9}} = 0.0110 \text{ (Equation B.77)}$$

$$\gamma_4 = \frac{1890{\times}10^3 \times 57 \times 210}{6.9{\times}10^9} = 3.28 \text{ (Equation B.79)}$$

and the final shrinkage curvature is then calculated as (Equation B.79):

$$\kappa^*_{E.sh} = 0.0110 \times 3.28 \times 600{\times}10^{-6}/210 = 103{\times}10^{-9} \ \text{mm}^{-1}$$

At section B: $e = -34$ mm, $\gamma_3 = 0.0110$ and $\gamma_4 = -1.96$ and the final shrinkage curvature is:

$$\kappa^*_{B.sh} = -0.0110 \times 1.96 \times 600{\times}10^{-6}/210 = -62{\times}10^{-9} \ \text{mm}^{-1}$$

At section C: $e = -51$ mm, $\gamma_3 = 0.0110$ and $\gamma_4 = -2.93$ and the final shrinkage curvature is:

$$\kappa^*_{C.sh} = -92{\times}10^{-9} \ \text{mm}^{-1}$$

The final shrinkage deflection for member BC is determined as (Equation 4.26):

$$\Delta^*_{E.sh} = \frac{9000^2}{96}[-62 + 103 - 92] \times 10^{-9} = 0.0 \text{ mm}$$

As the calculated deflection is small, we shall undertake a second assessment using the model of Gilbert and Ranzi (2011) given in Section 5.8 of Chapter 5 and by Equations 5.37 and 5.38 for an uncracked section.

At section E:

With $d_o = 153$ mm and, for the wide beam, $p_o = A_p/(bd_o) = 3550/(9000 \times 153) = 0.00258$, Equation 5.39 gives:

$$k_r = (100 \times 0.00258 - 2500 \times 0.00258^2)\left(\frac{2 \times 153}{210} - 1\right) = 0.110$$

and, by Equation 5.37, the final shrinkage curvature is then calculated as (Equation B.78):

$$\kappa^*_{E.sh} = 0.110 \times 600{\times}10^{-6}/210 = 314{\times}10^{-9} \ \text{mm}^{-1}$$

At section B:

With d_o = 139 mm, p_o = 0.0028 and k_r = 0.084, the final shrinkage curvature is $\kappa^*_{B.sh} = -240\times10^{-9}\ \text{mm}^{-1}$

At section C: $\kappa^*_{C.sh} = -342\times10^{-9}\ \text{mm}^{-1}$

The final shrinkage deflection for member BC is determined as:

$$\Delta^*_{E.sh} = \frac{9000^2}{96}[-240 + 3140 - 342]\times 10^{-9} = 2.2\ \text{mm}$$

As for the model of Section 4.6, the estimated shrinkage warping deflection is small. We take the larger of the two values for further calculations.

For a cantilever with a parabolic distribution of curvature, the deflection at the free end may be obtained from:

$$\Delta = \frac{L^2}{6}[\kappa_S + 2\kappa_M] \tag{12.19}$$

where κ_S is the curvatures at the fixed support and κ_M is at the mid-span. Assuming a uniform shrinkage curvature along the member length of $-342\times10^{-9}\ \text{mm}^{-1}$, the shrinkage deflection at the free end of the cantilever is:

$$\Delta^*_{A.sh} = \frac{2500^2}{6}[-342\times 3]\times 10^{-9} = -1.1\ \text{mm (upwards)}$$

Total wide beam deflections

The estimated total deflections for the wide beam are:

Mid-span of BC: $\Delta_{E.tot} = 2.1 + 4.2 + 2.2 = 8.5\ \text{mm}$ (downwards)

Cantilever: $\Delta_{A.tot} = -2.2 - 4.4 - 1.1 = -7.7\ \text{mm}$ (upwards)

Strip and mid-panel deflections

The strip deflections are estimated from Equation 12.11, with the ratio $M_{cs}/M_{wb} = 0.7$. The average second moment of area of the column strip is taken as the vales of I_g at the supports (10.1×10^9 mm^4) and at mid-span (3.5×10^9):

Column strip: $I_{cs} = 6.8\times10^9 \text{ mm}^4$

Middle strip: $I_{ms} = 3.5\times10^9 \text{ mm}^4$

Wide beam: $I_{wb} = (6.8 + 3.5)\times10^9 = 10.3\times10^9 \text{ mm}^4$

The strip deflections at E are thus:

$$\Delta_{E.cs} = 8.5 \times 0.7 \times 10.3/6.8 = 9.0 \text{ mm}$$

$$\Delta_{E.ms} = 8.5 \times 0.3 \times 10.3/3.5 = 7.5 \text{ mm}$$

and the mid-panel deflection is (Equation 12.9):

$$\Delta_{E.mp} = 9.0 + 7.5 = 16.5 \text{ mm}$$

The deflection at E on the column line is span/1000 and on the diagonal is span/770, which are satisfactory. At A, the deflection is span/320, which is also satisfactory.

10. Anchorage zone reinforcement

The critical region for anchorage zone cracking is behind the closely spaced (400 mm) anchorages at column lines. The slab cable anchorages have effective bearing areas approximately 80 mm deep by 225 mm wide.

Horizontal cracking

The prestress force at transfer is 5273 kN (Table 12.7); $5273/9 = 586$ kN/tendon. The closest spacing of tendons is in the column regions and is 400 mm. Taking a 400 mm wide 'equivalent prism', the stress on the prism boundary is:

$$\sigma_o = \frac{586 \times 10^3}{200 \times 400} = 7.0 \text{ MPa}$$

The transverse tensile force is (Equation 12.17):

$$T = 0.25P(1 - h_d/D) = 0.25 \times 586(1 - 80/210) = 91 \text{ kN}$$

Limiting the stress in the bursting reinforcement to 150 MPa, the required area of reinforcement is:

$$A_s = 91 \times 10^3/150 = 610 \text{ mm}^2$$

and may be provided by four 2-legged N10 stirrups centred around the anchorage, or by an appropriately sized rectangular helix surrounding the anchorage.

Vertical cracking

From Equation 12.18, the maximum tensile stress is determined as:

$$\sigma_{y.max} = \sigma_o(1 - h_s/S)$$
$$= 7.0(1 - 225/400) = 3.1 \text{ MPa}$$

The average compressive stress from the orthogonal prestress is:

$$\sigma = \frac{5273 \times 10^3}{210 \times 9000} = 2.8 \text{ MPa}$$

The difference between the bursting tensile stress and the average orthogonal compressive stress from the y-direction prestress is 3.1 – 2.8 = 0.3 MPa (tensile). We thus conclude that vertical cracking will not occur behind the anchorages.

12.8 References

Aalami, B.O. and Barth F.G., 1988, *Restraint Cracks and their Mitigation in Unbonded Post-tensioned Building Structures*, Post-tensioning Institute, Phoenix, Arizona, 49 pp.

C&CAA, 2003, *Guide to long-span concrete floors*, Cement and Concrete Association of Australia, 2nd Edition, 40 pp.

Cross, E., 2007, Post-tensioning in Building Structures, *Concrete in Australia,* Vol 33, Issue 4.

Eurocode 2, 2004, EN 1992-1-1:2004, *Design of concrete structures - Part 1-1 : General rules and rules for buildings*, European Committee For Standardization, Brussels, Belgium, 225 pp.

FIP, 1980, *Recommendations for the Design of Flat Slabs in Post-Tensioned Concrete (using Unbonded and Bonded Tendons)*, Federation Internationale de la Precontraite.

Foster, S J, Kilpatrick, A E, and Warner, R F, 2021, *Reinforced concrete basics*, 3rd Ed, Pearson, Melbourne, Australia, 589 pp.

Gilbert R.I. and Ranzi, G. 2011, *Time dependent behaviour of concrete structures*, Spon Press, Oxon, UK, 426 pp.

Gerber, L.L. and Burns, N.H., 1971, Ultimate strength tests of post-tensioned flat plates, Journal, *Prestressed Concrete Institute*, V. 16, N. 6, Nov-Dec, pp. 40-58.

Nilson, A.H. and Walters D.B., 1975, Deflection of Two-Way Floor Systems by the Equivalent Frame Method, *ACI Journal*, Vol 72, No 5.

Yettram, A.L. and Robbins, K., 1969, Anchorage-Zone Stresses in Axially Post-tensioned Members of Uniform Rectangular Section, *Magazine of Concrete Research*, Vol 21, No 7.

APPENDIX A

Analysis of uncracked sections

In Chapter 4 the stresses and strains in uncracked prestressed sections were calculated using the gross concrete section and ignoring the presence of reinforcing steel and prestressing ducts. This simplified approach will be adequate for most routine design calculations, but a more accurate analysis is sometimes required. The elastic analyses presented here take account of the reinforcing steel in the section, as well as the tendon duct and the grout in the duct.

A.1 Uncracked post-tensioned section with reinforcement

We consider the cross-section of a post-tensioned member as in Figure A.1(a) with tensile and compressive reinforcing steel at depths d_{st} and d_{sc}. The stressed tendon, of area A_p at depth d_p, is contained within a duct. The area of the duct, A_d, is larger than A_p and its centroid is at depth d_d, which will be slightly larger than d_p. Using the transformed area concept to take account of the reinforcing steel, we introduce additional concrete areas of $A_{sc}(E_s/E_c - 1)$ and $A_{st}(E_s/E_c - 1)$ at depths d_{sc} and d_{st}. The concrete section with transformed steel areas is shown in Figure A.1(b).

During the prestressing operation and before the duct is grouted, the equivalent transformed concrete area is A_E with centroid at depth d_E and with second moment of area I_E. We have:

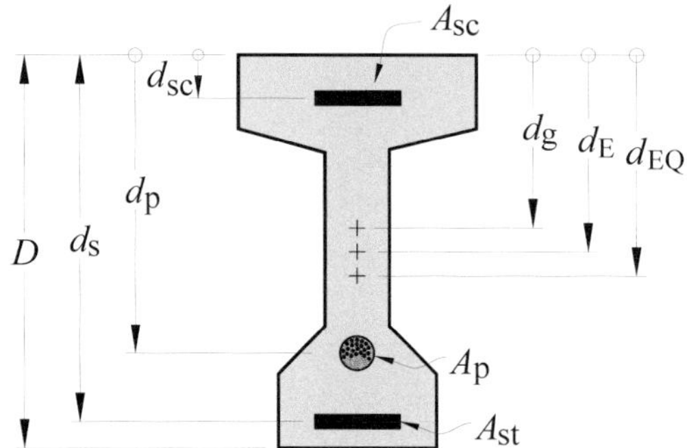

(a) Uncracked post-tensioned section with tendon, duct and reinforcement

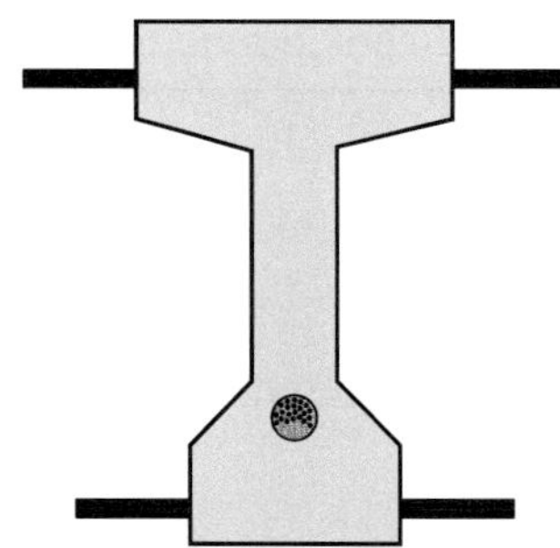

(b) Concrete section with transformed steel areas

Figure A.1 Uncracked post-tensioned section, with tendon, duct and reinforcing steel

$$A_E = A_g + (E_s/E_c - 1)(A_{sc} + A_{st}) - A_d \quad \text{(A.1)}$$

$$d_E = \frac{A_g d_g + (E_s/E_c - 1)(A_{sc} d_{sc} + A_s d_{st}) - A_d d_d}{A_g + (E_s/E_c - 1)(A_{sc} + A_{st}) - A_d} \quad \text{(A.2)}$$

$$I_E = I_g + A_g(d_g - d_E)^2 - A_d(d_d - d_E)^2 + I_{st} + I_{sc} \quad \text{(A.3)}$$

In Equation A.3 I_{st} and I_{sc} are the contributions of the tensile and compressive steels, and their values are:

$$I_{st} = (E_s/E_c - 1)A_{st}(d_s - d_E)^2 \tag{A.4}$$

$$I_{sc} = (E_s/E_c - 1)A_{sc}(d_{sc} - d_E)^2 \tag{A.5}$$

The term A_g is the gross area of the concrete section (including the duct area A_d), d_g is the depth to the centroid of A_g and I_g is the second moment of A_g about an axis at depth d_g.

In the cross-section just after prestressing, the force in the tendon is P_o and the moment due to self-weight and any superimposed load which comes into play at transfer is M_G. The compressive stress in the concrete at depth d below the top fibre is:

$$\sigma_c = \frac{P_o}{A_E} + \frac{P_o(d_p - d_E) - M_G}{I_E}(d - d_E) \tag{A.6}$$

The compressive stresses in the bottom and top concrete fibres due to the prestressing force P_o and the moment M_G are thus:

$$\sigma_{cb} = \frac{P_o}{A_E} + \left[\frac{P_o(d_p - d_E) - M_G}{I_E}\right](D - d_E) \tag{A.7}$$

$$\sigma_{ca} = \frac{P_o}{A_E} + \left[\frac{P_o(d_p - d_E) + M_G}{I_E}\right](-d_E) \tag{A.8}$$

The compressive stresses in the reinforcing steel areas A_{st} and A_{sc} are:

$$\sigma_{st} = \left\{\frac{P_o}{A_E} + \left[\frac{P_i(d_p - d_E) - M_G}{I_E}\right](d_{st} - d_E)\right\} \cdot \frac{E_s}{E_c} \tag{A.9}$$

$$\sigma_{sc} = \left\{\frac{P_o}{A_E} + \left[\frac{P_i(d_p - d_E) - M_G}{I_E}\right](d_{sc} - d_E)\right\} \cdot \frac{E_s}{E_c} \tag{A.10}$$

In the above equations a negative value for stress indicates tension.

It is assumed that the duct is grouted just after transfer so that when a live load moment M_Q is subsequently applied, the entire section acts in a composite manner. The equivalent area of this composite section is:

$$A_{EQ} = A_g + A_s' + A_p' + A_d' \tag{A.11}$$

where:

$$A_s' = \left(\frac{E_s}{E_c} - 1\right)(A_{sc} + A_{st}) \tag{A.12}$$

$$A_p' = \left(\frac{E_p}{E_c} - 1\right)(A_p) \tag{A.13}$$

$$A_d' = \left(\frac{E_d}{E_c} - 1\right)(A_d - A_p) \tag{A.14}$$

The term E_d is the modulus of elasticity of the hardened grout in the duct, which will usually be less than E_c in which case the term in Equation A.14 is negative. The grout is of course not prestressed and if it is cracked by the effect of M_Q then E_d should be taken as zero. The depth to the centroidal axis of the composite section is:

$$d_{EQ} = \frac{M_g + M_s + M_p + M_d}{A_g + A_s' + A_p' + A_d'} \tag{A.15}$$

where:

$$M_g = A_g d_g ; \quad M_s = (E_s/E_c - 1)(A_{sc} d_{sc} + A_{st} d_{st})$$

$$M_p = A_p' d_p ; \quad M_d = A_d' d_d$$

The second moment of area is:

$$I_{EQ} = I_G + I_s + I_p + I_d \tag{A.16}$$

where:

$$I_G = I_g + A_g(d_g - d_{EQ})^2$$

$$I_s = \left(\frac{E_s}{E_c} - 1\right)(A_{sc}(d_{sc} - d_{EQ})^2 + A_{st}(d_{st} - d_{EQ})^2)$$

$$I_p = \left(\frac{E_p}{E_c} - 1\right)A_p(d_p - d_{EQ})^2$$

$$I_d = \left(\frac{E_d}{E_c} - 1\right)(A_d - A_p)(d_d - d_{EQ})^2$$

The increment in compressive stress in the concrete at any depth d due to moment M_Q is:

$$\sigma_{cQ} = \frac{M_Q}{I_{EQ}}(d_{EQ} - d) \tag{A.17}$$

The stress increments in the steel, the prestressing tendon and the grout are determined by calculating the concrete stress σ_{cQ} at the required level and multiplying it by the appropriate modular ratio, i.e. by E_s/E_c, E_p/E_c or E_d/E_c.

The total stress is obtained by adding the stress due to prestress and M_G to the stress increment due to M_Q. In the case of the grout, the stress due to prestress and M_G is zero.

It is emphasised that this elastic analysis applies only to short-term loading. It ignores the creep and shrinkage strains that develop after prestressing. Nevertheless, the equations give an accurate estimate of the stress increments, (a) due to the application of prestress (Equations A.6 to A.8) and moment M_G, and (b) due to the live load moment M_Q (Equation A.17).

To make an approximate correction for time effects it was common practice in allowable stress design calculations to decrease E_c arbitrarily by a factor of about 2.0. The terms $(E_p/E_c - 1)$ and $(E_s/E_c - 1)$ were thus replaced by $(2E_p/E_c)$ and $(2E_s/E_c)$, respectively. The step-by-step method described in Appendix B provides a more satisfactory, more rational means of evaluating time effects.

EXAMPLE A.1

Refined elastic analysis, uncracked post-tensioned section

We make a refined analysis for a rectangular section similar to the one used in Example 4.1 of Chapter 4, except that it contains tensile steel reinforcement. The area of the tensile reinforcement is $A_{st} = 2700$ mm^2 at depth $d_{st} =$ 720 mm. The section is 800 mm deep and 400 mm wide, is post-tensioned by a cable with cross-sectional area $A_p = 1000$ mm^2 at a depth of 650 mm. The prestressing force is $P_i = 1200$ kN. The duct containing the cable has a 50 mm internal diameter. Its area is 1960 mm^2 and its centroid is at depth 660 mm. A moment of $M_G = 350$ kNm acts on the section just after prestressing.

Other data:

$$E_c = 30\times10^3 \text{ MPa};\ E_p = 195\times10^3 \text{ MPa};\ E_s = 200\times10^3 \text{ MPa}$$

From Example 4.1:

$$A_g = 320\times10^3 \text{mm}^2;\ I_g = 17.07\times10^9 \text{mm}^4.$$

From Equations A.1 to A.3:

$$A_E = 333.3\times10^3 \text{mm}^2;\ d_E = 413.2 \text{ mm};\ I_E = 18.45\times10^9 \text{ mm}^4.$$

SOLUTION

Stresses just after transfer:

Due to prestress alone, the stresses are:

$$\sigma_{cbp} = \frac{1.2\times10^6}{333.3\times10^3} + \frac{1.2\times10^6(800-413.2)(650-413.2)}{18.45\times10^9}$$

$$= 3.60 + 5.96 = 9.56 \text{ MPa (compression)}$$

$$\sigma_{cap} = \frac{1.2\times10^6}{333.3\times10^3} - \frac{1.2\times10^6(650-413.2)(-413.2)}{18.45\times10^9}$$

$$= 3.60 - 6.36 = -2.76 \text{ MPa (tension)}$$

This may be compared with the results from the simplified calculation method used in Example 4.1, which gave the stresses due to prestress alone as σ_{cbp} = +10.78 MPa and σ_{cap} = - 3.28 MPa.

Due to M_G = 350 kNm, the stress increments are:

$$\sigma_{cbG} = \frac{350\times10^6\times(413.2-800)}{18.45\times10^9} = -7.34 \text{ MPa}$$

$$\sigma_{caG} = \frac{350\times10^6\times(413.2)}{18.45\times10^9} = 7.84 \text{ MPa}$$

Hence the total stresses after transfer and after the application of M_G are:

$$\sigma_{cbpG} = 9.56 - 7.34 = 2.22 \text{ MPa}$$

$$\sigma_{capG} = -2.77 + 7.84 = 5.07 \text{ MPa}$$

At zero moment (due to prestress alone) the curvature is:

$$\kappa_o = \frac{-2.77 - 9.56}{30000} \cdot \frac{1}{800} = -0.52 \times 10^{-6} \text{mm}^{-1}$$

The negative sign indicates camber. With M_G acting the total curvature is:

$$\kappa = \frac{-2.27 + 5.05}{30000} \cdot \frac{1}{800} = 0.12 \times 10^{-6} \text{mm}^{-1}$$

which is a positive curvature.

After grouting:

In order to estimate the cracking moment, we assume that E_d is zero because the grout will crack well before the extreme concrete fibre cracks. From Equation A.11 we obtain $A_{EQ} = 340.5 \times 10^3 \text{mm}^2$; also $d_{EQ} = 418.2$ mm and $I_{EQ} = 18.84 \times 10^9 \text{mm}^4$.

The total moment M_{cr} that will bring the prestressed section to the point of cracking must produce a tensile stress increment in the bottom fibre of $\sigma_{cbp} + f_{cf}'$, where f_{cf}' is the tensile stress required to cause cracking, and includes an allowance for prior restrained shrinkage. Taking f_{cf}' as 3.20 MPa, the stress increment is 9.56 + 3.20 = 12.76 MPa and hence:

$$M_{cr} = \frac{12.76 \times 18.84 \times 10^9}{(800 - 418.2)} = 630 \text{ kNm}$$

The additional live-load moment required to cause cracking is:

$$M_Q = 630 - 350 = 280 \text{ kNm}$$

The total increment in curvature after prestressing is:

$$\Delta\kappa = \frac{12.76}{30 \times 10^3} \times \frac{1}{(800 - 418.2)} = 1.11 \times 10^{-6} \text{mm}^{-1}$$

A.2 Uncracked pretensioned section with reinforcement

In the case of a pretensioned member, an initial prestressing force of P_i is required in the tendon just prior to transfer, in order to achieve the smaller force after transfer of P_o. Also there is no duct. To evaluate short-term effects in a pretensioned section with reinforcement the quantities A_E, d_E and I_E are calculated from Equations A.1 to A.3, however with the grout ignored, i.e. with A_d set equal to A_p and E_d set to zero. At the level of the centroid of the prestressing strands the compressive stress and strain in the concrete after transfer are:

$$\sigma_{cp} = \frac{P_o}{A_E} + \left[\frac{P_o(d_p - d_E) - M_G}{I_E}\right](d_p - d_E) \tag{A.18}$$

$$\varepsilon_{cp} = \sigma_{cp}/E_c \tag{A.19}$$

We have:

$$P_i = P_o + \varepsilon_{cp} E_p A_p \tag{A.20}$$

which gives:

$$P_i = P_o\left(1 + \frac{E_p A_p}{E_c A_E} + \frac{E_p A_p}{E_c I_E}(d_p - d_E)^2\right) - \frac{E_p A_p}{E_c I_E} M_G (d_p - d_E) \tag{A.21}$$

The stresses in the concrete and steel due to P_o and M_Q can be calculated for the pretensioned member using Equations A.6 to A.8. In determining the effect of the live load moment M_Q, Equation A.17 can be used. Again, in evaluating A_{EQ}, d_{EQ} and I_{EQ}, the grout terms are set to zero by taking $A_d = A_p$ and $E_d = 0$.

As already mentioned, the effects of prior creep and shrinkage have not been considered in this analysis.

APPENDIX B

Creep and shrinkage in uncracked flexural members

Three alternative methods of analysis, of varying accuracy and complexity, are presented, with numerical examples, for evaluating the effects of creep and shrinkage in uncracked prestressed concrete members. A very simple order-of-magnitude procedure is also described. These methods allow long-term deflections and deferred prestress losses to be evaluated. The effects of non-uniform shrinkage and creep are briefly discussed at the end of this appendix.

B.1 Introductory note

The sustained loads that act on prestressed concrete flexural members are not usually large enough to induce flexural cracking. Even if cracking occurs under an early short-term overload, the prestress is usually sufficient to close the cracks following the overloading. Attention is therefore restricted here to the effects of creep and shrinkage in uncracked members.

B.2 Order-of-magnitude estimates of long-term deformations and prestress losses

Simple (but very approximate) estimates of long-term losses and deformations in a cross-section due to creep and shrinkage can be made by assuming that the concrete deforms freely under creep and shrinkage, unrestrained by the presence of the steel and tendon. This allows estimates to be made, with minimal calculation, of the long-term deformations and loss of prestress, and the redistribution of forces in the concrete, steel and tendon.

Simple estimates of creep effects

We consider creep in the beam section shown in Figure B.1, which is subjected to a prestressing force P at eccentricity e, and a positive sustained moment M_G. Both are applied at time t_0. If we ignore the presence of the tendon and any steel in the section, so that P and M_G act only on the concrete, then creep occurs freely in each concrete fibre under constant stress.

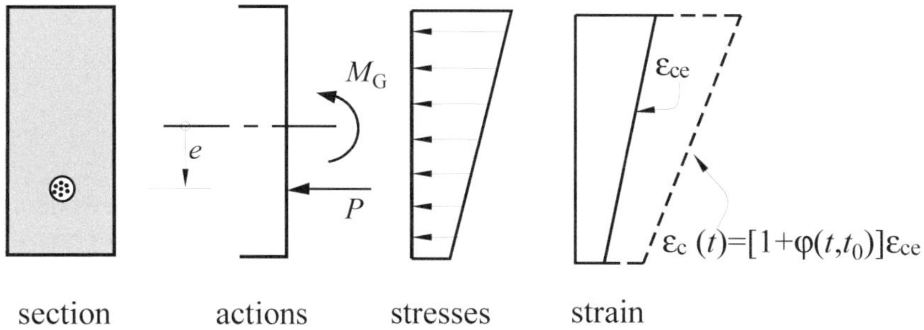

Figure B.1 Free creep strains, uncracked section, ignoring steel and tendon

In any fibre, the creep strain, $\varepsilon_{cc}(t)$, is obtained by multiplying the initial elastic strain, ε_{ce}, by the creep function (or creep coefficient) $\varphi(t, t_0)$, which is defined in Chapter 2. This gives:

$$\varepsilon_{cc}(t) = \varphi(t, t_0)\varepsilon_{ce} \tag{B.1}$$

Creep strain, like elastic strain, is therefore linearly distributed down the section, as shown in Figure B.1. The total strain, elastic plus creep, at any fibre is $\varepsilon_c(t) = (1 + \varphi(t, t_0))\varepsilon_{ce}$. At any time t greater than t_0, the creep curvature, $\kappa_c(t)$, is obtained from the initial elastic curvature, κ_0, and $\varphi(t, t_0)$:

$$\kappa_c(t) = \varphi(t, t_0)\kappa_0 \tag{B.2}$$

The elastic curvature κ_0 depends on the total moment acting on the concrete M and the elastic stiffness of the section. Ignoring the presence of the steel and tendon, the elastic stiffness is $E_c I_g$ and we have:

$$\kappa_o = \frac{M}{E_c I_g} \qquad \text{(B.3)}$$

where I_g is the second moment of area of the gross concrete section. The moment M consists of the positive applied moment M_G and the negative moment due to prestress, $M_p = Pe$. M is positive if M_G is larger in magnitude than M_p.

The total (elastic plus creep) curvature $\kappa(t)$ at time t can thus be obtained from the elastic curvature and a multiplying factor, $R(t, t_o)$:

$$\kappa(t) = R(t, t_o)\kappa_o = (1 + \varphi(t, t_o))\kappa_o \qquad \text{(B.4)}$$

At the end of the creep process, at time t^*, the long-term value of the creep function is φ_o^*, and the multiplying factor is $R_o^* = 1 + \varphi_o^*$. The long-term creep curvature is $\kappa_{co}^* = \varphi_o^* \kappa_o$ and the total long-term curvature is

$$\kappa^* = \kappa_o + \kappa_{co}^* = R_o^* \kappa_o \qquad \text{(B.5)}$$

The final long-term deflection curve for the entire member, $y^*(x)$, can also be obtained from the initial elastic deflection curve, $y_o(x)$, using the multiplying factor R_o^*:

$$y^*(x) = R_o^* \, y_o(x) \qquad \text{(B.6)}$$

The maximum long-term deflection in the member is similarly obtained from the maximum initial deflection, Δ_o, and R_o^*:

$$\Delta^* = R_o^* \, \Delta_o \qquad \text{(B.7)}$$

The initial deflections, $y_o(x)$ and Δ_o, are due to sustained load and prestress only and do not include any short-term deflections due to live load. Creep data in Chapter 2 can be used to evaluate φ_o^* and hence R_o^*.

If sustained load is applied in increments at different times, the short-term and long-term curvatures and deflections can be evaluated for each load increment and superposed, but allowance needs to be made (in the choice of values of φ_o^*) for the decrease in long-term creep with increasing age at first loading, t_o.

A rough, conservative estimate of the long-term loss of prestress in the tendon due to concrete creep, $\Delta X^*_{p.c}$, can be obtained from the long-term creep strain in the concrete at the depth of the tendon, $\varepsilon^*_{cc.p}$:

$$\Delta X^*_{p.c} = A_p E_p \varepsilon^*_{cc.p} \tag{B.8}$$

The creep strain $\varepsilon^*_{cc.p}$ can be evaluated from the free creep strains at the extreme top and bottom fibres, which are in turn obtained from the extreme fibre elastic strains, $\varepsilon_{ce.a}$ and $\varepsilon_{ce.b}$, as:

$$\varepsilon^*_{cc.p} = \varepsilon^*_{cc.a}\left(\frac{D-d_p}{D}\right) + \varepsilon^*_{cc.b}\frac{d_p}{D} = \varphi^*_o\left\{\varepsilon_{ce.a}\left(\frac{D-d_p}{D}\right) + \varepsilon_{ce.b}\frac{d_p}{D}\right\} \tag{B.9}$$

The term d_p is the depth of the tendon in the section, and D is the total depth. The loss of prestress in the concrete, $\Delta X^*_{c.c}$, is equal to $\Delta X^*_{p.c}$ if there is no reinforcement in the section. If reinforcing steel is present, $\Delta X^*_{c.c}$ will be much larger than $\Delta X^*_{p.c}$ because creep induces a progressive transfer of compressive force from the concrete to any adjacent bonded reinforcement. An estimate of the long-term compressive forces in the tensile and compressive layers of reinforcement, $\Delta X^*_{st.c}$ and $\Delta X^*_{sc.c}$, can be made, as for the tendon in Equation B.8, from the adjacent concrete creep strains. The loss of prestress in the concrete is then estimated as:

$$\Delta X^*_{c.c} = \Delta X^*_{p.c} + \Delta X^*_{st.c} + \Delta X^*_{sc.c}\,. \tag{B.10}$$

Simple estimates of shrinkage effects

In design, shrinkage is usually assumed to occur uniformly throughout a cross-section (even though this is rarely the case, as is explained below in Section B.6). However, shrinkage is partially restrained by the steel reinforcement and prestressing tendon in the section, and this induces shrinkage curvature if the steel and tendon are not uniformly distributed in the section. Nevertheless, the shrinkage curvature is usually quite small, and an order-of-magnitude estimate is rarely needed.

The loss of prestress due to shrinkage can be extremely high, and a rough estimate can be obtained using the final long-term uniform shrinkage strain ε^*_{sh} (or ε_{cs}, in AS 3600 terminology), in an expression similar to Equation B.8:

$$\Delta X^*_{p.sh} = A_p E_p \varepsilon^*_{sh} \tag{B.11}$$

Shrinkage of the concrete also results in a significant transfer of compressive forces to the reinforcing steel. These forces, $\Delta X^*_{st.sh}$ and $\Delta X^*_{sc.sh}$, can be estimated from the shrinkage strains at the levels of the steels, in the same way as for creep, and then added to $\Delta X^*_{p.sh}$ to estimate the total shrinkage loss of prestress in the concrete:

$$\Delta X^*_{c.sh} = \Delta X^*_{p.sh} + \Delta X^*_{st.sh} + \Delta X^*_{sc.sh} \tag{B.12}$$

Discussion

These simple calculations ignore the restraint provided by the steel and tendon and so tend to *overestimate* losses and long-term deformations. If the section contains only a prestressing tendon of small area, the errors will be small, but if substantial amounts of reinforcement are present, the errors can be large. The moment ratio M_G/M_p also has an effect: errors are larger when this ratio is small, but smaller when it is larger. Approximate calculations are included in the numerical examples that follow, and comparisons with the results of the more accurate calculation methods show how the accuracy of the simplified approach varies with the section details and other input data.

B.3 One-step analysis with age-adjusted effective modulus

B.3.1 One-step creep analysis

In a one-step analysis, the creep strains are allowed to increase freely between t_0 and t^*, as in the order-of-magnitude calculation in Section B.2. This implies that strain compatibility in the section is progressively lost because there is no change in the strains in the steel and tendon. To complete the analysis, compatibility is therefore restored at time t^* by applying an appropriate equilibrating force system to the section which consists of a tensile force increment, $\Delta X^*_{c.c}$, acting on the concrete at an initially unknown depth d_{xc}, and compressive force increments acting on the tendon and the tensile and compressive steel areas: $\Delta X^*_{p.c}$, $\Delta X^*_{st.c}$ and $\Delta X^*_{sc.c}$. An example of this one step analysis is presented in part (b) of Example B.1

A more general one-step analysis is now presented for a section which is symmetric about its vertical axis (Figure B.2). The centroid of the concrete is at depth d_g below the top fibre and the total depth is D. The tendon area, A_p, is at depth d_p. A layer each of tensile steel and compressive steel, with areas A_{st} and A_{sc}, are at depths d_{st} and d_{sc}. A moment M_G is applied together with the prestress at time t_o.

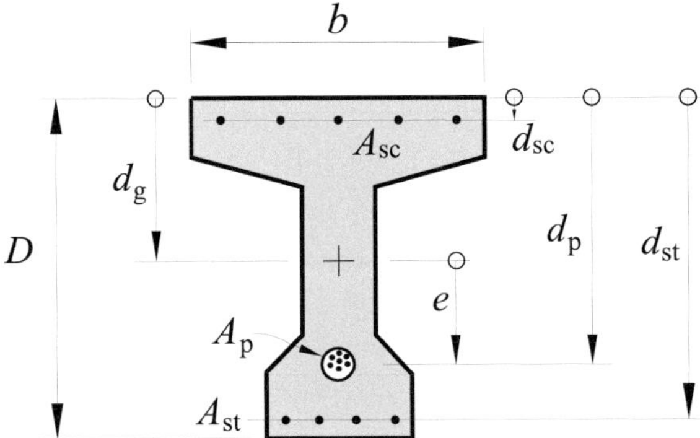

Figure B.2 Cross-section for analysis

The concrete strains in the section are assumed to be linearly distributed, as is shown in Figure B.3 for the case of a rectangular section. In the extreme fibres a and b at the top and bottom of the section, the initial compressive elastic strains at t_o are $\varepsilon_{ce.a}$ and $\varepsilon_{ce.b}$. These can be evaluated as explained in Chapter 4. The initial elastic curvature is $\kappa_o = (\varepsilon_{ce.a} - \varepsilon_{ce.b})/D$. The long-term compressive free creep strains in the extreme fibres at time t^* are:

$$\varepsilon^*_{cc.a} = \varphi^*_o \varepsilon_{ce.a} \tag{B.13}$$

$$\varepsilon^*_{cc.b} = \varphi^*_o \varepsilon_{ce.b} \tag{B.14}$$

At the steel and tendon levels, the long-term creep strains in the concrete are:

$$\varepsilon^*_{cc.st} = \varepsilon^*_{cc.a} \times \frac{D - d_{st}}{D} + \varepsilon^*_{cc.b} \times \frac{d_{st}}{D} \tag{B.15}$$

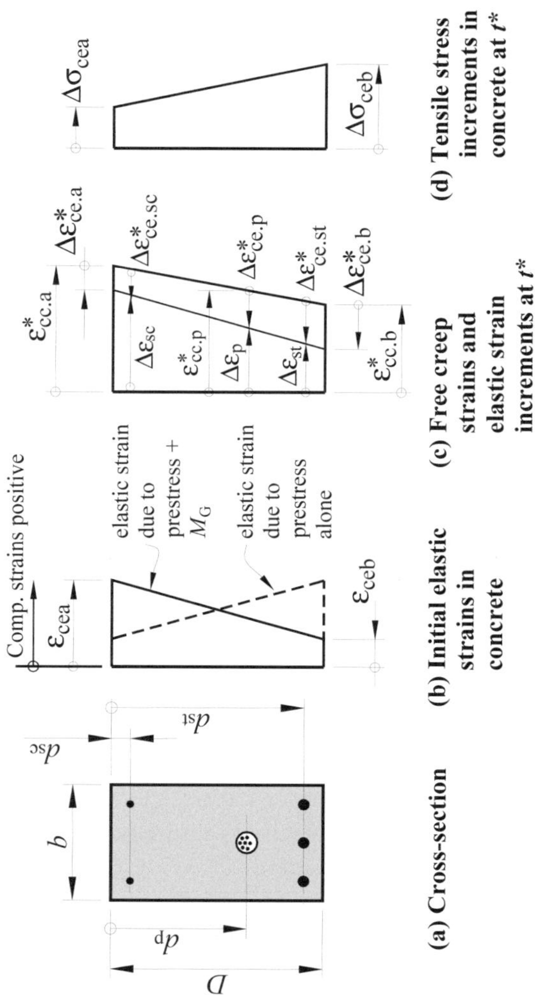

Figure B.3 Creep in an uncracked section with eccentric tendon

$$\varepsilon^*_{\text{cc.sc}} = \varepsilon^*_{\text{cc.a}} \times \frac{D - d_{\text{sc}}}{D} + \varepsilon^*_{\text{cc.b}} \times \frac{d_{\text{sc}}}{D} \tag{B.16}$$

$$\varepsilon^*_{\text{cc.p}} = \varepsilon^*_{\text{cc.a}} \times \frac{D - d_{\text{p}}}{D} + \varepsilon^*_{\text{cc.b}} \times \frac{d_{\text{p}}}{D} \tag{B.17}$$

There are two equilibrium requirements for the equilibrating force system:

Force equilibrium: $$\Delta X^*_{\text{c.c}} = \Delta X^*_{\text{p.c}} + \Delta X^*_{\text{st.c}} + \Delta X^*_{\text{sc.c}} \tag{B.18}$$

Moment equilibrium: $$\Delta X^*_{\text{c.c}} d_{\text{xc}} = \Delta X^*_{\text{p.c}} d_{\text{p}} + \Delta X^*_{\text{st.c}} d_{\text{st}} + \Delta X^*_{\text{sc.c}} d_{\text{sc}} \tag{B.19}$$

The elastic tensile force increment applied to the concrete at time t^*, $\Delta X^*_{\text{c.c}}$, induces elastic tensile strain increments in the concrete, which at the top and bottom fibres, *a* and *b*, are:

$$\Delta\varepsilon^*_{\text{ce.a}} = \frac{\Delta X^*_{\text{c.c}}}{A_{\text{g}} E^*_{\text{c}}} - \frac{\Delta X^*_{\text{c.c}}(d_{\text{xc}} - d_{\text{g}})}{I_{\text{g}} E^*_{\text{c}}} d_{\text{g}} \tag{B.20}$$

$$\Delta\varepsilon^*_{\text{ce.b}} = \frac{\Delta X^*_{\text{c.c}}}{A_{\text{g}} E^*_{\text{c}}} + \frac{\Delta X^*_{\text{c.c}}(d_{\text{xc}} - d_{\text{g}})}{I_{\text{g}} E^*_{\text{c}}} (D - d_{\text{g}}) \tag{B.21}$$

The compressive force increments in the tendon and steels induce elastic strain increments of $\Delta\varepsilon_{\text{p}}$, $\Delta\varepsilon_{\text{st}}$ and $\Delta\varepsilon_{\text{sc}}$, which are related to the compressive force increments by the relevant elastic stiffness terms, such as $E_{\text{p}}A_{\text{p}}$ for the tendon. From Figure B.3 the compatibility requirements are seen to be as follows:

At the tendon level: $$\varepsilon^*_{\text{cc.p}} = \Delta\varepsilon_{\text{p}} + \Delta\varepsilon^*_{\text{ce.p}} \tag{B.22}$$

At the tensile steel level: $$\varepsilon^*_{\text{cc.st}} = \Delta\varepsilon_{\text{st}} + \Delta\varepsilon^*_{\text{ce.st}} \tag{B.23}$$

At the compressive steel level: $$\varepsilon^*_{\text{cc.sc}} = \Delta\varepsilon_{\text{sc}} + \Delta\varepsilon^*_{\text{ce.sc}} \tag{B.24}$$

In these expressions the left hand term is the strain incompatibility, which is equal to the free concrete compressive creep at the relevant level of tendon or steel. The first term on the right side is the elastic compressive strain increment in the tendon or steel, and the second right hand side term is the elastic tensile strain increment induced at t^* in the concrete at this same level.

To proceed algebraically, we express the four force increments and the depth d_{xc} in terms of the extreme fibre elastic strain increments $\Delta\varepsilon^*_{ce.a}$ and $\Delta\varepsilon^*_{ce.b}$. Substituting these expressions into the two equilibrium equations, we then obtain two simultaneous equations in $\Delta\varepsilon^*_{ce.a}$ and $\Delta\varepsilon^*_{ce.b}$. Solving, we evaluate these elastic strain increments, and hence the force increments, the stress and strain increments in concrete, steels and tendon, and the prestress losses due to creep and the creep curvature.

To evaluate the tensile increment $\Delta X^*_{c.c}$, we first write an expression for the elastic strain increment in the concrete at the depth of the concrete section centroid, d_g, in terms of the extreme fibre elastic strain increments:

$$\Delta\varepsilon^*_{ce.g} = \Delta\varepsilon^*_{ce.a} \times \frac{D - d_g}{D} + \Delta\varepsilon^*_{ce.b} \times \frac{d_g}{D} = \frac{\Delta X^*_{c.c}}{A_g E^*_c} \qquad \text{(B.25)}$$

Rearranging, we have

$$\Delta X^*_{c.c} = A_g E^*_c \left(\Delta\varepsilon^*_{ce.a} \times \frac{D - d_g}{D} + \Delta\varepsilon^*_{ce.b} \times \frac{d_g}{D} \right) \qquad \text{(B.26)}$$

Equations for the three compressive force increments are similarly obtained:

$$\Delta X^*_{p.c} = A_p E_p \times \left(\varepsilon^*_{cc.p} - \left(\Delta\varepsilon^*_{ce.a} \frac{(D - d_p)}{D} + \Delta\varepsilon^*_{ce.b} \frac{d_p}{D} \right) \right) \qquad \text{(B.27)}$$

$$\Delta X^*_{st.c} = A_{st} E_s \times \left(\varepsilon^*_{cc.st} - \left(\Delta\varepsilon^*_{ce.a} \frac{(D - d_{st})}{D} + \Delta\varepsilon^*_{ce.b} \frac{d_{st}}{D} \right) \right) \qquad \text{(B.28)}$$

$$\Delta X^*_{sc.c} = A_{sc} E_s \times \left(\varepsilon^*_{cc.sc} - \left(\Delta\varepsilon^*_{ce.a} \frac{(D - d_{sc})}{D} + \Delta\varepsilon^*_{ce.b} \frac{d_{sc}}{D} \right) \right) \qquad \text{(B.29)}$$

To evaluate d_{xc} we subtract Equations B.19 from B.18, and substitute the expression for $\Delta X^*_{c.c}$ in Equation B.26. This gives:

$$d_{xc} = d_g + \left(\frac{I_g}{A_g}\right) \frac{(\Delta\varepsilon^*_{ce.b} - \Delta\varepsilon^*_{ce.a})}{(\Delta\varepsilon^*_{ce.a}(D - d_g) + \Delta\varepsilon^*_{ce.b} d_g)} \tag{B.30}$$

Substituting Equations B.26 to B.30 into the equilibrium equations (Equations B.18 and B.19), we obtain two simultaneous equations in $\Delta\varepsilon^*_{ce.a}$ and $\Delta\varepsilon^*_{ce.b}$:

$$a_{11}\Delta\varepsilon^*_{ce.a} + a_{12}\Delta\varepsilon^*_{ce.b} = c_1 \tag{B.31}$$

$$a_{21}\Delta\varepsilon^*_{ce.a} + a_{22}\Delta\varepsilon^*_{ce.b} = c_2 \tag{B.32}$$

or:

$$\begin{bmatrix} a_{11} & a_{12} \\ a_{21} & a_{22} \end{bmatrix} \cdot \begin{bmatrix} \Delta\varepsilon^*_{ce.a} \\ \Delta\varepsilon^*_{ce.b} \end{bmatrix} = \begin{bmatrix} c_1 \\ c_2 \end{bmatrix} \tag{B.33}$$

Solving for the elastic strain increments, $\Delta\varepsilon^*_{ce.a}$ and $\Delta\varepsilon^*_{ce.b}$, we obtain:

$$\Delta\varepsilon^*_{ce.a} = \frac{a_{22}c_1 - a_{12}c_2}{a_{11}a_{22} - a_{21}a_{12}} \tag{B.34}$$

$$\Delta\varepsilon^*_{ce.b} = \frac{a_{11}c_2 - a_{21}c_1}{a_{11}a_{22} - a_{21}a_{12}} \tag{B.35}$$

which are positive if tensile. The non-dimensional terms a_{ij} and c_i are:

$$a_{11} = \frac{D - d_g}{D} + n_p p_p \frac{D - d_p}{D} + n_{st} p_s \frac{D - d_{st}}{D} + n_{sc} p_s \frac{D - d_{sc}}{D} \tag{B.36}$$

$$a_{12} = \frac{d_g}{D} + n_p p_p \frac{d_p}{D} + n_{st} p_s \frac{d_{st}}{D} + n_{sc} p_s \frac{d_{sc}}{D} \tag{B.37}$$

$$c_1 = n_p p_p \varepsilon^*_{cc.p} + n_{st} p_s \varepsilon^*_{cc.st} + n_{sc} p_s \varepsilon^*_{cc.sc} \tag{B.38}$$

$$a_{21} = \frac{d_g}{D}\left(\frac{D-d_g}{D}\right) - i_g + n_p p_p d_p \frac{D-d_p}{D} + n_s p_{st} \frac{d_{st}}{D}\frac{D-d_{st}}{D} + n_s p_{sc} \frac{d_{sc}}{D}\frac{D-d_{sc}}{D} \quad \text{(B.39)}$$

$$a_{22} = \left(\frac{d_g}{D}\right)^2 + i_g + n_p p_p \left(\frac{d_p}{D}\right)^2 + n_s p_{st}\left(\frac{d_{st}}{D}\right)^2 + n_s p_{st}\left(\frac{d_{sc}}{D}\right)^2 \quad \text{(B.40)}$$

$$c_2 = \varepsilon^*_{cc.p} n_p p_p \frac{d_p}{D} + \varepsilon^*_{cc.st} n_s p_{st} \frac{d_{st}}{D} + \varepsilon^*_{cc.sc} n_s p_{sc} \frac{d_{sc}}{D} \quad \text{(B.41)}$$

The following non-dimensionalised geometric cross-sectional properties and material stiffness properties are used in the above expressions:

$$n_p = \frac{E_p}{E^*_c};\ n_s = \frac{E_s}{E^*_c};$$

$$p_p = \frac{A_p}{A_g};\ p_{st} = \frac{A_{st}}{A_g};\ p_{sc} = \frac{A_{sc}}{A_g};\ \text{and } i_g = \frac{I_g}{A_g D^2}.$$

Note that Equations B.38 and B.41 for c_1 and c_2 include the relevant free creep strain increments at various levels in the section, which are evaluated from $\varepsilon^*_{cc.a}$ and $\varepsilon^*_{cc.b}$, as in Equation B.9.

The resulting elastic increment in curvature at t^* is:

$$\Delta\kappa^*_c = -\frac{(\Delta\varepsilon^*_{ce.a} - \Delta\varepsilon^*_{ce.b})}{D} \quad \text{(B.42)}$$

which is negative if $\Delta\varepsilon^*_{ce.a}$ is greater than $\Delta\varepsilon^*_{ce.b}$. The final creep-induced curvature at t^* is:

$$\kappa^*_c = \kappa^*_{co} + \Delta\kappa^*_c \quad \text{(B.43)}$$

Note that κ^*_{co} is the free creep curvature: $\kappa^*_{co} = \varphi^*_o \kappa_o$, and the correction $\Delta\kappa^*_c$ accounts for the restraining effect of the tendon and reinforcement on the free creep deformations. In regards to signs, both $\Delta\varepsilon^*_{ce.a}$ and $\Delta\varepsilon^*_{ce.b}$ are positive when tensile, and so the correction $\Delta\kappa^*_c$ is positive when $\Delta\varepsilon^*_{ce.a} \leq \Delta\varepsilon^*_{ce.b}$.

With the elastic strain increments in the concrete at t^* determined, the tensile force increment in the concrete, $\Delta X^*_{c.c}$, which is also the loss of prestress in the concrete, is evaluated from Equation B.26. The other force increments are evaluated using Equations B.27 to B.29. The loss of prestress in the tendon, $\Delta X^*_{p.c}$, is only equal to the loss of prestress in the concrete, $\Delta X^*_{c.c}$, if there is no reinforcing steel in the section. The above equations provide an estimate of the effects of long-term creep in the section. The calculations are best carried out using spreadsheets.

B.3.2 Age-adjusted effective modulus modification

In the above analysis, the simplifying assumption has been made that the concrete stresses remain constant over the entire time interval between t_o and t^*, and that the elastic strain components, $\Delta\varepsilon^*_{ce.a}$ and $\Delta\varepsilon^*_{ce.b}$, and the corresponding elastic stress components, $\Delta\sigma^*_{ce.a}$ and $\Delta\sigma^*_{ce.b}$, are suddenly applied at time t^*. In fact these increments in stress and strain build up gradually from time t_o to t^*. To take account of these gradual increases, we can use the age-adjusted effective modulus for the concrete in lieu of the elastic modulus E^*_c at time t^*. The age-adjusted effective concrete modulus is defined in Equation 2.47 as:

$$E_{R\chi}(t, t_o) = \frac{E_c(t_o)}{1 + \chi(t, t_o)\varphi(t, t_o)} \tag{B.44}$$

In creating the age-adjusted effective modulus, Bazant used the "ageing coefficient", $\chi(t, t_o)$, to modify the effective modulus.

To employ this modification here, we replace E^*_c by the long-term age-adjusted effective modulus, $E_{R\chi}(t^*, t_o)$, in the relevant equations in Section B.3.1 where the final elastic increments $\varepsilon^*_{cc.a}$ and $\varepsilon^*_{cc.b}$ are evaluated. To do this it is convenient to use the term:

$$k^* = \frac{E_{R\chi}(t^*, t_o)}{E_c^*} \tag{B.45}$$

so that the modification to the equations consists simply of using $k^* E_c^*$ in lieu of E_c^*. To evaluate k^*, we rewrite Equation B.45 as follows:

$$k^* = \frac{E_{R\chi}(t^*, t_o)}{E_c(t)} \times \frac{E_c(t)}{E_c^*} = \frac{1}{1 + \chi(t^*, t_o)\varphi(t^*, t_o)} \left(\frac{E_c(t)}{E_c^*}\right)$$

The ageing coefficient at time t^*, which we can write as χ_o^*, has a value less than unity when the sustained stress *increases* with time. This is the situation we are considering, as the tensile stress increments in the extreme fibres of concrete increase from zero towards final values of $\Delta\sigma_{ce.a}^*$ and $\Delta\sigma_{ce.b}^*$ at t^*. A typical value of 0.8 is often used for such cases, for loading first applied at 28 days. This value was suggested by Bazant in his original work in 1972.

More recently, values for the ageing coefficient have been proposed by Gilbert and Ranzi (2011). Their recommended long-term values are:

For concrete loaded at an early age, $t_o < 20$ days:

with near-constant external load, $\chi_o^* = 0.65$;

with near-constant deformation (relaxation), $\chi_o^* = 0.8$.

For concrete loaded at later ages, $t_o > 20$ days:

with near-constant external load, $\chi_o^* = 0.75$;

with near-constant deformation (relaxation), $\chi_o^* = 0.85$.

The age t_o (at which sustained load is first applied) has an important effect on the value of k^*. Maximum values of k^* occur for aged concrete, i.e. when t_o is large. For example, for $t_o = 180$ days, and taking typical values of $E_c(180)/E_c^* = 1.0$, $\chi_o^* = 0.75$ and $\varphi_o^* = 1.0$ we obtain $k^* = 1/1.75$, or about 0.7. Much smaller values for k^* occur when the concrete is very young. For example, for $t_o = 7$ days, and taking $\chi_o^* = 0.65$, $\varphi_o^* = 4.0$ and $(E_c(14))/E_c^* = 1.3$ we obtain $k^* = 0.22$.

For sustained loading at 28 days, and with values $\chi_o^* = 0.75$, $\varphi_o^* = 3.0$ and $E_c(28)/E_c^* = 1.1$, we obtain $k^* = 0.28$. A value of around 0.3 will be used in the numerical examples that follow. If creep becomes a critical design issue, a range of values of k^* can of course be used in a sensitivity check on the final predicted results.

In summary, to allow approximately for the gradual increase in the elastic tensile concrete stresses and strains over time, we replace the long-term elastic modulus, E_c^*, by $k^* E_c^*$ in Equations B.18 to B.43.

EXAMPLE B.1

ONE-STEP ANALYSIS OF CREEP IN A REINFORCED PRESTRESSED RECTANGULAR CROSS-SECTION

A one-step analysis is used with the effective modulus correction to estimate the long-term loss of prestress and increase in curvature due to creep in a reinforced, prestressed section. The rectangular section has $b = 400$ mm and $D = 800$ mm. The tendon eccentricity is $e = 250$ mm.

The constant sustained moment is $M_G = 400$ kNm, applied at 28 days at the time of prestressing. For comparison purposes, the calculations are repeated in Example B.2, for the section without steel reinforcement, and in Examples B.3 and B.4 taking into account shrinkage and creep.

Other data:

$A_p = 1000$ mm^2; $E_p = 195 \times 10^3$ MPa; $P_o = 1200$ kN

$A_{st} = A_{sc} = 1350$ mm^2; $d_{st} = 750$ mm; $d_{sc} = 50$ mm.

$\varphi_o^* = 3.0$; $E_{co} = 30 \times 10^3$ MPa; $E_c^* = 33 \times 10^3$ MPa.

$A_c = 320 \times 10^3$ mm^2; $d_g = 400$ mm; $I_g = 17.07 \times 10^9$ mm^4

SOLUTION

Initial conditions

The elastic concrete compressive stresses in the top a and bottom b fibres at time t_o due to the prestress and the moment M_G are determined as explained in Section 4.2.2 of Chapter 4. The top and bottom fibre elastic strains are 203×10^{-6} and 47×10^{-6}, respectively.

The initial elastic negative curvature due to prestress alone is:

$$\kappa_{op} = -\frac{P_o e}{E_c I_g} = -\frac{1200\times10^3 \times 250}{30\times10^3 \times 17.07\times10^9} = -586\times10^{-9}\text{mm}^{-1}$$

The positive elastic curvature due to moment M_G is:

$$\kappa_{oG} = +\frac{400\times10^6}{30\times10^3 \times 17.07\times10^9} = +781\times10^{-9}\text{mm}^{-1}$$

The nett initial curvature is:

$$\kappa_o = \kappa_{op} + \kappa_{oG} = +195\times10^{-9}\text{ mm}^{-1}$$

Free creep strains and curvature

The free creep strains, calculated using Equations B.13 and B.14, are: $\varepsilon^*_{cc.a} = 609\times10^{-6}$ and $\varepsilon^*_{cc.b} = 141\times10^{-6}$. The long-term free creep curvature is the initial elastic curvature multiplied by φ^*_o:

$$\kappa^*_{co} = 3.0 \times 195\times10^{-9} = 585\times10^{-9}\text{mm}^{-1}$$

At the level of the tendon the free creep strain is $\varepsilon^*_{cc.p} = 228\times10^{-6}$. At the steel levels we have $\varepsilon^*_{cc.st} = 170\times10^{-6}$ and $\varepsilon^*_{cc.sc} = 580\times10^{-6}$.

(a) Order-of-magnitude estimates

Estimates of the long-term free creep curvature and the total long-term curvature are then obtained as:

$$\kappa_{co}^{*} = 3 \times 195{\times}10^{-9}\,\mathrm{mm}^{-1} = 585{\times}10^{-9}\,\mathrm{mm}^{-1}$$

$$\kappa^{*} = \kappa_{o} + \kappa_{co}^{*} = 4 \times 195{\times}10^{-9} = 780{\times}10^{-9}\,\mathrm{mm}^{-1}$$

To estimate the long-term prestress loss in the tendon, we have $\varepsilon_{cc.p}^{*} = 228{\times}10^{-6}$ and, hence, $\Delta X_{p.c}^{*} = A_p E_p \varepsilon_{cc.p} = 45$ kN, which is under 4 per cent of the initial prestress. Calculation of the compressive forces transferred from the concrete to the steel and tendon can be calculated similarly, as previously explained.

(b) One-step creep analysis

Evaluation of elastic strain increments at t^*

The following values for the a_{ij} and c_i parameters are obtained using Equations B.36 to B.41 in a spreadsheet calculation:

$$a_{11} = 0.5290;\ a_{12} = 0.5406;\ c_1 = 23.396{\times}10^{-6}$$

$$a_{21} = 0.1725;\ a_{22} = 0.3681;\ c_2 = 4.175{\times}10^{-6}$$

From Equations B.34 and B.35, we obtain:

$$\Delta\varepsilon_{ce.a}^{*} = 39.96{\times}10^{-6};\quad \Delta\varepsilon_{ce.b}^{*} = 4.175{\times}10^{-6}$$

Force increments

The force increments are obtained from Equations B.26 to B.29:

$$\Delta X_{p.c}^{*} = 43\ \mathrm{kN};\ \Delta X_{c.c}^{*} = 233\ \mathrm{kN}$$

$$\Delta X_{sc.c}^{*} = 146\ \mathrm{kN};\ \Delta X_{st.c}^{*} = 44\ \mathrm{kN}$$

As a check on equilibrium, $43 + 146 + 44 = 233 = \Delta X_{c.c}^{*}$ and thus okay.

The loss of prestress in the tendon (43 kN) is small, but in the concrete it is more than five times as large, at about 20 per cent of P_o. The difference is due to the transfer of the forces $\Delta X^*_{st.c}$ and $\Delta X^*_{sc.c}$ from the concrete to the bar reinforcements.

Creep curvature correction increment $\Delta\kappa^*_c$

At t^*, when the extreme fibre elastic strain increments $\Delta\varepsilon^*_{ce.a}$ and $\Delta\varepsilon^*_{ce.b}$ are applied, the increment in curvature is:

$$\Delta\kappa^*_c = -\frac{(39.96 - 4.175) \times 10^{-6}}{800} = -44.7 \times 10^{-9}\text{mm}^{-1}$$

$$\kappa^*_c = \kappa^*_{co} + \Delta\kappa^*_c = (585 - 44.7)\times 10^{-9} = 542\times 10^{-9}\text{mm}^{-1}$$

The correction is negative because the tensile strain increment in the top fibre, $\Delta\varepsilon^*_{ce.a}$, is higher than in the lower fibre. This is in turn due to the compressive steel present in the upper fibres, coupled with the high compressive concrete stresses in the top fibres due to the large moment M_G.

(c) Age-adjusted effective modulus adjustment

We repeat the above calculations, using a long-term age-adjusted effective modulus value $k^* E^*_c = 9600$ MPa to allow for the gradual increase in the concrete stress with time. The initial conditions and the free creep calculations are unchanged.

Strain increments at t^*

The a_{ij} and c_i parameters become:

$$a_{11} = 0.5983;\ a_{12} = 0.6355;\ c_1 = 7.894\times 10^{-5}$$

$$a_{21} = 0.1862;\ a_{22} = 0.4493;\ c_2 = 2.823\times 10^{-5}$$

The values for the strain increments (which now include a creep component and are somewhat higher) at t^* are now:

$$\Delta\varepsilon^*_{ce.a} = 116.5\times 10^{-6};\ \Delta\varepsilon^*_{ce.b} = 14.53\times 10^{-6}$$

Loss of prestress

The force increments, obtained from Equations B.26 to B.29, are:

$$\Delta X^*_{p.c} = 38 \text{ kN}; \ \Delta X^*_{c.c} = 204 \text{ kN};$$

$$\Delta X^*_{sc.c} = 127 \text{ kN}; \ \Delta X^*_{st.c} = 40 \text{ kN}$$

Creep curvature correction $\Delta\kappa^*_c$

The elastic curvature correction to be made at t^* is:

$$\Delta\kappa^*_c = -\frac{\Delta\varepsilon^*_{ce.a} - \Delta\varepsilon^*_{ce.b}}{D} = -\frac{(116.51 - 14.53)\times10^{-6}}{800}$$

$$= -128\times10^{-9}\text{mm}^{-1}$$

With this negative correction the corrected creep curvature is:

$$\kappa^*_c = \kappa^*_{co} + \Delta\kappa^*_c = (585 - 127.5)\times10^{-9} = 457\times10^{-9}$$

(d) Discussion:

Use of the age-adjusted effective modulus has increased the final creep curvature correction, $\Delta\kappa^*_c$, from -44.7×10^{-9} to -128×10^{-9} mm^{-1}. This is a large difference, but the change in the final value of the creep curvature, κ^*_c, is only from 542×10^{-9} to 457×10^{-9} mm^{-1}, which is about 20 per cent. The order-of-magnitude estimate of 585×10^{-9} mm^{-1} is conservative and an over-estimate by about 30 per cent. In this example, of this 30 per cent, the correction to restore compatibility makes up about 10 per cent and the correction to allow for varying stress about 20 per cent.

Estimates of the loss of prestress in the concrete with and without the age-affected modulus correction are 204 kN and 233 kN, respectively. The error due to ignoring the change in stress during creep is about 15 per cent in this case. The order-of-magnitude calculation was only made for loss of prestress in the tendon, which was 45 kN. The estimates without and with the age-adjusted modulus were 43 kN and 38 kN, suggesting a slight overestimate in the rough estimate. It should be noted however that the absolute value for the tendon is small.

EXAMPLE B.2

ONE-STEP ANALYSIS OF CREEP IN A PRESTRESSED BEAM SECTION WITHOUT LONGITUDINAL BAR REINFORCEMENT

In this example, Example B.1 is repeated using the same cross-section, but without bar reinforcement for flexure; that is, $A_{st} = A_{sc} = 0$.

SOLUTION

(a) Order-of-magnitude estimates

The order of magnitude estimates for deformations and prestress loss in the tendon are unchanged from that of the previous example.

(b) One-step analysis

For the cross-section without reinforcement, the initial conditions are unchanged because we are using the elastic properties of the plain concrete section. However, values for the a_{ij} and c_i parameters change and the elastic corrections at t^* now become:

$$\Delta\varepsilon^*_{ce.a} = -3.550\times10^{-6};\quad \Delta\varepsilon^*_{ce.b} = 11.66\times10^{-6}$$

The force increments are:

$$\Delta X^*_{p.c} = 43\text{ kN};\quad \Delta X^*_{c.c} = 43\text{ kN};$$

$$\Delta X^*_{sc.c} = 0\text{ kN};\quad \Delta X^*_{st.c} = 0\text{ kN}$$

The loss of prestress in the concrete is now equal to that in the tendon (which is still about 4 per cent) and so much less than in Example B.1. This demonstrates how the presence of reinforcement magnifies the loss of prestress in the concrete due to creep.

Creep curvature correction at *t**

As is to be expected, the curvature correction:

$$\Delta\kappa_c^* = -\frac{\Delta\varepsilon_{ce.a}^* - \Delta\varepsilon_{ce.b}^*}{D} = -\frac{(-3.55 - 11.66)\times10^{-6}}{800}$$

$$= -19.0\times10^{-9}\text{mm}^{-1}$$

is less than one-half that in Example B.1. The corrected long-term creep curvature is then:

$$\kappa_c^* = \kappa_{co}^* + \Delta\kappa_c^* = (585 + 19)\times10^{-9} = 604\times10^{-9}\text{mm}^{-1}$$

and is about 4 per cent larger than in the previous example.

(c) Age-adjusted effective modulus modification

Using the age-adjusted effective modulus with $k^*E_c^* = 9600$ MPa, in lieu of $E_c^* = 33000$ MPa, the initial conditions are unaltered.

Strain increments at *t**

Using the reduced modulus value $k^*E_c^* = 9600$ MPa, the a_{ij} and c_i parameters become:

$$a_{11} = 0.5983;\ a_{12} = 0.6355;\ c_1 = 7.894\times10^{-5}$$

$$a_{21} = 0.1862;\ a_{22} = 0.4493;\ c_2 = 2.823\times10^{-5}$$

The values for the strain increments (including a creep component) at t^* are:

$$\Delta\varepsilon_{ce.a}^* = -11.02\times10^{-6};\ \Delta\varepsilon_{ce.b}^* = 36.2\times10^{-6}$$

The creep curvature correction is positive and is:

$$\Delta\kappa_c^* = -(-11.02 - 36.2)\times10^{-6}/800 = 59\times10^{-9}\text{mm}^{-1}$$

The corrected creep curvature is:

$$\kappa_c^* = \kappa_{co}^* + \Delta\kappa_c^* = (585 + 59)\times10^{-9} = 645\times10^{-9}\text{mm}^{-1}$$

Loss of prestress

The force increments are:

$$\Delta X_{p.c}^* = 39 \text{ kN}; \ \Delta X_{c.c}^* = 39 \text{ kN}$$

$$\Delta X_{sc.c}^* = 0 \text{ kN}; \ \Delta X_{st.c}^* = 0 \text{ kN}$$

(d) Discussion

With the reinforcement removed, the loss of prestress in the concrete has fallen from 17 per cent to around 3 per cent, and the loss of prestress in the tendon is now equal to that in the concrete.

As in Example B.1, the difference between the curvature corrections without and with the age-adjusted modulus adjustment is again large, $+19\times10^{-9}$ versus $+59\times10^{-9}$ mm^{-1}; nevertheless, the difference in the final κ_c^* values is much smaller: 565×10^{-9} and 645×10^{-9} mm^{-1}, a difference of about 6 per cent. The order-of-magnitude estimate of curvature is much closer in this example, 604×10^{-9} mm^{-1}, an error of only 10 per cent, which is because the section does not have any reinforcement.

Use of the age-adjusted modulus correction reduces the loss of prestress in the concrete from 43 to 39 kN. The order-of-magnitude estimate, 45 kN, is about 12 per cent in error, but the absolute magnitudes are all very small.

The results of Examples B.1 and B.2 highlight the important effect of reinforcement on prestress loss. Without reinforcement, the loss of prestress in the concrete is equal to that in the tendon and is relatively small. The loss of prestress in the concrete increases significantly when reinforcement is added to the section. On the other hand, the presence of the reinforcement decreases the long-term curvature due to creep.

B.3.3 Analysis for combined creep and shrinkage

When uniform shrinkage develops in the section of a member, it is partially restrained by the tendon and the reinforcing steel in the same way that creep is partially restrained. If the tendon and reinforcement are unsymmetrically placed (which they usually are), a shrinkage-induced strain gradient develops over time. As we have seen, the additional curvature and the resulting increase in deflection are referred to as **shrinkage warping**. The tendon and steel are usually below the section centroid, and restrain the compressive creep increments in the lower fibres, so that the shrinkage-induced curvature, $\kappa_{sh}(t)$, is positive and the deflection is downward.

In reality, both shrinkage and creep develop simultaneously in the section over time. A simple modification can be made to the one-step analysis in Section B.3.1 to allow for the effect of shrinkage. It is only necessary to evaluate the long-term free shrinkage strains in the top and bottom fibres, $\Delta\varepsilon^*_{csh.a}$ and $\Delta\varepsilon^*_{csh.b}$, and add these to the free creep strain increments, $\Delta\varepsilon^*_{cc.a}$ and $\Delta\varepsilon^*_{cc.b}$ in Equations B.13 and B.14. These are then the total inelastic long-term strains and can be used in lieu of the creep strains in the analysis, which proceeds otherwise unchanged.

A one-step analysis for shrinkage alone could be made, but it would ignore the tensile stress that progressively builds up over time in the concrete due to restraint provided by the steel and tendon, and hence an induced tensile creep. The age-adjusted effective modulus could be used in the shrinkage analysis; however, two separate analyses would be then used, instead of one for the combined effects.

EXAMPLE B.3

ONE-STEP ANALYSIS FOR COMBINED SHRINKAGE AND CREEP IN A PRE-STRESSED, REINFORCED SECTION

The one-step analysis in Example B.1 is repeated, but in this example a final uniform shrinkage strain $\varepsilon_{cs} = 800\times10^{-6}$ is included. The details of the section are unchanged.

SOLUTION

(a) Order-of-magnitude estimates

There is no curvature increment due to the free uniform shrinkage. The additional loss of prestress in the tendon due to shrinkage is obtained from the inelastic strain in the concrete at the level of the tendon. The additional shrinkage strain increment is $800{\times}10^{-6}$, and so the additional loss of prestress in the tendon due to shrinkage is conservatively estimated as:

$$\begin{aligned}\Delta X^*_{\text{p.sh}} &= A_\text{p}E_\text{p}(800{\times}10^{-6}) = 1000\times 195{\times}10^3\times 800{\times}10^{-6}\\ &= 156{\times}10^3\ \text{N}\quad(156\ \text{kN})\end{aligned}$$

The additional compressive forces in the steels are estimated as:

$$\begin{aligned}\Delta X^*_{\text{st.sh}} = \Delta X^*_{\text{sc.sh}} &= 1350\times 200{\times}10^3\times 800{\times}10^{-6}\\ &= 216{\times}10^3\ \text{N}\quad(216\ \text{kN})\end{aligned}$$

Thus, the loss of prestress in the concrete is:

$$\Delta X^*_{\text{c.sh}} = 156 + 2\times 216 = 588\ \text{kN}$$

(b) One-step analysis

Free creep and shrinkage strains

The free creep strains in the extreme fibres at time t^* are increased by the shrinkage strains of $800{\times}10^{-6}$:

$$\varepsilon^*_{\text{cc.a}} + \varepsilon^*_{\text{csh.a}} = (609+800){\times}10^{-6} = 1409\times 10^{-6}$$

$$\varepsilon^*_{\text{cc.b}} + \varepsilon^*_{\text{csh.b}} = (141+800){\times}10^{-6} = 941\times 10^{-6}.$$

These values are used in lieu of $\varepsilon^*_{\text{cc.a}}$ and $\varepsilon^*_{\text{cc.b}}$ in the subsequent calculations, which otherwise follow those in Example B.1. Since the shrinkage is uniform in the section, the free inelastic curvature is the same as in Example B.1:

$$\kappa^*_{\text{co}} + \kappa^*_{\text{sh}} = ((1409-941)\times 10^{-6})/800 + 0 = 585\times 10^{-9}\ \text{mm}^{-1}$$

Elastic strain increments at *t**

The values of the a_{ij} parameters do not change from those in Example B.1, but the c_i parameters depend on the free inelastic strains and become:

$$c_1 = 79.07\times10^{-6};\ c_2 = 40.89\times10^{-6}$$

The elastic strain increments at t^* (tensile positive) become:

$$\Delta\varepsilon^*_{ce.a} = 69.03\times10^{-6};\quad \Delta\varepsilon^*_{ce.b} = 78.73\times10^{-6}$$

The force increments (now due to creep plus shrinkage) are obtained from Equations B.27 to B.30 as:

$$\Delta X^*_{p.csh} = 186\text{ kN};\ \Delta X^*_{c.csh} = 780\text{ kN}$$

$$\Delta X^*_{sc.csh} = 354\text{ kN};\ \Delta X^*_{st.csh} = 241\text{ kN}$$

Curvature correction

The elastic curvature correction at t^* is obtained using the values of the extreme fibre elastic strain increments:

$$\Delta\kappa^*_c = -\frac{\Delta\varepsilon^*_{ce.a} - \Delta\varepsilon^*_{ce.b}}{D} = -\frac{(69.03 - 78.73)\times10^{-6}}{800}$$

$$= 8.875\times10^{-9}\text{mm}^{-1}$$

In this case the correction is small and positive and the total inelastic curvature becomes:

$$\kappa^*_c = \kappa^*_{co} + \Delta\kappa^*_c = (585 + 9)\times10^{-9} = 594\times10^{-9}\text{mm}^{-1}$$

(c) Age-adjusted effective modulus adjustment

Evaluation of strain increments at *t**

The a_{ij} terms do not change, but the c_i parameters are now:

$$c_1 = 256.9\times10^{9};\ c_2 = 136.6\times10^{8}$$

The elastic strain increments at t^* are therefore somewhat higher:

$$\Delta\varepsilon^*_{\text{ce.a}} = 218.2\times10^{-6}; \quad \Delta\varepsilon^*_{\text{ce.b}} = 220.5\times10^{-6}$$

Curvature correction

The curvature correction at t^* is very small:

$$\Delta\kappa^*_{\text{c}} = -\frac{\Delta\varepsilon^*_{\text{ce.a}} - \Delta\varepsilon^*_{\text{ce.b}}}{D} = -\frac{(218.2 - 220.5)\times10^{-6}}{800}$$

$$= 2.3\times10^{-9}\text{mm}^{-1}$$

The corrected creep curvature becomes:

$$\kappa^*_{\text{c}} = \kappa^*_{\text{co}} + \Delta\kappa^*_{\text{c}} = (585 + 2.3)\times10^{-9} = 587\times10^{-9}\text{mm}^{-1}$$

Loss of prestress

The force increments due to long-term creep and shrinkage are now:

$$\Delta X^*_{\text{p.csh}} = 158 \text{ kN}; \quad \Delta X^*_{\text{c.csh}} = 672 \text{ kN}$$

$$\Delta X^*_{\text{sc.csh}} = 313 \text{ kN}; \quad \Delta X^*_{\text{st.csh}} = 203 \text{ kN}$$

(d) Discussion

These results and those from Example B.1 show that the loss of prestress in the concrete is greatly increased when shrinkage is included, from 204 kN to 780 kN. Likewise, the loss of prestress in the tendon increases from 38 kN to 186 kN. The differences in prestress loss, caused by shrinkage, are 576 kN and 148 kN. The order-of-magnitude estimates for the prestress losses are 588 kN and 156 kN and are very close in this example. In Example B.1, the curvature correction due to creep is negative, but due to combined creep and shrinkage it becomes almost zero, indicating that the correction due to shrinkage is positive and nearly as large in magnitude as that due to creep. The correction due to shrinkage is positive because the areas of steel and tendon are higher in the lower fibres.

EXAMPLE B.4

ONE-STEP ANALYSIS FOR COMBINED SHRINKAGE AND CREEP IN A PRE-STRESSED SECTION WITHOUT LONGITUDINAL BAR REINFORCEMENT

Example B.2 is repeated, taking account of the same shrinkage as in B.3.

SOLUTION

(a) One-step-analysis

Without tensile and compressive steel reinforcement, the elastic strain increments at t^*, allowing for shrinkage, are larger than in Example B.2 but smaller than in B.3. They are:

$$\Delta\varepsilon^*_{ce.a} = -15.98\times10^{-6};\ \ \Delta\varepsilon^*_{ce.b} = 52.50\times10^{-6}$$

The curvature increment is positive because the tendon is below the section centroid, but is larger than in Example B.2 because of shrinkage:

$$\Delta\kappa^*_c = -((-15.98 - 52.50)\times10^{-6}/800) = 85.59\times10^{-9}\ \text{mm}^{-1}$$

The force increments in the steel areas are of course zero, and the loss of prestress in the concrete is larger than in Example B.2, but smaller than in B.3:

$$\Delta X^*_{p.csh} = \Delta X^*_{c.csh} = 193\ \text{kN}$$

The corrected final curvature is:

$$\kappa^*_c = \kappa^*_{co} + \Delta\kappa^*_c = (585 + 85.59)\times10^{-9} = 671\times10^{-9}\ \text{mm}^{-1}$$

(b) Age-adjusted effective modulus adjustment

Using the age-adjusted effective modulus, the strain increments applied at t^* are found to be:

$$\Delta\varepsilon^*_{ce.a} = -49.59\times10^{-6};\ \ \Delta\varepsilon^*_{ce.b} = 162.9\times10^{-6}$$

Curvature correction

The curvature correction at t^* is:

$$\Delta\kappa_c^* = -\frac{\Delta\varepsilon_{ce.a}^* - \Delta\varepsilon_{ce.b}^*}{D} = -\frac{(-49.59 - 162.9)\times10^{-6}}{800}$$

$$= 265.7\times10^{-9}\text{mm}^{-1}$$

The corrected creep curvature is:

$$\kappa_c^* = \kappa_{co}^* + \Delta\kappa_c^* = (585 + 266)\times10^{-9} = 851\times10^{-9}\text{mm}^{-1}$$

Loss of prestress

The force increments are:

$$\Delta X_{p.csh}^* = 177 \text{ kN}; \ \Delta X_{c.csh}^* = 177 \text{ kN}$$

$$\Delta X_{sc.csh}^* = 0 \text{ kN}; \ \Delta X_{st.csh}^* = 0 \text{ kN}$$

(c) Discussion

With the reinforcement removed, the prestress loss due to joint creep and shrinkage is now 177 kN in the concrete (as well as the tendon), which is far less than the loss of 672 kN in Example B.3. On the other hand, the loss is far greater than that obtained in Example B.2 where only creep was considered. The long-term curvature κ_c^* has increased significantly from 587×10^{-9} to 851×10^{-9} mm^{-1}.

B.3.4 Long-term behaviour patterns

There are some inherent patterns of long-term behaviour that become apparent from the numerical examples. Long-term behaviour is strongly influenced by (a) the amount of reinforcement in the section and (b) the relative magnitudes of the positive sustained moment M_G and the negative prestressing moment, $M_p = Pe$.

When M_G is small relative to M_p, the initial compressive concrete stresses and the initial elastic strains are a maximum in the bottom fibres and a minimum in the top fibres, and possibly even tensile. The compressive free creep strains are therefore maximum in the bottom fibres and minimum in the top fibres, so that the free creep curvature is negative. However, the compressive creep strains in the concrete in the lower fibres induce compressive strain increments (and hence compressive stress) in any adjacent bonded reinforcing steel and a decrement in tension in the tendon. There is a progressive transfer of compressive force from the concrete to the steel and tendon and so a loss of compressive prestress in the concrete and a loss of tensile prestress in the tendon. On the other hand, the presence of the "tensile" steel and tendon in the lower fibres restrains free creep in this region and significantly reduces the deformations that would otherwise occur in the section. In the upper fibres, the initial concrete compressive stress is small and so there is little creep there. There is some transfer of compressive force away from the concrete, but this is minor and does not play an important role in the creep process when M_G is small relative to M_p.

When M_G is large relative to M_p, the initial compressive stresses and strains are maximum in the top fibres and minimum in the bottom fibres, and the initial curvature is positive. The subsequent free creep strains are also maximum in the top fibres and minimum in the bottom fibres, and the free creep curvature is positive. In this case, the transfer of compressive force from the concrete is primarily to the top fibre 'compressive' steel reinforcement. The roles of the reinforcement have reversed, and there is now little transfer of compressive force to the 'tensile' steel.

If there is little or no reinforcement at all in the section and only a small area of tendon, the losses will be relatively small, and the creep-induced curvature will be only slightly smaller than what would occur under free creep conditions.

Uniform shrinkage in the section is restrained by the tendon and the steel. As these are located primarily in the lower fibres, they provide restraint there to the compressive shrinkage strains and the shrinkage curvature is positive. If the long-term shrinkage strain is large, the decrease in compressive force in the concrete (the prestress loss) due to shrinkage can become very large.

B.4 Step-by-step analysis

In the one-step analysis of time effects presented in Section B.3.1, the creep strains are evaluated on the incorrect assumption of constant concrete sustained stress. Instead of introducing the age-adjusted modulus to allow for the changes in concrete stress that actually occurs over time, a multi-step analysis can be undertaken in which the stresses and strains in the section are updated at a number of intermediate time points, t_1, t_2, etc., between t_o and t^*.

Alternative formulations are possible, but we use a free creep and free shrinkage approach, similar to the analyses in Section B.3. In the i-th time step, between times $t_{(i\text{-}1)}$ and t_i, inelastic creep and shrinkage strains develop freely in the concrete under the conditions existing at $t_{(i\text{-}1)}$. Strain incompatibilities therefore develop between the concrete and the adjacent tendon and steel. To restore compatibility at time t_i a correction is made by introducing a self-equilibrating incremental force system to the section which consist of a tensile force increment in the concrete and compressive force increments in the tendon and steels. For the section in Example B.2, the forces are ΔX_c^i, ΔX_p^i, ΔX_st^i and ΔX_sc^i. The concrete force increment induces an elastic, linear tensile strain increment in the concrete with extreme fibre strains of $\Delta\varepsilon_\text{ca}^i$ and $\Delta\varepsilon_\text{cb}^i$. Compressive strain increments are induced in the tendon and steels. The stress and strain increments, together with the force increments at t_i are evaluated. The reduction in concrete stress during the time step is thus evaluated, and the stresses are updated and become the initial conditions at the start of the new time step for the next round of calculations. The analysis is of course approximate, but the accuracy increases with the number of time steps used.

Although we only consider here the analysis of individual cross-sections, this step-by-step procedure can be extended to treat a number of cross-sections along a member, and hence to evaluate the time-varying behaviour of the entire member. A further extension can be made to treat an entire structural system made up of various individual members. Changes in sustained load can also be allowed for. The analysis then becomes a computer simulation of the time-varying behaviour of the structural system. Even the step-by-step analysis of a single cross-section requires intensive computations, and is best carried out using a computer program. Information on wider structural applications of the step-by-step method is available in the literature (Kawano and Warner, 1996; Gilbert and Ranzi, 2011).

The intermediate time points, t_1, t_2 ... t_i ... $t_{(n-1)}$ are best chosen so that roughly the same amount of inelastic strain occurs within each time step. The steps thus vary in size. If the last time point (t^*) is at infinity, then the final step is infinitely large. In practice, t^* is often taken to be around 30 years. The time points should also be chosen to allow for any changes in external sustained load. Although the use of a large number of steps improves accuracy, a relatively small number will often be adequate for design calculations.

The multi-step analysis commences with an initial elastic analysis to evaluate the conditions of stress and strain at time t_o, which is usually when prestress is first applied to the member. Within any single time step, the calculations are the same as those in the one-step analysis, as described in Section B.3.1.

B.4.1 Step-by-step section analysis for creep

It is important to observe how the free creep calculation changes from step to step. During the i-th time step, creep is induced not only by the initial compressive stress applied at t_o, but also by (i -1) tensile stress increments that are applied at the end of each intermediate time step. The free creep in the i-th step requires i separate calculations, with the results added algebraically. The i-th free creep increment in the top fibre, $\Delta\varepsilon^i_{cc.a}$, is thus made up of a compressive creep strain increment due to σ_{coa}, *minus* (i – 1) tensile creep increments due to the tensile stress increments $\Delta\sigma^1_{c.a}$, $\Delta\sigma^2_{c.a}$ ... $\Delta\sigma^{i-1}_{c.a}$:

$$\Delta\varepsilon^i_{cc.a} = \Delta\varepsilon^i_{cc.a}(\sigma_{coa}) - \Delta\varepsilon^i_{cc.a}(\Delta\sigma^1_{c.a}) - \ldots - \Delta\varepsilon^i_{cc.a}(\Delta\sigma^{(i-1)}_{c.a}) \quad \text{(B.46)}$$

To demonstrate how the step-by-step analysis is carried out, a simple unreinforced member with axial prestress is considered below in Examples B.5 and B.6. Although the overall process is unchanged in a step-by-step analysis, the calculations within each time step are more complex but similar to those used in the one-step analysis described in Section B.3.3.

B.4.2 Step-by-step analysis for combined shrinkage and creep

In reality, shrinkage always occurs together with creep, and it is easier to treat the two effects together rather than separately, as we have seen in Section B.3.2. To undertake a joint step-by-step analysis for creep and uniform shrinkage, the shrinkage increments in the extreme *a* and *b* fibres within each time step are evaluated and added to the free creep increments. The calculations then proceeds as explained in Section B.4.1.

EXAMPLE B.5

ONE-STEP CREEP ANALYSIS FOR A SYMMETRIC PRESTRESSED SECTION WITHOUT LONGITUDINAL BAR REINFORCEMENT

To provide a simple example of the step-by-step procedure we consider the effect of creep in the axially prestressed member of length *L*, shown in Figure B.4. The section is rectangular and contains no steel reinforcement. The time-varying behaviour of this member was discussed qualitatively in Section 1.3 of Chapter 1. We will use a three-step analysis in Example B.6 following, but we first use a single-step analysis to simplify the explanation.

The concrete section is 400 mm square, and the initial prestressing force, $P_o = 1600$ kN, is applied at 14 days. We have: $A_c = 160{\times}10^3$ mm^2 and $A_p = 1430$ mm^2. We use the following creep and elastic properties:

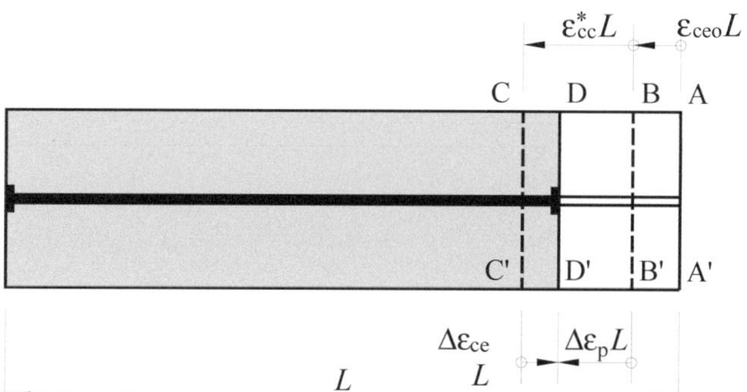

Figure B.4 One-step analysis of creep, axial prestress

Other data: $\varphi(t^*, 14) = 3.0$; $E_{co} = 28\times10^3$ MPa ; $E_p = 195\times10^3$ MPa
$E_c^* = E_c(t^*) = 32\times10^3$ MPa

In Figure B.4, the left end of the member is fixed, while the right end is free to move from its original location AA'. Just after prestressing at time t_o, the immediate elastic shortening of the concrete is $(\varepsilon_{ceo} L)$ and the right end has moved to BB'. In a one-step analysis, free creep occurs between t_o and t^*, and the total free creep strain is ε_{cc}^*. This results in a further shortening of the concrete of $(\varepsilon_{cc}^* L)$ and the right end in Figure B.4 is now at CC'. However, the right end of the tendon has remained at BB'. To restore compatibility we apply a compressive force increment $\Delta X_{p.c}^*$ to the tendon and a tensile force increment of equal magnitude, $\Delta X_{c.c}^*$, to the concrete. The right end of the concrete moves back from CC' by $\Delta\varepsilon_{ce} L$ to DD', and the tendon shortens by $\Delta\varepsilon_p L$ and its end moves from BB' to DD', thus restoring compatibility.

Solution

(a) Order-of-magnitude estimates

The initial elastic concrete compressive strain is $\varepsilon_{ceo} = 360\times10^{-6}$ and with $\varphi_o^* = 3$, the free concrete creep is $\varepsilon_{cc}^* = 1080\times10^{-6}$, which is a conservative estimate of the final axial compressive strain. A conservative estimate of the loss of prestress is:

$$\Delta X_{c.c}^* = \Delta X_{p.c}^* = 1430 \times 195\times10^3 \times 1080\times10^{-6}$$
$$= 301\times10^3 \text{ N} \quad (301 \text{ kN})$$

which is about 19 per cent of the original prestress.

(b) one-step analysis

Time step: The single time step is infinitely large, extending from t_o = 14 days to t^*.

Initial elastic analysis: The initial elastic stress σ_{co} and elastic strain ε_{ceo} are:

$$\sigma_{co} = P_o/A_c = 1600 \times 10^6/160000 = 10 \text{ MPa}$$

$$\varepsilon_{ceo} = \sigma_{co}/E_{co} = 10/28{\times}10^3 = 360{\times}10^{-6}$$

The initial stress in the tendon is: $\sigma_{po} = P_o/A_p = 1119$ MPa.

Free creep

With the concrete uncoupled from the tendon, the total free creep at t^* is:

$$\varepsilon^*_{cc} = \varphi(t^*, t_o)\varepsilon_{ceo} = 3.0 \times 360{\times}10^{-6} = 1080{\times}10^{-6}$$

Restoration of compatibility

To restore compatibility at t^* we apply the tensile force $\Delta X^*_{c.c}$ to the concrete and the compressive force $\Delta X^*_{p.c}$ to the tendon. From Figure B.4 we see that for compatibility we require $\Delta\varepsilon_{ce} + \Delta\varepsilon_p = \varepsilon^*_{cc}$. Using the instantaneous stiffness properties of the materials at t^* we have:

$$\Delta\varepsilon_{ce} = \Delta X^*_{c.c}/(A_c E^*_c);\ \Delta\varepsilon_p = \Delta X^*_{p.c}/(A_p E_p)$$

and hence:

$$\Delta X^*_{c.c} = \Delta X^*_{p.c} = \frac{A_p E_p}{A_p E_p + A_c E^*_c} A_c E^*_c \varepsilon^*_{cc};\ \Delta\varepsilon_{ce} = \frac{A_p E_p}{A_p E_p + A_c E^*_c}\varepsilon^*_{cc}$$

Substituting values:

$$A_p E_p = 1430 \times 195000 = 0.279 \times 10^9\ \text{N}$$

$$A_c E^*_c = 160000 \times 32000 = 5.12 \times 10^9\ \text{N}$$

$$\Delta\varepsilon_{ce} = \frac{0.279}{5.12 + 0.279} \times 1080{\times}10^{-6} = 55{\times}10^{-6}$$

$$\Delta X^*_{c.c} = \Delta X^*_{p.c} = \frac{0.279}{5.12 + 0.279} \times 1080{\times}10^{-6} \times 5.12{\times}10^9$$
$$= 286{\times}10^3\ \text{N}\ \ (286\ \text{kN})$$

The prestressing force at time t^* is:

$$P^* = P_o - \Delta X^*_{p.c} = 1600 - 286 = 1317 \text{ kN}$$

The proportional loss of prestress due to creep is: $286/1600 = 0.18$; i.e.,18 per cent. The loss of prestress in the concrete in this unreinforced cross-section is of course equal to that in the tendon.

(c) Discussion

This one-step analysis using E^*_c ignores the progressive increase in the tensile stress increments in the concrete over time. As previously explained, a correction can be made by replacing E^*_c by the age-adjusted modulus $k^* E^*_c$, as in earlier examples. Instead, we use a three-step analysis in Example B.6.

The order-of-magnitude estimate of creep loss was 302 kN, less than 5 per cent different to the result obtained here. The results are close because the section does not contain reinforcement.

EXAMPLE B.6

THREE-STEP CREEP ANALYSIS FOR A SYMMETRIC PRESTRESSED SECTION WITHOUT LONGITUDINAL BAR REINFORCEMENT

Example B.5 is repeated using a three-step analysis. The time intervals between four time instants, t_o, t_1, t_2, and t^*, are the three time steps. These are shown in Figure B.5, together with the decrements in prestressing force at the end of each time interval, and the relevant creep functions.

Other data:

$$E_{co} = 28\times10^3 \text{ MPa}; \; E_{c1} = 30\times10^3 \text{ MPa};$$

$$E_{c2} = 31\times10^3 \text{ MPa}; \; E^*_c = 32\times10^3 \text{ MPa}$$

SOLUTION

(a) Order-of-magnitude estimates

The order-of-magnitude estimates are unchanged and are as given in Example B.5.

(b) Step-by-step analysis

Choice of time steps

With initial prestressing at $t_o = 14$ days, and the end of the process at t^*, we choose t_1 and t_2 so that about the same amount of free creep occurs in the each time interval. With $\varphi(t^*, 14) = 3.0$, we require $\varphi(t_1, t_{14}) = 1.0$ and $\varphi(t_2, t_{14}) = 2.0$.

From the creep data in Chapter 2 and with a theoretical thickness of t_h = 200 mm, equal creep increments are obtained with times after first loading of approximately 40 and 200 days. To obtain the corresponding ages of the concrete we add 14 days, and so obtain: $t_1 = 54$ days, and $t_2 = 214$ days to give:

$$\varphi(t^*, 14) = 3.0; \quad \varphi(214, 14) = 2.0; \quad \varphi(54, 14) = 1.0$$

In the following calculations we will need specific values of the creep function at t_1 and t_2. Using creep data from Chapter 2 we obtain:

$$\varphi(t^*, 54) = 2.0; \quad \varphi(214, 54) = 1.8; \quad \varphi(t^*, 214) = 1.9$$

Initial elastic analysis

From Example B.5, the initial conditions (time t_o) are:

$$P_o = 1600 \text{ kN}; \ \sigma_{co} = 10 \text{ MPa}; \ \varepsilon_{ceo} = 360\times10^{-6};$$

$$\sigma_{po} = 1119 \text{ MPa}; \ \varepsilon_{po} = 5700\times10^{-6}.$$

Step 1(a) of time analysis: Free creep between 14 and 54 days

The free creep strain increment in the first step is:

$$\Delta\varepsilon_{cc}(54, 14) = \varphi(54,14)\times\varepsilon_{ceo} = 1.0\times360\times10^{-6} = 360\times10^{-6}$$

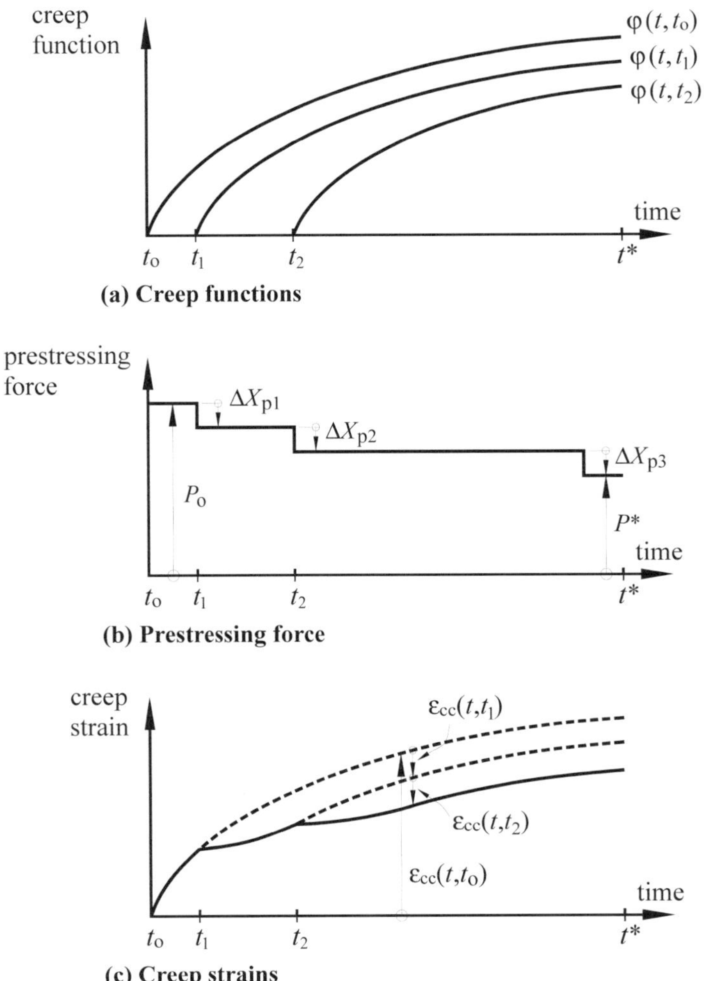

Figure B.5 Three-step analysis of creep

This is the total free creep strain increment $\Delta\varepsilon_{cc}^{1}$ in the first time step. It is also the strain incompatibility between concrete and tendon at 54 days. It is convenient here to evaluate the creep strain increments due to σ_{co} for the second and third time steps as they will be needed in later calculations:

$$\Delta\varepsilon_{cc.o}(214, 14) = \varphi(214,14) \times \varepsilon_{ceo} = 2.0 \times 360\times10^{-6} = 720\times10^{-6}$$

$$\Delta\varepsilon_{cc.o}(t^*, 14) = \varphi(t^*, 14) \times \varepsilon_{ceo} = 3.0 \times 360\times10^{-6} = 1080\times10^{-6}$$

Step 1(b) of time analysis: Restoration of compatibility at 54 days

To complete Step 1, we restore compatibility at $t_1 = 54$ days using a compressive force increment $\Delta X_{p.c}^{1}$ in the tendon and an equal tensile force increment $\Delta X_{c.c}^{1}$ in the concrete. For consistency in this and future examples we will refer to all the step changes in forces, strains and stresses as 'increments', but we will indicate whether the increment is tensile or compressive. The elastic strain increments induced in the concrete and tendon, $\Delta\varepsilon_{ce}^{1}$ and $\Delta\varepsilon_{p}^{1}$, are tensile and compressive, respectively. Together they add up to the value $\Delta\varepsilon_{cc}^{1}$:

$$\Delta\varepsilon_{cc}^{1} = \Delta\varepsilon_{ce}^{1} + \Delta\varepsilon_{p}^{1}$$

With:

$$\Delta\varepsilon_{ce}^{1} = (\Delta X_{c.c}^{1})/(E_{c}^{1}A_{c})\,;\;\; \Delta\varepsilon_{p}^{1} = (\Delta X_{p.c}^{1})/(E_{p}A_{p})$$

we have:

$$\Delta X_{c.c}^{1} = \Delta\varepsilon_{cc}^{1}E_{p}A_{p}/(1 + n_{p1}p_{p})$$

where $n_{p1} = E_{p}/E_{c}^{1}$, and E_{c}^{1} is the elastic modulus of the concrete at 54 days. Substituting values we obtain $\Delta X_{c.c}^{1} = 94.9$ kN. The elastic stress and strain increments in the concrete at 54 days are tensile and have the values $\Delta\sigma_{c}^{1} = 0.6$ MPa, and $\Delta\varepsilon_{ce}^{1} = 20\times10^{-6}$.

Step 2(a) of time analysis: Free creep between 54 and 214 days

In Step 2, free creep of the concrete occurs due both to the sustained compressive stress σ_{co} and the tensile increment $\Delta\sigma_c^1$. The free compressive creep caused by σ_{co} is:

$$\varepsilon_{cc.o}(214, 14) - \varepsilon_{cc.o}(54,14) = (720 - 360)\times 10^{-6} = 360\times 10^{-6}$$

The free tensile creep at 214 days due to $\Delta\sigma_c^1$ is:

$$\Delta\varepsilon_{cc}^1(214, 54) = (\varphi(214, 54))\Delta\varepsilon_{ce}^1 = 1.8\times 20\times 10^{-6} = 36\times 10^{-6}$$

The total free compressive creep strain during Step 2 is:

$$\Delta\varepsilon_{cc}^2 = (360 - 36)\times 10^{-6} = 320\times 10^{-6}$$

We will also calculate the free creep strain due to $\Delta\sigma_c^1$ that occurs in step 3, $\Delta\varepsilon_{cc}^3(t^*, 54)$, as it will be needed in the step 3 calculation. With $\varphi(t^*, 54) =$ 2.0 we obtain the tensile value:

$$\Delta\varepsilon_{cc}^3(\Delta\sigma_c^1) = 2.0\times 20\times 10^{-6} = 40\times 10^{-6}$$

Step 2(b) of time analysis: Restoration of compatibility at 214 days

We apply a tensile force $\Delta X_{c.c}^2$ at 214 days to the concrete and an equal compressive force $\Delta X_{p.c}^2$ to the tendon to restore compatibility. We have:

$$\Delta\varepsilon_{ce}^2 = (\Delta X_{c.c}^2)/(E_c^2 A_p);\ \Delta\varepsilon_p^2 = (\Delta X_{p.c}^2)/(E_p A_p)$$

With:

$$\Delta X_{c.c}^2 = \Delta\varepsilon_{ce}^2 E_p A_p/(1 + n_{p2}p_p)$$

we obtain $\Delta X_{c.c}^2 = 84.5$ kN. The strain and stress increments are:

$$\Delta\varepsilon_{ce}^2 = 17\times 10^{-6} \text{ and } \Delta\sigma_c^2 = 0.5 \text{ MPa} \quad \text{(tensile)}$$

Step 3(a) of time analysis: Free creep between 214 days and t*

During Step 3, free creep occurs in the concrete under the compressive stress σ_{co} and the tensile increments $\Delta\sigma_c^1$ and $\Delta\sigma_c^2$. The free creep due to σ_{co} is:

$$\begin{aligned}\varepsilon_{cc}^3(\sigma_{co}) &= (\varphi(t^*, 14) - \varphi(214, 14))\varepsilon_{ceo} \\ &= (3-2)360\times10^{-6} \\ &= 360\times10^{-6}\end{aligned}$$

The values $\Delta\varepsilon_{cc}(214, 54) = 36\times10^{-6}$ and $\Delta\varepsilon_{cc}(t^*, 54) = 40.0\times10^{-6}$ were obtained in previous calculations. The free tensile creep in step 3 due to $\Delta\sigma_{c1}$ is thus:

$$\Delta\varepsilon_{cc}^3(\Delta\sigma_c^1) = \varepsilon_{cc}(t^*, 54) - \Delta\varepsilon_{cc}(214, 54) = 4\times10^{-6}$$

For $\Delta\sigma_c^2$, with previously obtained values, the free tensile creep is:

$$\Delta\varepsilon_{cc}^3(\Delta\sigma_c^2) = \varphi(t^*, 214)\times\Delta\varepsilon_{ce}^2 = 1.9\times17\times10^{-6} = 30\times10^{-6}$$

The sum of the free creep strain increments in Step 3 is:

$$\Delta\varepsilon_{cc}^3 = (360-4-32)\times10^{-6} = 324\times10^{-6} \quad \text{(compressive)}$$

Step 3(b) of time analysis: Restoration of compatibility at t*

To restore compatibility we require:

$$\Delta\varepsilon_{cc}^3 = \Delta\varepsilon_{ce}^3 + \Delta\varepsilon_p^3$$

and, as previously:

$$\Delta\varepsilon_{ce}^3 = \Delta X_{c.c}^3/(A_c E_c^3); \quad \Delta\varepsilon_{p.c}^3 = \Delta X_p^3/(A_p E_p)$$

The final force increment to be applied at time t^* to restore compatibility at the end of the last step is:

$$\Delta X_{c.c}^3 = \varepsilon_{cc}^3 E_p A_p/(1 + n_{p3}p_p)$$

Substituting values, we obtain $\Delta X^{3}_{c.c} = 84.6\ \text{kN}$.

Final conditions at end of process

The total loss in tensile force in the tendon is:

$$\Delta X^{*}_{p.c} = \Delta X^{1}_{p.c} + \Delta X^{2}_{p.c} + \Delta X^{3}_{p.c} = 94.9 + 84.5 + 84.6 = 264\ \text{kN}$$

The final prestressing force is: $P^{*} = 1600 - 264 = 1336\ \text{kN}$ so that the per cent loss is $(264/1600\times10^{3})\times 100 = 16.5$ per cent, which is the same in the concrete.

(b) Discussion

The difference in results from Examples B.5 and B.6 is quite small. This is to be expected because there is no reinforcing steel in the section and the area of prestressing steel is relatively small. As also expected, the one-step analysis slightly overestimates the inelastic deformation and the prestress loss, as compared with the three step analysis. The losses are 18 per cent versus 17 per cent. Although the three-step analysis gives better accuracy, the amount of calculation is much higher. In this particular example, little would be gained by going to a larger number of steps.

B.5 Approximate closed form equations for losses and deformations

The one-step and the step-by-step methods of analysis are both computationally intensive, and it is of advantage to have simpler, if less accurate, alternatives available for estimating the long-term effects of creep and shrinkage. Closed-form, approximate equations for long-term deformations and loss of prestress can be obtained by introducing simplifications and approximations into the one-step analysis.

B.5.1 Section with prestress only

The equations for the one-step analysis in Section B.3.2 become much simpler if there is no reinforcing steel in the section. The initial elastic analysis and the free creep calculations remain unchanged, but the final elastic analysis at t^* is much simplified. With $A_{st} = A_{sc} = 0$, the force equilibrium requirement at t^* is $\Delta X^*_{c.c} = \Delta X^*_{p.c}$. As the force increments both act at the same depth d_p, the requirement of moment equilibrium is automatically satisfied. Also, there is only one strain compatibility requirement, namely $\Delta\varepsilon_p = \varepsilon^*_{cc.p} - \Delta\varepsilon^*_{ce.p}$. This allows closed-form equations to be derived.

The elastic strain increment in the tendon at t^* is $\Delta\varepsilon_p = \Delta X^*_{p.c}/(E_p A_p)$ and the elastic strain increment required in the concrete at this level is:

$$\Delta\varepsilon^*_{ce.p} = \Delta\sigma^*_{ce.p}/E^*_c$$

which can be expressed in terms of the force increment, $\Delta X^*_{c.c}$, as:

$$\Delta\varepsilon^*_{ce.p} = -\frac{\Delta X^*_{c.c}}{A_g E^*_c} - \frac{\Delta X^*_{c.c} e^2}{I_g E^*_c} \tag{B.47}$$

Substituting the expressions for $\Delta\varepsilon_p$ and $\Delta\varepsilon^*_{ce.p}$ into the strain compatibility requirement and rearranging, and noting that $\Delta X^*_{p.c}$ and $\Delta X^*_{c.c}$ are equal in value, we obtain the following expression for the tensile force increment in the concrete, which is also the loss of compressive prestress in the concrete and the loss of tensile prestress in the tendon:

$$\Delta X^*_{c.c}(A_p) = \gamma_1 A_g k^* E^*_c \varepsilon^*_{cc.p} \tag{B.48}$$

The notation $\Delta X^*_{c.c}(A_p)$ is introduced here to indicate that this loss is due to the presence of the tendon. Note that the elastic modulus E^*_c has been replaced by the age-adjusted effective modulus $k^* E^*_c$ to take account of changes in the concrete stress during the creep process, as previously explained in Section B.3.2. The non-dimensional parameter γ_1 is a function of the cross-sectional properties:

$$\gamma_1 = \frac{1}{1 + \dfrac{A_g k^* E^*_c}{A_p E_p} + \dfrac{e^2 A_g}{I_g}} = \frac{1}{1 + \dfrac{1}{p_p n^*_p} + \dfrac{f^2}{i_g}} \tag{B.49}$$

where A_g and I_g are the area and second moment of area, respectively, of the concrete section, and e is the eccentricity of the tendon measured from the centroid of the concrete. The second, alternative, expression for γ_1 in Equation B.49 uses the following non-denationalised terms:

$$n_p^* = E_p/k^* E_c^* ; p_p = A_p/A_g ; i_g = I_g/(A_g D^2) ; \text{and } f = (e/D)$$

In Equation B.48, $\varepsilon_{cc.p}^*$ is the final free-creep strain in the concrete at the tendon level. It is obtained from the final free creep strains at the extreme fibres:

$$\varepsilon_{cc.p}^* = \varepsilon_{cc.a}^* \times \frac{D - d_p}{D} + \varepsilon_{cc.b}^* \times \frac{d_p}{D} \quad \text{(B.50)}$$

To evaluate the long-term curvature we determine the tensile strain increments $\Delta\varepsilon_{ce.a}^*$ and $\Delta\varepsilon_{ce.b}^*$ in the concrete at the top and bottom fibres due to $\Delta X_{c.c}^*$:

$$\Delta\varepsilon_{ce.a}^* = \frac{\Delta X_{c.c}^*}{A_g E_c^*} - \frac{\Delta X_{c.c}^* e}{I_g E_c^*} d_g \quad \text{(B.51)}$$

$$\Delta\varepsilon_{ce.b}^* = \frac{\Delta X_{c.c}^*}{A_g E_c^*} + \frac{\Delta X_{c.c}^* e}{I_g E_c^*}(D - d_g) \quad \text{(B.52)}$$

The term d_g is the depth to the centroid of the concrete section. The creep curvature correction is $\Delta\kappa_c^* = (\Delta\varepsilon_{ce.b}^* - \Delta\varepsilon_{ce.a}^*)/D$. Substituting and rearranging, and using the notation $\Delta\kappa_c^*(A_p)$ to indicate that this correction is associated with the tendon area, we have:

$$\Delta\kappa_c^*(A_p) = \gamma_1\gamma_2(\varepsilon_{cc.p}^*/D) \quad \text{(B.53)}$$

The non-dimensional parameter γ_1 is given above in Equation B.49 and γ_2 is another non-dimensional function of the section properties:

$$\gamma_2 = (A_g eD)/I_g \quad \text{(B.54)}$$

The curvature correction term $\Delta\kappa_c^*(A_p)$ is positive if the tendon is located below the centroid of the concrete. The nett long-term curvature, $\kappa_c^*(A_p)$, is obtained by adding the curvature increment at the end of the process to the free creep curvature:

$$\kappa_c^*(A_p) = \kappa_{co}^* + \Delta\kappa_c^*(A_p) \tag{B.55}$$

EXAMPLE B.7

CALCULATION OF PRESTRESS LOSS AND CREEP CURVATURE FOR A PRE-STRESSED SECTION WITHOUT LONGITUDINAL BAR REINFORCEMENT USING SIMPLIFIED EQUATIONS

The calculations in Example B.2 for a section with no reinforcement are undertaken using the simplified expressions for $\Delta X_c(A_p)$ and $\Delta\kappa_c^*(A_p)$.

SOLUTION

Substituting section property values into Equations B.49 and B.54 we obtain $\gamma_1 = 0.0570$ $\gamma_2 = 3.750$ and from Equation B.48:

$$\Delta X_{c.c}^*(A_p) = \Delta X_{p.c}^*(A_p) = 39 \text{ kN}$$

From Equations B.53 and B.55 we obtain $\Delta\kappa_c^*(A_p) = 61.01\times10^{-9}$ mm^{-1} and:

$$\kappa_c^*(A_p) = (585.9 + 61.01) \times 10^{-9} = 647\times10^{-9} \text{ mm}^{-1}$$

The results from Example B.2 were as follows:

Without the age-adjusted effective modulus correction:

$$\Delta X_{c.c}(A_p) = \Delta X_{p.c}(A_p) = 43\text{kN}$$

$$\Delta\kappa_c^*(A_p) = 1.9\times10^{-9} \text{ mm}^{-1}; \; \kappa_c^*(A_p) = 605\times10^{-9} \text{ mm}^{-1}$$

With the age-adjusted effective modulus correction:

$$\Delta X_{c.c}(A_p) = \Delta X_{p.c}(A_p) = 39 \text{ kN} \; ; \; \Delta\kappa_c^*(A_p) = 59\times10^{-9}\text{ mm}^{-1}$$

$$\kappa_c^*(A_p) = (585.9 + 59)\times10^{-9} = 645\times10^{-9}\text{ mm}^{-1}$$

Discussion

The simplified equations should give identical results to those from Example B.2 when $A_{sc} = A_{st} = 0$. Minor differences are due to round off.

B.5.2 Section with prestress and tensile reinforcement

A simplification is introduced to take account of the presence of tensile reinforcement in the section. Provided the tensile steel is reasonably close to the prestressing tendon, we can replace the tendon area A_p by an 'equivalent' combined steel-tendon area A_{eq}, located at depth d_{eq}, where:

$$A_{eq} = A_p + A_{st} \tag{B.56}$$

$$d_{eq} = \frac{A_p d_p + A_{st} d_{st}}{A_{eq}} \tag{B.57}$$

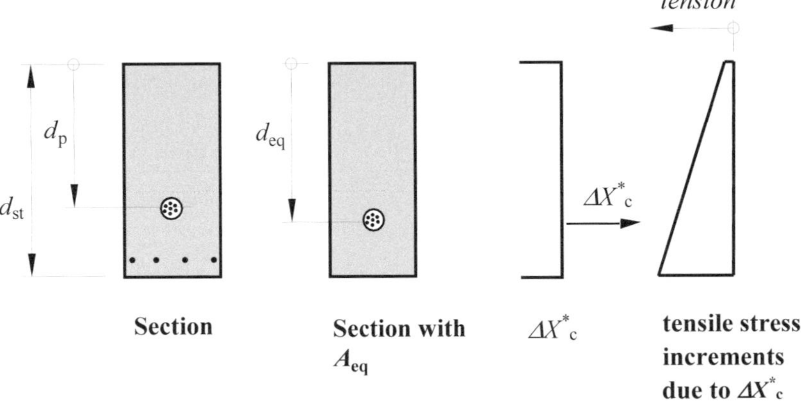

Figure B.6 Equivalent tendon area A_{eq}

This is shown in Figure B.6 for the case of a rectangular section, although the approximation is generally applicable. Given the approximations being made, the use of an equivalent elastic modulus, E_{eq}, is not warranted. Replacing A_p and d_p by A_{eq} and d_{eq}, the analysis in Section B.5.1 is repeated. Rewriting the equations and using the terms $\Delta X_c^*(A_{eq})$ and $\Delta\kappa_c^*(A_{eq})$ to indicate that we are now dealing with a section with tensile reinforcement, and an equivalent area A_{eq}, we obtain the following expression for the prestress loss:

$$\Delta X_{c.c}^*(A_{eq}) = \gamma_3 A_g k^* E_c^* \varepsilon_{cc.eq}^* \tag{B.58}$$

The term $\varepsilon_{cc.eq}^*$ is the free creep strain at the depth d_{eq}. It is obtained as

$$\varepsilon_{cc.eq}^* = \varepsilon_{cc.a}^* \times \frac{D - d_{eq}}{D} + \varepsilon_{cc.b}^* \times \frac{d_{eq}}{D} \tag{B.59}$$

The non-dimensional parameter γ_3 is:

$$\gamma_3 = \frac{1}{1 + \dfrac{A_g k^* E_c^*}{A_{eq} E_{eq}} + \dfrac{e_{eq}^2 A_g}{I_g}} \tag{B.60}$$

The curvature correction is:

$$\Delta\kappa_c^*(A_{eq}) = \gamma_3 \gamma_4 \frac{\varepsilon_{cc.eq}^*}{D} \tag{B.61}$$

where γ_3 is as above and γ_4 is:

$$\gamma_4 = (A_g e_{eq} D)/I_g \tag{B.62}$$

These equations do *not* apply to compressive reinforcement because the steel and tendon areas have been assumed to be close together.

In regard to signs, in the positive moment region of a beam the initial elastic curvature due to prestress is negative, and positive due to M_G. The free creep curvature κ_c^* always has the same sign as the initial curvature, and hence will be positive provided M_G is greater than $M_p = Pe$. Irrespective of whether the initial elastic curvature (and hence the free creep curvature) is positive or negative, the correction to the creep curvature, $\Delta\kappa_c^*$, is positive provided the equivalent area is below the centroid of the concrete area.

EXAMPLE B.8

CALCULATION OF PRESTRESS LOSS AND CREEP CURVATURE FOR A PRE-STRESSED SECTION WITH LONGITUDINAL TENSILE BAR REINFORCEMENT, USING SIMPLIFIED EQUATIONS

In this example, Example B.7 is repeated but with flexural tensile bar reinforcement included in the section. The initial analysis and the free creep curvature calculation are unchanged. We take $E_{eq} = E_s$. With $A_{st} = 1350$ mm^2 (as in Examples B.1 and B.3) we have $A_{eq} = 2350$ mm^2 and $d_{eq} = 707.4$ mm.

SOLUTION

Using Equations B.60 and B.62 we obtain:

$$\gamma_3 = 0.03873 \text{ and } \gamma_4 = 4.6117 .$$

Substituting in Equation B.58:

$$\Delta X^*_{c.c}(A_{eq}) = \Delta X^*_{eq.c}(A_{eq}) = 79.7\text{kN} .$$

Substituting in Equation B.61:

$$\Delta\kappa_c^*(A_{eq}) = 43.5\times10^{-9} \text{ mm}^{-1}$$

which gives $\kappa_c^* = 629.4\times10^{-9} \text{ mm}^{-1}$.

For comparison purposes, calculations for this cross-section with tensile reinforcement were made using a one-step analysis with the age-adjusted effective modulus modification. The results are as follows:

$$\Delta X^*_{p.c} = 34.9 \text{ kN} ; \ \Delta X^*_{st.c} = 45.1 \text{ kN}$$

$$\Delta X^*_{sc.c} = 0 \text{ kN} ; \ \Delta X^*_{c.c} = 80.0 \text{ kN}$$

$$\Delta \kappa^*_c = 115.7 \times 10^{-9} \text{ mm}^{-1} ; \ \kappa^*_c = 701.7 \times 10^{-9} \text{ mm}^{-1}$$

Discussion

The creep loss in the concrete, 80 kN, as obtained from the one-step analysis, is almost identical with that obtained from the closed form equation. However, the curvature correction term from the closed form equations is less than a half of the value from the other method. Nevertheless the final long-term curvatures differ by only about 11 per cent.

For this cross-section the order-of-magnitude estimate of creep curvature, $\kappa^*_{co} = 586 \times 10^{-9} \text{ mm}^{-1}$, underestimates the age-adjusted effective modulus calculation by about 17 per cent, which is adequate for preliminary design estimates.

B.5.3 Allowing for compressive reinforcement

To take account of compressive reinforcement, we make the further simplifying assumption that corrections for the compressive and tensile reinforcement can be made independently of each other and then superposed. Of course, this is not strictly correct. The application of a tensile force ΔX^*_c to the concrete at depth d_{eq} introduces strain increments into the upper fibres, and at depth d_{sc} this increment will not usually be zero. Nevertheless, if the tensile and compressive steel areas are in just two layers, near the top and bottom fibres respectively, the top-fibre strains due to $\Delta X^*_{c.c}$ will be small, and the interaction will be limited and the error will be contained.

Using this simplification, we consider the situation shown in Figure B.7. Ignoring the tensile steel and tendon, the free creep strain at depth d_{sc}, $\varepsilon^*_{cc.sc}$, is evaluated from $\varepsilon^*_{cc.a}$ and $\varepsilon^*_{cc.b}$ as:

$$\varepsilon^*_{cc.sc} = \varepsilon^*_{cc.a} \times \frac{D - d_{sc}}{D} + \varepsilon^*_{cc.b} \times \frac{d_{sc}}{D} \tag{B.63}$$

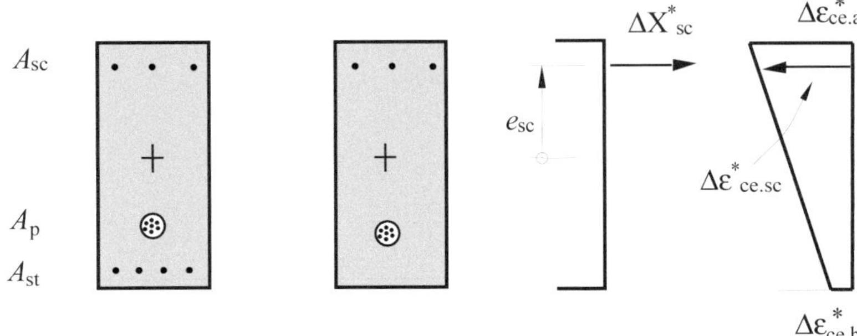

Figure B.7 Correction for compressive steel

which is also the strain incompatibility between the compressive steel and the adjacent concrete. Applying the compressive force ΔX^*_{sc} to the compressive steel and an equal tensile force $\Delta X^*_{c.c}$ to the concrete, also at depth d_{sc}, we obtain equations in the same format as those derived previously in Section B.5.2. The eccentricity of $\Delta X^*_{c.c}$, relative to the centroid of the concrete section, is:

$$e_{sc} = d_{sc} - 0.5D \tag{B.64}$$

which will always be negative.

The tensile elastic strain increment in the concrete at depth d_{sc}, due to $\Delta X^*_{c.c}$, is $\Delta\varepsilon^*_{ce.sc}$ and the compressive strain increment in the compressive steel, due to the incremental force ΔX^*_{sc}, is $\Delta\varepsilon_{sc}$. The concrete elastic strain increment and compressive steel increment are:

$$\Delta\varepsilon^*_{ce.sc} = \frac{\Delta X^*_{c.c}}{E^*_c A_c} + \frac{\Delta X^*_{c.c} e^2_{sc}}{E^*_c I_c} \tag{B.65}$$

$$\Delta\varepsilon_{sc} = \frac{\Delta X^*_{sc}}{E_s A_{sc}} \tag{B.66}$$

To restore compatibility we require:

$$\Delta\varepsilon_{sc} = \varepsilon^*_{cc.sc} - \Delta\varepsilon^*_{ce.sc} = \varepsilon^*_{cc.sc} - \frac{\Delta X^*_{c.c}}{E^*_c A_c} - \frac{\Delta X^*_{c.c} e^2_{sc}}{E^*_c I_c} \qquad \text{(B.67)}$$

To allow for the change in concrete stress during the creep process, we replace E^*_c by the adjusted reduced modulus, $k^* E^*_c$. Rearranging, and noting that the force increments are equal in magnitude, we obtain a simplified expression for $\Delta X^*_{c.c}$, which is the loss in prestress in the concrete due to the presence of the compressive steel. Writing this as $\Delta X^*_{c.c}(A_{sc})$ we obtain:

$$\Delta X^*_{c.c}(A_{sc}) = \gamma_5 A_g k^* E^*_c \varepsilon^*_{cc.sc} \qquad \text{(B.68)}$$

The term γ_5 is a non-dimensional function of the section details and material properties, and is comparable with the terms γ_1 and γ_3:

$$\gamma_5 = \frac{1}{1 + \dfrac{A_g k^* E^*_c}{A_{sc} E_{sc}} + \dfrac{e^2_{sc} A_g}{I_g}} \qquad \text{(B.69)}$$

With $\Delta X^*_{c.c}(A_{sc})$ determined, we evaluate the strain increments that this force induces in the top and bottom concrete fibres, $\Delta\varepsilon^*_{ce.a}$ and $\Delta\varepsilon^*_{ce.b}$, and hence the creep curvature correction. Writing the curvature correction as $\Delta\kappa^*_c(A_{sc})$ we have:

$$\Delta\kappa^*_c(A_{sc}) = \gamma_5 \gamma_6 \frac{\varepsilon^*_{cc.sc}}{D} \qquad \text{(B.70)}$$

where:

$$\gamma_6 = \frac{A_g e_{sc} D}{I_g} \qquad \text{(B.71)}$$

Discussion

This treatment for the effect of A_{sc} is inherently less accurate than the treatment for A_{st}. The assumption that corrections can be made independently for the tensile and compressive reinforcement cannot be applied to more complex sections, for example when reinforcing steel is distributed down the side faces of the section or even when there are more than two levels of reinforcement. For such cross-sections, the simplified equations are not appropriate and losses and creep curvatures need to be evaluated using the step-by-step method or a one-step analysis with the age-adjusted effective modulus.

Intuitively, we can see that the correction for the effect of the compressive steel will not be important when M_G is small relative to M_p, and the compressive stresses in the upper fibres are small, but it will be quite important when M_G is large.

The similarities and differences among the various γ terms become clear when the relevant equations are compared. In regard to signs, the term γ_5 is always positive, as are γ_3 and γ_1, but while γ_4 and γ_2 are both always positive, γ_6 is always negative. This is because the eccentricity e_{sc} is negative. The curvature correction due to the presence of the compressive steel, $\Delta\kappa_c^*(A_{sc})$, is therefore negative, which is intuitively clear.

EXAMPLE B.9

Calculation of Prestress Loss and Creep Curvature for a Prestressed Section with Tensile and Compressive Flexural Bar Reinforcement Using Simplified Equations

The simplified equations are used to evaluate the effect of adding compressive steel of area $A_{sc} = 1350\ \text{mm}^2$ (also used in Examples B.1 and B.3) to the section previously considered in Example B.8.

SOLUTION

From Equations B.69 and B.71: $\gamma_5 = 0.0236$ and $\gamma_6 = -5.25$

From Equation B.68: $\Delta X^*_{c.c}(A_{sc}) = 144.4$ kN

From Equation B.70: $\Delta\kappa^*_c(A_{sc}) = -89.8\times10^{-9}$ mm^{-1}

From Example B.8, the values for the contributions of A_{eq} were:

$$\Delta X^*_{c.c}(A_{eq}) = 79.7 \text{ kN kN; } \Delta\kappa^*_c(A_{eq}) = 43.5\times10^{-9} \text{ mm}^{-1}$$

Superposing the results, we obtain for the doubly reinforced section:

$$\Delta X^*_{sc.c} = 144.4 \text{ kN}; \ \Delta X^*_{eq.c} = 79.7 \text{ kN}; \ \Delta X^*_{c.c} = 224.1 \text{ kN}$$

$$\Delta\kappa^*_c = (43.5 - 89.8)\times10^{-9} = -46.3\times10^{-9} \text{ mm}^{-1}$$

$$\kappa^*_c = (586 - 46.3)\times10^{-9} = 539.7\times10^{-9} \text{ mm}^{-1}$$

Comparing these results with those obtained in Example B.1, we find that the loss of prestress in the concrete, as obtained here (224 kN), is about 10 per cent greater than that obtained using the one-step, age-adjusted effective modulus method.

The difference in the curvature calculations is greater. The value of the curvature correction $\Delta\kappa^*_c$ from the simplified equations is almost the same as the one from Example B.3, but is a little more than a third of the value from Example B.1. Nevertheless, the final long-term curvature obtained using the simplified equations is about 18 per cent larger than the value obtained from Example B.1, and on the conservative side.

B.5.4 Allowing for time-varying sustained loads

In previous sections of this appendix we have restricted attention to the case where the constant sustained moment M_G is applied at the same time as the prestress. In practice, the sustained load may vary over time. Increments in dead load can be applied to a member at various critical stages in the construction sequence, while further permanent loads can be introduced when the structure is brought into service.

We consider a cross-section that is subjected to an applied moment that changes in steps over time. The initial sustained moment M_G is applied with the prestress at time t_o, but additional moments, M_1, M_2 ... M_i ... M_n, occur at times t_1, t_2 ... t_i ... t_n. The effects of the moment increments can be calculated separately and independently, using any appropriate method, and then superposed on those for M_G. However, an approximation can be used to simplify the analysis of creep in a member under step-wise increasing sustained load (Warner, 2014c).

We use the concept of an equivalent, constant, sustained moment M_{eq}, which is applied with the prestress at time t_o. This equivalent moment has by definition the same effect as the moment sequence so that M_{eq} can be used in lieu of M_G in the previously derived formulae. This avoids a complex sequence of separate calculations for each moment increment.

The problem is to evaluate M_{eq}. For a specific moment increment M_i, applied at t_i, there is an equivalent moment $M_{i.eq}$, which, if applied at time t_o, would yield the same long-term free creep as M_i. An approximate expression for $M_{i.eq}$ is:

$$M_{i.eq} = \frac{\varphi_i^* E_{co}}{\varphi_o^* E_{ci}} M_i \tag{B.72}$$

where E_{ci} is the elastic modulus of the concrete at time t_i, and φ_i^* is the final value of the creep function for initial loading at t_i. We will take $M_{i.eq}$ as an approximation for the equivalent moment that would produce the same long-term effect as M_i, applied at t_i. The adequacy of this simplifying assumption is best checked by means of comparison computations, which have been carried out to a limited extent (Warner, 2014c).

For the moment sequence $M_G + M_1 + M_2$ etc., the equivalent moment thus becomes:

$$M_{eq} = M_G + \sum_{i=1}^{n} M_{i.eq} \tag{B.73}$$

It can be seen that M_{eq} depends on the shape of the three-dimensional surface of the creep function $\varphi(t, t_o)$ and specifically on values of this function for the particular load application times t_o, t_1, t_2, etc.

A single calculation can be made using M_{eq}, thus avoiding the sequence of calculations for individual moments. The free creep due to M_{eq} is:

$$\kappa^*_{c,tot} = \varphi^*_o \times \frac{M_{eq}}{E_{co} I_g} \tag{B.74}$$

Equations B.72 and B.73 are approximations. In practice, however, the precise history of sustained loading on a particular member during its life time will rarely be known in advance with any good degree of accuracy by the designer. In this light, the approximation is reasonable.

EXAMPLE B.10

CREEP CALCULATIONS FOR TIME-VARYING SUSTAINED LOAD

We consider a prestressed cross-section with a time-varying moment history that is made up of an initial sustained moment M_G applied at the time of prestressing at t_o = 14 days and two additional dead-load moments subsequently applied at 54 days and 214 days. The relevant data are as follows:

At t_o = 14 days: M_G =250 kNm; φ^*_o = 3.0; $E_{co} = 28\times10^3$ MPa

At t_1 = 54 days: M_1 =200 kNm; φ^*_1 = 2.0; $E_{c1} = 30\times10^3$ MPa

At t_2 = 214 days: M_2 =100 kNm; φ^*_2 = 1.8; $E_{co} = 31\times10^3$ MPa

SOLUTION

To evaluate the equivalent moment M_{eq} which, applied at time $t_o = 14$ days, is to be used in a single creep calculation in lieu of three separate calculations, we use Equation B.72:

$$\text{For } M_1\text{: } M_{1,eq} = \frac{\varphi_1^*}{\varphi_o^*}\frac{E_{co}}{E_{c1}}M_1 = \frac{2.0}{3.0}\times\frac{28\times10^3}{30\times10^3}\times 200 = 124.4 \text{ kNm}$$

$$\text{For } M_2\text{: } M_{2,eq} = \frac{\varphi_2^*}{\varphi_o^*}\frac{E_{co}}{E_{c2}}M_1 = \frac{1.8}{3.0}\times\frac{28\times10^3}{31\times10^3}\times 100 = 54.2 \text{ kNm}$$

Hence $M_{eq} = 250 + 124.4 + 54.2 = 428.6$ kNm, to be applied at $t_o = 14$ days. The moment M_{eq} replaces the total sustained moment of 550 kNm, which is applied in steps over 214 days.

B.5.5 Simplified formulae for long-term losses and deformations due to shrinkage

Simplified formulae for the long-term effects of shrinkage can be derived using a similar approach to that for creep. We consider a concrete section as in Figure B.2 with total concrete area A_g and overall depth D. The depth to the centroid of the concrete section from the top fibre is d_g. We initially allow free, uniform shrinkage to occur in the concrete, without any restraint from the steel or tendon. The result is a uniform shortening of the concrete, the total shrinkage strain being ε_{sh}^* (or in AS 3600 terminology, ε_{cs}). Compatibility of strains in the section is progressively lost as the concrete shrinks.

We again use the idea of an equivalent steel-tendon area A_{eq} located at depth d_{eq} below the top surface of the section. However, since the shrinkage strain ε_{cs} is uniformly distributed over the section it is valid now to include the compressive reinforcement A_{sc} in A_{eq} as well as A_{st} and A_p. This was not possible in the creep analysis. Equations B.56 and B.57 are therefore modified to include A_{sc} as follows:

$$A_{eq} = A_p + A_{st} + A_{sc} \text{ and } d_{eq} = (A_p d_p + A_{st} d_{st} + A_{sc} d_{sc})/A_{eq}$$

We apply a resultant compressive force $\Delta X^*_{eq.sh}$ to the equivalent steel area at depth d_{eq}, and an equal tensile force $\Delta X^*_{c.sh}$ to the concrete at the same level in order to restore compatibility.

The force $\Delta X^*_{eq.sh}$ produces a compressive strain increment $\Delta\varepsilon_{eq}$ in the equivalent steel area at depth d_{eq}, while the tensile strain increment in the concrete at this depth is $\Delta\varepsilon^*_{ce.eq}$. Compatibility is restored provided these add up to ε^*_{sh}:

$$\Delta\varepsilon_{eq} + \Delta\varepsilon^*_{ce.eq} = \Delta\varepsilon^*_{sh} \quad \text{(B.75)}$$

The resulting equations are similar to those for the creep analysis for a section with tendon and steel. We obtain:

$$\Delta X^*_{c.sh} = \gamma_3 A_g k^* E^*_c \varepsilon^*_{sh} \quad \text{(B.76)}$$

$$\gamma_3 = \frac{1}{1 + \dfrac{A_g k^* E^*_c}{A_{eq} E_{eq}} + \dfrac{e^2_{eq} A_g}{I_g}} \quad \text{(B.77)}$$

Here, expression B.77 is similar to B.70, except that the equivalent area A_{eq} now also contains the compressive steel area. As the free shrinkage is uniformly distributed in the section, there is no initial curvature and the total shrinkage curvature is equal to the final elastic strain increments, and is:

$$\kappa^*_{sh} = \gamma_3 \gamma_4 \frac{\varepsilon^*_{sh}}{D} \quad \text{(B.78)}$$

where:

$$\gamma_4 = \frac{A_c e_{eq} D}{I_c} \quad \text{(B.79)}$$

Equation B.78 is similar in form to Equation B.61, but A_{eq} now includes the compressive steel area A_{sc} as well as A_{st} and A_p.

In a single-span beam the equivalent area A_{eq} will lie in the lower part of the section over the major length of span, i.e. with e_{eq} positive, so that the shrinkage curvature is also positive and the deflection of the beam due to shrinkage warping is downward. If the equivalent area lies in the upper part of a cross-section, with e_{eq} negative, the term γ_4 is negative and so also is the shrinkage curvature. When the curvature is positive in some regions of the beam and negative in other regions, integration of the curvatures will be needed to determine whether the deflection is positive or negative, and its value.

Tensile stresses develop gradually in the concrete as shrinkage proceeds, so that secondary tensile creep strains also develop. This can also be allowed for, approximately, by using the effective modulus term $k^* E_c^*$ in Equation B.77 for the calculation of γ_3.

EXAMPLE B.11

USE OF SIMPLIFIED FORMULAE TO EVALUATE SHRINKAGE EFFECTS

The long-term shrinkage effects in the cross-section of the doubly reinforced beam of the previous examples are calculated here using Equation B.76 to B.79. The final shrinkage strain is taken to be $\varepsilon_{sh}^* = 0.0008$ and is assumed uniform over the concrete section.

SOLUTION

With $A_{eq} = 3700$ mm:

$$d_{eq} = \frac{(1350 \times 50 + 1350 \times 750 + 1000 \times 650)}{3700} = 467.6\text{mm}$$

$$e_{eq} = 467.6 - 400 = 67.6 \text{ mm}$$

we evaluate γ_3 and γ_4 from Equations B.77 and B.79 as:

$$\gamma_3 = 0.1889; \quad \gamma_4 = 1.014$$

Prestress loss due to shrinkage

From Equation B.76: $\Delta X^*_{c.sh} = \Delta X^*_{eq.sh} = \gamma_3 A_c k^* E^*_c \varepsilon_{cs} = 471$ kN

Shrinkage curvature

Since the free shrinkage strain is uniform, the free curvature due to shrinkage is zero and the total shrinkage curvature is given by Equation B.78:

$$\kappa^*_{sh} = 191.4 \times 10^{-9} \text{ mm}^{-1}$$

Recalculation for section without reinforcement: $A_{sc} = A_{st} = 0$

When the calculations are redone for zero steel areas we obtain:

$$A_{eq} = 1000 \text{ mm}^2;\ d_{eq} = 650 \text{ mm};\ e_{eq} = 250 \text{ mm};$$

$$\gamma_3 = 0.0564;\ \gamma_4 = 3.75$$

$$\Delta X^*_{c.sh} = 140 \text{ kN};\ \kappa^*_{sh} = 211 \times 10^{-9} \text{ mm}^{-1}$$

Discussion

In Example B.3 the loss of concrete prestress in the same cross-section due to shrinkage and creep was found to be 780 kN. Due to creep alone it was evaluated in Example B.1 to be 233 kN. The difference, 547 kN, can be attributed to shrinkage. The value determined here, 471 kN, is about 14 per cent less. Without the reinforcement in the section, the loss of prestress reduces to 140 kN, while the curvature increases to 211×10^{-9} mm^{-1}.

B.6 Non-uniform shrinkage and creep

A crucial assumption underlying the creep and shrinkage provisions in AS 3600 is that the shrinkage and creep properties of the concrete do not vary within a member, either vertically (down the cross-section), laterally, or along its length. The same values of the long-term shrinkage strain, ε_{cs}, or ε^*_{sh}, and of the creep function, φ^*_o, are applied to all fibres in each cross-section of a member.

This uniformity assumption has also been made in this appendix, but it is in fact a gross simplification of real behaviour. Shrinkage and creep strains develop in concrete in a far more complex manner than the uniformity assumption implies. The processes of creep and shrinkage in a mass of concrete are extremely complex and are not yet fully understood, but local changes in the moisture content are known to influence the time-dependent strains. In particular, creep and shrinkage are influenced by progressive drying of the concrete at free surfaces open to the atmosphere, and also by the movement of moisture from the interior mass of concrete towards the free surfaces. This inevitably results in non-uniform shrinkage strains within a concrete member.

Shrinkage strains are relatively large at and near the free (relatively dry) surfaces of an unloaded mass of concrete, but attenuate to small values in the still-moist core that is remote from the free surfaces. Shrinkage over the cross-section is non-uniform and non-linear. Local stresses therefore develop that are comparable to the residual stresses that build up in the sections of a steel member during its manufacture. Large tensile stresses develop in the concrete regions at and near the free surfaces where shrinkage is greatest, while there are smaller equilibrating compressive stresses in the larger inner core region. This is why surface cracks appear in many concrete members, even in compressive members carrying considerable compression.

In a rectangular beam section with dimensions b and D of comparable size, and with all surfaces open to the atmosphere, the shrinkage strains vary three-dimensionally, with maximum values at the surfaces. For a slab-like member, with $b >> D$, the non-linear distributions near the sides do not significantly affect overall flexural behaviour, and the distribution of the free shrinkage strains can be approximated as two-dimensional and non-linear, with maximum strains at the top and bottom surfaces and minimum values near mid-depth. A simplified, piece-wise linear two-dimensional approximation is shown in Figure B.8 for the situation where the atmospheric conditions are the same at the top and bottom surfaces.

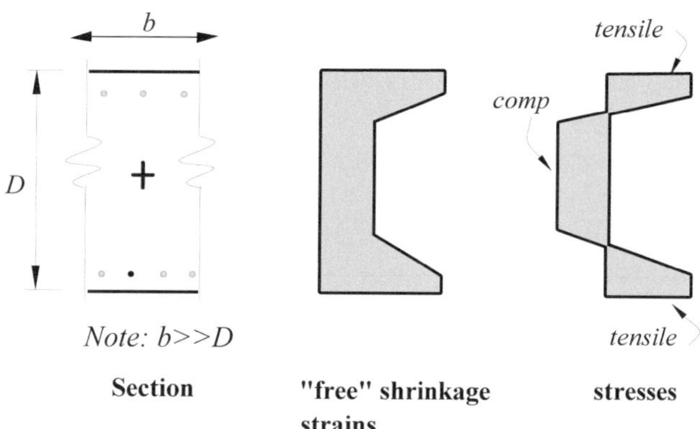

Figure B.8 Residual stresses in concrete slab section due to non-uniform shrinkage

Such a free strain distribution could not exist over a finite length of the member without tearing of the material. In fact, additional elastic strains, with accompanying elastic 'residual' stresses, develop in the concrete and these are imposed on the 'free' shrinkage strains. The stresses are self-equilibrating, as indicated in Figure B.8. By representing the section as a number of thin, discrete, horizontal layers an analysis can be undertaken to evaluate the imposed stresses and strains that result from an assumed initial free shrinkage strain pattern. A section analysis using a segmental representation can be used for a beam section, while the three-dimensional distribution of strains in a beam section can be dealt with by representing the section as a grid of fibres. The analyses can allow for the presence of the steel and tendon in the section and can be introduced into a step-by-step time analysis to evaluate the long-term, non-uniform shrinkage effects in entire members and structures.

Creep, like shrinkage, is also affected by local moisture content, and by loss of moisture at the free concrete surfaces, with moisture movement within the concrete. The previous comments relating to non-uniform shrinkage apply also to creep, with minor modifications. Creep in concrete members in most practical situations is non-uniform. A layered (or two-dimensional grid) analysis can be used in much the same way as for shrinkage, with varying creep function values, $\varphi(t_j, t_i)$, allocated as appropriate to the different thin layers or fibres in the idealised cross-section.

Existing methods of analysis can thus be used to investigate non-uniform creep and shrinkage. However, the main problem facing designers is the lack of empirical, quantitative information on how creep and shrinkage vary throughout cross-sections and from region to region along real members.

The assumption of uniform creep and shrinkage in design calculations is a gross simplification which makes the pursuit of precise values of long-term stresses and strains pointless. In this light, the use of simple but approximate methods of analysis will often be preferred to complex analysis.

B.7 References

Bazant, Z. P. & Wittmann, F. H.(Editors), 1982, *Creep and shrinkage in concrete structures*, John Wiley & Sons, New York

Bazant, Z. P, & Carol, I. (Editors), 1988, *Mathematical modelling of creep and shrinkage of concrete*, John Wiley and Sons, New York, N.Y.

Bazant, Z. P., Qiang, Y., Hubler, M. H., Kristek, V. & Bittner, Z., 2011, *Wake-up call for creep myth about size effect and black holes in safety: What to improve in fib model code draft*, Proceedings, fib Symposium, ISBN 978-80-87158-29-6.

Gilbert, R. I. & Ranzi, G., 2010, *Time-dependent behaviour of concrete structures*, Spon Press, Oxon, UK, 426 pp.

Kawano, A. & Warner, R. F., 1996, Model formulations for numerical creep calculations for concrete, *ASCE Journal of Structural Engineering*, Vol 122, No 3, March.

Rebentrost, M. & Warner, R. F., 2002, Influence of bond on the behaviour of prestressed concrete beams, Proceedings, 3rd International Symposium, fib-ACI, *Bond in Concrete*, Budapest University of Technology and Economics, ISBN 9634207146.

Trost, H., 1967, Auswirkungen des Superpositionsprinzips auf Kriech- und Relaxationsprobleme (Application of the superposition principle to creep and shrinkage problems in concrete and prestressed concrete), *Beton und Spanbeton*, Vol 62, No 10, pp 230- 38, No 11, pp 261-69.

Warner, R.F., Rangan, B.V., Hall, A.S., and Faulkes, K.A., 1998, *Concrete Structures*, Longman Cheshire, Pearson Education, Melbourne.

Warner, R. F., 2014a, Simplified analysis of creep in prestressed concrete beams, *Australian Journal of Structural Engineering*, Vol 15, No 1, pp89-96.

Warner, R. F., 2014b, Analysis of shrinkage in prestressed concrete slabs and beams, *Australian Journal of Structural Engineering*, Vol 15, No 2, pp151-159.

Warner, R. F., 2014c, Creep in prestressed concrete beams under time-varying sustained loading, Australian Journal of Structural Engineering, Vol 15, No 3, pp 260-268.

Warner, R. F., 2015, Simple design formulae for evaluating creep effects in prestressed concrete members, Australian Journal of Structural Engineering, Vol 16, No 2, pp 137-149.

APPENDIX C

Effects of prior creep and shrinkage on flexural strength

The effects of creep and shrinkage on the behaviour of prestressed concrete beams under sustained load can be evaluated using the step-by-step method of analysis described in Appendix B. This appendix investigates the extent to which prior creep and shrinkage affect (a) the behaviour under short-term live load and (b) the flexural strength. It is shown that prior creep and shrinkage have no significant effect on flexural strength provided the member is ductile. Furthermore, prior creep and shrinkage strains in a member do not significantly affect the short term response of the member to live load increments. In practice, this means that the different deflection components can be calculated separately and added together to obtain the total deflection. Creep and shrinkage must of course be considered when losses are determined, and their contribution to total deflection must always be allowed for.

C.1 Introduction

In Chapter 5, the methods of analysis used to determine the response of cracked flexural members to short-term loading ignored the presence of large creep and shrinkage strains that develop progressively over time. In fact, creep and shrinkage cause significant stress redistribution in cross-sections, especially when significant amounts of steel reinforcement are present. Large additional deformations and deflections also occur as the result of creep and shrinkage. The presence of prior creep and shrinkage strains and stresses was also ignored in Chapter 6 in the analyses to determine moment capacity.

The important practical question is whether or not it is acceptable to ignore the effects of prior creep and shrinkage in routine design calculations. The analyses in this appendix suggest that it is safe to ignore the effects of prior creep and shrinkage on moment capacity in ductile sections, and also when the short-term response to working loads is calculated.

C.2 Short-term service load application

When prior creep and shrinkage strains are allowed for, the neutral axes of strain and stress no longer coincide in a cross-section subjected to a short-term live load moment. The complexity of the analysis increases considerably. In the following we will retain the term d_n to denote the depth of the neutral axis of strain, and introduce the term d_{ne} for the depth of the neutral axis of stress (and hence also of elastic strains).

In Figure C.1(b) the initial elastic strains at transfer are represented by line AB, and YA and ZB are the top fibre and bottom fibre concrete strains, respectively. Under the effects of creep and shrinkage in the period up to the time of loading, the total strains in the section increase, while the prestressing force decreases somewhat. In Figure C.1(c), the total strain just prior to loading is represented by line A' B'. Although the total strains in the top and bottom fibres have increased to YA' and ZB', respectively, the elastic components have decreased to YA" and ZB", because of the loss of prestress. The inelastic components are represented in Figure C.1(c) by A"A' and B"B'.

It is convenient to rearrange the elastic and inelastic strains in this diagram to that shown in Figure C.1(d), where the creep and shrinkage strain components, represented by line Y'Z', are plotted from the vertical zero reference line for total strain, YZ. In this rearrangement the elastic strains are added to the inelastic strains, so that, with Y'Z' as the reference line, A'B' defines the elastic strains.

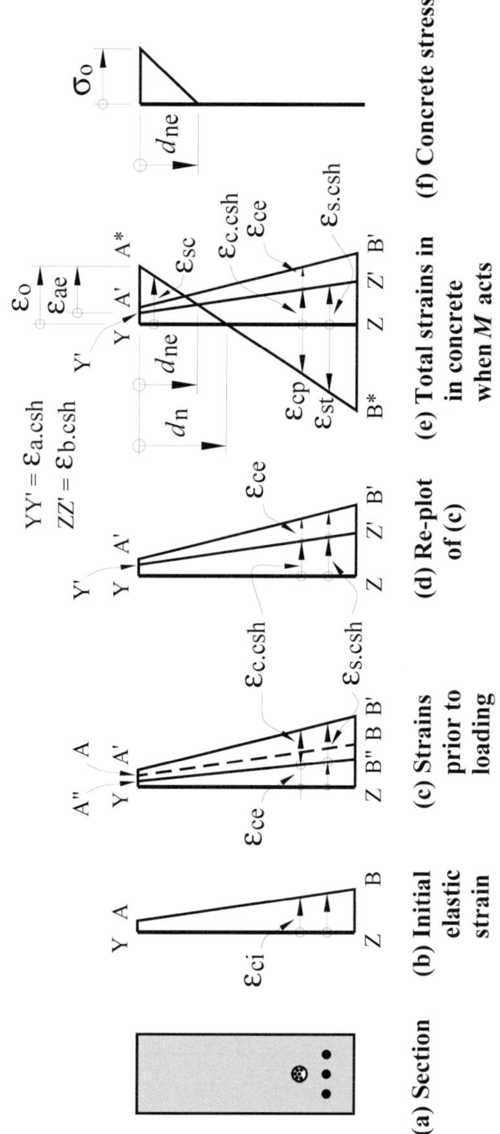

Figure C.1 Cracked section analysis – effect of creep and shrinkage

In Figure C.1(e), YZ is the zero reference line and A^*B^* represents the total strain in the cracked section when moment *M* acts. The neutral axis of *strain* is at depth d_n. The interval between line Y'Z' and A^*B^* is the elastic concrete strain. The concrete compressive stress is proportional to elastic strain, and so the point of zero elastic strain at depth d_{ne} is also the point of zero concrete stress. In other words, the neutral axis of *stress* is at depth d_{ne}. In Figure C.1(e) the lines Y'Z' and A'B' are plotted from the reference line YZ, as follows:

YY' = $\varepsilon_{a.csh}$ (creep and shrinkage strain in top fibre)

Y'A' = ε_{ap} (elastic strain in top fibre due to prestress)

ZZ' = $\varepsilon_{b.csh}$ (creep and shrinkage strain in bottom fibre)

Z'B' = ε_{bp} elastic strain in bottom fibre due to prestress)

In order to find the total strain line (line A^*B^* in Figure C.1(e)), we have to use a trial and error approach. We choose trial values of d_n and ε_o and then determine the stresses and forces acting in the section. The trial values are progressively modified until force equilibrium and moment equilibrium in the cross-section are satisfied.

The total compressive strain in the top fibre is YA* = ε_o, the elastic component being Y'A* = $\varepsilon_{ae} = \varepsilon_o - \varepsilon_{a.csh}$. In the bottom fibre the total tensile strain is ZB* = ε_b, where:

$$\varepsilon_b = \varepsilon_o \cdot \frac{D - d_n}{d_n} \tag{C.1}$$

The corresponding neutral axis of stress, at depth d_{ne}, may be obtained from the geometry of Figure C.1(e) as follows:

$$d_{ne} = D \cdot \frac{Y'A^*}{Y'A^* + ZZ' + ZB^*} \tag{C.2}$$

The absolute values of the terms in the denominator are all added together, even though ZB^* is tensile. Using strains, we obtain the following expression for d_{ne}:

$$d_{ne} = D \cdot \frac{\varepsilon_{ae}}{\varepsilon_{ae} + \varepsilon_{b} + \varepsilon_{b.csh}} \tag{C.3}$$

The elastic top fibre strain ε_{ae} and top fibre stress $\sigma_o = E_c\varepsilon_{ae}$, and hence the concrete compressive force, C, can be determined in the manner already used in Chapter 5:

$$C = 0.5\sigma_o b_w d_{ne} \tag{C.4}$$

The force C acts at depth d_w:

$$d_w = \frac{d_{ne}}{3} \tag{C.5}$$

However, when the stresses and forces in the reinforcing and prestressing steel are determined, the total strains (and hence the neutral axis of strain d_n) have to be used:

$$\varepsilon_{st} = \varepsilon_o \cdot \frac{d_s - d_n}{d_n} \tag{C.6}$$

$$\varepsilon_{cp} = \varepsilon_o \cdot \frac{d_p - d_n}{d_n} \tag{C.7}$$

$$\varepsilon_p = \varepsilon_{c.csh} + \varepsilon_{pe} + \varepsilon_{ce} + \varepsilon_o \cdot \frac{d_p - d_n}{d_n} \tag{C.8}$$

$$\varepsilon_{sc} = \varepsilon_o \cdot \frac{d_n - d_{sc}}{d_n} \tag{C.9}$$

The steel forces can be obtained from the stresses given by Equations C.8 and C.9. A check can now be made to see whether force equilibrium (Equation 5.5) is satisfied. Adjustments of the assumed value of the strain ε_o can be made until force equilibrium is satisfied. The moment M which corresponds to the value of d_n is then calculated by means of Equation 5.6. The procedure can then be repeated for a sequence of neutral axis positions.

To apply this method of analysis to a particular cross-section, we first need to evaluate the stresses and strains (elastic, creep and shrinkage) in the section just before the external moment acts. A step-by-step analysis of the uncracked section, as explained in Appendix B, can be used to obtain these strains; alternatively a visco-elastic analysis can be undertaken.

The quantitative effect of prior creep and shrinkage can be observed by undertaking comparative numerical analyses. In Example C.1 below, the cracked section analysis already used in Example 5.3 in Chapter 5 is repeated, taking account of prior creep and shrinkage. The results of the two sets of analyses are summarised and compared in Table C.1 for a bending moment of 279 kNm.

TABLE C.1 Comparison of results for *M* = 279 kNm

Calculation method:	Elastic analysis	Elastic analysis, but with prior creep and shrinkage
σ_p	1174	1127
σ_{st}	56	48
σ_{co}	13	14
κ	0.91×10^{-6}	2.39×10^{-6}

Note: In comparing the values in the second and third columns, it should be remembered that the elastic analysis values (Column 2) were based on σ_{pe} = 1100 MPa, whereas the calculations for Column 3 used a slightly different value, of 1038 MPa.

Clearly the differences in stresses are marginal and of no practical significance from a design point of view. The total curvature is of course much larger than the elastic curvature. This comparison supports the view that prior creep and shrinkage are not of practical design importance in regard to the stresses, internal forces and short term deformation increment resulting from a short-term load application. They are of course of great importance when total deformations and deflections are calculated.

EXAMPLE C.1

Cracked Section Analysis with Allowance for Prior Creep and Shrinkage

In this example we repeat the cracked section analysis in Example 5.3, but this time taking account of pre-existing creep and shrinkage in the section.

Calculations are only presented for a chosen neutral axis depth of strain of $d_n = 600$ mm. Trial and error calculations are carried out to obtain the total top fibre strain, ε_o, that corresponds to $d_n = 600$ mm.

Solution

Prior creep and shrinkage strains:

The strains and stresses in the section, following an extended period of sustained moment, have been determined by a viscoelastic analysis and are as follows:

The sum of the creep and shrinkage strains (excluding elastic strains) in the concrete are:

top fibre: $\varepsilon_{a.csh} = +0.000999$;

bottom fibre: $\varepsilon_{b.csh} = +0.000302$

The elastic concrete strains are:

top fibre: $\varepsilon_{a.e} = +0.000201$

bottom fibre: $\varepsilon_{b.e} = -0.000075$

The concrete stresses are:

top fibre: $\sigma_a = +6.43$ MPa

bottom fibre: $\sigma_b = -2.40$ MPa

At the tendon level: $\sigma_{ce} = -0.93$ MPa .

The stress in the tendon at the time of loading is $\sigma_{pe} = 1038$ MPa, having reduced to this value from an original value of $\sigma_{po} = 1100$ MPa.

Calculations for $d_n = 600$ mm:

For each calculation with $d_n = 600$ mm, and a trial ε_o value, we have:

$$\varepsilon_{a.e} = \varepsilon_o - 0.000999 \,; \quad \varepsilon_{b.e} = \varepsilon_o/4 + 0.000302 \,; \quad \varepsilon_{st} = \varepsilon_o/6$$

$$\varepsilon_p = 0.005190 + 0.000418 - 0.000029 + 0.04\varepsilon_o$$

$$= 0.005579 + 0.04\varepsilon_o$$

$$d_{ne} = \frac{\varepsilon_{ae}}{\varepsilon_{ae} + \varepsilon_{be}} D$$

$$C_w = 0.5 \times 32{\times}10^3 \times 250 \times \varepsilon_{a.e} \times d_{ne} = 4.0{\times}10^6 \varepsilon_{a.e} d_{ne}$$

$$\text{Error} = C_w - T_s - T_p$$

The calculations are given in Table C.2. Taking the values in the last row, the bending moment is:

$$M = 457 \times 0.625 + 65 \times 0.7 - 520 \times 0.298/3 = 279 \text{ kNm}$$

at a curvature of:

$$\kappa = \varepsilon_o/d_n = 0.001435/600 = 2.39{\times}10^{-6} \text{ mm}^{-1}$$

TABLE C.2 Calculations for trial values of ε_o

ε_o me	ε_{ae} me	ε_{be} me	d_{ne} mm	C_w kN	ε_{st} me	T_s kN	ε_p me	T_p kN	ΣH kN
1500	501	677	319	639	250	67	5639	457	-115
1400	401	652	286	459	233	63	5635	456	60
1435	436	661	298	520	239	65	5636	457	2

C.3 Effect on flexural strength

In the flexural strength analyses in Chapter 6, pre-existing creep and shrinkage strains in the cross-section have been ignored. No serious error is to be expected from this simplification, particularly if the prestressing and reinforcing steels are at yield when M_u is reached. Nevertheless, account can easily be taken of the prior creep and shrinkage strains in the section up to the time of loading to failure. The method is similar to that used above in Section C.2 for the cracked section analysis.

In Figure C.2, which is similar to Figure C.1(e), line A^*B^* represents the total strain line when M_u acts; however, the stress-related strains are measured from reference line Y'Z' since the region YY'-Z'Z represents the inelastic (creep and shrinkage) strains present in the section. The limiting strain ε_u is represented by $Y'A^*$. The depth d_{ne} to the neutral axis of stress is now as shown in Figure C.2, and the ratio of the neutral axes of stress and strain is as follows:

$$\frac{d_{ne}}{d_n} = \frac{\varepsilon_u}{\varepsilon_o} = \frac{\varepsilon_u}{\varepsilon_u + \varepsilon_{a.csh}} \tag{C.10}$$

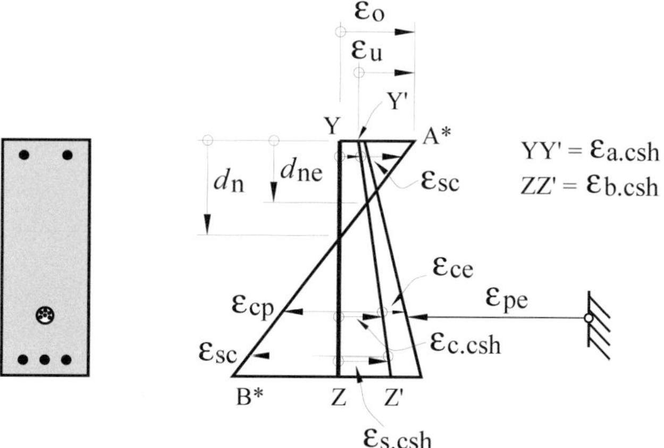

Figure C.2 Strains at M_u, including inelastic strains

To obtain this ratio the inelastic strains in the section first have to be evaluated. Methods to do this have been explained in Section C.2. The compressive force in the concrete and its line of action are then:

$$C_c = 0.85\ f_c'\ b\,\gamma\, d_{ne} \tag{C.11}$$

$$d_{sc} = 0.5\,\gamma\, d_{ne} \tag{C.12}$$

In the calculation of M_u the depth d_{ne} has to be used in Equations 6.18 and 6.19. However, the total strains need to be considered when the check on steel yielding is made, i.e.

$$\varepsilon_{pu} = \varepsilon_{pe} + \varepsilon_{ce} + \varepsilon_o \cdot (d_p - d_n)/d_n \tag{C.13}$$

$$\varepsilon_{su} = \varepsilon_o \cdot (d_s - d_n)/d_n \tag{C.14}$$

It will be noted that ε_o is the sum of ε_u (assumed to be 0.003) plus the creep and shrinkage strains present in the top fibre, $\varepsilon_{a.csh}$.

Calculations for a variety of cross-sections show that the presence of the inelastic strains has a negligible effect on moment capacity.

C.4 Concluding remarks

Creep and shrinkage induce large additional deformations and strains in a flexural member during its in-service life and a significant redistribution of internal stresses occurs in critical sections. There is a large increase in stress in the compressive reinforcement and a corresponding decrease in the compressive stresses in the concrete above the neutral axis.

Nevertheless, these changes do not have any significant effect on the moment capacity of a member provided it is ductile. In practical design it is therefore reasonable to ignore the inelastic strains present in the section, as well as any stress redistributions that may have taken place under prior sustained load, in order to calculate the moment capacity.

The deflections that occur due to creep and shrinkage must always be taken into account in serviceability design calculations. However, the increase in deflection that occurs under short-term live load is not affected to any significant degree by the stress changes and inelastic strains already present. This means that for practical design it is appropriate to calculate separately and independently the short-term deflections due to prestress and sustained load, the additional long-term deflections due to creep and shrinkage, and the short-term deflection due to live load, and to add these components together to estimate the total deflections.

APPENDIX D

Elastic deflections and end rotations for single-span beams

TABLE D.1 Deflections and end rotations for various load cases

Load Case	δ	θ
(a) 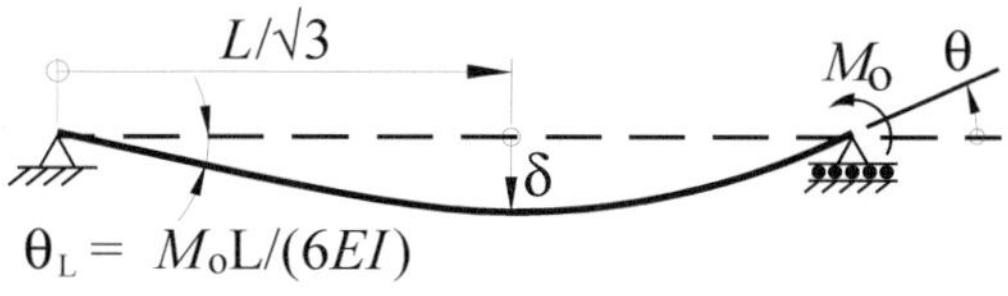 $L/\sqrt{3}$; M_o; θ; δ; $\theta_L = M_o L/(6EI)$	$\dfrac{M_o L^2}{9\sqrt{3}\,EI}$	$\dfrac{M_o L}{3EI}$
(b) 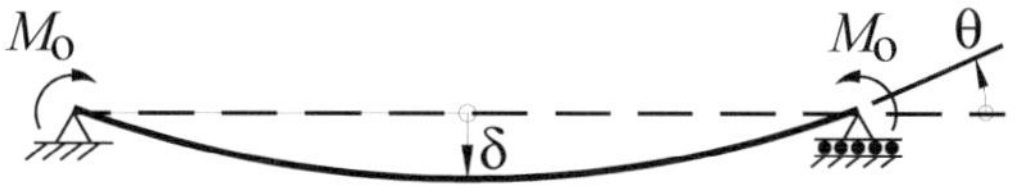 M_o; M_o; θ; δ	$\dfrac{M_o L^2}{8EI}$	$\dfrac{M_o L}{2EI}$
(c) 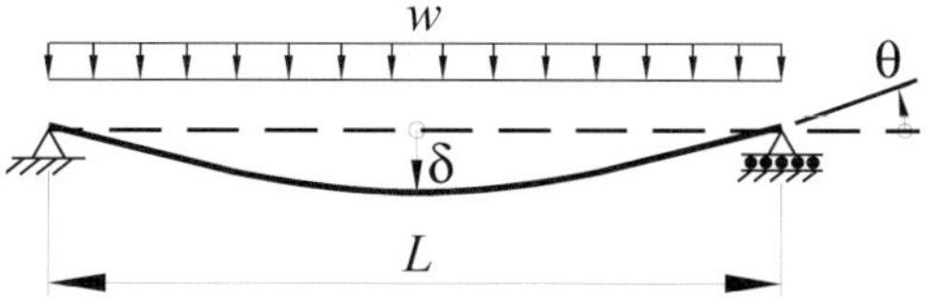 w; θ; δ; L	$\dfrac{5}{384}\dfrac{wL^4}{EI}$	$\dfrac{1}{24}\dfrac{wL^3}{EI}$

TABLE D.1 Deflections and end rotations for various load cases

Load Case	δ	θ
(d) 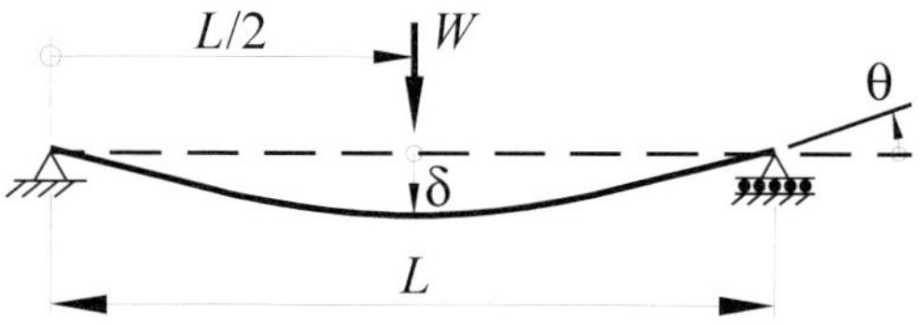 L/2, W, θ, δ, L	$\frac{1}{48}\frac{WL^3}{EI}$	$\frac{1}{16}\frac{WL^2}{EI}$
(e) 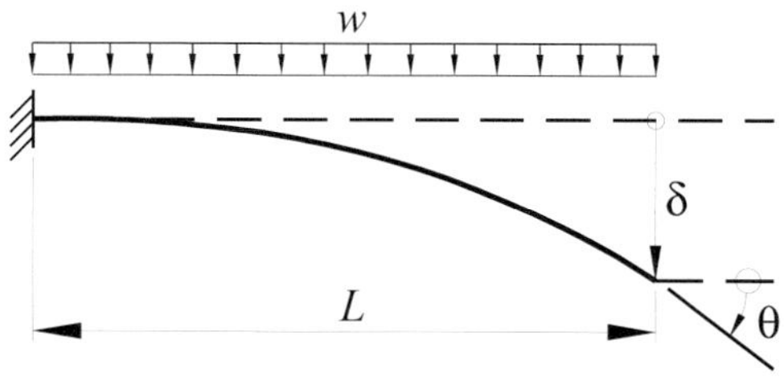 w, δ, L, θ	$\frac{1}{8}\frac{wL^4}{EI}$	$\frac{1}{6}\frac{wL^3}{EI}$
(f) 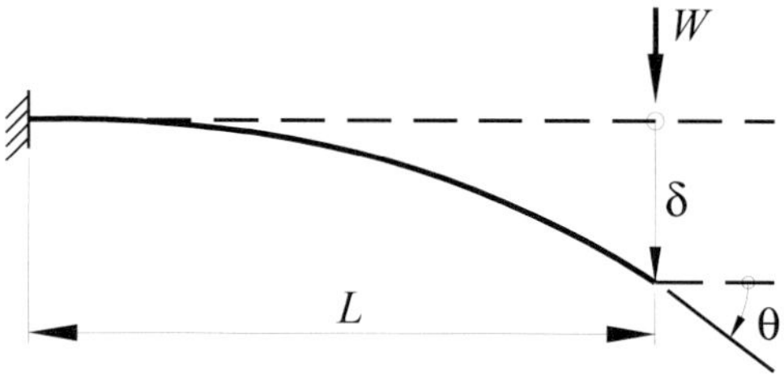 W, δ, L, θ	$\frac{1}{3}\frac{WL^3}{EI}$	$\frac{1}{2}\frac{WL^2}{EI}$
(g) 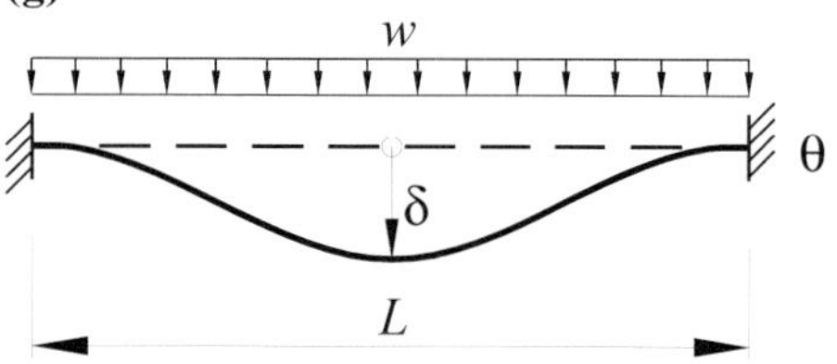 w, θ, δ, L	$\frac{1}{384}\frac{wL^4}{EI}$	0

TABLE D.1 Deflections and end rotations for various load cases

Load Case	δ	θ
(h) 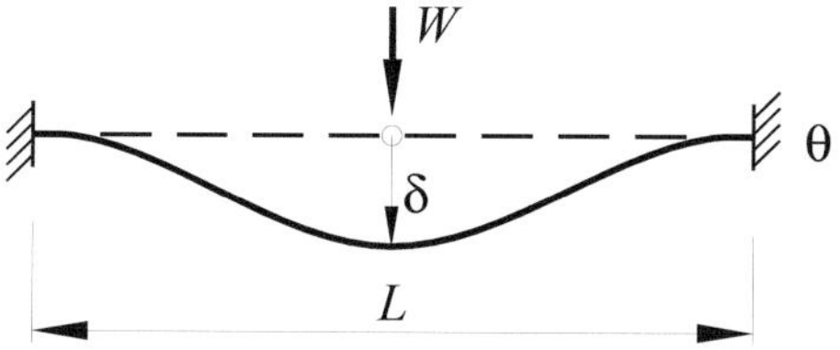	$\frac{1}{192}\frac{WL^3}{EI}$	0
(i) 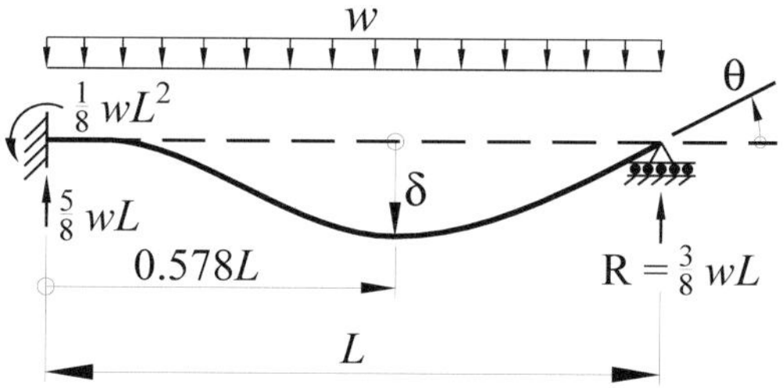	$\frac{1}{185}\frac{wL^4}{EI}$	$\frac{1}{48}\frac{wL^3}{EI}$
(j) 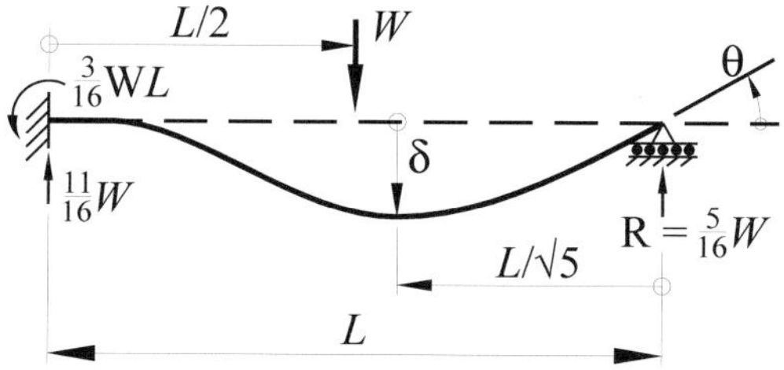	$0.00932\frac{WL^3}{EI}$	$\frac{1}{32}\frac{WL^2}{EI}$

INDEX

E

F

H

I

L

M

N

P

R

S

T

U

W